Cambridge IGCSE™ and O Level

Geography

Fourth Edition

Paul Guinness
Garrett Nagle

Endorsement indicates that a resource has passed Cambridge International Education's rigorous quality-assurance process and is suitable to support the delivery of their syllabus. However, endorsed resources are not the only suitable materials available to support teaching and learning, and are not essential to achieve the qualification. For the full list of endorsed resources to support this syllabus, visit www.cambridgeinternational.org/endorsedresources

Any example answers to questions taken from past question papers, practice questions, accompanying marks and mark schemes included in this resource have been written by the authors and are for guidance only. They do not replicate examination papers. In examinations the way marks are awarded may be different. Any references to assessment and/or assessment preparation are the publisher's interpretation of the syllabus requirements. Examiners will not use endorsed resources as a source of material for any assessment set by Cambridge International Education.

While the publishers have made every attempt to ensure that advice on the qualification and its assessment is accurate, the official syllabus, specimen assessment materials and any associated assessment guidance materials produced by the awarding body are the only authoritative source of information and should always be referred to for definitive guidance.

Our approach is to provide teachers with access to a wide range of high-quality resources that suit different styles and types of teaching and learning.

For more information about the endorsement process, please visit www.cambridgeinternational.org/endorsedresources

Third-party websites and resources referred to in this publication are not endorsed.

Cambridge International Education material in this publication is reproduced under licence and remains the intellectual property of Cambridge University Press & Assessment.

Author dedications

Garrett Nagle: To Angela

Paul Guinness: To Mary

Maps use the boundaries shown on the UN world map. The presentation of these boundaries does not imply an opinion on or endorsement of their legal status.

Every effort has been made to trace all copyright holders, but if any have been inadvertently overlooked the publishers will be pleased to make the necessary arrangements at the first opportunity.

Although every effort has been made to ensure that website addresses are correct at time of going to press, Hachette Learning cannot be held responsible for the content of any website mentioned in this book. It is sometimes possible to find a relocated web page by typing in the address of the home page for a website in the URL window of your browser.

Hachette UK's policy is to use papers that are natural, renewable and recyclable products and made from wood grown in well-managed forests and other controlled sources. The logging and manufacturing processes are expected to conform to the environmental regulations of the country of origin.

To order, please visit www.hachettelearning.com or contact Customer Service at education@hachette.co.uk/ +44 (0)1235 827827.

ISBN: 978 1 0360 1083 6

© Garrett Nagle and Paul Guinness 2025

Third edition published in 2018.
This edition published in 2025 by
Hachette Learning (a trading division of Hodder & Stoughton Limited),
An Hachette UK Company
Carmelite House
50 Victoria Embankment
London EC4Y 0DZ
www.hachettelearning.com

The authorised representative in the EEA is Hachette Ireland, 8 Castlecourt Centre, Castleknock Road, Castleknock, Dublin 15, D15 YF6A, Ireland

Impression number   10 9 8 7 6 5 4

Year   2028 2027 2026

All rights reserved. Apart from any use permitted under UK copyright law, no part of this publication may be reproduced or transmitted in any form or by any means, electronic or mechanical, including photocopying and recording, or held within any information storage and retrieval system, without permission in writing from the publisher or under licence from the Copyright Licensing Agency Limited. Further details of such licences (for reprographic reproduction) may be obtained from the Copyright Licensing Agency Limited, www.cla.co.uk

Cover photo © Sergey Nivens - stock.adobe.com

Illustrations by Integra Software Services Pvt Ltd

Typeset in India by Integra

Printed and Bound in Great Britain by Bell & Bain Ltd, Glasgow

A catalogue record for this title is available from the British Library.

Contents

	Introduction	v
	How to use this book	vi
	Physical geography	1

TOPIC 1 Changing river environments
- **1.1** The main hydrological characteristics and processes that operate in rivers and drainage basins — 2
- **1.2** The main landforms associated with these processes — 12
- **1.3** Rivers present opportunities and hazards for people — 18

TOPIC 2 Changing coastal environments
- **2.1** The physical processes that shape the coast — 35
- **2.2** The main landforms associated with these processes — 39
- **2.3** Coasts present opportunities and hazards for people — 47

TOPIC 3 Changing ecosystems
- **3.1** The characteristics of the Antarctic ecosystem — 67
- **3.2** The threats to the Antarctic ecosystem and how they can be managed — 72
- **3.3** The characteristics of the tropical rainforest ecosystem — 76
- **3.4** The threats to the tropical rainforest ecosystem and how they can be managed — 80

TOPIC 4 Tectonic hazards
- **4.1** The structure of the Earth and the distribution of earthquakes and volcanoes — 87
- **4.2** The processes and features associated with earthquakes and volcanoes — 93
- **4.3** The impact of tectonic hazards — 98
- **4.4** Managing the impacts of tectonic hazards — 102

TOPIC 5 Climate change
- **5.1** The natural and human causes of climate change — 110
- **5.2** The impacts of climate change at a range of geographic scales — 115
- **5.3** The responses to climate change — 118

Human geography — 123

TOPIC 6 Changing populations
- **6.1** Populations grow and decline — 124
- **6.2** Population structures change over time — 134
- **6.3** The causes and impacts of international migration — 142

TOPIC 7 Changing towns and cities
- **7.1** Where people live — 157
- **7.2** The opportunities and challenges of urbanisation — 161
- **7.3** The management of urban growth — 168

TOPIC 8 Development
- **8.1** Measuring development — 178
- **8.2** The world is developing unevenly — 184
- **8.3** Achieving sustainable development — 191

CONTENTS

TOPIC 9   Changing economies
- **9.1** Changing employment structures   208
- **9.2** The impact of globalisation and the role of transnational corporations   218
- **9.3** Tourism is a growing industry   232

TOPIC 10   Resource provision
- **10.1** How our food is produced   247
- **10.2** The global patterns of food supply and demand   251
- **10.3** The challenges of food supply   258
- **10.4** How our energy is produced   270
- **10.5** The global patterns of energy supply and demand   272
- **10.6** The impacts of energy production   279

Geographical skills   291

Command words   325

Glossary   326

Acknowledgements   339

Index   341

Introduction

This book has been written to help you obtain the knowledge, understanding and skills you need as you study Cambridge IGCSE™, IGCSE (9–1) and O Level Geography (0460/0976/2217) for examination from 2027. It will also increase your awareness of geography from a local to a global scale, using examples and case studies from around the world. Geography is about people and places, so we hope that you will use your own area as much as possible to add to the material in this book.

We would encourage you to keep up to date with geographical events – one way is through listening to the news or reading about events in newspapers or on the internet. Geography is happening every day, everywhere – so think about your own geographical location and new geographical events.

This book has been written to support the syllabuses for examination from 2027. It includes a number of activities to help you succeed with written assessments and coursework. Helpful support with essential maths skills is also provided.

Other resources published to support this Student Book include:

» a Revision and Study Guide, which provides a condensed version of the course, and includes common errors (misconception in geography) and exam-style questions and answers.
» a Workbook, which provides a series of questions relating to all themes (both the Revision and Study Guide and Workbook have not been through the endorsement process for the Cambridge Pathway).

Below are details of the assessment for the Cambridge IGCSE, IGCSE (9–1) and O Level syllabuses. Be prepared – knowing what to expect from the syllabus you are following will help you be more successful in your studies. Make sure you also use your teachers' experience – they are an excellent resource waiting to be tapped. Good luck and enjoy your geographical studies.

Scheme of assessment for Cambridge IGCSE

The information in this section is taken from the Cambridge International Education syllabus. You should always refer to the appropriate syllabus document for the year of examination to confirm the details and for more information. The syllabus document is available on the website: www.cambridgeinternational.org.

All candidates take Paper 1, Paper 2 and either Component 3 or Paper 4. Papers 1, 2 and 4 consist of combined question papers and answer booklets where candidates answer in the spaces provided.

Paper 1 Physical Geography (1 hour 45 minutes; 75 marks): You are required to answer structured questions containing short answer and extended response items. Some items are based on source material. It is worth 36 per cent of the total marks.

Paper 2 Human Geography (1 hour 45 minutes; 75 marks): Candidates answer structured questions containing short answer and extended response items. Some items are based on source material. It is worth 36 per cent of the total marks.

Either

Component 3 Coursework (school-based assessment; 60 marks): Teachers set one school-based assignment of 1800–2200 words. This component is worth 28 per cent of the total marks.

Or

Paper 4 Geographical Investigations (1 hour 30 minutes; 60 marks): Candidates answer two questions containing short answer and extended response items. Some items are based on source material. It is worth 28 per cent of the total marks.

Scheme of assessment for O Level

All candidates take three papers.

Paper 1 Physical Geography (1 hour 45 minutes; 75 marks): Candidates are required to answer structured questions containing short answer and extended response items. Some items are based on source material. It is worth 36 per cent of the total marks.

Paper 2 Human Geography (1 hour 45 minutes; 75 marks): Candidates answer structured questions containing short answer and extended response items. Some items are based on source material. This paper is worth 36 per cent of the total marks.

Paper 3 Geographical Investigations (1 hour 30 minutes; 60 marks): Candidates answer two questions containing short answer and extended response items. Some items are based on source material. It is worth 28 per cent of the total marks.

How to use this book

To make your study of Cambridge IGCSE, IGCSE (9–1) and O Level Geography as rewarding and successful as possible, this textbook, which is endorsed for the Cambridge Pathway, offers the following important features:

Maps
Clear, detailed maps include useful information.

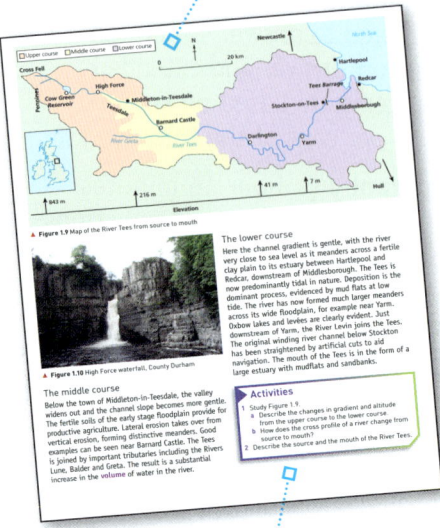

Tables, diagrams
Valuable information and statistics are summarised in numerous diagrams, charts and graphs. Tables provide data drawn from up-to-date sources in an easy-to-digest format.

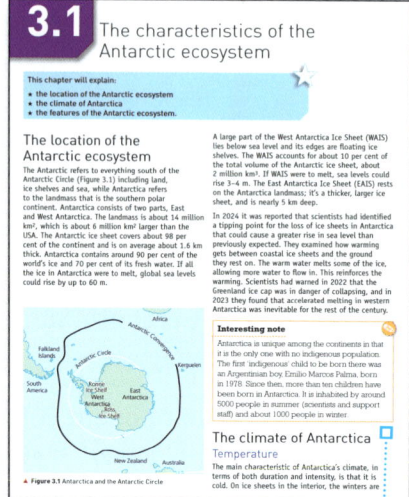

Activities
Carefully designed activities support your learning and check your progress on each topic.

Interesting notes
Fascinating facts are scattered throughout the book.

Photos
Vibrant photos with informative captions provide visual stimuli along with lots of useful information.

Detailed specific examples
Up-to-date, in-depth examples with analysis questions designed to help you understand and think about concepts in the real world.

Practice questions
Practice questions provide essential practice at answering questions to the required standard. All practice questions and sample answers have been written by the authors.

Glossary
Key words, highlighted in bold in the text, are defined in the glossary.

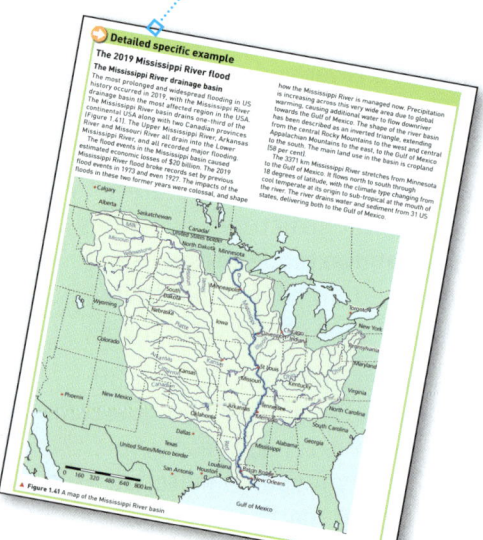

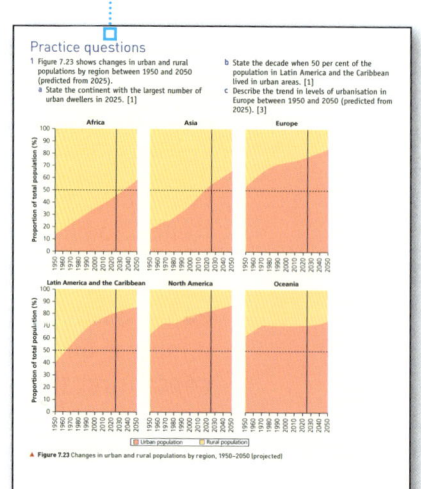

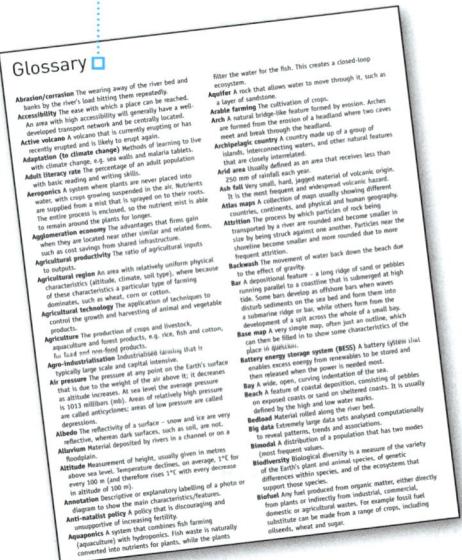

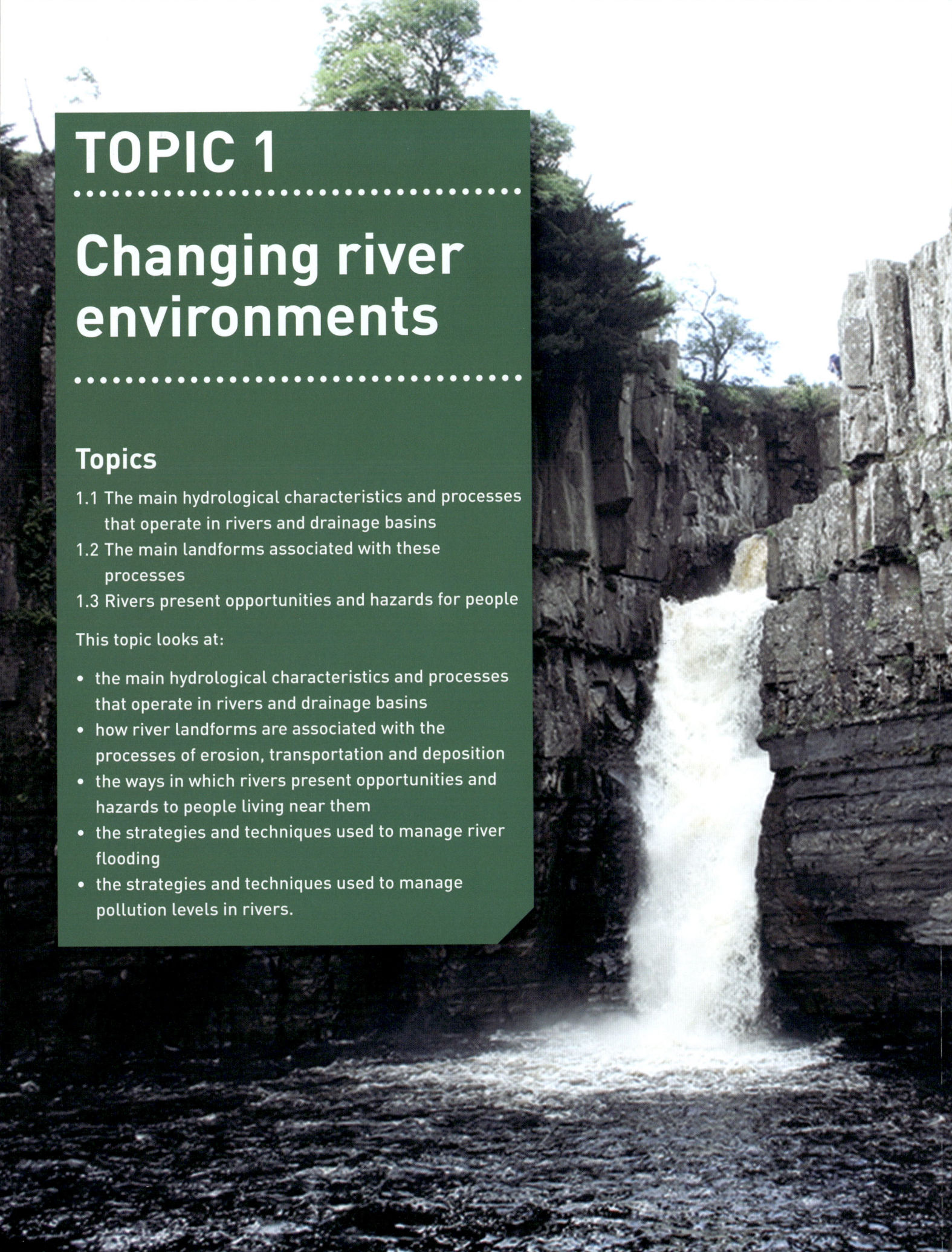

TOPIC 1

Changing river environments

Topics

1.1 The main hydrological characteristics and processes that operate in rivers and drainage basins
1.2 The main landforms associated with these processes
1.3 Rivers present opportunities and hazards for people

This topic looks at:

- the main hydrological characteristics and processes that operate in rivers and drainage basins
- how river landforms are associated with the processes of erosion, transportation and deposition
- the ways in which rivers present opportunities and hazards to people living near them
- the strategies and techniques used to manage river flooding
- the strategies and techniques used to manage pollution levels in rivers.

1.1 The main hydrological characteristics and processes that operate in rivers and drainage basins

This chapter will explain:

★ the characteristics of rivers and drainage basins
★ how the drainage basin operates within the water cycle
★ the processes that operate in a drainage basin
★ the processes that operate within a river.

The characteristics of rivers and drainage basins

A **drainage basin** (or catchment area) is the area drained by a river and its tributaries. Some drainage basins are very small (such as less than 10 km²). The largest drainage basins are huge, however – the Mississippi River and its tributaries drain over one-third of the USA (Table 1.1).

▼ Table 1.1 The drainage basins of some of the world's major rivers

River	Continent	Length (km)	Area of drainage basin (km²)	Average discharge (m³/sec)
Amazon	South America	6387	6,144,727	219,000
Nile	Africa	6690	3,254,555	5100
Mississippi/ Missouri	North America	6270	3,202,230	16,200
Yangtze	Asia	6211	1,800,000	31,900

Drainage basins have a number of distinct features (Figure 1.1):

» The boundary of a drainage basin is called the **watershed**.
» The point where a river begins is known as its **source**.
» A river reaches the sea at its **mouth**.
» A **tributary** joins the main river at a **confluence**.
» A main river and all its tributaries form a **channel network** or river system.

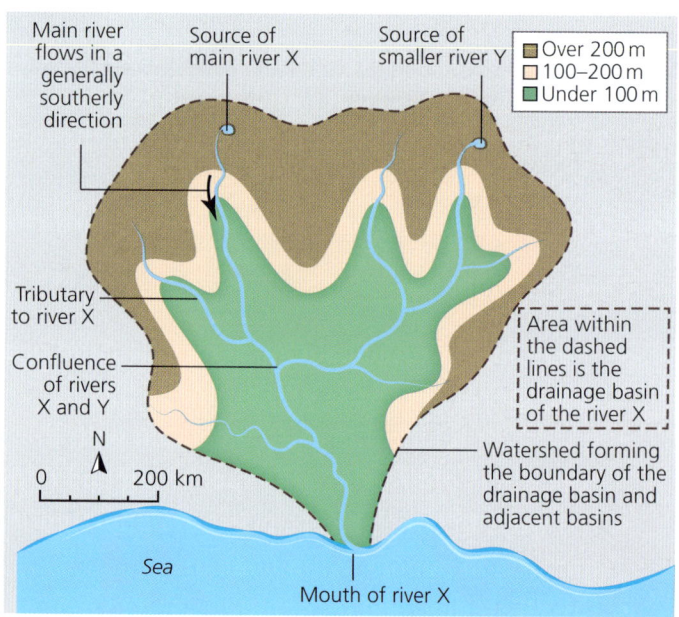

▲ Figure 1.1 The features of a drainage basin

A watershed (Figure 1.2) is a ridge of high land that forms the boundary between one drainage basin and other adjacent basins.

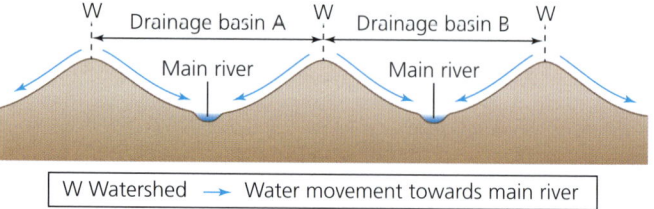

▲ Figure 1.2 A cross-section showing drainage basins and watersheds

The source of a river

A river is a large, natural stream of flowing water. The place where a river begins may be:

» an upland lake – the Mississippi River, the largest river in North America, begins as a stream from Lake Itasca in the US state of Missouri
» a melting glacier (Figure 1.3) – the Gangotri Glacier in the Himalaya mountains is the source of the River Ganges in Asia

▲ **Figure 1.3** A meltwater stream emerging from the Fox Glacier in the Southern Alps, New Zealand

» a spring in a boggy upland area, where the soil is so saturated that recognisable surface flow begins – the source of the Danube River is a spring in such an area of the Breg River, in the Black Forest in Germany
» a spring at the foot of an escarpment at the boundary between permeable and impermeable rock (Figure 1.4) – there are many such springs at the foot of the North Downs and South Downs, in southeast England.

When small streams begin to flow, they act under gravity, following the fastest route downslope. As they take the lowest path in the local landscape, water is added to them from tributaries, **groundwater flow**, **throughflow** and **overland flow** (surface runoff).

Channel networks

Some main rivers have a large number of tributaries, meaning that no place in the drainage basin is very far from a river. Such an area is said to have a high **drainage density**. The Amazon River, for example, receives water from more than a thousand tributaries. Where a main river has few tributaries, the drainage density is low. Channel networks often form a distinct pattern, which is due to the structure of rocks in the drainage basin.

▲ **Figure 1.4** Water issuing from a spring at the boundary of permeable and impermeable rock, Malham Cove, Yorkshire, UK; this is the source of Malham Beck

The mouth of a river

A river mouth occurs where a river empties into another body of water – a larger river, a lake or a sea or ocean. The great majority of rivers drain into a sea or ocean, but some drain into lakes that may be far from a coastline. For example, the River Volga, the longest river in Europe (at approximately 3685 km), flows into the inland Caspian Sea. Deltas sometimes form at the mouth of a river where the strength of tides and currents is insufficient to clear the large-scale sediment arriving from further upstream. The biggest delta in the world is the Ganges delta in Bangladesh and India.

The long profile

The **long profile** of a river is a longitudinal section of the course of the river, drawn along the river from source to mouth. It is expressed graphically as a curve, with the idealised form being a concave-upwards curve. Figure 1.5 shows how rivers change along their course from upstream to downstream:

» Discharge, width, depth, speed of flow/velocity and load quantity all increase.
» Load particle size, channel bed roughness and the river's gradient all decrease.

1.1 HYDROLOGICAL CHARACTERISTICS AND PROCESSES IN RIVERS AND DRAINAGE BASINS

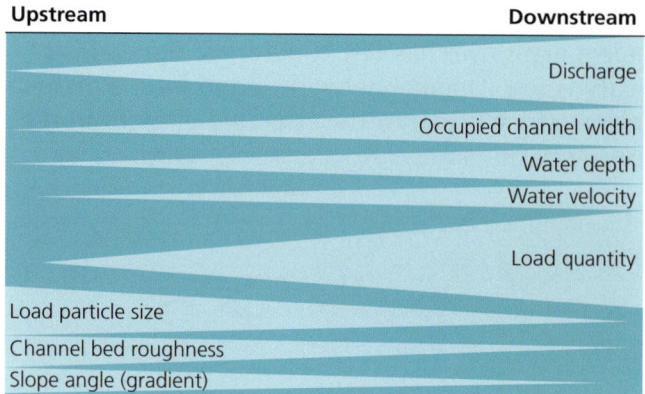

▲ Figure 1.5 Changes in a river from upstream to downstream

Channel shape

The efficiency of a stream's shape is measured by its **hydraulic radius** – that is, the cross-sectional area divided by the **wetted perimeter** (Figure 1.6). The wetted perimeter is the total length of the bed and the bank sides that are in contact with the water in the channel. The higher the ratio, the more efficient the stream and the smaller the frictional loss. The ideal form is semi-circular.

Figure 1.6 shows two channels with the same cross-sectional area, but with different shapes and hydraulic radii:

» Stream A: With a larger hydraulic radius, this stream has a smaller amount of water in contact with the wetted perimeter. This results in less friction and reduced energy loss, and therefore greater velocity.
» Stream B: With a smaller hydraulic radius, this stream has a larger amount of water in contact with the wetted perimeter. This results in greater friction and more energy loss, and therefore reduced velocity.

Stream A is more efficient than stream B.

There is a close relationship between the characteristics of the channel in which water is flowing and its **velocity** and **discharge**. These characteristics include the depth, width and channel roughness.

» River velocity is the speed at which water is flowing, in metres per second (m/s).
» Discharge (Figure 1.7) is the volume of water that passes through a section of the river per unit of time, expressed in cubic metres per second (m³/s). The discharge of a river is calculated by multiplying the river's cross-sectional area by its velocity.

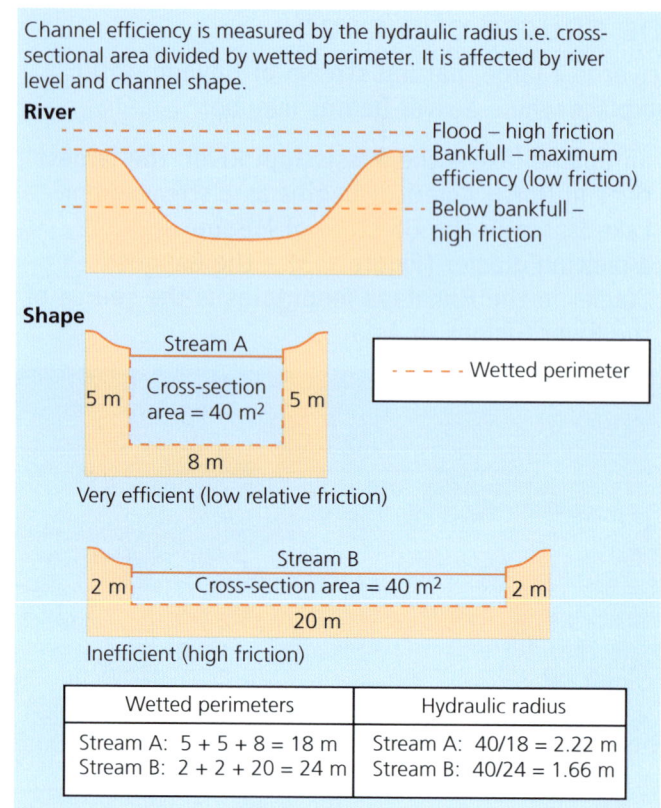

▲ Figure 1.6 The cross-sectional area and wetted perimeter

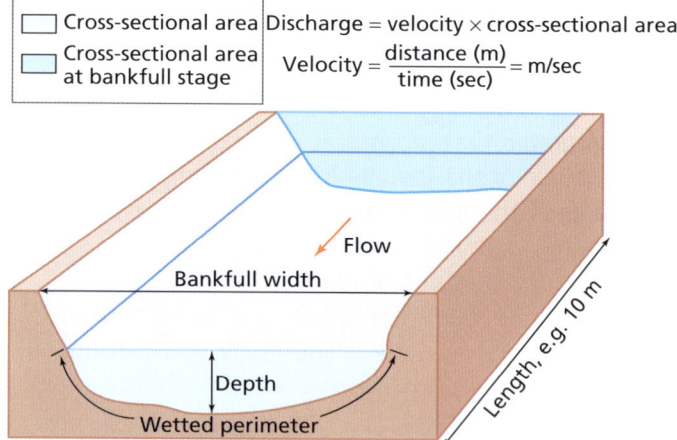

▲ Figure 1.7 Discharge

Channel roughness

Channel roughness describes how rough (uneven) the bed of a river is. Channel roughness causes friction, which slows down the velocity of the river water. Friction is caused by irregularities in the river bed, such as boulders, pebbles, potholes and vegetation, and by contact between the water and the river bed and banks.

If a river bed was equally smooth from its source to its mouth, it would be reasonable to expect that the velocity of the river would be greater near the source because of the steeper gradients associated with upland source regions. However, in reality the reverse is true, due to the very high degree of channel roughness in upland areas compared to the relatively smooth channel beds usually found in the lowland sections of rivers.

Discharge normally increases downstream, as do width, depth and velocity. The increase in channel width downstream is normally greater than the increase in channel depth. Large rivers, with a higher width-to-depth (w/d) ratio, are more efficient than smaller rivers with a lower width-to-depth ratio, because less energy is spent in overcoming friction. Thus, the carrying capacity increases and a lower gradient is required to transport the load. Although river gradients decrease downstream, the load carried is smaller and therefore easier to transport.

> ### Activities
> 1. Describe three different sources of rivers.
> 2. What is the long profile of a river?
> 3. Explain the terms:
>   a. Wetted perimeter
>   b. Hydraulic radius.
> 4. What is channel roughness and how does it affect the efficiency of a river?

The Bradshaw model

The **Bradshaw model** is a geographical model that suggests how a river's characteristics change from the source to the mouth of the river. Figure 1.8 shows the generalised ways in which the long and cross profiles of a river change as the river's gradient decreases. The long profile is sub-divided into three sections:

» Upper course
» Middle course
» Lower course.

The characteristics of each of these three sections are distinctly different, as are the physical processes operating in them. Look back at Figure 1.5 (page 4) to remind yourself of how factors such as width, depth, velocity and discharge change from upstream to downstream.

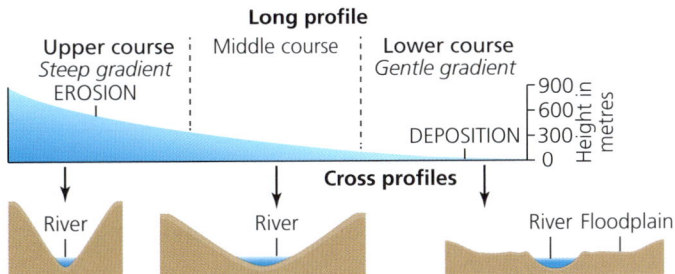

▲ **Figure 1.8** The long and cross profiles of a river

The River Tees is a reasonable exemplification of the Bradshaw model. The Tees is one of the major rivers in northeast England; it drains an area of about 1800 km². The source of the Tees is at Cross Fell, on the eastern side of the Pennine Mountains (Figure 1.9). The river rises at a height of over 750 m, flowing 160 km (channel length) eastwards to the mouth of the river in the North Sea. The Tees exhibits most of the classic processes and landforms of the upper, middle and lower courses of rivers.

The upper course

The upper course of the Tees is mainly an area of moorland, where the main land use is sheep farming. Annual precipitation can rise to over 2000 mm per year on the highest land. Precipitation decreases significantly eastwards towards the North Sea. The river channel is shallow and narrow. The bed is uneven, with sizeable angular boulders in places. There is much friction and the water flows more slowly here than further downstream, where the channel is wider, deeper and less uneven. High levels of friction upstream can cause considerable turbulence.

Vertical erosion has created a steep channel gradient and steep valley sides. These features, combined with impermeable rock, result in the river reacting quickly to rainfall. Impressive waterfalls are evident at Cauldron Snout and High Force, along with clear examples of interlocking spurs. High Force (Figure 1.10) is the UK's largest waterfall, at 21 m high. A deep plunge pool has been eroded at the base of the waterfall. At High Force, a bed of hard rock (dolerite) overlies softer rock (sandstone and shale). Look at Figure 1.21 (page 13) to understand what happens when this occurs. As the waterfall has eroded upstream, it has left behind an impressive gorge downstream of High Force. Rapids are also in evidence in this section of the river.

1.1 HYDROLOGICAL CHARACTERISTICS AND PROCESSES IN RIVERS AND DRAINAGE BASINS

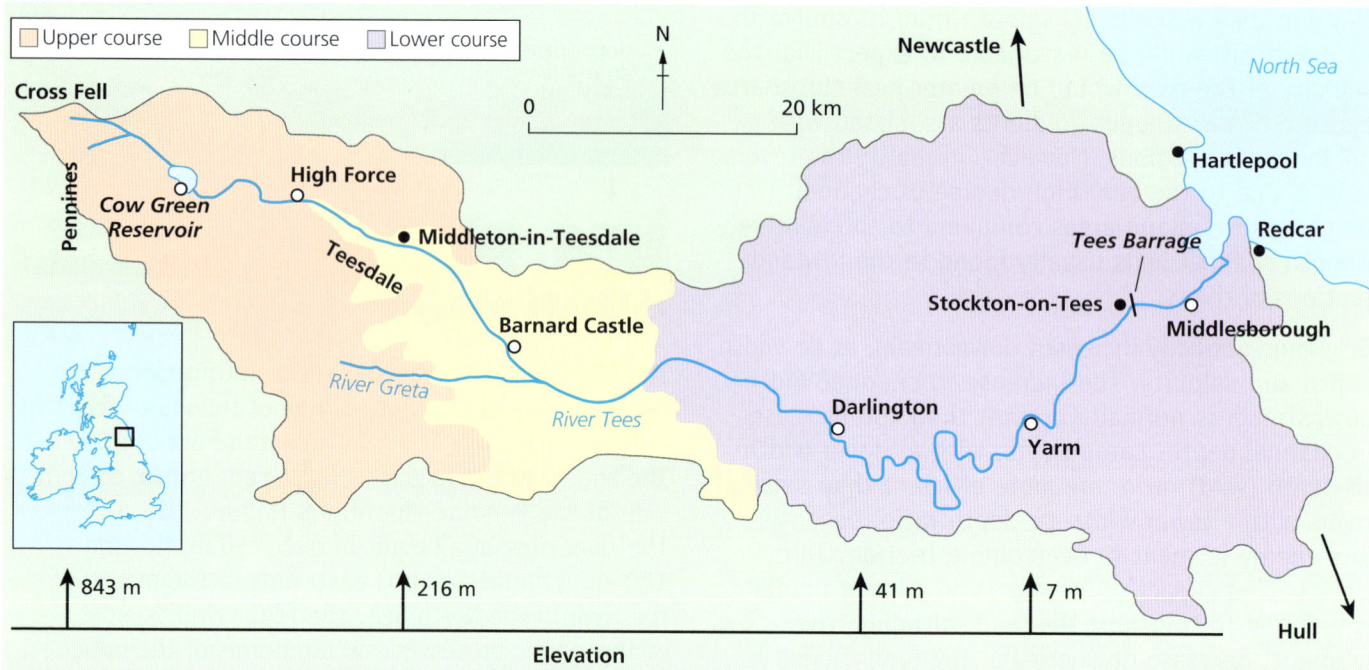

▲ **Figure 1.9** Map of the River Tees from source to mouth

▲ **Figure 1.10** High Force waterfall, County Durham

The middle course

Below the town of Middleton-in-Teesdale, the valley widens out and the channel slope becomes more gentle. The fertile soils of the early stage floodplain provide for productive agriculture. Lateral erosion takes over from vertical erosion, forming distinctive meanders. Good examples can be seen near Barnard Castle. The Tees is joined by important tributaries including the Rivers Lune, Balder and Greta. The result is a substantial increase in the **volume** of water in the river.

The lower course

Here the channel gradient is gentle, with the river very close to sea level as it meanders across a fertile clay plain to its estuary between Hartlepool and Redcar, downstream of Middlesborough. The Tees is now predominantly tidal in nature. Deposition is the dominant process, evidenced by mud flats at low tide. The river has now formed much larger meanders across its wide floodplain, for example near Yarm. Oxbow lakes and levées are clearly evident. Just downstream of Yarm, the River Levin joins the Tees. The original winding river channel below Stockton has been straightened by artificial cuts to aid navigation. The mouth of the Tees is in the form of a large estuary with mudflats and sandbanks.

> ### Activities
> 1 Study Figure 1.9.
>   a Describe the changes in gradient and altitude from the upper course to the lower course.
>   b How does the cross profile of a river change from source to mouth?
> 2 Describe the source and the mouth of the River Tees.

How the drainage basin operates within the water cycle

Hydrology is the study of water. The Earth's water is constantly recycled in a **closed system** called the **hydrological cycle** (water cycle). A closed hydrological system means that the volume of water in the hydrosphere today is the same as has always been present in the Earth's atmosphere system.

Figure 1.11 shows that water can be held for varying periods of time in different **stores**, namely:

» in oceans and seas
» on land as rivers, lakes and reservoirs
» in bedrock as groundwater
» in the atmosphere as water vapour and clouds.

Over 97 per cent of the world's water is stored in oceans and seas. These water bodies make up about 70 per cent of the surface of the Earth. This water is, of course, saline. The rest of the world's water (less than 3 per cent) is fresh. Of this fresh water, more than 68 per cent is held as ice and snow, with most of this in Antarctica and Greenland, and just over 30 per cent as groundwater. Only about 0.3 per cent of fresh water is found in rivers, lakes and surface reservoirs.

Just 0.001 per cent of the world's water is held in the atmosphere at any one time. This amounts to only about 10 days' supply of average rainfall around the world. This means that without transfers in the hydrological cycle (Figure 1.12), the world would run short of fresh water very quickly.

» Antarctica covers an area of almost 14 million km² and contains 30 million km³ of ice. This equates to around 61 per cent of all fresh water on Earth. The Antarctic ice sheet holds an amount of water such that if it were to melt, the sea level would rise by 70 m. The Greenland ice sheet covers 1.7 million km², which is about 70 per cent of the surface of Greenland.

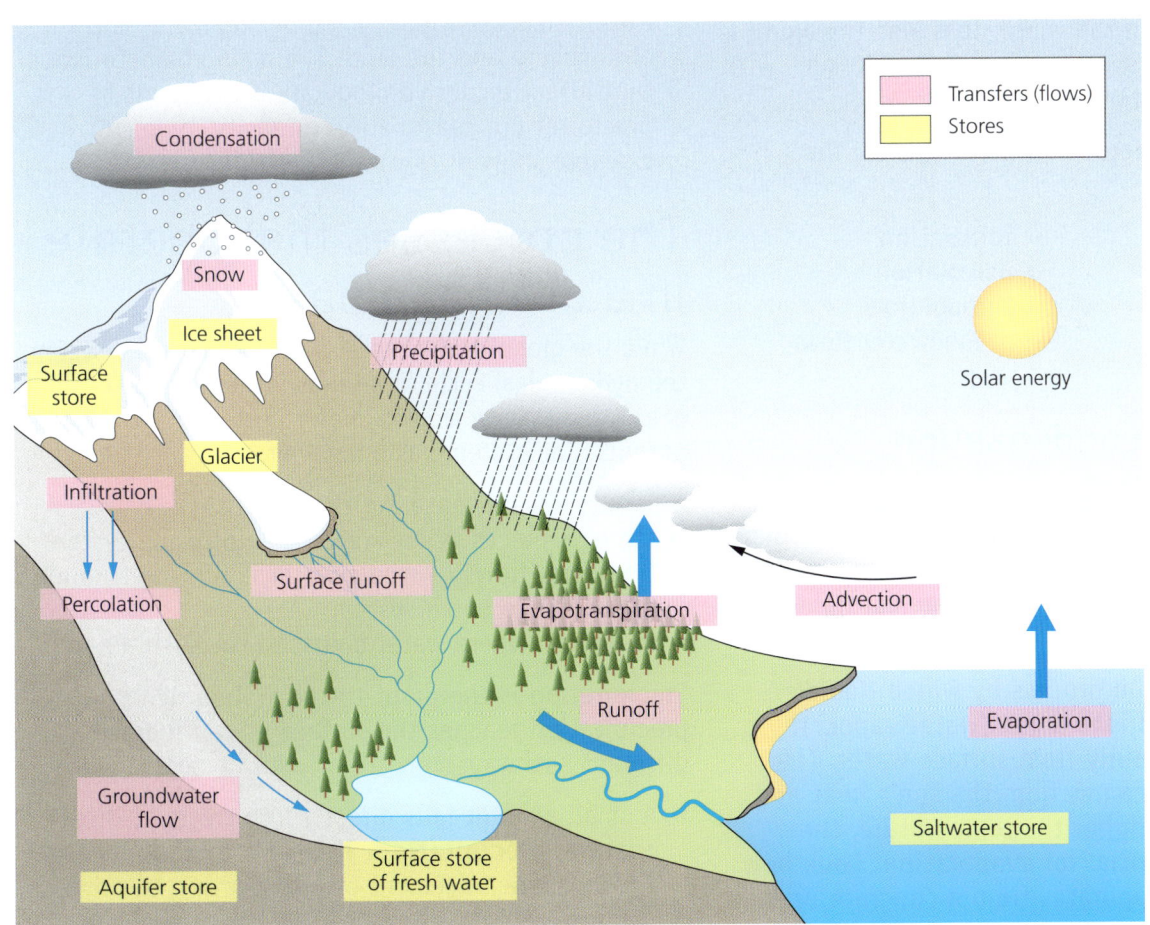

▲ **Figure 1.11** Processes, stores and transfers in the hydrological cycle

1.1 HYDROLOGICAL CHARACTERISTICS AND PROCESSES IN RIVERS AND DRAINAGE BASINS

▲ **Figure 1.12** Ice melting (a transfer) at the edge of Antarctica, by far the world's largest mass of ice

» About 30 per cent of the Earth's fresh water is held as groundwater. At over 1.7 million km³, the Great Artesian Basin in Australia underlies 22 per cent of the country and is arguably the largest groundwater **aquifer** in the world.

» Lake Baikal in eastern Russia is the largest-volume fresh water lake in the world. It is also the world's deepest lake. It covers an area of 31,500 km², with a maximum depth of 1637 m.

Transfers of water occur between stores by the following processes:

» evaporation
» condensation
» precipitation
» transpiration
» overland flow
» infiltration
» percolation
» throughflow
» groundwater flow.

Evaporation, condensation and precipitation

These are the three main processes in the hydrological cycle. Water exists in three states – liquid, solid and vapour – and these three states are constantly interchanging.

» **Evaporation** is the process by which liquid water is changed into a gas (water vapour). It takes place mainly from surface water. The energy required comes from the sun's heat and from wind. The higher the temperature, the greater the potential for evaporation. Look how quickly water evaporates from a concrete or tarmac surface on a very hot day compared with a cooler day. Evaporation is also faster on a windy day than on a calm day. Evaporation from water surfaces on land would not be enough to keep rivers and lakes full and to provide the human population with enough drinking water. Fortunately, large amounts of water evaporated from the seas and oceans are carried by air masses on to land, where condensation and precipitation take place.

» **Condensation** is the process by which water vapour changes into water droplets. It happens when water vapour is cooled to a level known as the dew point. This is when clouds begin to form. The extent of cloud cover at any point in time is a good indication of the intensity of condensation in the atmosphere. Clouds are tiny water droplets suspended in air, while rain droplets (precipitation) are much larger. This larger size enables rain droplets to overcome rising currents in the air in order to reach the ground surface.

» **Precipitation** occurs when water in any form falls from the atmosphere to the surface. This is mainly as rain, snow, sleet and hail. Thus, water is constantly recycled between the sea, the atmosphere and the land. The main characteristics that affect local hydrology are the amount of precipitation, seasonality, intensity, type (for example snow or rain) and variability.

The processes that operate in a drainage basin

While the global hydrological system is a closed system, the hydrological cycle of an individual drainage basin is an open system as it is open to external inputs and outputs. The system has a range of:

» inputs – water entering the system
» stores – places where water is held in the system
» transfers (flows) – where water is flowing through the drainage basin system
» outputs – where water is lost to the system.

Precipitation is the input to the system. When precipitation reaches the surface, it can follow different pathways:

» A small amount falls directly into rivers as **direct channel precipitation**. This adds to **channel flow**, which is the movement of water within the river channel.
» The rest falls on to vegetation or the ground.

▲ **Figure 1.13** A waterlogged field, near the town of Navan, Republic of Ireland – the soil in this field was already saturated when heavy rain fell in the night before the photograph was taken

If heavy rain has fallen previously and all the air pockets in the soil are full of water, the soil is said to be **saturated** (Figure 1.13). Because the soil is unable to take in any more water, the rain flows on the surface under the influence of gravity or remains on a flat surface in a waterlogged state. Flowing surface water is called **surface runoff** or overland flow.

If the soil is not saturated, rainwater will soak into it through the process of **infiltration**. It then moves vertically down through the soil and rock by the process of **percolation**. If the rock below the soil is **permeable** (meaning it allows water into it), the rainwater will continue to soak down deeper into the rock. This water will eventually come to **impermeable** rock (which does not allow water into it). The underground water level will build up towards the surface from here. This water does not remain stationary but flows downslope under gravity. The upper level of underground water is the **water table**. Water contained in rocks is known as **groundwater**, and water on the move in rocks is called groundwater flow. Rock that holds groundwater is known as an aquifer.

A spring occurs when underground water emerges at the surface.

» This happens where a permeable rock such as limestone covers an impermeable rock such as clay. Rainwater that can percolate into the permeable rock is unable to penetrate the impermeable rock below. This water will emerge at the surface as a spring, provided the water table is above the surface level.
» This also happens when the water table in a normally dry area reaches the surface during a period of unusually heavy rain. Such springs generally flow for only a short period of time.

Rainwater can be intercepted by vegetation. **Interception** is greatest in summer, when trees and plants have greater leaf coverage.

» Some rainwater will be stored on leaves and then evaporated directly into the atmosphere.
» The remaining intercepted water will either drip to the ground from leaves and branches or it will trickle down tree trunks or plant stems (**stemflow**) to reach the ground.

Vegetation takes in moisture through its root system. It loses some of this into the air by **transpiration**. Surface water is also lost by evaporation. The combination of these two processes is known as **evapotranspiration**.

In some countries precipitation is fairly regular throughout the year. However, in other countries there may be distinct wet and dry seasons. In these areas, rivers may dry up completely for many months. In **deserts**, small river channels may be dry for most of the year.

> ### Activities
> 1 Study Figure 1.11 (page 7).
>   a List three stores and three transfers (flows) in the hydrological cycle.
>   b Draw a labelled diagram to show the relationship between evaporation, condensation and precipitation.
> 2 Why is the global hydrological system a closed system, while the hydrological cycle of an individual drainage basin is an open system?
> 3 Explain the differences between:
>   a overland flow and groundwater flow
>   b infiltration and percolation.
> 4 What is evapotranspiration?

The processes that operate within a river

Energy is needed for transfers to occur. Around 95 per cent of a river's energy is used to overcome **friction**. The remaining 5 per cent or so is used to erode the

1.1 HYDROLOGICAL CHARACTERISTICS AND PROCESSES IN RIVERS AND DRAINAGE BASINS

river channel and transport this material downstream. The amount of energy in a river is determined by:

» the amount of water in the river
» the speed at which it is flowing.

Figure 1.14 shows that in the upper course of a river, near the source, a river's channel is shallow and narrow, and the river bed is often strewn with boulders and very uneven. There is a lot of friction so the water flows more slowly here than it does further downstream in the middle course and lower course, where the channel is wider, deeper and less uneven. Figure 1.15 shows a river in its upper course – notice the steep gradient, the boulders in the river and the amount of 'white water'. The latter is a good indication of a high level of friction.

▲ **Figure 1.15** A river in its upper course – British Columbia, Canada

Erosion

There are four processes of **erosion** that take place in a river:

» **Hydraulic action**: The sheer force of the river water removing loose material from the bed and banks of the river.
» **Abrasion/corrasion**: The wearing away of the river bed and banks as a result of the river's load hitting them repeatedly.
» **Attrition**: In swirling water, rocks and stones collide with each other and with the bed and banks. Over time, the sharp edges become smooth and the rocks and stones become smaller in size (Figure 1.16).
» **Solution**: Some rocks, such as limestone, which is soluble in slightly acidic water, dissolve slowly in river water.

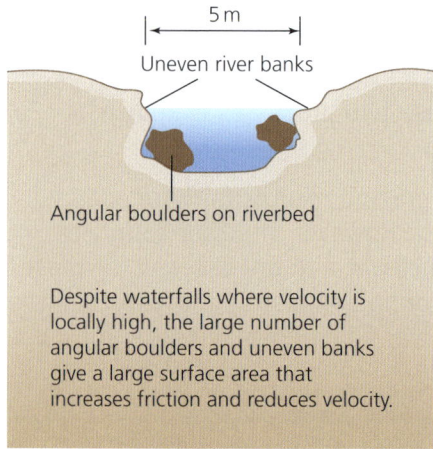
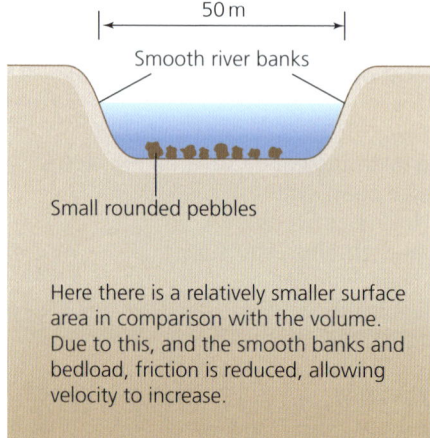

▲ **Figure 1.14** Velocity and discharge in the upper and lower courses of a river

▲ **Figure 1.16** Rocks in a dry section of river bed that have been rounded and reduced in size by attrition – Kyrgyzstan, Central Asia

Along the course of a river there are two main types of erosion that take place:

- **Vertical erosion** (downward): This takes place in the upper course of the river near the source, where the river cuts down into its bed and deepens the valley.
- **Lateral erosion** (sideward): This takes place in the middle and lower courses and widens the valley.

Most erosion occurs when discharge is high and rivers are in **flood**.

Transportation

There are four processes by which a river can transport its **load** (Figure 1.17):

- Solution: In areas of calcareous rock (limestone), material is carried in solution as dissolved load.
- **Suspension**: The smallest particles (silt and clay) are carried in suspension by the moving water.
- **Saltation**: Larger particles (sand, gravel, very small stones) are transported in a series of 'hops' or bounces.
- **Traction**: Large stones are pushed along the bed by the process of traction.

Parts of the load that are moved by traction when the discharge of the river is low may be transported by saltation when the discharge is high.

Deposition

Deposition takes place when a river does not have enough energy to carry its load. This can happen when:

- the gradient decreases
- the discharge falls during a dry period
- the current slows down on the inside of a meander
- the river enters a lake or the sea.

When a river loses energy, the large, heavy material known as the **bedload** is deposited first. Lighter material is carried further downstream. The gravel, sand and silt that are deposited is called **alluvium**. This is spread over the floodplain. The load transported by solution is carried out to sea with much of the clay and the lightest suspended particles.

Table 1.2 gives examples of the factors affecting the processes of erosion, **transportation** and deposition.

▼ **Table 1.2** Examples of factors affecting processes

Factor	Effect on erosion, transportation and deposition
Climate	• Heavy rainfall leads to higher discharge, which leads to increased action of river processes. • Higher temperature leads to increased evaporation, which leads to lower discharge and reduced action of river processes.
Slope	• Steep slopes result in fast-flowing rivers with strong erosive power. • Gentle slopes encourage deposition.
Geology	• Rivers erode valleys made of soft rock at a rapid rate. • Very porous (e.g. chalk) and permeable (e.g. carboniferous limestone) rocks may lack surface river flow for all or part of the year.
Altitude	• Snowmelt and melting glaciers have a big impact on river regimes and processes.
Aspect	• South-facing slopes (in the Northern Hemisphere) have higher rates of evaporation and transpiration, which can affect discharge.

Activities

1. a Define friction.
   b How much of a river's energy is used to overcome friction?
2. Describe and explain the differences between the two diagrams in Figure 1.14 (page 10).
3. List the four processes of:
   a erosion   b transportation.
4. Under what conditions is deposition likely to occur?

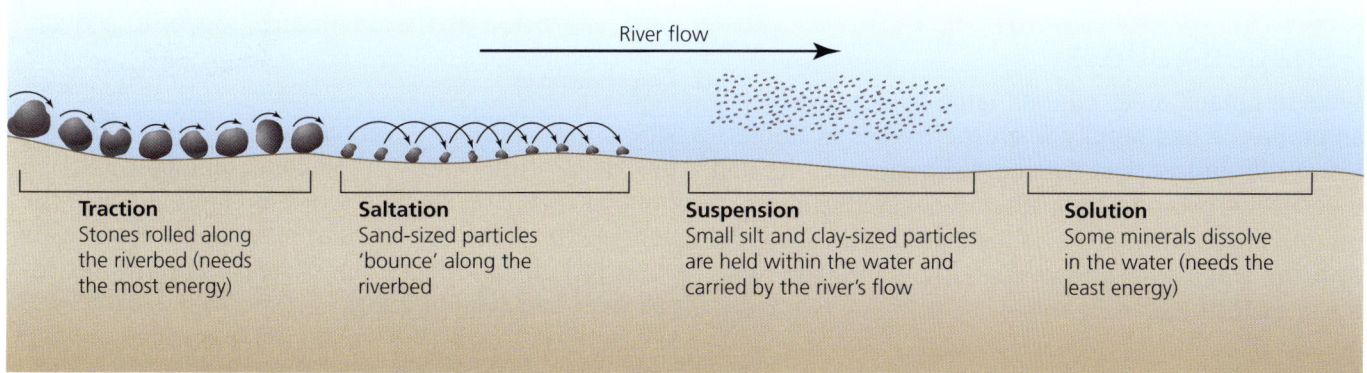

▲ **Figure 1.17** The processes of transportation

1.2 The main landforms associated with these processes

This chapter will explain:

★ the characteristics and formation of upland landforms
★ the characteristics and formation of lowland landforms

The characteristics and formation of upland landforms

The characteristic river landforms in upland areas (Figures 1.18 and 1.19) are a steep **V-shaped valley**, a steep gradient, interlocking spurs, potholes, waterfalls, rapids and gorges. In the upper course, much of the river's energy is needed to overcome friction. The rest is used to transport the load.

▲ **Figure 1.19** The interlocking spurs of the Waitaki River, South Island, New Zealand

▲ **Figure 1.18** Meandering river in its upper course – a snow-covered mountain landscape in Switzerland

V-shaped valleys

Rivers in upland areas contain large boulders, which can erode the bed rapidly when the river is in flood. This results in the river cutting downward into its bed by vertical erosion to form steep V-shaped valleys. Soil and loose rock on the valley sides are washed down the steep slopes by overland flow into the river. This adds to the load.

Interlocking spurs

Rivers begin to meander in the upper course. Erosion is concentrated on the outside banks of these small meanders. This eventually produces **interlocking spurs** that alternate on each side of the river. Figure 1.19 shows that interlocking spurs are ridges of high land that project towards a river at right angles. They decrease in height towards the river. Interlocking spurs are eroded further down a river's course when lateral erosion takes over from vertical erosion as the dominant process in a river.

Potholes

Where the river bed is very uneven, pebbles carried by fast, swirling water can become temporarily trapped by obstacles. The swirling currents cause the pebbles to rotate in a circular movement, eroding (by abrasion) circular depressions known as **potholes** in the river bed (Figure 1.20).

The characteristics and formation of upland landforms

▲ Figure 1.20 Potholes on the River Wharfe, Yorkshire, UK

Waterfalls and rapids

Waterfalls are the most spectacular feature of the upper course, but they can also be found in the middle course. They occur when:

- there is a sudden change in the course of the river due to differences in rock hardness along the valley (Figure 1.21)
- a fault line has created an escarpment over which the river flows
- glaciation has left a tributary valley hanging high above the main valley
- a steep drop at the edge of a plateau has been formed by uplift of the land
- a lava flow has crossed the path of a river, so it pours over its edge as a waterfall.

Figure 1.21 shows what happens when a river flows from a band of hard, resistant rock on to a band of softer, less resistant rock. Waterfalls can form

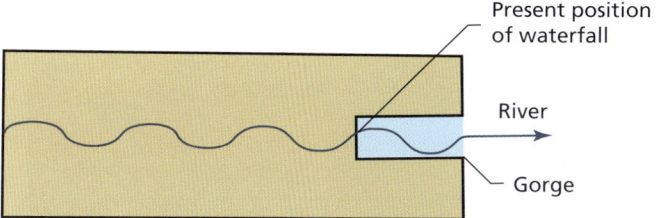

▲ Figure 1.22 The formation of a gorge of recession

when the hard rock is horizontal, vertical or dipping upstream. The lower, softer rock is eroded more quickly, causing the hard rock to overhang. The undercutting is caused by corrosion and hydraulic action. The overhang steadily becomes larger until a critical point is reached. When this occurs, the overhang collapses. The rocks that crash down into the plunge pool are swirled around by the currents. This increases erosion, making the plunge pool deeper. The rocks in the plunge pool are eroded mainly by attrition.

This process, beginning with the collapse of a layer of hard rock, is repeated time after time. As a result, the waterfall retreats upstream (Figure 1.22), leaving behind a steep-sided **gorge**. The Niagara Falls (Figure 1.23) is a group of three waterfalls at the southern end of Niagara Gorge, which is 11 km long. The Falls span the border between the province of Ontario, Canada and the state of New York, USA.

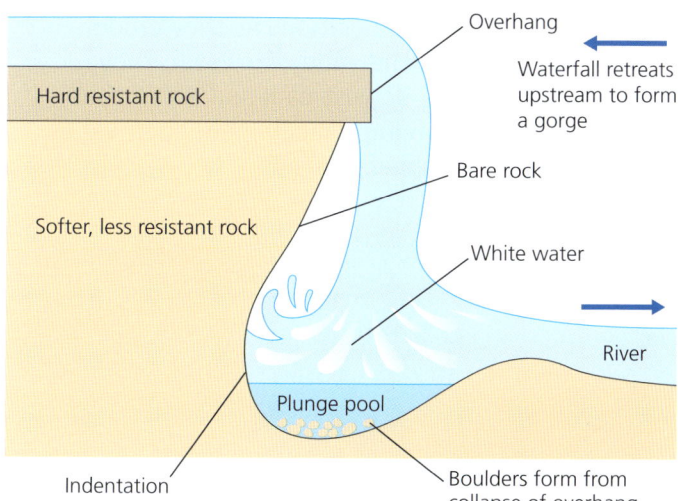

▲ Figure 1.21 The formation of a waterfall

▲ Figure 1.23 The Niagara Falls, Ontario, Canada

13

1.2 THE MAIN LANDFORMS ASSOCIATED WITH THESE PROCESSES

▲ Figure 1.24 Section of rapids along the River Wharfe, Yorkshire, UK

Sometimes very thin alternating bands of hard and soft rock cross the course of a river. This creates an uneven river bed and a zone of turbulent water known as **rapids** (Figure 1.24). Rapids can also form when a layer of hard rock dips gently downstream. The tallest waterfall in the world is the Angel Falls (979 m) located in the rainforest of Canaima National Park, Venezuela.

> ### Activities
> 1 Which river landforms might you expect to find in an upland area?
> 2 Explain the formation of interlocking spurs.
> 3 Draw a fully labelled diagram to explain the formation of a waterfall.

The characteristics and formation of lowland landforms

As more tributaries join the main river downstream, the volume of water increases. In lowland areas, if the rock is permeable, throughflow and groundwater flow also add water to the river.

Meanders and meander migration

In lowland areas, lateral erosion takes over from vertical erosion as the most important process. As a result of this, meanders (wide bends in the river's course) become larger (Figure 1.25).

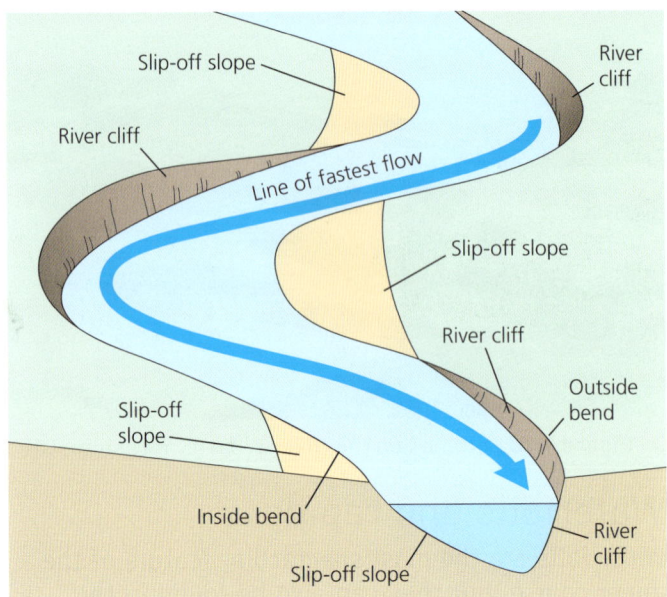

▲ Figure 1.25 The cross-section of a meander

» The current is fastest and most powerful on the outside bank of a meander, particularly on the downstream section. Erosion is relatively rapid. The outside bank is **undercut**. Again, the emphasis is on the downstream section. Eventually it collapses and retreats, causing the meander to spread further across the valley.
» If the meander has already reached the side of the valley, erosion on the outside bend may create a very steep slope or **river cliff**.
» The current on the inside of the meander is much slower. As the river slows, it drops some of its load and deposition occurs (Figure 1.26). This builds up to form a gently sloping **slip-off slope** (Figure 1.27).
» This results in the water being shallow on the inside of the meander and deep on the outside.

Due to the power of lateral erosion, meanders slowly change their shape and position. As they erode sideways, they widen the valley. However, they also move or migrate downstream and erode the interlocking spurs, giving a much more open valley compared with the upper course.

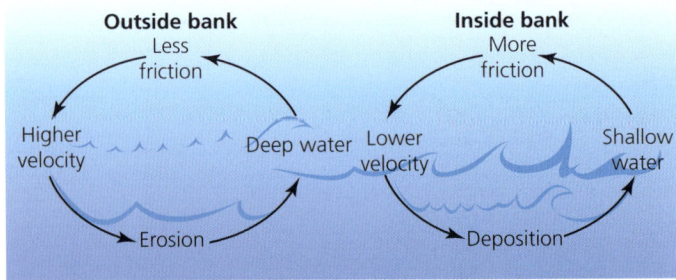

▲ Figure 1.26 The processes operating on the inside and outside banks of a meander

14

The characteristics and formation of lowland landforms

▲ **Figure 1.27** Meander and slip-off slope on the Marsyangdi River in the Annapurna Himalayan Range, Nepal

Oxbow lakes

» As a river flows towards its mouth, meanders become more pronounced and the valley becomes wider and flatter (see the left of Figure 1.28).
» As erosion continues to cut into the outside bends of a meander, a **meander neck** may form (see the middle of Figure 1.28).
» Eventually, when the river is in flood, it may cut right across the meander neck and shorten its course (see the right of Figure 1.28).
» For a while water will flow both along the old meander route and along the new straight course. However, because the current will slow down at the entry and exit points of the meander, deposition will occur.
» Over time, the meander will be cut off from the new straight course and leave an **oxbow lake**.

Floodplains and levées

A **floodplain** is a wide area of almost flat land on both sides of a river (Figures 1.29 and 1.30). It is formed by the movement of meanders, explained above. When discharge is high, the river is able to transport a large amount of material in suspension. At times of exceptionally high discharge, the river overflows its banks and floods the low-lying land around it. The increase in friction as the river water surges across the floodplain reduces velocity and causes the material carried in suspension to be deposited on the floodplain as alluvium. Alluvium is deposits of clay, silt and sand left by floodwater.

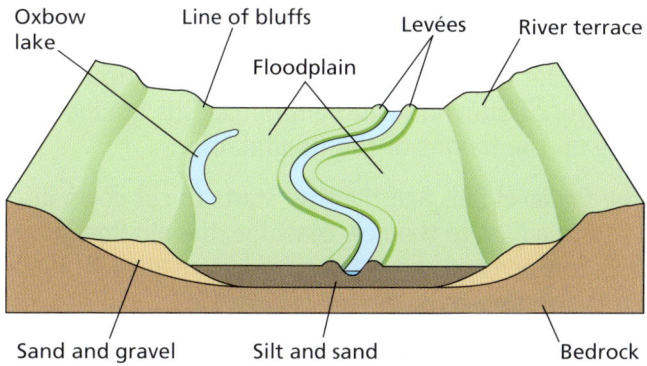

▲ **Figure 1.29** A cross-section of a river floodplain

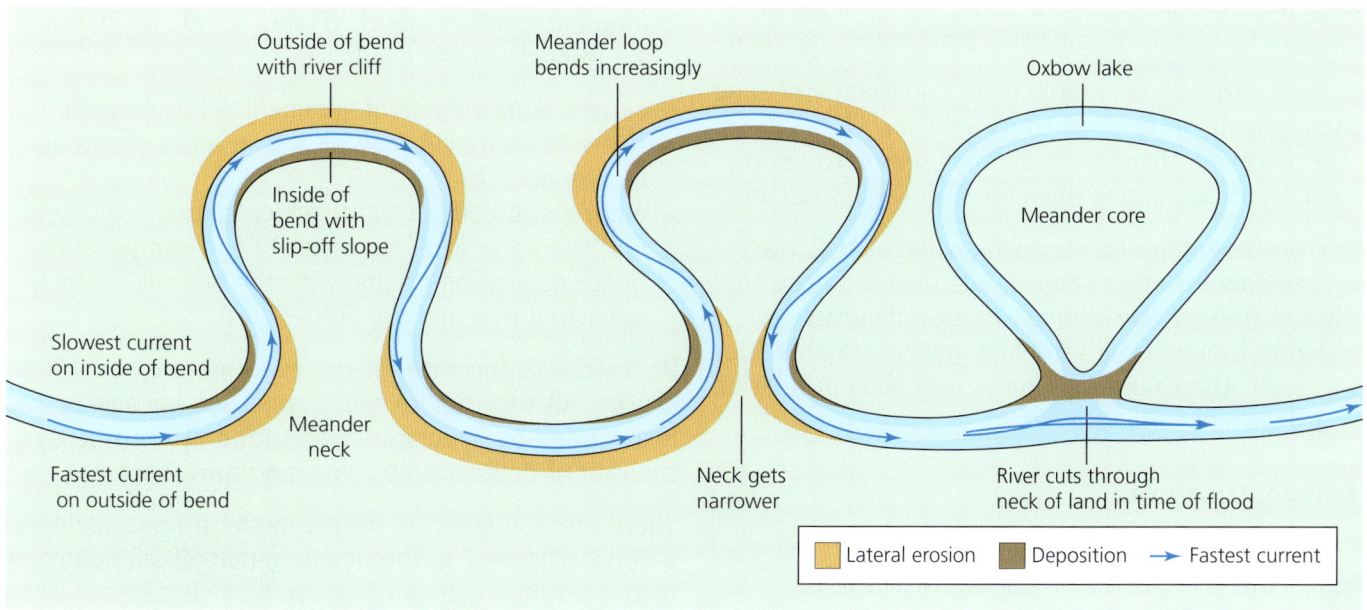

▲ **Figure 1.28** The formation of an oxbow lake

1.2 THE MAIN LANDFORMS ASSOCIATED WITH THESE PROCESSES

▲ **Figure 1.30** The floodplain of the River Tuul, south of Ulaanbaatar, Mongolia

The heaviest or coarsest material will be dropped nearest to the river. This can form natural embankments alongside the river called **levées** (Figure 1.29). The lightest material will be carried towards the valley sides. Each time a flood occurs, a new layer of alluvium will be formed. This gradually builds up the height of the floodplain.

If the sea level falls and the river starts to cut down into its bed to adjust to the new coastline, the old floodplain will be left perched above the new river channel. This is then known as a **river terrace** (Figure 1.29).

Braided channels

Braiding is when a river divides for various distances into two or more channels (Figure 1.31). The channels are separated by islands of sediment. Braiding occurs when:

» a river carries a very large load, particularly of sand and gravels, in relation to its velocity
» the discharge changes rapidly from season to season and deposition occurs when the river current slackens.

The river then deposits so much sediment that the river channel becomes choked. The river is forced to split and find its way through its own deposits. The banks formed from sand and gravels are unstable. As a result, the channel becomes very wide in relation to its depth.

> **Interesting note**
>
> Seven of the world's rivers are more than 5000 km long – the Nile, Amazon, Yangtze, Mississippi-Missouri, Yenisei, Yellow and Ob.

▲ **Figure 1.31** Braided channel with vegetation established on a central island – Kyrgyzstan, Central Asia

> **Activities**
>
> 1 What are the causes and consequences of meander migration?
> 2 With reference to Figure 1.25 (page 14), explain the processes operating on the inside and outside banks of a meander.
> 3 How are oxbow lakes formed?
> 4 Describe the formation of a floodplain.

Deltas

A **delta** is a flat, low-lying deposit of sediment that is found at a river's mouth (Figure 1.32). For deltas to be formed, a river needs to:

» carry a large volume of sediment – for example, rivers in semi-**arid** regions and in areas of intense human activity
» enter a still body of water, which causes velocity to fall; the water loses its **capacity** and **competence**, hence deposition occurs, with the heaviest particles deposited first and the lightest last.

Deposition is increased if the water is salty, as this causes salt particles to group together, become heavier and be deposited. Vegetation also increases the rate of deposition by slowing down the water.

There are a number of stages in the formation of a delta. The first is the development of sandbanks in the original mouth of the river. This causes the river to divide, and then there is a period of

The characteristics and formation of lowland landforms

repeated subdivision until there are a large number of distributaries flowing towards the sea. Each of the channels develops its own set of levées.

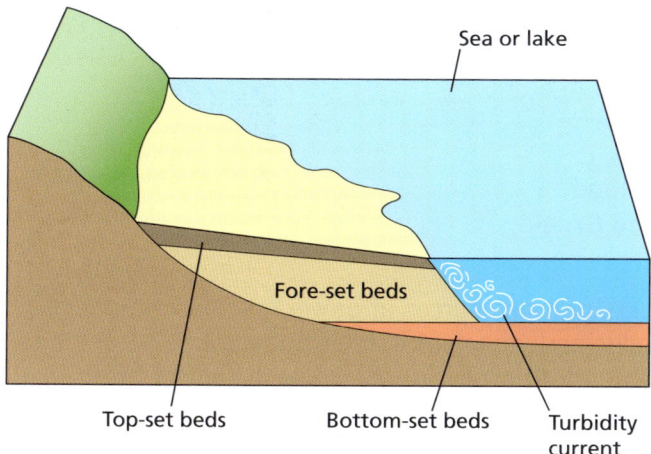

▲ **Figure 1.32** The formation of a delta

The Rhône delta

The Rhône river (Figure 1.33) divides into two main distributaries 4 km north of Arles. The east branch, the Grand Rhône, is the larger of the two, and carries 85 per cent of the Rhône's water into the Mediterranean.

» At Arles the river is just 2 m above sea level and takes almost 50 km to reach the sea.
» The delta is criss-crossed by numerous small islands, abandoned channels and active levées.
» Most settlements and transport routes are located close to the river, where the land is slightly higher. Further away from the river, the land is lower, swampy and frequently covered with water.
» Deposition by the river is estimated to be about 17 million m³ each year, or about 50 tonnes every minute. As the Mediterranean Sea has a very small tidal range, there are no currents to carry away these deposits.

> ### Activities
> 1 Looking at Figure 1.33, what is the length of the Grand Rhône between Arles and the sea?
> 2 How wide is the Rhône delta when it meets the sea?
> 3 What is the map evidence that:
>   a conservation
>   b tourism
>   are important in the Rhône delta?

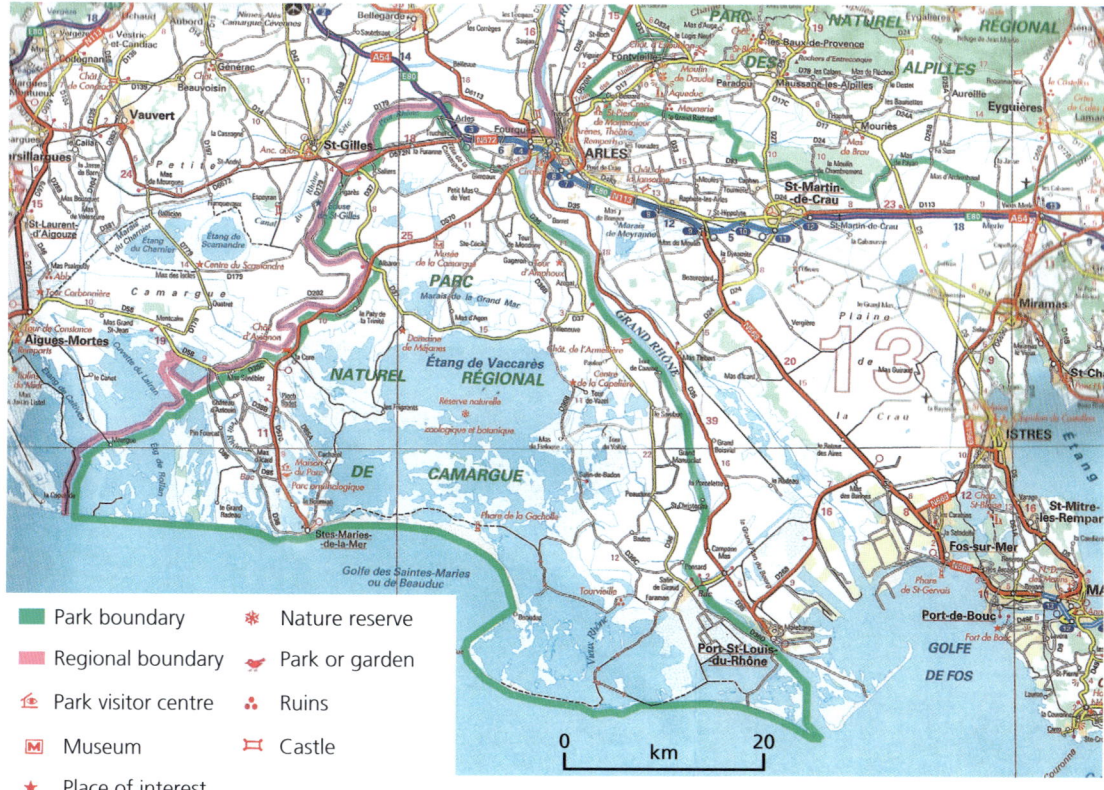

▲ **Figure 1.33** Map extract of the Rhône delta

1.3 Rivers present opportunities and hazards for people

This chapter will explain:

★ the opportunities of living near a river
★ the hazards of living near a river: floods and pollution
★ detailed specific example: the 2019 Mississippi River flood
★ detailed specific example: restoring Shanghai's Suzhou Creek.

The opportunities of living near a river

The land surrounding rivers, particularly the lower courses of rivers, is often very densely populated. This is because of:

» the significant supply of water that can be obtained from a large river
» the fertile soils usually found on floodplains, which allow for highly productive agriculture
» the opportunities for fishing, which provides a substantial additional source of food
» the use of the river as an important transport route
» the large areas of flat land for construction of homes and infrastructure.

The emergence of the world's first cities some 5500 years ago is the result of the advantages listed above. The areas that first experienced this important socioeconomic change towards larger urban areas were:

» Mesopotamia – the valleys of the Tigris and Euphrates Rivers (modern-day Iraq)
» the lower Nile valley (Egypt, Figure 1.34)
» the floodplains of the River Indus (Pakistan).

Rivers provide sources of power, mainly in terms of producing hydroelectricity at a range of scales, along with opportunities for the development of tourism, which can be a major source of employment. Many major river infrastructure projects have multiple purposes (Figure 1.35).

Locations along or near rivers frequently provide attractive environments in which to live. This fact is often reflected in higher house prices close to rivers compared with further away. However, in many areas, particularly those that have been subject to recent floods, houses have proved difficult to sell and also to insure.

▲ **Figure 1.35** A system of locks along the River Danube – part of a multipurpose river scheme providing hydroelectricity, flood control and improved navigation

▲ **Figure 1.34** Arable farming in the lower Nile valley, with the pyramids in the background

The hazards of living near a river

Floods

Floods are a natural feature of all rivers. For most of the time a river is contained within its channel, but at other times it may burst its bank and a flood occurs. Floods bring advantages such as water and fertile alluvium, which allow farmers to grow crops. But the problem is that they may bring too much water and too much silt. The results can be devastating, as the many disastrous floods throughout the history of China show. Rapid river bank erosion during floods can cause population displacement and socioeconomic impacts. For example, the flooding of the Meghna River in Bangladesh caused major disruptions during the 1990s and 2000s. The frequency and intensity of floods has increased significantly in many countries due to climate change, and this is a trend that is likely to continue.

Hazards associated with flooding can be divided into primary, secondary and tertiary effects.

Primary effects:

- Lives being lost, particularly when flooding is severe and occurs with little warning.
- Rural areas being inundated, resulting in crop loss and livestock destruction.
- Urban areas being inundated, impacting housing, industry and services infrastructure. Underground transit systems can be particularly affected.
- Flood control and other river infrastructure being damaged – bridges, levées, flood walls.
- Pollutants being transported and distributed over a wide area.

Secondary effects:

- Drinking water supply systems becoming polluted, especially if sewage treatment plants are flooded.
- Disease spreading due to contaminated water and other factors.
- Electricity and other energy-supply services being disrupted.
- Transportation systems being impacted, which may result in food shortages.

Tertiary (long-term) effects:

- Wildlife habitats being destroyed.
- The routes of river channels may change, leaving old channels dry and infrastructure abandoned.
- The quality of some farmland may be adversely impacted, leading to farming being abandoned or severely curtailed.
- Land, housing and all types of infrastructure may become more expensive to insure.

Floods account for about one-third of all natural catastrophes. They cause more than half the fatalities and are responsible for one-third of the economic losses.

The natural causes of flooding

The initial causes of floods are natural. However, human interference intensifies many floods (Figure 1.36). A flood is a high flow of water that overtops the banks of a river. The primary causes of floods are mainly the results of external climatic forces. The secondary causes of floods tend to be drainage-basin specific. Most floods in Western Europe, for example, are associated with deep depressions (low-pressure systems) in autumn and winter, which are both long in duration and wide in areal coverage. By contrast, in India up to 70 per cent of the annual rainfall occurs in just 100 days, in the summer southwest **monsoon**. Elsewhere, melting snow may be responsible for widespread flooding.

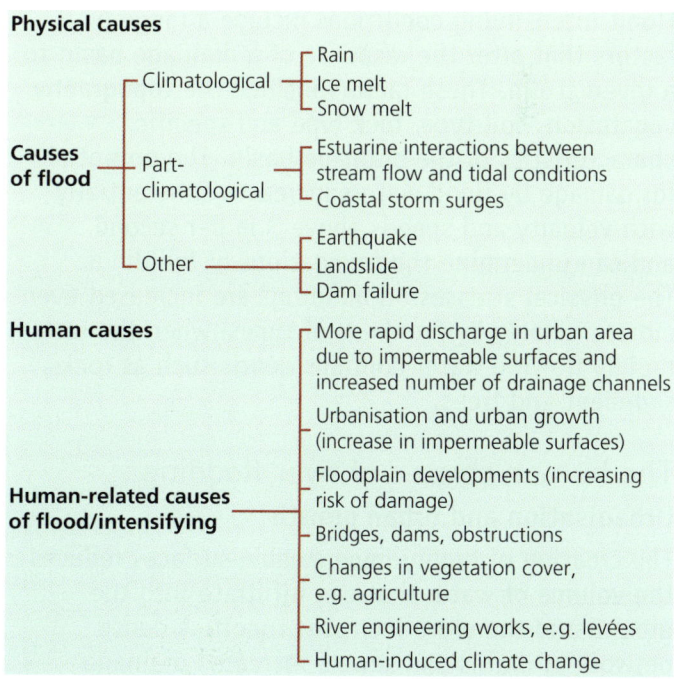

▲ Figure 1.36 The natural and human causes of floods

Heavy rainfall is the major cause of flooding in most countries, particularly long periods of rain that completely saturate the soil. When soil is saturated, surface runoff (overland flow) is at its

1.3 RIVERS PRESENT OPPORTUNITIES AND HAZARDS FOR PEOPLE

most significant, resulting in the extremely rapid movement of water into river systems. Snowmelt can be an important contributing factor in some areas, as unusually rapid snowmelt can sometimes be the major cause of floods.

Floods caused primarily by heavy rainfall are sometimes referred to as slow-building floods, as opposed to **flash floods** that impact much more suddenly. Flash floods are usually the result of a short period of unusually intensive rainfall. They are especially common in arid and semi-arid areas, and sometimes occur after a period of **drought** when land has been baked hard by the sun, preventing infiltration. Small river basins are particularly subject to flash floods.

In low-lying areas of active floodplains and river estuaries, flood water can spread out easily over a large area, but the gradient is too small to allow the water to leave the area quickly. In Bangladesh, 110 million people are living relatively unprotected on the floodplains of the Ganges, Brahmaputra and Meghna Rivers. Floods caused by the monsoon regularly cover 20–30 per cent of the flat delta. In very serious floods, up to half of the country may be flooded.

Flood-intensifying conditions include a range of factors that alter the response of a drainage basin to a given storm. These factors include the topography, vegetation, soil type, rock type and specific characteristics of the drainage basin. The potential for damage by flood waters increases substantially with velocity and speeds above 3 m per second, and can undermine the foundations of buildings. The physical stresses on buildings are increased even more, probably by hundreds of times, when the rough, rapidly flowing water contains debris such as rocks, sediment and trees.

The human causes of river flooding

Urbanisation and urban growth

The creation of highly impermeable surfaces reduces the volume of water that can infiltrate into the ground and therefore increases runoff. A dense network of drains and sewers increases drainage density so runoff is transferred quickly into river channels. Increased storm runoff means many sewerage systems cannot cope with the resulting peak flow. Natural river channels are often constricted by bridge supports or riverside facilities, reducing their carrying capacity.

Floodplain development

The increasing development of floodplains over the past century for housing and economic activity at the expense of farming and forestry, has significantly increased the proportion of land covered by impermeable surfaces, putting a higher percentage of people at risk of flooding in many countries.

The build-up of river debris from human activities

Debris from industrial, agricultural and domestic activities can significantly decrease the flow rates of river channels. Such debris can also pollute water channels and impact ecosystems.

Deforestation and poor agricultural practices

Deforestation, particularly in upland areas, has increased the rapid movement of water into lowland areas, often overwhelming the capacity of the physical landscape to absorb such large amounts of water. Poor agricultural practices, such as removing hedgerows and leaving large areas bare of vegetation, increase the problem of the rapid movement of water into river channels.

Human-induced climate change

A warmer atmosphere can hold more moisture and subsequently release greater volumes of precipitation. Higher temperatures can cause more 'rain-on-snow' events, with warm rains causing faster and often earlier snowmelt. Hurricanes and other storm events are predicted to become more frequent.

> ### Activities
> 1. Discuss two physical causes of flooding.
> 2. How has urban and economic development on floodplains increased the risk of flooding?
> 3. Outline the causes and consequences of a major flood that has occurred anywhere in the world in the last few years.

The impacts of river flooding

Floods cause more than US$40 billion in damage worldwide every year. Floods are projected to increase for many world regions. The USA, Australia, South Africa and India were among countries affected by serious flooding in 2022 (Table 1.3).

The hazards of living near a river

▼ **Table 1.3** Examples of high-impact flood events in 2022

Event	Effects
Pakistan floods	• Extremely heavy monsoon rains impacted more than 33 million people. • The month of July experienced 181 per cent of average precipitation and August 243 per cent. • The number of deaths exceeded 1700 and there were economic losses of nearly US$30 billion.
India floods	• Monsoon flooding and landslides caused 700 deaths.
Bangladesh floods	• Floods affected 7 million people and 141 people died.
Sahel floods	• Nigeria, Niger, Chad and southern Suden all suffered floods due to heavy rain at the end of the monsoon season (Oct-Dec). • Deaths exceeded 600 in Nigeria and there were economic losses of US$4.2 billion.

Climate change is altering the frequency, severity and location of flooding. A higher average global temperature is increasing the amount and intensity of rainfall during precipitation events. This is likely to amplify the severity of flooding, although floods could become rarer in some regions.

Flooding can seriously affect the quality of groundwater and surface water resources, along with the quantity of safe water that can be delivered to water users. Damaged infrastructure can result in severe interruptions to water supply services. The most frequent problems are:

» scarcity of safe drinking water
» disruption of **water treatment** facilities
» the outbreak of disease as a consequence.

The reduction of groundwater quality can be caused by pollutants being transported below the surface on groundwater recharge. Flooding can affect **wellfields**, with floodwater contaminating damaged wells. A wellfield is the land above and surrounding the wells that have been drilled into an aquifer. Significant interruptions to treatment facilities may occur, along with damage to distribution systems. Severe flooding can result in the interruption of abstraction from artificial reservoirs and the deterioration of stored water due to turbidity (murkiness due to more sediment).

Flood risk

Flood risk (Figure 1.37) is a combination of:

» hazard – the probability of occurrence
» exposure – a measure of the number of people or things that may be affected
» vulnerability – a measure of the potential for people and property to be affected.

Flood risk is dependent on there being:

» a source of flooding, in most cases a river
» a route, such as a river valley, for the flood water to take (a pathway)
» a receptor (people and property) that will be affected by the flood.

Therefore, the impact of a flood event will be considerably lower in a sparsely populated region (low exposure) and/or where people are able to evacuate quickly (low vulnerability), compared to in a densely populated region (high exposure) and/or where evacuation would be problematic (high vulnerability). This concept is known as the source-pathway-receptor model. In most countries there will be a system of **flood impact assessment**.

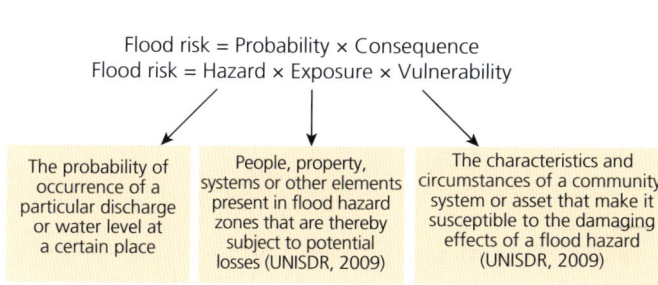
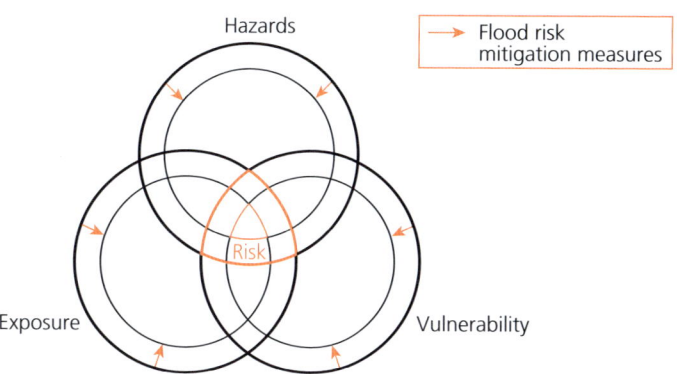

▲ **Figure 1.37** The concept of flood risk and its reduction, including Venn diagram

1.3 RIVERS PRESENT OPPORTUNITIES AND HAZARDS FOR PEOPLE

The strategies to manage river flooding: prediction and prevention

Prediction

In recent decades, flood forecasting and warning has become more accurate, due mainly to advances in weather satellites and the use of radar. This is particularly the case in developed countries. There is much less effective flood forecasting in most developing countries. However, there are interesting exceptions, such as Bangladesh. Most floods in Bangladesh originate in the Himalayas, giving authorities about 72 hours' warning. Disastrous floods, however, can often come as a surprise.

In the USA, National Weather Service forecasters rely on a network of almost 10,000 river gauges to monitor the discharge of rivers across the country. According to the **US Geological Survey**, flood prediction requires several types of data:

- the amount of rainfall occurring
- the type of storm producing the precipitation
- the rate of change in the discharge of the river/channel network
- the characteristics of the drainage basin.

The task then is to convey information as quickly as possible about the immediacy and severity of the flood risk to the people who are likely to be affected. For example, using the UK government website (https://gov.uk), allows you to check if you are:

- at immediate risk of flooding
- at risk of flooding in the next five days
- in an area that's likely to flood in the future.

Prevention

Traditionally, floods have been managed by methods of '**hard engineering**', such as dams, reservoirs, levées, straightened channels and flood-relief channels (Figure 1.38). Although hard engineering can be effective in reducing the risk in the locations selected, it may cause unexpected effects elsewhere in the drainage basin. Such negative consequences include increased sedimentation, bed and bank erosion, decreased water quality and loss of habitats. Levées are the most common form of river engineering. Over 4500 km of the Mississippi River has levées.

▲ **Figure 1.38** Hard engineering structures: (a) river defences on the River Thames, London; (b) levées in Zermatt, Switzerland

More recently, '**soft engineering**' measures have increasingly come to the fore. These techniques focus on working with natural processes and features rather than on attempting to control them. They include catchment management plans, river restoration and wetland conservation. **Flood abatement** involves reducing the amount of runoff in a drainage basin. This can be achieved through actions such as:

- reforestation, to slow down the movement of water in a drainage basin
- reseeding sparsely vegetated areas, to reduce evaporation losses and control soil erosion
- treatment of slopes by contour ploughing or terracing (Figure 1.39), which reduces surface flow
- clearance of sediment and other debris from streams, to increase river discharge
- preservation of natural water storage zones, such as lakes
- construction of small water- and sediment-holding areas
- comprehensive protection of vegetation from wildfires, **overgrazing** and clear-cutting of forests.

The hazards of living near a river

Flood diversion refers to the practice of allowing certain areas, such as wetlands and floodplains, to be flooded to a greater extent. Natural flooding may be increased through the use of flood-relief channels (diversion spillways) to direct more water into these areas during times of flood.

▲ Figure 1.39 Terracing in mountainous terrain in Nepal – terracing increases agricultural production and reduces surface runoff

Land-use zoning

Hydrological responses to rainfall strongly depend on the local characteristics of the soil, such as water storage capacity and infiltration rates. The type and density of vegetation cover and the characteristics of the built environment are also very important in the hydrological response to rainfall.

Land-use zoning segregates land use on a floodplain into different areas by type of use. Most land-use zoning practices date from the mid-twentieth century, although earlier examples can be found. Carefully planned land-use zoning can reduce the number of premises and people at risk of flooding. For example, parks and gardens can be planned and designed to absorb and contain significant amounts of water during and after heavy rainfall. On a larger scale, **catchment management** of the whole catchment area to optimise the functioning of the catchment can use a source control approach that seeks to keep as much rainwater as possible retained where it falls, thus reducing flood peaks.

Floodplain land that floods regularly (once a year) could be used for pastoral agriculture, as animals can be moved to higher ground when there is a risk of flooding. Alternatively, such land could be used for recreational purposes, where there would be minimal damage to infrastructure. In many rapidly expanding urban areas, unplanned and poorly planned development has often outpaced the construction and improvement of drainage infrastructure.

Hazard-resistant building design

Hazard-resistant design has also become increasingly important. Hazard-resistant design (for example flood-proofing) includes any adjustments to buildings and their contents that help reduce losses. Table 1.4 shows adjustments to building and site design that should be considered when building in flood-risk areas. Incorporating flood-resistant construction measures during the construction and renovation of homes not only prevents damage, but also reduces post-flood stress.

▼ Table 1.4 Four levels of adjustments to building and site design in flood-risk areas

Flood avoidance	Constructing a building and site in a way that minimises the chance of it flooding. For example: • Building it higher than the flood level. • Moving to an area that is not at risk of flooding.
Flood resistance	Constructing a building in a way that stops floodwater coming into the building and causing damage. For example: • Using specialist flood doors and windows. • Installing automatic anti-flood airbricks.
Flood resilience	Constructing a building in a way that reduces the impact of flooding should water get into the building, so that: • permanent damage is avoided • structural integrity is preserved • the clear up and drying is made easier. For example: • Ensuring that electrics and sockets are positioned above likely flood levels. • Installing tiles instead of fitted carpets.
Flood repairable	Constructing a building in a way that, should water get into the building, any damage caused by the flood can be repaired easily or replaced. (This is also a type of flood resilience.) For example: • Using treated timber for flooring rather than untreated timber. • Installing 'marine ply' kitchen units.

1.3 RIVERS PRESENT OPPORTUNITIES AND HAZARDS FOR PEOPLE

Sustainable drainage systems

Sustainable drainage systems (SuDS) are natural drainage methods for the drainage of surface water to nearby watercourses in urban areas. They provide an alternative to the direct channelling of surface water by pipes and sewers. They aim to:

» lower flow rates in watercourses
» increase surface water storage capacity
» improve water quality by reducing the transport of pollutants to water environments.

Figure 1.40 shows the different methods used in SuDS to decrease flow rates in watercourses and to improve water quality:

1 Source-control methods, such as intercepting runoff water on roofs for subsequent reuse or for storage and subsequent evapotranspiration (for example, green roofs), to reduce the volume of water entering watercourses.
2 Pre-treatment steps, such as vegetated swales (channels) or filter trenches that remove pollutants.
3 Retention systems that delay the discharge of surface water to watercourses by providing storage within ponds, retention basins or wetlands.
4 Infiltration systems, such as infiltration trenches and soakaways, that allow water to soak into the ground.

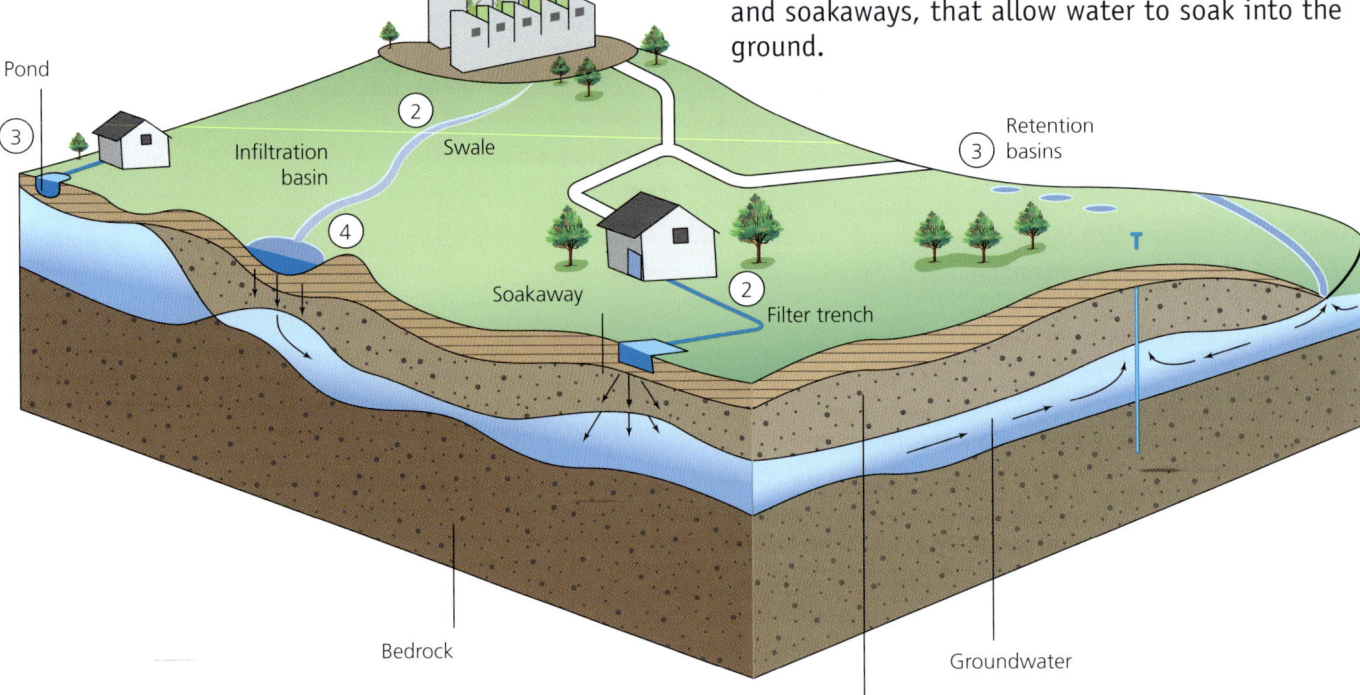

▲ **Figure 1.40** Sustainable drainage systems

Activities

1 Write a brief analysis of Figure 1.37 (page 21).
2 Which types of data does the US Geological Survey use for flood prediction?
3 Describe three methods of hard engineering.
4 a Define soft engineering.
   b List four types of soft engineering.
5 Explain why appropriate land-use zoning can reduce flood risk.
6 What are sustainable drainage systems?

Detailed specific example

The 2019 Mississippi River flood

The Mississippi River drainage basin

The most prolonged and widespread flooding in US history occurred in 2019, with the Mississippi River drainage basin the most affected region in the USA. The Mississippi River basin drains one-third of the continental USA along with two Canadian provinces (Figure 1.41). The Upper Mississippi River, Arkansas River and Missouri River all drain into the Lower Mississippi River, and all recorded major flooding.

The flood events in the Mississippi basin caused estimated economic losses of $20 billion. The 2019 Mississippi River flood broke records set by previous flood events in 1973 and even 1927. The impacts of the floods in these two former years were colossal, and shape how the Mississippi River is managed now. Precipitation is increasing across this very wide area due to global warming, causing additional water to flow downriver towards the Gulf of Mexico. The shape of the river basin has been described as an inverted triangle, extending from the central Rocky Mountains to the west and central Appalachian Mountains to the east, to the Gulf of Mexico to the south. The main land use in the basin is cropland (58 per cent).

The 3371 km Mississippi River stretches from Minnesota to the Gulf of Mexico. It flows north to south through 18 degrees of latitude, with the climate type changing from cool temperate at its origin to sub-tropical at the mouth of the river. The river drains water and sediment from 31 US states, delivering both to the Gulf of Mexico.

▲ Figure 1.41 A map of the Mississippi River basin

1.3 RIVERS PRESENT OPPORTUNITIES AND HAZARDS FOR PEOPLE

The causes and duration of the 2019 floods

A lethal combination of heavy spring rain, rising temperatures and melting snow overwhelmed waterways in the drainage basin. The 12 months from May 2018 to April 2019 were the wettest year-long period in records dating back to 1895. In the parts of the basin subject to heavy annual snow accumulation, snow cover has generally decreased. This is due to:

- warmer temperatures causing earlier melting
- more precipitation falling as rain rather than snow.

There is also increasing evidence to suggest that the persistent jet stream behaviour behind these record-setting precipitation events is most likely due to human-induced climate change. Red River Landing is located on the Mississippi River near the confluence of the Mississippi and Red Rivers, and Figure 1.42 shows the number of days above flood stage at Red River Landing for the floods of 1927, 1973, 2011 and 2019.

There were three major periods of flooding between March and September 2019. Many areas in the basin were flooded more than once.

- In March, part of the US Midwest and the northern states were impacted by a 'bomb cyclone' that precipitated huge amounts of snow and rain in a short time period.
- Peak flooding from spring to July was the most devastating event.
- A final period of flooding occurred in late summer/early autumn.

However, the US Army Corps of Engineers had been working since early November 2018 to manage flood waters and prevent flooding in communities across the Mississippi River Valley. While communities upriver had experienced devastating flooding, those in and around Louisiana had been largely protected by the Mississippi River and tributaries system.

Between January and June 2019, precipitation was as high as 150–250 per cent above normal in much of the drainage basin. For much of the region it was the wettest period in 124 years. By early July 2019:

- the Mississippi River had been above flood stage for 235 days, the longest period in recorded history

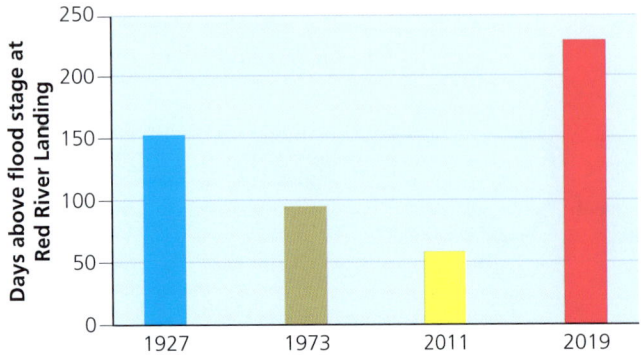

▲ **Figure 1.42** The number of days above flood stage at Red River Landing for the floods of 1927, 1973, 2011 and 2019

- nearly 210 trillion gallons of water had flowed down the Mississippi River since the beginning of the year – 64 per cent above the 10-year average.

For the first time in its almost 100-year history, the Bonnet Carre Spillway was opened twice in back-to-back years to relieve pressure on levées and prevent devastating flooding over a large area. In 2019, the Bonnet Carre Spillway was open for 123 days.

The effects of the flooding

- The floods caused nearly 20 million acres of farmland to go unplanted.
- Around 14 million people were displaced for some period of time.
- More than 550,000 acres in the Mississippi delta, almost half of it farmland, was impacted by flooding and stagnant water.
- St Louis Harbour was closed for 38 consecutive days between May and June 2019. From March to June, an estimated 6.3 million tonnes of grains worth almost $1 billion went unshipped due to disruptions to barge traffic.
- The flooding led river authorities to believe many river passages unsafe due to strong currents and obstructions along the river bed. They closed multiple locks, suspended traffic on certain sections, set tow size restrictions and in some areas limited passage under bridges to daylight hours.
- In Louisiana, the seafood industry suffered severe disruption. The influx of a much higher than normal level of fresh water impacted the Gulf of Mexico ecosystem and decimated species such as shrimp and oyster.

Flood protection measures

Although flood protection measures have steadily improved through a number of phases along the Mississippi, and in more recent times have involved soft engineering as well as hard engineering, the USA was taken by surprise by both the scale and duration of the 2019 floods. Hydrologists argue that lessons must be learned, stressing the urgent need to consider managing the Mississippi River more holistically to deal with the more frequent and intense floods expected in the future. Communities across coastal Louisiana also fear water coming up from the Gulf of Mexico.

Figure 1.43 shows the reaches of the Mississippi River and the alterations made to the natural river over time. The headwater reach extends from the river's source at Lake Itasca to Coon Rapids, Minnesota. Through much of this reach, the river flows through wetlands and forest. Alterations in the impounded reaches focused on making the river navigable for commercial shipping and on flood control to allow development of the floodplain for agriculture and urban areas. The process of impoundment began in the 1930s. Levée construction disconnected a substantial portion of the historic floodplain from the river.

The magnitude and range of alteration is greatest in the free-flowing reaches of the Mississippi River. This section of the river begins just upstream of its confluence with the Missouri River. Dikes (also called wing dams or wing dikes) are used to direct the flow of water to maintain the alignment and depth of the navigation channel. Dikes are usually constructed in groups of three or more (dike fields), with each successive downstream dike extending further into the channel. Revetted banks, which are reinforced with concrete or other supportive material, protect the outside of meanders from erosion. Levées dominate the river banks and have been incrementally raised over time in response to large flood risks. As in the impounded reach, the levées disconnect the river from much of the historic floodplain. In urban areas along the river the levées are often replaced by floodwalls, which use less land. Dredging occurs when required and is used in various stretches of the river's course all of the time.

There are of course many other examples of dams and other alterations on the tributaries that feed into the Mississippi River. These include on the Missouri, Ohio and Tennessee Rivers.

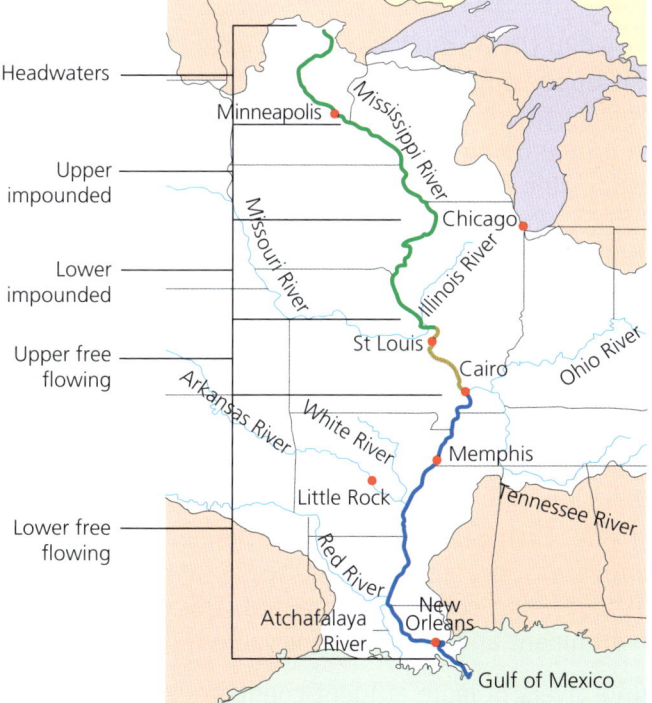

▲ **Figure 1.43** The reaches of the Mississippi River

Sustainable management strategies

Restoring floodplains is an important strategy for communities attempting to manage annual flooding along the Mississippi River. As the river proceeds downstream from its headwaters, an increasing proportion of the floodplain is disconnected from the river. Disconnected floodplain refers to the portion of the historic floodplain that is now separated from the river by levées.

Flood preventative land-use zoning is widely recognised as an important soft engineering technique. The main problem is that, without huge expense and disruption, this process can only function at a slow speed. Areas of land on floodplains have been purchased by federal and local governments and by NGOs to prevent further development and to implement nature-based solutions for flood control. Protecting existing wetlands through zoning is often an important first step. Differences of opinion between the federal government and local governments over land use in floodplains have been a regular occurrence over the years.

Pairing levées with nature-based solutions is viewed as the best way forward by many hydrologists working along the Mississippi River. Nature-based solutions mean that levées are better prepared to resist floods. The concept of 'levée setbacks' is gaining increasing attention. This involves relocating levées further away from a river and creating space for nature-based solutions between the river and the relocated levée.

The Nature Conservancy has played a significant role in promoting nature-based solutions:

- In the Atchafalaya River basin in Louisiana, a million acres of wetland provide a critical natural habitat. The Nature Conservancy is working to ensure the natural benefits of flooding by protecting the existing floodplain forest, planting new trees and restoring natural hydrology where possible. It has shaved the tops off sections of canal levées and cut notches in river banks to allow floodwaters to once again flow through the wetlands.
- In Iowa, The Nature Conservancy is working with farmers to promote in-field practices such as prairie strips. These are sections of native prairie vegetation planted within crop fields to slow runoff. The planting of cover crops has also been encouraged. Cover crops are planted during the off-season to hold soil in place and break up ground compacted by farm machinery. Additional practices, including no-till farming or only partially tilling fields, create a more absorbent soil surface that holds in moisture during the drier months and reduces runoff when conditions are wet.
- In Missouri, flooding periodically affects almost every county. Communities are advised on the benefits of nature-based solutions to address flood risk. These include planting trees to replace areas of turf grass on floodplains (as the latter has a very low level of water percolation when the land is sloping), and building bio-retention ponds to hold some of the runoff from flooding.

> ## Activities
> 1 Describe the area covered by the Mississippi River basin.
> 2 Discuss the causes and consequences of the 2019 floods.
> 3 Explain two traditional methods of flood protection used along the Mississippi River.
> 4 Outline two nature-based strategies that have been used in the Mississippi River basin.

1.3 RIVERS PRESENT OPPORTUNITIES AND HAZARDS FOR PEOPLE

River pollution

The human causes of river pollution

Pollution is sometimes referred to as the 'invisible water crisis'. The composition of surface and underground water has an impact on **water quality**, and this depends on two broad factors:

- Natural conditions in the drainage basin
- Human intervention.

Human intervention is mainly in the form of **pollution**. Water pollution comes from a number of sources (Figure 1.44), including:

- contamination by agricultural runoff, particularly from factory farming, where livestock are kept on impermeable surfaces as opposed to grazing in fields
- industrial pollution of rivers and other water bodies
- urban runoff carrying pollutants from cars, factories and other sources
- untreated sewage, which has increased in many areas as population has grown.

▲ **Figure 1.44** A polluted river in Christchurch, New Zealand

Countries in all income groups are showing signs of risks related to water quality. Low levels of wastewater treatment is the main cause of poor water quality in low-income countries (LICs), while in high-income countries (HICs) runoff from agriculture is a more serious problem. A serious constraint in the efforts to improve water quality is that detailed water quality data remains sparse, largely due to weak monitoring and reporting capacity.

In 2019, the World Bank identified three types of pollution:

- Pollutants of poverty: Developing countries have lacked the finance to provide adequate water treatment facilities.
- Pollutants of growing prosperity: The use of nitrogen as a fertiliser has risen more than 700 per cent since 1960. Much of this growth has occurred in Asia.
- Emerging pollutants: Microplastic pollution has grown rapidly throughout the world's fresh water sources. Emerging pollutants are largely associated with developed economies.

Water pollution is forecast to intensify over the next 20 years or so and to present a serious threat to sustainable development. The list of contaminants of concern is increasing and exposure to pollutants is likely to increase significantly in low- and lower-middle-income countries. In these countries, pollution will be driven by population growth, economic growth and the lack of wastewater treatment. The increase in pollution is projected to be particularly strong in some African countries.

The 2020 UN World Water Development Report found that opportunities were being missed to use water projects to cut **greenhouse gas** (GHG) emissions while improving access to clean water. Wastewater is responsible for about 5 per cent of all GHG emissions globally. Processing sewage can turn wastewater from a source of GHGs to a source of clean energy if the methane is captured and used in place of natural gas. Currently 80–90 per cent of wastewater is discharged to the environment without treatment.

The impacts of river pollution

Water pollution reduces the ability of a water source to provide the quality of water that it would otherwise provide. Each year, more than 80 per cent of the world's wastewater is released to the environment without being collected or treated. This pollutes the environment and wastes a renewable resource.

While rivers in more affluent countries have steadily become cleaner in recent decades, the reverse has been true in much of the developing world. A significant reason for this is the impact of globalisation, in terms of outsourcing production to developing countries. Rivers in Asia are the most polluted. For example, each day around 800 million litres of sewage are drained into the Yamuna River that flows through Delhi. For many people, the only alternative to using this water for drinking and cooking is to turn to water vendors, who sell tap water at greatly inflated prices. According to a 2018 survey by the Delhi government, 44 per cent of residents in Delhi's lowest-income areas rely on bottled water.

Sewage and runoff from farms frequently contain nutrients such as nitrogen and phosphorus. These nutrients cause excessive aquatic plant growth (**eutrophication**), which can have a range of adverse ecological effects.

Poor water quality can have a very direct and massive impact on people's lives. Most seriously it can result in waterborne diseases, which can have significant health implications. Apart from health, the economic wellbeing of individuals and families is also likely to be adversely affected. If large numbers of people are affected by waterborne diseases, this will impact on national productivity. Diseases related to poor quality water include:

- bilharzia, where snails transmit flatworms to people, causing internal organ damage
- malaria and yellow fever, both caused by mosquitoes which breed around water
- cholera, which causes extreme diarrhoea.

Although most people in developed countries think their water supplies are clean and healthy, there is growing concern about traces of potentially dangerous medicines that may be contaminating tap water. Easily dissolved in water, these remain highly toxic when leaving the body and are hard to destroy in water treatment plants. Traces of medicines in water come from manufacturing plants, the livestock industry and domestic households. Many of the more than 4000 prescription medicines used for animal and human health ultimately find their way into the environment, including Bisphenol-A (BRA), antibiotics and opiates.

> **Activities**
> 1. a Define pollution.
>   b List three sources of pollution.
> 2. Comment on the types of pollution (related to income) identified by the World Bank.
> 3. Name three diseases related to poor water quality.

The strategies and techniques used to manage river pollution

River pollution management strategies can be grouped into three categories:

- Reducing human activities that produce pollution that ends up in rivers.
- Reducing the release of pollution into rivers.
- Removing pollution from rivers and restoring ecosystems.

Reducing human activities that produce pollution

Pollutants draining into rivers from agricultural activities are a major problem in many drainage basins. Agriculture is the single largest cause of river pollution in the UK and many other countries. The more intensive the agricultural practices are, such as intensive poultry units (IPUs), the higher the level of pollutants washed into rivers. Farming using organic principles ('green' farming) substantially reduces water pollution, using methods including the following:

- No-till and low-till farming: By disturbing the soil less, there is less runoff from rainwater. This reduces both sedimentation and pollution from runoff.
- Reducing the use of chemical inputs: Sustainable farming aims to reduce as far as possible the use of artificial fertilisers, insecticides and weed killers. Fertiliser pollution increases the level of nitrates and phosphates in river water, resulting in the growth of algae that form a bloom over the water surface.
- Preventing animal waste from leaching into groundwater and causing contamination.
- Agroforestry: Combining farmland with the planting of trees, shrubs and hedges. It can also involve the use of riparian buffer strips – vegetation planted alongside a river to act as a buffer for runoff.
- Less intensive animal rearing: Meaning that animals will not require the routine use of antibiotics.

Point-source pollution from industrial premises located beside rivers can be a heavy source of water pollution. This depends very much on government regulations and monitoring. In general, the combination of regulation and monitoring is more extensive in HICs, which have greater resources to establish effective monitoring systems. At the domestic/household level, limiting the use of detergents containing phosphates is an effective way to reduce a household's pollution footprint.

An increasing awareness of how salt harms fresh water sources has seen some local authorities in the USA and elsewhere significantly cut back on the use of road salt to counter potential freezing conditions on road surfaces. Such local authorities are now using 'salting' more carefully and selectively. Road salt is generally the dominant source of chloride in regions that experience a significant level of freezing in winter. Unlike other pollutants, chloride doesn't break down in water over time, and levels of chloride have increased by more than a third since the late 1980s across the entire Upper Mississippi River basin.

1.3 RIVERS PRESENT OPPORTUNITIES AND HAZARDS FOR PEOPLE

Other reasons for the increased level of chloride in river water include:

- salt from water softeners
- potassium chloride fertiliser.

Reducing the release of pollution into rivers

Wastewater treatment is a vital process that helps to ensure water supply systems remain safe and sustainable. Pollution from unmanaged wastewater remains a pressing global challenge. Wastewater treatment involves removing pollutants from wastewater through physical, chemical or biological processes. The more efficient these processes are, the cleaner the river water becomes. The infrastructure required for large-scale effective wastewater treatment is expensive. Therefore it is not surprising that there is a high correlation between the GDP per capita of a country and the quality and extent of its wastewater systems.

However, cost is not the only obstacle. Wastewater treatment processes contribute as much to global GHG emissions as the global aviation industry. Conserving water by reducing per capita water use in a practical way, without compromising personal health, would relieve pressure on existing systems and reduce GHG emissions.

When stormwater is not properly managed, it can pick up pollutants from the surfaces it flows over. Effective stormwater management therefore reduces the amount of pollutants contaminating rivers, by methods including:

- permeable pavements to allow stormwater to infiltrate through porous surfaces into the soil and groundwater
- roadside curb 'cuts' to allow road runoff to be directed into pervious areas
- vegetated filter strips – bands of dense vegetation through which runoff is directed
- constructed wetlands designed to operate in a similar way to natural wetlands.

Altering human activities to reduce the volume and range of pollutants entering rivers is normally achieved by a combination of government legislation and education campaigns. Raising both public and commercial awareness of the causes and effects of river pollution can result in significant benefits.

Removing pollution from rivers

A range of remedial actions can be employed to revive a polluted river:

- The clearing and stabilisation of river banks to minimise further direct pollution from an area.
- Physical pollution, much of it in the form of litter, can end up in rivers from a variety of sources. Volunteer organisations in a wide variety of locations around the world carry out regular litter picks. A significant proportion of such river litter is in the form of plastic pollution (Figure 1.45). A high proportion of plastic reaching the sea comes from rivers.
- Physically removing algae blooms, which have been created by excessive nutrients (for example fertiliser runoff) being washed into a river, may be an essential activity in the process of returning a river to a healthy state.
- Pumping air into the water where the environmental state of a river is extremely poor, using floating aerators. This can prove effective in combination with other clean-up activities. Oxygen is essential for the balance and function of fresh water ecosystems. Aeration methods can be top down (surface aeration) or bottom up (benthic aeration). Some water bodies require constant aeration, while others require only emergency aeration during high-risk periods.
- Mechanical dredging to remove contaminated sediments, which will also have the benefit of increasing river flow.

▲ **Figure 1.45** Plastic pollution along the banks of the lower Danube River, Romania

> ### Activities
> 1 Describe two changes that could be made to agricultural practices to reduce polluted runoff into rivers.
> 2 How can good stormwater management reduce river pollution?

Detailed specific example

Restoring Shanghai's Suzhou Creek

Location, causes and consequences

Suzhou Creek (Figure 1.46) is a part of the Yangtze River drainage basin. Its source is Taihu Lake, from where it winds 125 km, of which almost 53 km flows through Shanghai, before its confluence with the larger Huangpu River. Shanghai is one of the world's largest cities. It is located at the mouth of the Yangtze River and bounded to the east by the East China Sea.

▲ Figure 1.46 Suzhou Creek flowing through central Shanghai

Beginning in the early 1900s, the waters of Suzhou Creek deteriorated, mainly due to increasing population and the expansion of industrial activities. Domestic sewage and industrial wastewater were discharged directly into the river, gradually polluting the water quality. By the late 1970s the entire river was heavily polluted. Suzhou's fish and shrimp populations became extinct in the 1980s. By this time it had become the most polluted water body in Shanghai. There was a high level of visual pollution and frequent complaints about the foul odour released by the mix of pollutants. Algal blooms in early summer had become a common occurrence.

▼ Table 1.5 The increasing population of Shanghai

Year	Population
2024	29.9 million
2010	20.3 million
1990	8.6 million
1970	6.1 million

An important transport route for centuries, Suzhou Creek had become overloaded with boats transporting a wide array of products. Apart from its transport function, the waterway also provided vital flood control and drainage functions for the area, and water for farmland irrigation and nearby factories. Over a long period, the river had deteriorated due to:

- raw industrial pollution
- urban wastewater
- spills and waste from boats
- human waste.

The situation had become so bad that the river was nicknamed the 'black and stink'. At this point, Suzhou Creek:

- failed to meet China's lowest national water quality standard (Class V)
- had become a public health hazard due to the risks of diseases such as cholera, typhoid and dysentery.

Low-income communities were most vulnerable to the hazards posed by the river as they lived, and their children played, near the creek.

Restoration in three phases

In an attempt to tackle this major water pollution problem, in 1996 the Shanghai Municipal Government and the Asian Development Bank (ADB) entered a long-term partnership to restore the creek. The Economic and Social Development Plan for Shanghai was adopted to begin the 12-year Suzhou Creek Rehabilitation Project. The first phase of this project was launched in 1998 and completed in 2003. It involved:

- the reduction of sewage discharge into the river by developing systems to divert sewage away from the creek
- the construction of a pumping station to flush the creek and input oxygen into the dead water body
- the installation of a water lock between the Huangpu River and Suzhou Creek
- the construction of a wastewater treatment plant, with a capacity of 400,000 m^3 a day
- the installation of solid waste collection wharves.

The second phase followed on immediately from the first phase and was completed in 2005. Its objectives were:

- to maintain and improve the water quality (building on the improvements in the first phase)
- to extend the cleaning to its six tributaries
- to construct embankments to develop large areas of green space along the river banks.

The third phase (2006–08) continued the emphasis on improving riverside accessibility for surrounding residents. Public use of riverfront space, appropriate distribution of a riverfront greenbelt and regulation of riverfront buildings were included in the Suzhou Creek Landscape Planning. This final phase emphasised the social and environmental benefits of the project to local residents.

The combination of these measures removed wastewater from the creek and kept it out. Overall, the project benefited about 3 million people living in proximity to the creek. It improved the environment of a considerable area of Shanghai by:

- improving sanitation services
- reducing serious risks to public health
- providing greater access to parks and green spaces along the river banks.

1.3 RIVERS PRESENT OPPORTUNITIES AND HAZARDS FOR PEOPLE

Public surveys conducted after completion of the project showed that satisfaction with the city's environment had increased from 12 per cent in 2000 to 71 per cent in 2003. Satisfaction with water quality had risen from 12 per cent in 2000 to 76 per cent in 2003.

Finance and project partners

The initiation and ultimate success of the project depended on collaboration between a number of partner organisations. The total cost of the project was US$876 million, comprised of:

- a US$300 million loan from the Asian Development Bank
- US$325.4 million from the State Development Bank
- US$132.5 million from the Ministry of Finance
- US$62.7 million from the Shanghai Municipal Government
- US$55.4 million from district and county governments.

Sustainable management

The Chinese government has stressed environmental protection and sustainable management as a national strategy under its 'Ecological Civilisation' programme. UN HABITAT has also recognised the regeneration of Suzhou Creek as a success.

Major sustainability measures include:

- Education and legislation to reduce pollutants entering the river system.
- A 'River Chief' system of responsibility at all levels of government has been established for the management and protection of rivers.
- Maintaining a sufficient level of investment to monitor water quality and maintain the necessary quality of infrastructure. Continuous monitoring of pollution levels in the drainage basin as a whole is vitally important for successful sustainability.

> ### Activities
>
> 1. Why is Suzhou Creek regarded as an important waterway?
> 2. How did the river become so polluted?
> 3. Describe the three phases in the restoration of the river.
> 4. Suggest why so many organisations were involved in the restoration project.

Practice questions

1 Figure 1.47 shows the inflows and outflows on the Nile water and rainfall graphs for the source areas.

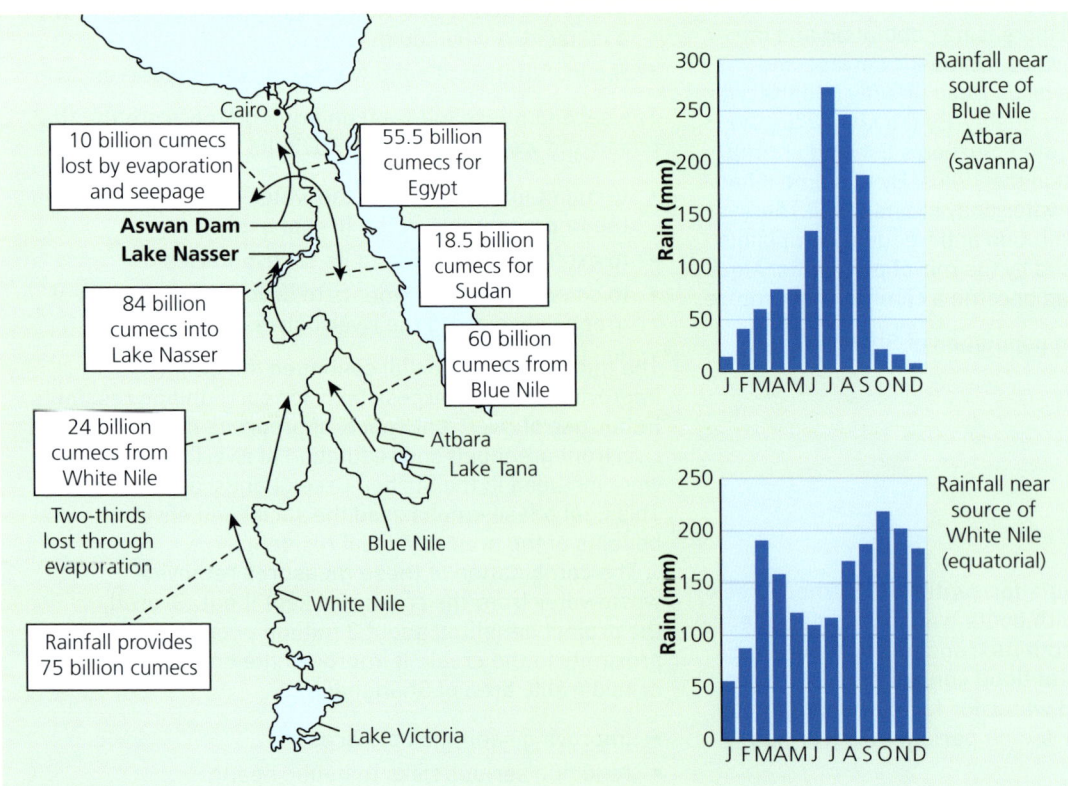

▲ Figure 1.47 The inflows and outflows on the Nile water and rainfall graphs for the source areas

a State the amount of water that:
  i initially comes from the White Nile [1]
  ii joins the Blue Nile. [1]
b Describe the rainfall pattern for the source of the Blue Nile. [3]
2 Figure 1.48 shows the changes in the flow of water in the Nile before and after the building of the Grand Ethiopian Renaissance Dam (GERD), completed in 2020.
a State the maximum flow in the Nile:
  i without the GERD [1]
  ii with the GERD. [1]
b Compare the flow of the Blue Nile without the GERD and with the GERD. [3]

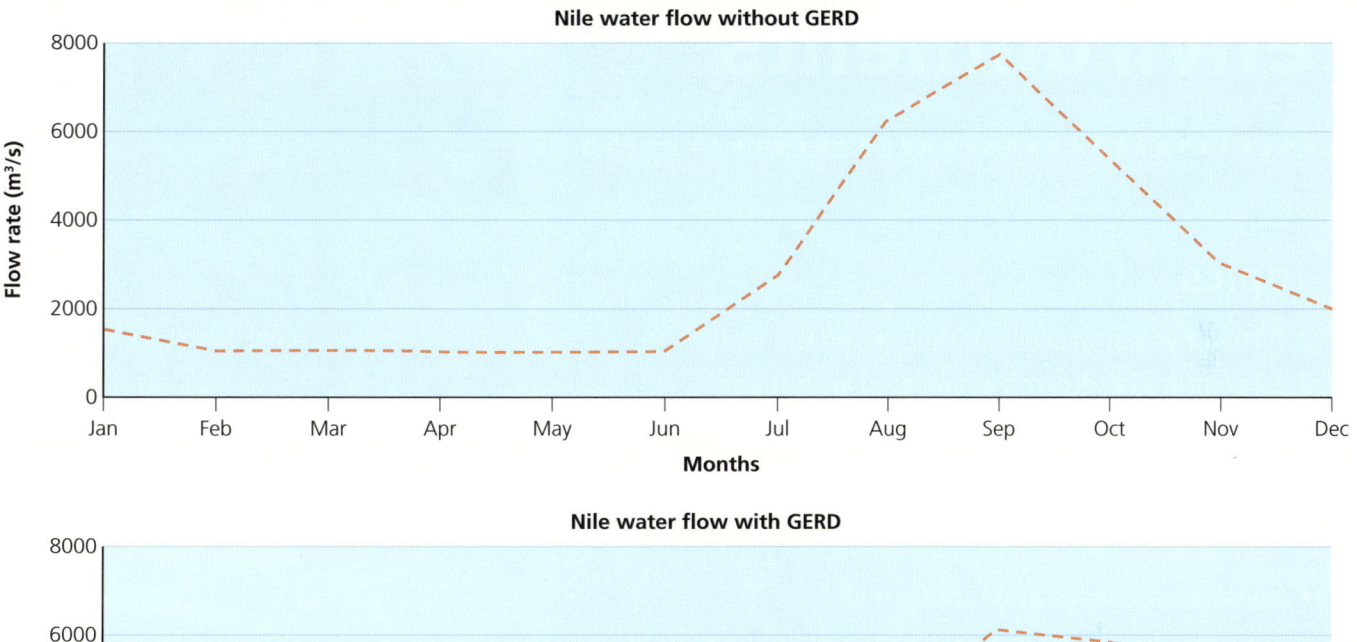
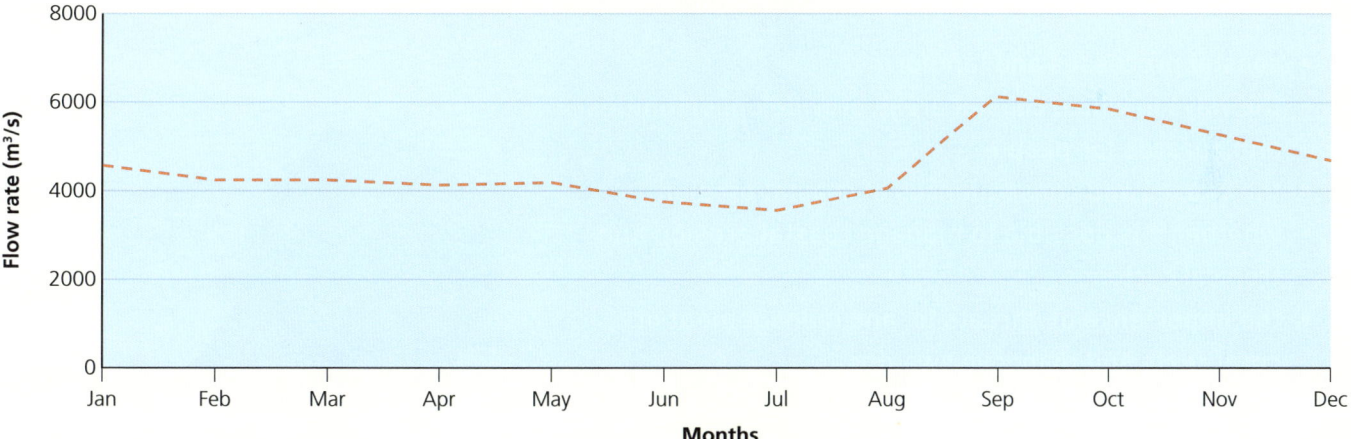

▲ Figure 1.48 The changes in the flow of water in the Nile before and after the building of the GERD, completed in 2020

TOPIC 2

Changing coastal environments

Topics

2.1 The physical processes that shape the coast
2.2 The main landforms associated with these processes
2.3 Coasts present opportunities and hazards for people

This topic looks at:

- coastal processes and landforms
- the ways in which coastal areas provide opportunities and hazards for people
- the causes and consequences of coastal erosion
- whether it is possible to manage coastal erosion and tropical cyclones
- the importance of, threats to, and management of coral reefs and mangove swamps.

2.1 The physical processes that shape the coast

This chapter will explain the causes and consequences of coastal erosion:

★ transportation
★ deposition
★ longshore drift
★ types of wave
★ wave refraction.

▲ **Figure 2.1** Jolly Harbour, Antigua

The factors that affect coastal processes and coastal landforms include:

» waves and currents, including longshore drift
» local geology – that is, rock type, structure and strength
» changes in **sea level**
» human activity and the increased use of coastal engineering.

All of these factors interact and produce a unique set of processes that occur at the coast. These processes go on to produce different types of landform for every coastal area.

Coastal erosion

There are many types of erosion carried out by **waves**:

» Hydraulic action occurs as waves hit or break against a cliff face. Any air trapped in cracks is put under great pressure. As the wave retreats, this build-up of pressure is released with explosive force (Figure 2.2). This is especially important in well-jointed rocks such as limestone, sandstone and granite, and in weak rocks such as clays and glacial deposits. Hydraulic action makes the most impact during storms.
» Corrasion (abrasion) is the process of a breaking wave hurling materials, such as pebbles or shingle, against a cliff face. It is similar to abrasion in a river.

▲ **Figure 2.2** Hydraulic action on a coral coastline

2.1 THE PHYSICAL PROCESSES THAT SHAPE THE COAST

» Attrition is the process by which eroded material, such as broken rock, is worn down to form smaller, rounder beach material.
» Corrosion (solution) occurs on limestone and chalk. Calcium carbonate, a salt found in these rocks, dissolves slowly in acidic water.

Transportation

Transportation in coastal systems is generally divided into two types:

» Bedload: Grains transported by bedload are moved through continuous contact (traction), dragging or discontinuous contact (saltation or bouncing) with the sea floor. In traction, sediments slide or roll along the sea floor – a slow form of transport. Weak currents may transport sand, and strong currents may transport pebbles and boulders. In saltation, the grains of sediment bounce along the sea bed. Moderate currents may transport sand, whereas strong currents may transport pebbles and gravel.
» Suspended load: Grains are carried by turbulent flow and generally held up by the water. Suspension occurs when moderate currents are transporting sands. Some grains are permanently held in suspension (wash loads), and these typically consist of clay.

Deposition

Deposition is governed by sediment size and shape. In some cases, notably clay, sediments will stick together (flocculate), become heavier and be deposited. Deposition occurs when there is a decrease in wave energy or velocity. This may occur along a gently sloping shoreline where there is friction with the sea bed, or when waves meet an irregular, indented coastline, for example at the mouth of a river. Obstacles on a beach, such as **groynes**, will cause a wave to break and may lead to deposition updrift of the groyne.

Longshore drift

Longshore drift occurs when waves move up to the beach (**swash**) in one direction, but the waves draining back down the beach (**backwash**) take a different route (under the effect of gravity). The net movement is therefore *along* the shore, hence the term 'longshore drift' (Figure 2.3 and Figure 2.4). A wooden or concrete wall (groyne) may be built to prevent longshore drift moving sand or shingle away from the beach.

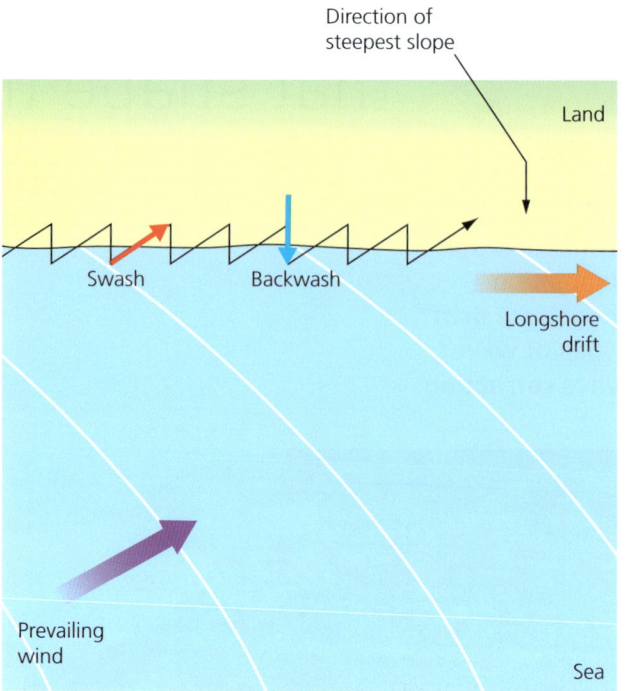

▲ **Figure 2.3** Longshore drift

▲ **Figure 2.4** Longshore drift has caused the build up of sediment on the downdrift (near side) of the stone groyne and its removal on the updrift (far side) of the groyne

Human activity and longshore drift in West Africa

The increase in coastal retreat in Ghana has been blamed on the construction of the Akosombo Dam on the Volta River. It is just 110 km from the coast and disrupts the flow of sediment from the Volta, stopping it from reaching the shore. Thus there is less sand to replace that which has already been washed away by longshore drift, so the coastline is retreating due to erosion by the Guinea Current. Towns such as Keta, 30 km east of the Volta estuary, have been destroyed as their protective beach has been removed (Figure 2.5).

Types of wave

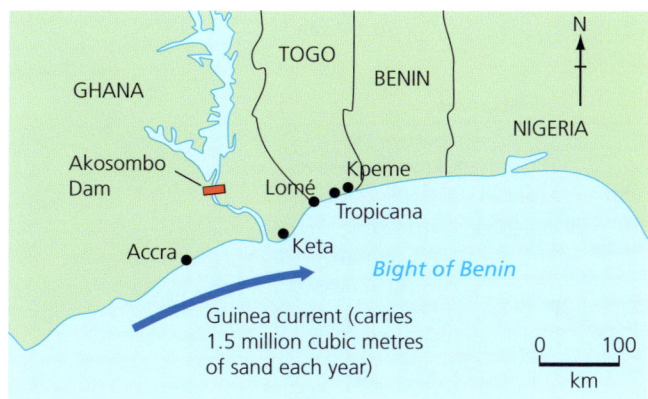

▲ Figure 2.5 Human activity and longshore drift in West Africa

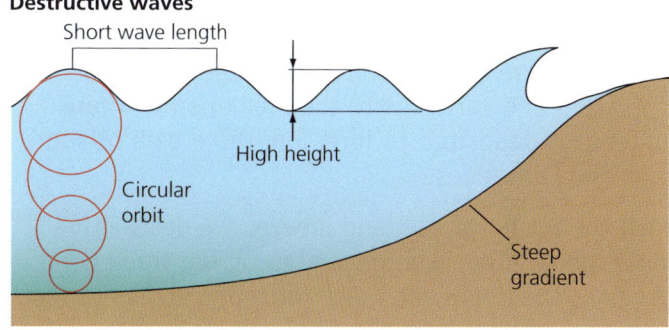

▲ Figure 2.6 Destructive and constructive waves

Types of wave

Wave length is the distance between two successive crests or troughs. Wave height is the distance between the trough and the crest. Wave frequency is the number of waves per minute. Wave velocity is the speed of a travelling wave, and is influenced by wind, **fetch** and depth of water. The fetch is the amount of open water over which a wave has passed. Waves are sometimes divided into **constructive** and **destructive waves** (Table 2.1 and Figure 2.6).

▼ Table 2.1 Destructive and constructive waves

Destructive waves (erosional waves)	Constructive waves (depositional waves)
Short wave length (less than 20 m)	Long wave length (up to 100 m)
High height (more than 1 m)	Low height (less than 1 m)
High frequency (10–12/minute)	Low frequency (6–8/minute)
Low period (one every 5–6 seconds)	High period (one every 8–10 seconds)
Backwash greater than swash	Swash greater than backwash
Steep gradient	Low gradient
Caused by local winds and storms	Caused by swell from distant storms
High-energy waves	Low-energy waves

Waves on the Palisadoes, Jamaica

Beaches are transitory features – they regularly change shape. In Jamaica, one of the most important factors for wave formation is the local sea–land breeze. The Palisadoes, on the south coast of Jamaica, has a tidal range of just 0.23 m.

Here, swell is generated by trade winds on a year-round basis, and occasionally by easterly waves and tropical cyclones (hurricanes) that originate far from the island. The sea breeze is most persistent in the summer months from May to August, and strongest in June. It most commonly approaches from an east-southeasterly direction (Figure 2.7). The breeze normally develops in the late morning, reaching velocities of 12 m/s by the early afternoon. Once the sea breeze declines, a land breeze develops from the northwest.

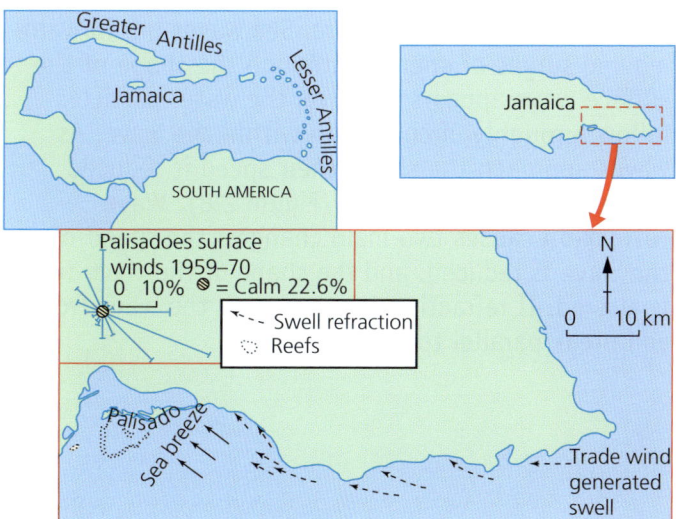

▲ Figure 2.7 Constructive and destructive waves on the Palisadoes, Jamaica

2.1 THE PHYSICAL PROCESSES THAT SHAPE THE COAST

The sea breeze is associated with an increase in wave and breaker height. Wave heights regularly exceed 1 m and may reach 5 m. The sea breezes provide a mechanism for shoreline erosion, caused by destructive waves during daylight hours. The change from constructive waves, where sediment is transported landwards, to destructive waves, where sediment is dragged seawards, is related to wave steepness. As steepness increases, erosion occurs.

When onshore winds occur, a return counter-current is formed, flowing seawards towards the break point. This current removes material from the front of the beach and carries it to the break point, where material is deposited to form longshore bars offshore from the Palisadoes.

With the return of the land breeze, the destructive waves decay. The swell generated by the trade wind becomes the dominant wave, returning sediment lost from the beach during the day.

> ### Activities
> 1 State the size of the tidal range of the Palisadoes.
> 2 Identify the direction in which the sea breeze blows.
> 3 Identify the direction in which the land breeze blows.
> 4 Suggest the impacts of:
>   a the sea breeze
>   b the land breeze
>   on wave activity.

Wave refraction

Waves result from friction between wind and the sea surface. Waves in the open, deep sea are different from those breaking on shore. Sea waves are forward-moving surges of energy. Although the shape of the surface wave appears to move, the water particles follow a roughly circular path within the wave. As waves approach the shore, their speed is reduced as they touch the sea floor (Figure 2.8). **Wave refraction** causes two main changes: the speed of the wave is reduced, and the shape of the wave front is altered. If refraction is completed, the wave fronts will break parallel to the shore.

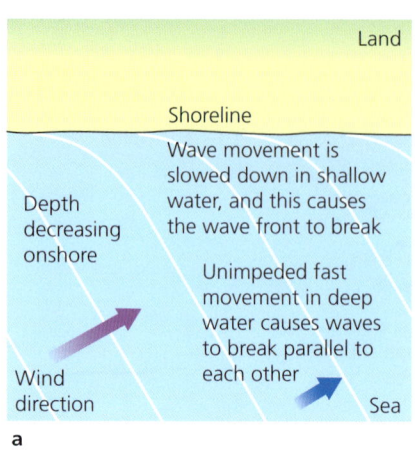

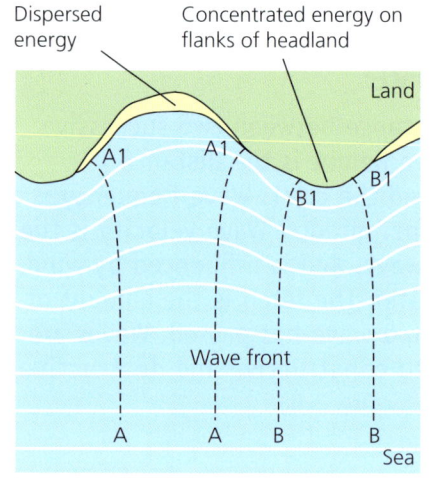

▲ Figure 2.8 Wave refraction

Wave refraction also distributes wave energy along a stretch of coast. On a coastline with alternating **headlands** and bays, wave refraction will concentrate destructive/erosive activity on the headlands, while deposition will tend to occur in the bays. Irregularities in the shape of the coastline mean that refraction is not always totally achieved. This causes longshore drift, which is a major force in the transport of material along the coast (Figure 2.3 on page 36).

> ### Activities
> 1 Define the following terms:
>   • Swash
>   • Fetch
>   • Wave refraction
>   • Backwash.
> 2 Describe the main differences between a destructive wave and a constructive wave.
> 3 Describe and explain the process of longshore drift.
> 4 Briefly describe how human activity affected the impact of longshore drift in West Africa.

2.2 The main landforms associated with these processes

This chapter will explain:

★ the main landforms associated with coastal processes
★ the formation and main characteristics of discordant and concordant coastlines.

Headlands and bays and their features

Hard rocks form headlands that protrude, whereas weaker rocks are eroded to form **bays** (Figures 2.9 and 2.10).

Wave refraction in the bay spreads wave energy around the bay, whereas it focuses wave energy on the flanks of the headlands. Bayhead beaches are formed when constructive waves deposit sand between two headlands, such as at Maracas Bay and Tyrico Bay in northern Trinidad.

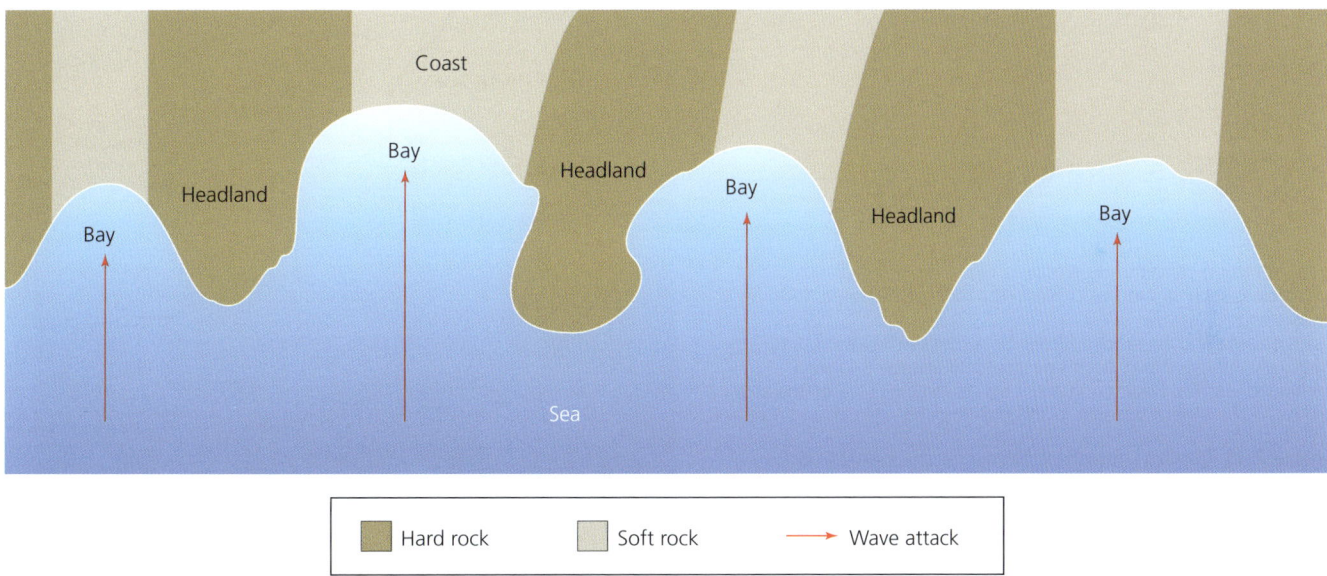

▲ **Figure 2.9** Headlands and bays

▲ **Figure 2.10** Headlands and bays in Praia de Rocha, Portugal

On a headland, erosion will exploit any weakness, creating, at first, a **cave** (Figure 2.11). Once the cave reaches both sides of the headland, an **arch** is formed (Figure 2.12). A collapse of the top of the arch forms a **stack**, and when the stack is eroded a **stump** is created (Figure 2.13). Where erosion opens up a vertical crack, allowing sea water to spout up at the surface, a blowhole is formed. The sandstone of the Cape Peninsula in South Africa has been attacked by the sea, forming steep vertical **cliffs** and small-scale features such as arches and stacks (Figure 2.14).

2.2 THE MAIN LANDFORMS ASSOCIATED WITH THESE PROCESSES

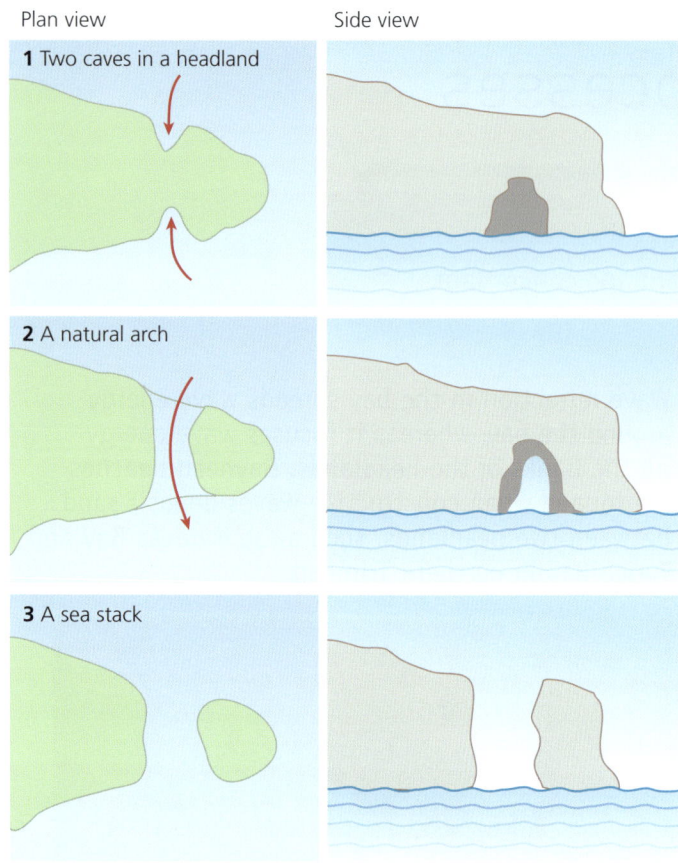

1. Wave refraction concentrates erosion on the sides of headlands. Weaknesses such as joints or cracks in the rock are exploited, forming caves.
2. Caves enlarge and are eroded further back into the headland until eventually the caves from each side meet and an arch is formed.
3. Continued erosion, weathering and mass movements enlarge the arch and cause the roof of the arch to collapse, forming a high standing stack.

▲ **Figure 2.11** The formation of caves, arches and stacks

▲ **Figure 2.12** An arch at Durdle Door, Dorset, UK

▲ **Figure 2.13** Stacks and stumps – part of the Twelve Apostles, Victoria, Australia

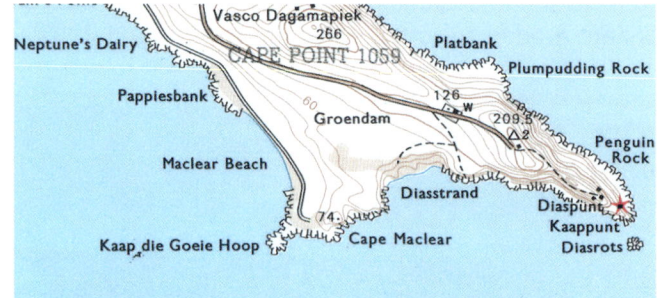

▲ **Figure 2.14** The Cape Peninsula, South Africa – showing Cape Maclear and the Cape of Good Hope in the background

Wave action is concentrated between the high water mark (HWM) and the low water mark (LWM). HWM is the level reached by the sea at high tide, while LWM is the level reached by the sea at low tide. The sea may undercut a cliff face, creating a notch and overhang (Figure 2.15). As erosion continues, the notch becomes deeper and eventually the overhang collapses, causing the cliff line to retreat. The base of the cliff is left behind as an increasingly longer platform. This is sometimes called a **wave-cut platform**, because it has been cut or eroded by wave action (Figure 2.16).

Beaches

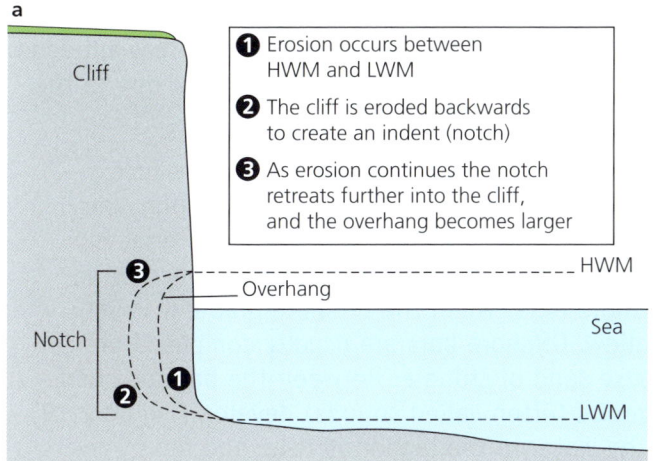

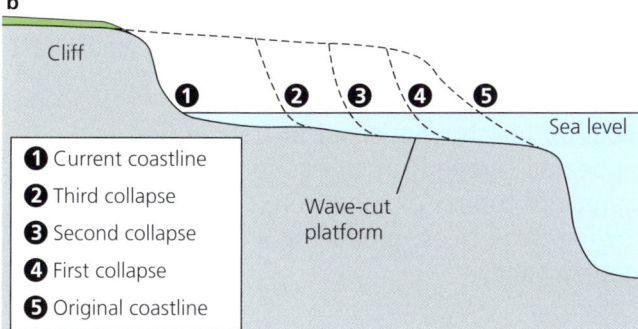

▲ Figure 2.15 The formation of a wave-cut platform

▲ Figure 2.16 A wave-cut platform, Cornwall, UK

Cliff profiles vary greatly, depending on the:

» rate of coastal erosion (cliff retreat)
» strength of the rock
» presence of joints and bedding planes.

In addition, cliffs change over time, from ones that are dominated by marine processes (like wave erosion), to ones that are protected from marine processes but affected by land-based processes.

> ### Activities
> 1. Explain the difference between attrition and corrasion (abrasion).
> 2. Explain why hydraulic action occurs in jointed rocks.
> 3. State the type of rocks likely to be affected by corrosion (solution).
> 4. Identify the types of erosion that are most likely to take place:
>   a during a storm
>   b on a beach
>   c on the face of a cliff.
> 5. In your own words, describe how a wave-cut platform may be formed.
> 6. Make a sketch of Figure 2.10 (page 39) and label the following features:
>   - headland
>   - bay
>   - stack
>   - beach.

Beaches

The best beach development occurs on a lowland coast (constructive waves) with a sheltered aspect/trend, composed of 'soft' rocks that provide a good supply of material, or where longshore drift supplies abundant material.

On Tenerife, the lack of beach material other than volcanic material (Figure 2.17) has led to one beach, Las Teresitas, being formed of sand imported from the Sahara Desert (Figure 2.18). An artificial barrier prevents the sand from being eroded by wave action.

▲ Figure 2.17 Volcanic beach, Las Teresitas, Tenerife

2.2 THE MAIN LANDFORMS ASSOCIATED WITH THESE PROCESSES

▲ Figure 2.18 An artificial beach with imported sand

The term '**beach**' refers to the accumulation of material deposited between low spring tides and the highest point reached by storm waves at high spring tides. A typical beach will have three zones: backshore, foreshore and offshore. The backshore is marked by a line of dunes or a cliff. Above the high water mark there may be a berm or shingle ridge. This is coarse material pushed up the beach by spring tides and aided by storm waves flinging material well above the level of the waves themselves. These are often referred to as storm beaches. The seaward edge of the berm is often eroded and irregular due to the creation of beach cusps.

The foreshore is exposed at low tide. The first material is deposited offshore. In this zone, the waves touch the sea bed and so the material is usually disturbed, sometimes being pushed up as offshore **bars**, when the offshore gradient is very shallow. Offshore bars are usually composed of coarse sand or shingle. Between the bar and shore, **lagoons** (often called sounds) develop (Figure 2.19). If the water in the lagoon is calm and fed by rivers, marshes and mudflats can be found. Bars can be driven onshore by storm winds and waves. A classic area is off the coast of the Carolinas in the southeast of the USA.

In 2017, a beach reappeared in Ireland that had disappeared (eroded) in storms in 1984 (Figure 2.20). Hundreds of tonnes of sand were deposited following freak tides, producing a 330 m stretch of golden sand.

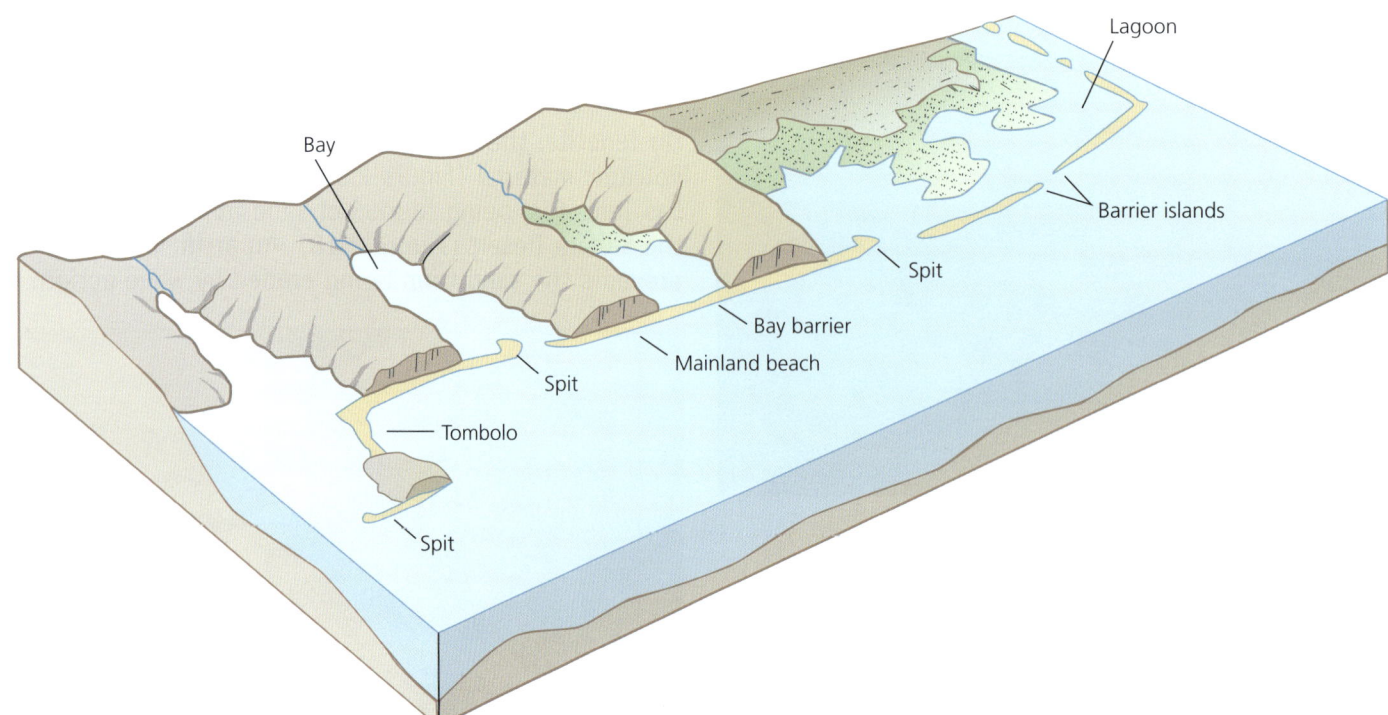

▲ Figure 2.19 The features of coastal deposition

Bars and spits

▲ **Figure 2.20** The beach at Achill Island, Ireland that was washed away by storms in 1984 (top image) and returned after 33 years (bottom image)

Bars and spits

Bars and spits are more localised features. They develop where:

» abundant material is available, particularly shingle and sand
» the coastline is irregular, for example where there is a variable geology
» there are estuaries and major rivers.

A **spit** is a beach of sand or shingle that is linked at one end to land. It is found where wave energy is reduced, for example along a coast where headlands and bays are common and near river mouths (in estuaries and rias).

Spits often become curved as waves undergo refraction (Figure 2.21). Cross-currents or occasional storm waves may assist this hooked formation. A good example is the sandspit in Walvis Bay, Namibia.

The main body of this spit is curved, but it has additional, smaller hooks, or recurves. Longshore drift moves sediment northwards along the coast. However, the coastline is very irregular here and there is a sudden change in the trend of the coastline. Consequently, refraction occurs, causing the waves to bend around eastwards.

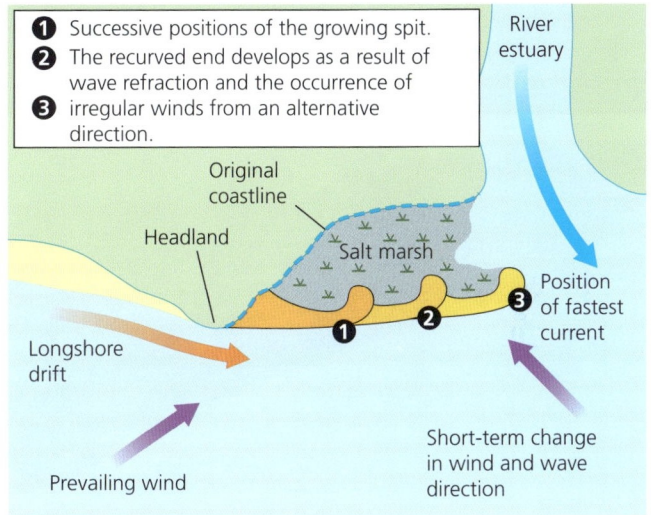

▲ **Figure 2.21** The development of a spit

On the seaward side, the slope to deeper water is very steep. Within the curve of the spit, the water is shallow and a considerable area of mudflat and salt marsh is exposed at low water. These salt marshes continue to grow as mud is trapped by the marsh vegetation.

Related features include bars. These are ridges that block off a bay or river mouth (Figure 2.22). There are many examples on the west coast of Antigua (Figure 2.23).

▲ **Figure 2.22** A bar at Slapton Ley, Devon, UK

2.2 THE MAIN LANDFORMS ASSOCIATED WITH THESE PROCESSES

▲ **Figure 2.23** The west coast of Antigua. (Top) Mosquito Cove, (centre) Jolly Harbour Marina and (bottom) Crab Hill Bay

> **Interesting note**
>
> The longest spit in the world is the 112 km Arabat Spit in the Sea of Azov, between Russia and Ukraine. There are thirteen spits in the Sea of Azov.

Tombolos are ridges that link the mainland to an island. Good examples include the Lumley area of Sierra Leone, and the Cape Verde Peninsula, Senegal. The Cape Peninsula in South Africa is a complex tombolo that has developed on a very large scale.

The Palisadoes, Jamaica: a spit or a tombolo?

The Palisadoes is one of the largest deposited coastal features in the Caribbean (Figure 2.24).

▲ **Figure 2.24** The Palisadoes tombolo

Located just south of Kingston in Jamaica, this 13 km-long feature has been formed and re-formed many times during its history. Scientists believe that it may be 4000 years old.

Longshore drift occurs from east to west on the south coast of Jamaica. The sediment comes from rivers, cliff erosion and offshore sediments. The Palisadoes is located at a sharp bend in the coastline. Longshore drift carries sediment westwards and extends the length of the spit. As it grew longer it linked up with a number of cays (small islands), turning the spit into a tombolo.

The region experiences tropical storms and hurricanes. These can seriously damage the coast. For example, in 2004 Hurricane Ivan eroded up to a metre off the 2 m-high sand dunes. Even under normal conditions, summer sea breezes cause powerful destructive waves, which are capable of eroding the seaward face of the beach, causing it to become steeper.

Sand dunes

> **Activities**
> 1. State:
>   a the age
>   b the length
>   of the Palisadoes.
> 2. Identify the direction in which longshore drift occurs on the Palisadoes.
> 3. Identify where the sediment that builds up the Palisadoes comes from.
> 4. Suggest the impact of hurricanes on the Palisadoes.

Sand dunes

Sand dunes are one of the most dynamic environments in physical geography. Important changes take place here in a very short space of time. Extensive sandy beaches are almost always backed by sand dunes because strong onshore winds can easily transport the sand that has dried out and been exposed at low water. The sand grains are trapped and deposited against any obstacle on land to form dunes (Figure 2.25). Dunes can be blown inland and can therefore threaten coastal farmland and even villages. The interaction of winds and vegetation helps form sand dunes.

On the beach, conditions are very windy, dry (much water just soaks into the sand) and salty. Few plants can survive these extreme conditions but some can, including sea couch and marram grass (Figure 2.26). These are adapted to tolerate water with a high salt content and high wind speeds, and they can survive burial by sand. In fact, marram grass needs to be buried by fresh sand in order to send out fresh shoots.

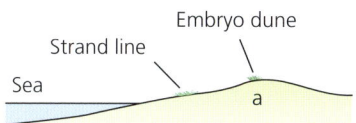

As the tide goes out, the sand dries out and is blown up the beach. A small embryo dune forms in the shelter behind the strand line.

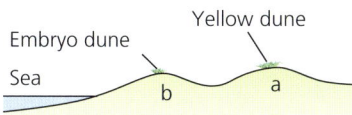

Sea couch grass colonises and helps bind the sand. Once the dune grows to over 1 m high, marram grass replaces the sea couch.

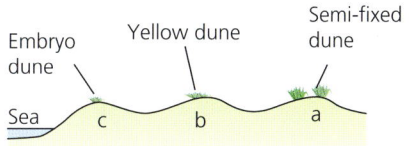

Once the yellow dune is over 10 m high, less sand builds up behind it and marram grass dies to form a thin humus layer. As the original dune a has developed, new embryo and yellow dunes have formed.

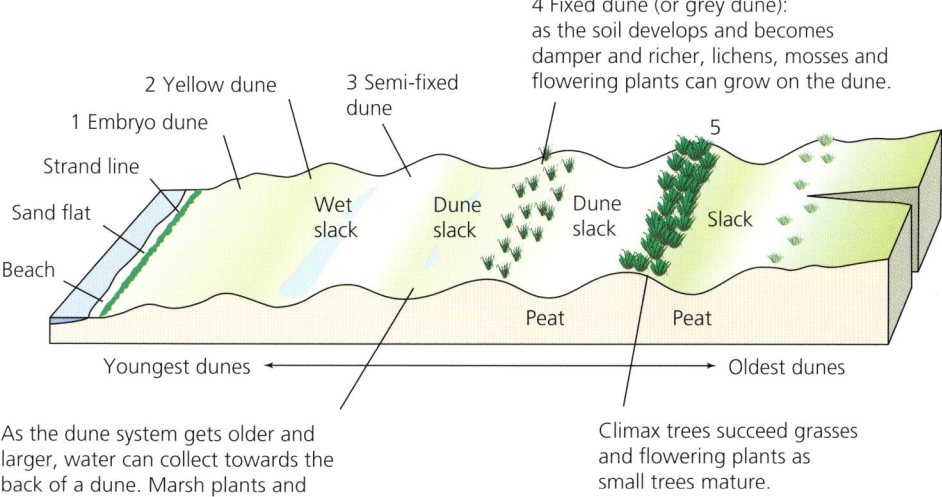

As the dune system gets older and larger, water can collect towards the back of a dune. Marsh plants and small willow trees can grow here.

Climax trees succeed grasses and flowering plants as small trees mature.

▲ **Figure 2.25** The formation of sand dunes

2.2 THE MAIN LANDFORMS ASSOCIATED WITH THESE PROCESSES

▲ Figure 2.26 Sand dune vegetation

The growth of new plants is called succession. Plants such as heather cannot tolerate the dry, windy, salty conditions of the beach, but they can survive in the less windy, moister, less salty dunes. They in turn alter the environment so that other species can invade and develop. Over a distance of just a few hundred metres on a sand dune, there may be as many as four or five different types of ecosystem.

> ### ▶ Activities
>
> Study Figure 2.25 (page 45), which shows the development of vegetation on a sand dune.
>
> 1 Contrast the conditions between the shoreline and inland.
> 2 Explain why deposition occurs on the sand dunes.
> 3 Suggest how human activities might affect the sand dunes and/or the salt marsh.

Once marram grass and sea couch are established on a beach, they reduce the wind speed. This helps trap fresh sand. As the sand builds up, these plants send out new shoots, trapping more sand and building up a dune. Increasingly, the presence of plants in the sand dune adds organic matter and moisture to the dune and allows other plants to grow, such as heather.

Discordant and concordant coastlines

Some coastlines have their geology parallel or at right angles to the shoreline. **Concordant coastlines** (also known as Pacific-type coastlines) tend to be straight and regular, with relief and geology parallel to the coastline (Figure 2.27), for example the south coast of the Isle of Purbeck or the coast of British Colombia. In contrast, **discordant coastlines** have relief and geology at right angles to the coast, for example the east coast of the Isle of Purbeck or northwest Spain.

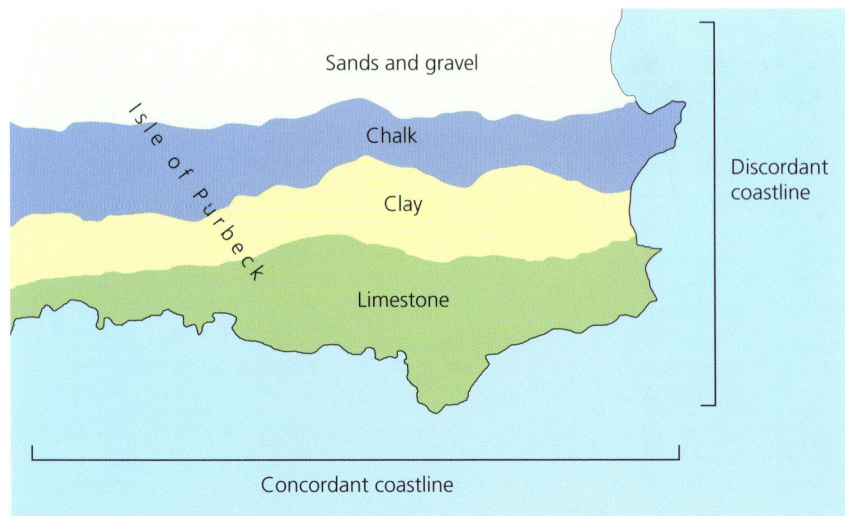

▲ Figure 2.27 Concordant and discordant coastlines

2.3 Coasts present opportunities and hazards for people

This chapter will explain:

★ the opportunities and hazards of living in coastal areas
★ the distribution and characteristics of tropical cyclones and how they can be managed
★ detailed specific example: the causes, impacts and potential solutions to coastal erosion in the USA's Eastern Seaboard
★ the extent to which coral reefs and mangroves can be managed so that people can benefit from them
★ detailed specific example: the Coral Triangle, Southeast Asia.

The opportunities of living near the coast

Coastal areas provide many advantages to people, including health, employment, trade, a place for recreation, tourism and investment, climate. For instance, many people living by the coast experience better health. Living near coastal areas can have a positive impact on mental health and may reduce stress. These areas have long been used for convalescence, holidays and physical activities, such as surfing, swimming, sailing and biking. In addition, the climate in coastal areas is generally milder than in inland areas, with cooler summers and warmer winters.

Coastal areas are also important for employment and trade. Employment opportunities include in the fishing and related industries, such as fish processing, as well as in tourism, with jobs in hotels, restaurants, boating, coastal fishing and water sports. Many coastal areas are important for the import and export of goods, and many ports have developed into major cities, such as London, New York, Shanghai and Singapore.

Coastal areas are also good places for investments (Figure 2.28). In the USA, for example, coastal regions attract residents with higher incomes, and there is a correlation between areas with higher incomes and lower crime rates. This may be because in coastal areas that rely on tourism, residents tend to realise the importance of ensuring the safety and security of visitors, and because residents in coastal areas generally wish to safeguard their investment.

▲ Figure 2.28 Tourism developments in St Lucia

While coastal areas offer many opportunities to people, their actions may cause problems. These are summarised in Table 2.2.

▼ Table 2.2 Relationships between human activities and coastal zone problems

Human activity	Agents/consequences	Coastal zone problems
Urbanisation and transport	• Land use changes, e.g. for ports, airports	• Loss of habitats and species diversity
	• Road, rail and air congestion	• Visual intrusion
	• Dredging and disposal of harbour sediments	• Lowering of groundwater table

2.3 COASTS PRESENT OPPORTUNITIES AND HAZARDS FOR PEOPLE

▼ Table 2.2 (Continued)

Human activity	Agents/consequences	Coastal zone problems
	• Water abstraction • Wastewater and waste disposal	• Saltwater intrusion • Water pollution • Human health risks • Eutrophication • Introduction of alien species
Agriculture	• Land reclamation • Fertiliser and pesticide use • Livestock densities • Water abstraction	• Loss of habitats and species diversity • Water pollution • Eutrophication • River channelisation
Tourism, recreation and hunting	• Development and land use changes, e.g. golf courses • Road, rail and air congestion • Ports and marinas • Water abstraction • Wastewater and waste disposal	• Loss of habitats and species diversity disturbance • Visual intrusion • Lowering of water table • Saltwater intrusion in aquifers • Water pollution • Eutrophication • Human health risks
Fisheries and aquaculture	• Port construction • Fish processing facilities • Fishing gear • Fish farm effluents	• Overfishing • Impacts on non-target species • Litter and oil on beaches • Water pollution • Eutrophication • Introduction of alien species • Habitat damage and change in marine communities
Industry (including energy production)	• Land use changes • Power stations • Extraction of natural resources • Process effluents • Cooling water • Windmills • River impoundment • Tidal barrages	• Loss of habitats and species diversity • Water pollution • Eutrophication • Thermal pollution • Visual intrusion • Decreased input of fresh water and sediment to coastal zones • Coastal erosion

The hazards of living near the coast

As well as opportunities, there are many hazards present in coastal areas. Rising sea levels are threatening to increase flooding in many low-lying coastal zones. Fresh water resources can become contaminated by saltwater intrusion, and some coastal ecosystems, such as salt marshes and wetlands, are being squeezed and eroded. Some coastal areas are at risk from **natural hazards** such as tropical cyclones and tsunamis. **Tropical cyclones** bring high wind speeds (over 119 km/hour) and heavy rain to tropical and sub-tropical coastal areas. **Tsunamis** create large waves that can devastate coastal regions, such as the Indian Ocean tsunami of 2004, in which over 230,000 people died.

Water pollution is another hazard to affect coastal areas. Some of this may be runoff from the land, for example of nitrate fertilisers, that leads to 'dead zones' in coastal areas due to the excessive growth of algae using up the oxygen and cutting off light to the lower zones. Some of it may be from the discharge of waste from ships. The discharge of poorly treated wastewater into coastal waters poses environmental and health risks. Moreover, the destruction of natural habitats, such as at Rodney Bay, St Lucia, to create an artificial lagoon (Figure 2.29), may lead to unforeseen impacts. At Rodney Bay currents became stronger, which increased erosion on nearby beaches, and fisheries declined as offshore waters became murkier.

▲ **Figure 2.29** Some tourist developments may lead to unforeseen environmental damage, such as at Rodney Bay, St Lucia

Hard and soft engineering strategies

Human pressures on coastal environments create the need for a variety of **coastal management strategies** (Table 2.3). Coastal defence protects against **coastal erosion** and flooding by the sea. Coastal management strategies may be long term or short term, sustainable or non-sustainable. Successful management strategies require a detailed knowledge of coastal processes. Rising sea levels, more frequent storm activity and continuing coastal development are likely to increase the need for coastal management.

Defence options include:

» 'do nothing'
» maintaining existing levels of coastal defence
» improving the coastal defence
» allowing retreat of the coast in selected areas.

Hard engineering structures

The effectiveness of sea walls depends on their cost and their performance. Their function is to prevent erosion and flooding, but much depends on:

» whether they are sloping or vertical
» whether they are permeable or impermeable
» whether they are rough or smooth
» what material they are made of (clay, steel or rock, for example).

In general, flatter, permeable, rougher walls perform better than vertical, impermeable, smooth walls.

Cross-shore structures such as groynes, breakwaters, piers and strongpoints have been used for decades. Their main function is to stop the drifting of material. Traditionally, groynes were constructed from timber, brushwood and wattle. However, modern cross-shore structures are often made from rock. They may be part of a more complex form of management that includes beach nourishment and offshore structures.

▼ Table 2.3 Hard engineering strategies and techniques for coastal management

Type of management	Aim/method	Strengths	Weaknesses
Hard engineering	To control natural processes		
Cliff base management	To stop cliff or beach erosion		
Sea walls	Large-scale concrete curved walls designed to reflect wave energy	• Easily made • Good in areas of high density	• Expensive • Life span about 30–40 years • Foundations may be undermined

2.3 COASTS PRESENT OPPORTUNITIES AND HAZARDS FOR PEOPLE

▼ **Table 2.3** (*Continued*)

Type of management	Aim/method	Strengths	Weaknesses
Revetments	Sloping wooden structures built on shorelines, cliff bases or in front of sea walls to absorb wave energy and reduce coastal erosion	• Easily made • Cheaper than sea walls	• Limited life span
Gabions	Rocks held in wire cages to absorb wave energy	• Cheaper than sea walls and revetments	• Small scale
Groynes	Barriers (usually wooden) placed at right angles to the shoreline designed to trap material transported by longshore drift. They can have a negative impact on coastal zones downdrift as they deprive them of sediment	• Relatively low cost • Easily repaired	• Cause erosion on downdrift side • Interrupt sediment flow
Rock armour	Large rocks at base of cliff to absorb wave energy	• Cheap	• Unattractive • Small scale • May be removed in heavy storms
Offshore breakwaters	To reduce wave power offshore	• Cheap to build	• Disrupt local ecology
Rock strongpoints	To reduce longshore drift	• Relatively low cost • Easily repaired	• Reduces sediment downdrift, which could lead to increased erosion downdrift
Cliff face strategies	To reduce the impacts of sub-aerial processes		
Cliff drainage	To remove water from rocks in the cliff	• Cost-effective	• Drains may become new lines of weakness • Dry cliffs may produce rockfalls
Vegetation	To increase interception and reduce overland runoff	• Relatively cheap	• May increase moisture content of soil and lead to landslides

▼ Table 2.3 (Continued)

Type of management	Aim/method	Strengths	Weaknesses
Cliff regrading	To lower slope angle to make cliff safer	• Useful on clay (most other measures are not)	• Uses large amounts of land – impractical in heavily populated areas

The sustainability of hard engineering strategies

As the table suggests, all forms of hard engineering can have negative impacts or last only a relatively short time. In that sense, they could be considered to be unsustainable. However, they may protect buildings and businesses and land for many decades, and so in that sense they are extremely important, even though they will need to be replaced or upgraded after a number of years.

Soft engineering strategies

Managed retreat allows nature to take its course – erosion in some areas, deposition in others. Benefits include less money being spent and the creation of natural environments. More soft engineering strategies are outlined in Table 2.4.

The sustainability of soft engineering strategies

Soft-engineering schemes would appear to be more sustainable than hard-engineering schemes as they use, in some cases, natural materials or waste materials. However, soft-engineering also allows land to be eroded and so such schemes may be very unpopular with landowners and homeowners.

The distribution and characteristics of tropical storms and how they can be managed

Tropical storms are some of the most dangerous natural hazards for people and for the environment. Damage is caused by high winds, floods and storm surges. In Asia, tropical storms are known as tropical cyclones, and in Northern America they are called hurricanes.

▼ Table 2.4 Soft engineering strategies and techniques for coastal management

Type of management	Aim/method	Strengths	Weaknesses
Soft engineering	Working with nature		
Offshore reefs	Waste materials, e.g. old tyres weighted down, to reduce speed of incoming waves	• Low technology and relatively cost-effective	• Long-term impacts unknown
Beach nourishment	Sand pumped from sea bed to replace eroded sand	• Looks natural	• Expensive • Short-term solution
Managed retreat	Coastline allowed to retreat in certain places	• Cost-effective • Maintains a natural coastline	• Unpopular • Political implications
'Do nothing'	Accept that nature will win	• Cost-effective	• Unpopular • Political implications
Red-lining	Planning permission withdrawn New line of defences set back from existing coastline	• Cost-effective	• Unpopular • Political implications

2.3 COASTS PRESENT OPPORTUNITIES AND HAZARDS FOR PEOPLE

> **Interesting note**
>
> Tropical storms are known by several different names in different parts of the world. 'Tropical cyclone' is the generic name; 'hurricanes' occur in the Atlantic, 'typhoons' in the western Pacific, and 'cyclones' in the Indian Ocean.

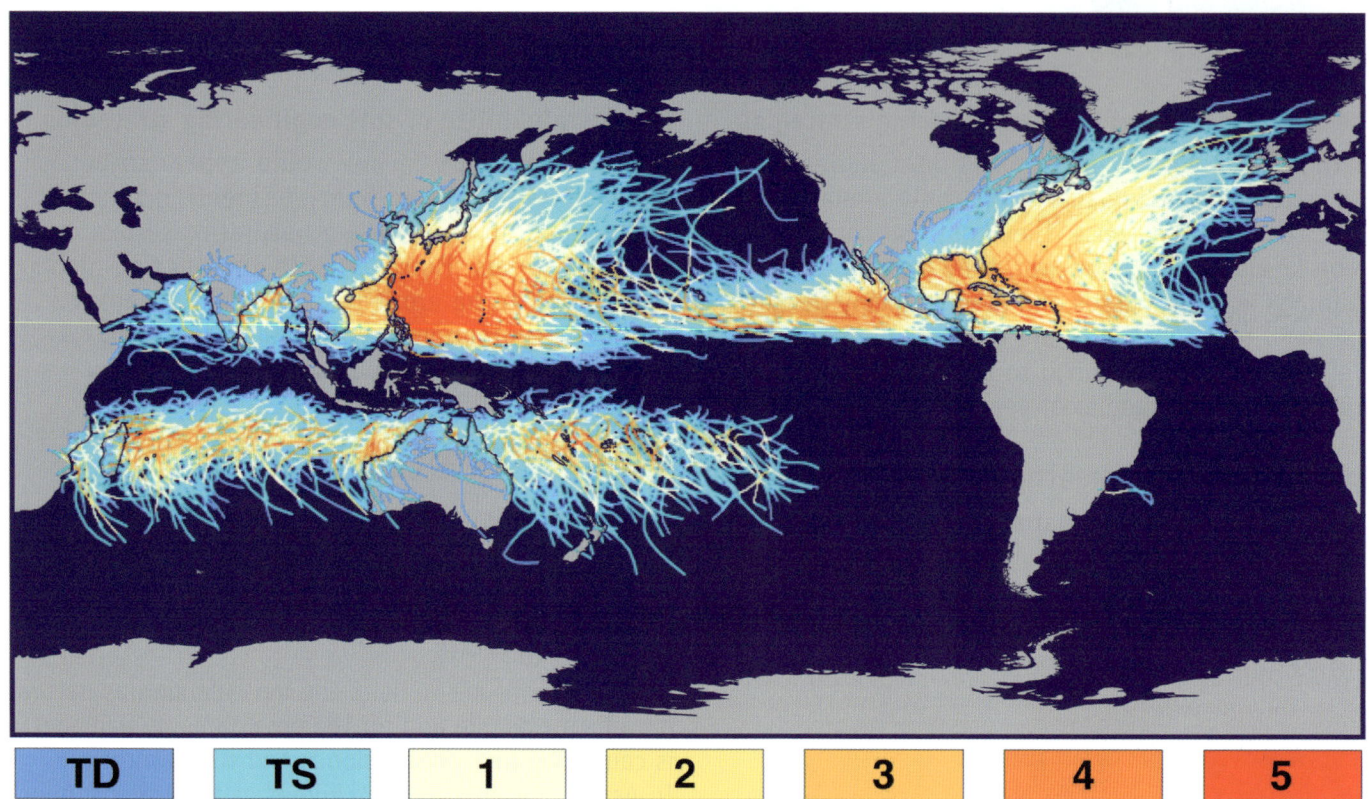

▲ **Figure 2.30** The distribution of tropical cyclones (TD = tropical depression, TS = tropical storm, 1–5 = tropical cyclone categories). The trajectories show the main paths of tropical cyclones 1851–2006.

Tropical cyclones are intense hazards that bring heavy rainfall, strong winds and high waves, and they cause other hazards such as flooding and mudslides. Tropical cyclones are also characterised by enormous quantities of water. This is due to their origin over moist tropical seas. High-intensity rainfall, with totals of up to 500 mm in 24 hours, invariably causes flooding. The path of a tropical cyclone is erratic, so it is not always possible to give more than 12 hours' warning. This is insufficient for proper evacuation measures.

Tropical cyclones develop as intense low-pressure systems over tropical oceans (Figure 2.30). Winds spiral rapidly around a calm central area known as the eye. The diameter of the whole hurricane may be as much as 800 km, although the very strong winds that cause most of the damage are found in a narrower belt up to 300 km wide. In a mature hurricane, pressure may fall to as low as 880 mb. This very low pressure, and the strong contrast in pressure between the eye and the outer part of the hurricane, is what leads to strong gale-force winds.

Tropical cyclones move excess heat from low latitudes to higher latitudes. They normally develop in the westward-flowing air just north of the equator (known as an easterly wave). They begin life as small-scale tropical depressions, localised areas of low pressure that cause warm air to rise. These trigger thunderstorms that persist for at least 24 hours and may develop into tropical storms, which have greater wind speeds of up to 118 km/hr. However, only about 10 per cent of tropical disturbances ever become tropical cyclones – storms with wind speeds above 118 km/hr.

The distribution and characteristics of tropical storms and how they can be managed

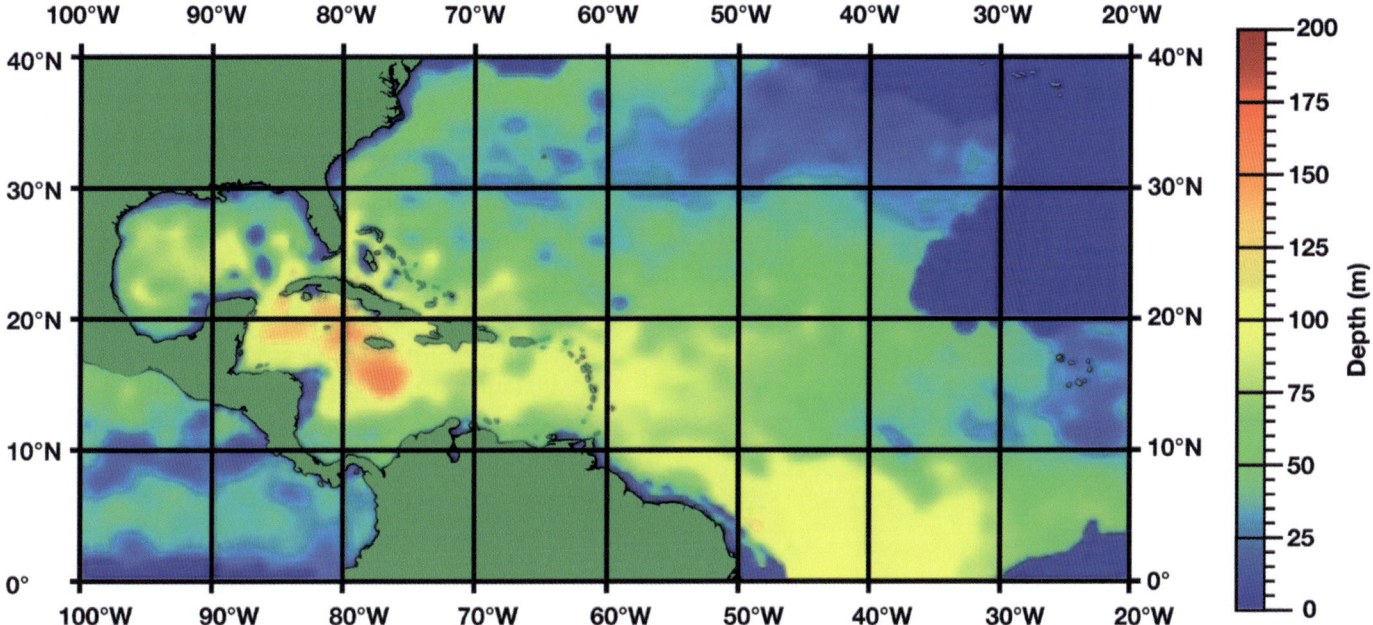

▲ **Figure 2.31** The depth of the Caribbean Sea at which temperatures remain over 27°C

For tropical cyclones to form, a number of conditions are needed:

» Sea temperatures must be over 27°C to a depth of 60 m (Figure 2.31) – warm water gives off large amounts of heat when it is condensed; this is the heat that drives the hurricane.
» The low-pressure area has to be sufficiently far away from the equator so that the Coriolis force (the force caused by the rotation of the Earth) creates rotation in the rising air mass. If it is too close to the equator, there is insufficient rotation and a hurricane will not develop.
» Conditions must be unstable: some tropical low-pressure systems develop into tropical cyclones, but not all of them, and scientists are unsure why some do but others do not.

The impacts of tropical cyclones

The **Saffir-Simpson Scale**, developed by the National Oceanic and Atmospheric Administration, assigns tropical cyclones to one of five categories of potential disaster (Table 2.5). The categories are based on wind intensity, and in order to be classified as a hurricane a tropical cyclone must have maximum sustained winds of at least 118 km/hr. The classification is used for tropical cyclones forming in the Atlantic and northern Pacific – other areas use different scales.

▼ **Table 2.5** The Saffir-Simpson Scale

Tropical cyclone category	Windspeed (km/hr)	Storm surge (metres above normal)	Pressure (mb)	Predicted damage
1	119–53	1.2–1.5	≥ 980	• Minimal damage • No real damage to building structures • Damage primarily to unanchored building structures, e.g. mobile homes • Some coastal road flooding and minor pier damage
2	154–77	1.6–2.5	965–79	• Moderate damage • Some damage to roofing material, doors and windows • Considerable damage to vegetation and piers

2.3 COASTS PRESENT OPPORTUNITIES AND HAZARDS FOR PEOPLE

▼ Table 2.5 (Continued)

Tropical cyclone category	Windspeed (km/hr)	Storm surge (metres above normal)	Pressure (mb)	Predicted damage
				• Coastal and low-lying escape routes flood 2–4 hours before arrival of the storm eye • Small craft in unprotected anchorages break moorings
3	178–209	2.6–3.6	945–64	• Extensive damage • Some structural damage to small residences and utility buildings • Flooding near the coast destroys smaller structures, with larger structures damaged by floating debris • Land below 1.5 m above mean sea level may be flooded 13 km or more inland
4	210–49	3.7–5.5	920–44	• Extreme damage • Some complete roof structure failures on small residences • Extensive damage to doors and windows • Land below 3 m above mean sea level may be flooded, requiring massive evacuation of residential areas as far as 10 km inland
5	⩾ 250	> 5.5	< 920	• Catastrophic damage • Complete roof failure on many residences and industrial buildings • Some complete building failures, with small utility buildings blown over or blown away • Complete destruction of mobile homes • Severe and extensive window and door damage • Low-lying escape routes are cut off by rising water 3–5 hours before the arrival of the centre of the storm • Major damage to lower floors of all structures located less than 4.5 m above mean sea level and within 500 m of the shoreline • Massive evacuation may be required of residential areas on low ground within 8–16 km of the shoreline

Other scales exist. The Hurricane Severity Index rates hurricanes (tropical cyclones) on their size and wind speed. For example, a large tropical cyclone that is less intense may cause more damage than a smaller one that is more intense, on account of its size. The National Oceanic and Atmospheric Administration uses the Accumulated Cyclone Energy (ACE), which calculates the estimated maximum sustained velocities at six-hourly intervals. Other scales include the Australian Tropical Cyclone Intensity Scale and the Indian Meteorological Society Scale of Hurricanes.

Predicting tropical cyclones can be difficult for several reasons:

» The unpredictability of hurricane paths makes the effective management of tropical cyclones difficult.
» The strongest storms do not always cause the greatest damage.
» The distribution of the population, for instance throughout the Caribbean Islands, may increase the risk associated with hurricanes.
» Hazard mitigation depends on the effectiveness of the human response to natural events and the protection strategies available (Figure 2.32).

» Developing countries continue to lose more lives to natural hazards as a result of inadequate planning and preparation.

Strategies to manage the impacts of tropical storms

▲ **Figure 2.32** Hurricane management strategies: (a) hurricane shelter; (b and c) steel shutters over windows

Tracking hurricanes

Most hurricanes are tracked by the US National Hurricane Center in Miami, USA, and by the Met Office in the UK. The National Hurricane Center in turn warns countries if they are at risk of a hurricane.

Information regarding hurricanes is received from a number of sources, including:

- » satellite images
- » aircraft that fly into the eye of the hurricane to record weather information
- » weather stations at ground level
- » radars that monitor areas of intense rainfall.

Planning and preparing for hurricanes

National governments, international agencies and meteorological societies can help people prepare for a hurricane. They provide advice on risk assessment, land-use control, floodplain management and reducing vulnerability.

- » Risk assessment: Risks from tropical cyclones can be shown in a hazard map. The information may be used to estimate the probability of hurricanes hitting a particular place. Weather records may show how often cyclones have struck, and their location and magnitude. In many places records go back for over a century.
- » Land-use zoning: This ensures that the most important facilities are placed in the safest areas. Policies regarding future development may need to take into account possible changes related to global climate change.
- » Floodplain management: A plan for floodplain management should aim to protect important buildings and infrastructure from river and coastal flooding.
- » Reducing vulnerability of structures and infrastructure:
  - New buildings should be designed to be wind and water resistant. Housing is particularly vulnerable to hurricanes. Hurricane Luis (1995) caused damage to 90 per cent of Antigua's houses, while Hurricane Gilbert (1988) made 800,000 people temporarily homeless in Jamaica. To limit damage to houses, owners are now encouraged to fix hurricane straps to roofs and put storm shutters over windows. Buildings can be raised above the ground to protect against floods and storm surges; for instance houses can be built on stilts to allow flood waters to pass away safely below them.
  - Communication and utility lines should be located away from coastal areas or installed underground.
  - Levées and coastal dikes should be regularly inspected for breaches due to erosion.
  - The planting of mangrove trees and other vegetation helps reduce breaking wave energy, and therefore reduces the impact of soil erosion and landslides.

Hurricane watches and warnings

- » A tropical storm watch is issued when tropical storm winds are expected within 36 hours.
- » A tropical storm warning is issued when there are risks of tropical storm winds within 24 hours.

2.3 COASTS PRESENT OPPORTUNITIES AND HAZARDS FOR PEOPLE

- A hurricane watch is issued when there is a threat of hurricane conditions within 24–36 hours.
- A hurricane warning is issued when hurricane conditions (winds of 119 km/hr or greater, or dangerously high water and rough seas) are expected within 24 hours.

Keeping safe during a hurricane

Before a hurricane arrives, households should be aware of where the emergency shelters are located, and have some essential supplies at home, such as a first aid kit, essential medical supplies, food and water, portable radio, torch and extra batteries. Windows should be protected with shutters or boarded with plywood. Gardens should be cleared of material that can be picked up and thrown by high winds.

During a hurricane, household members should stay inside, away from windows, skylights and glass doors; listen to the radio or television for hurricane progress reports; store valuables and personal papers in a waterproof container on the highest level of their home; and avoid using open flames as a source of light.

After the hurricane, it is recommended that people seek medical attention for the injured; assist in search and rescue if it is safe to do so; watch out for secondary hazards (fire, flooding, etc.); clean up debris; report damage to utilities; and assist in community response efforts.

The emergency relief offered after a hurricane can take many forms – including providing food supplies, clean water, blankets and medicines. Much of this is provided in hurricane shelters. In some communities, emergency electrical generators may be needed. The community normally becomes involved in the clean-up operation, and electricity and phone companies work to restore power lines and communications.

Long-term redevelopment

Long-term redevelopment may include construction of new buildings in areas away from the coastline and on high ground. Long-term reconstruction concentrates on:

- housing and community projects
- water supply and sanitation
- transport and communications
- agriculture, fisheries and small businesses
- schools
- government expenses.

> **Activity**
>
> To what extent is it possible to manage the impacts of tropical cyclones?

The impacts of Typhoon Haiyan

The Philippines experiences about 20 typhoons every year. In 2012 Typhoon Bopha killed more than 1100 people and caused over US$1 billion in damage.

At least 6000 people were killed in the central Philippine province of Leyte when Typhoon Haiyan, one of the strongest storms ever to make landfall, struck the Philippines in November 2013. The super-typhoon brought winds of up to 315 km/hr and tore roofs off buildings, turned roads into rivers full of debris, and knocked out electricity pylons.

About 70–80 per cent of the buildings in the area in the path of Haiyan in Leyte Province were destroyed. Tacloban, the provincial capital of Leyte, had a population of over 200,000. The storm surge caused sea waters to rise by over 6 m when the typhoon hit. Power was knocked out and there was no mobile phone signal, making communication possible only by radio.

With many provinces left without power or telecommunications, and airports closed in the hardest-hit areas such as Tacloban, it was impossible to know the full extent of the storm's damage – or to provide badly needed aid. Government figures showed that more than 4 million people had been directly affected. The World Food Programme mobilised some US$2 million in aid and aimed to deliver 40 tonnes of fortified biscuits to victims within days. Estimates of the economic cost are about US$15 billion. Many countries pledged aid to the Philippines, including the UK (US$131 m), Japan (US$52 m), Canada (US$40 m) and the USA (US$37 m).

Satellite images showed normally green patches of vegetation ripped up into brown squares of debris in Tacloban, where a local TV station broadcast images of huge storm surges, flattened buildings and families wading through flooded streets with their possessions held high above the water. Those living in the hardest-hit areas, such as the eastern Visayas, have some of the lowest incomes in the Philippines. Many have little or no savings, so the typhoon put an already vulnerable population at even greater risk of future food and job insecurity. On Bohol Island, where a 7.3 magnitude earthquake had killed some 200 people in the previous month, residents were successfully evacuated ahead of the storm. However, because the island's main power supply comes from neighbouring Leyte, residents were left without electricity or water. In Tacloban, the sheer force of the storm was just too much for some evacuation centres, which collapsed.

> **Activity**
>
> Describe the main impacts of Typhoon Haiyan.

 Detailed specific example

The causes, impacts and potential solutions to coastal erosion in the USA's Eastern Seaboard

Many beaches along the east coast of America have disappeared since 1900, such as that at Marshfield, Massachusetts. Nearly 40 per cent of Americans live along the coasts, where ageing buildings, roads and railways face structural damage from floods. As the sea level rises, the beaches and barrier islands (barrier beaches) that line the coasts of the Atlantic Ocean and the Gulf of Mexico, from New York to the Mexican border, are in retreat.

The problem is that much of the shore cannot retreat naturally because industries and properties worth billions of dollars have been built here. Many important cities and tourist centres, such as Miami, Atlantic City and Galveston (Texas), are sited on barrier islands. Consequently, many shoreline communities have built sea walls and other protective structures to shield them from the power of destructive waves.

- Relief: The flat topography of the coastal plains from New Jersey southward means that a small rise in sea level can allow the ocean to advance a long way inland.
- Changing sea levels: Much of the Northern American coast is sinking relative to the ocean, so local sea levels are rising faster than global averages (Figure 2.33). The level of tides along the coasts shows that subsidence varies between 0.5 and 19.5 mm a year. By contrast, the west coast, in particular in Alaska, is rising. A dominant cause of subsidence along the east coast is groundwater depletion.
- Erosion and tourism-related developments: Extensive coastal development has accelerated erosion as sea level rises, apartment blocks, resorts and second homes have developed rapidly along the shoreline. Erosion is evident at many places along the coasts of the Atlantic and the Gulf of Mexico. Major resorts such as Miami Beach and Atlantic City have pumped in dredged sand to replenish eroded beaches. Erosion threatens islands to the north and south of Cape Canaveral, although the Cape itself appears safe. Resorts built on barrier beaches in Virginia, Maryland and New Jersey have also suffered major erosion.
- Rates of erosion: Overall losses are not well known. Massachusetts loses about 26 hectares (ha) a year to rising seas. Nearly 10 per cent of that loss is from the island of Nantucket, south of Cape Cod. However, these losses are minimal compared with that of Louisiana, which is losing 40 ha of wetlands a day – about 15,500 ha a year.

There have been many attempts to manage the impacts of tropical storms along the USA's southeast coast. However, most homeowners are not covered by insurance for flooding and so have to take measures to protect their properties from flooding. Florida, for example, experiences flash floods from tropical cyclones. Heavy rain typically occurs when a storm makes landfall. The amount of rain can overwhelm drainage systems within minutes. Coastal flooding is also widespread. Low-lying areas and coastal developments are especially vulnerable. Storm surges can also add to flooding. Cities in Florida that are vulnerable to flooding include Miami, Pensacola, Tampa and Fort Lauderdale.

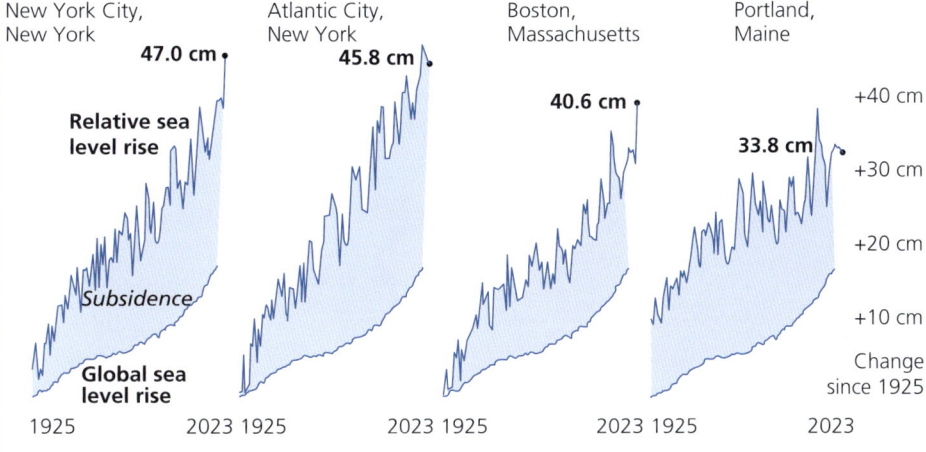

▲ **Figure 2.33** Sinking land and rising sea along the east coast of the USA

2.3 COASTS PRESENT OPPORTUNITIES AND HAZARDS FOR PEOPLE

Protective measures include the use of trap bags (a type of modified sand bag) to set up flood barriers, and levées. Sea walls have been built in many locations. However, these may give protection only up to a certain magnitude of event. With global climate change, storm heights may become greater and existing sea walls may prove ineffective. Temporary dams can be built using traditional sand bags. Sand bags and trap bags can be used to prevent beach erosion.

Following the devastation of Hurricane Andrew, a Category 5 Hurricane in 1992 that killed 65 people and caused over $27 billion worth of damage, Florida's Division of Emergency Management created the Hurricane Loss Mitigation Program to provide retrofits to residential, commercial and mobile homes. The Gulf State College Mobile Home Tie-Down Program inspects and provides tie-downs for mobile homes to protect them against the high wind speeds experienced in tropical storms.

According to the Canadian engineering company WSP, current rates of subsidence are increasing the risk of sea level change. Up to 2.1 million people are affected by subsidence annually in 867,000 properties. Hot spots include New York, Florida and Charleston, South Carolina. WSP invested in evacuation routes and access to hospitals. They have raised roads at Key Largo, Florida and Virginia Beach, Virginia. The North Carolina Department of Transport has managed the risk of flooding by raising parts on Interstate 40 and Highway 53.

> ### ▶ Activities
> 1. Explain why beaches on the Eastern Seaboard of the USA are retreating.
> 2. State the proportion of Americans who live along the coast.
> 3. Estimate by how much the east coast is subsiding each year.

The extent to which coral reefs and mangroves can be managed

Coral reefs and **mangroves** are other types of coastal ecosystems that people can benefit from (Figure 2.34).

Coral reefs

Coral reefs are generally located in tropical seas. They are calcium carbonate structures, made up of reef-building stony corals. Coral is limited to the depth that light can reach, so reefs develop in shallow water, ranging to depths of 60 m. This dependence on light also means that reefs are only found where the surrounding waters contain relatively small amounts of suspended material. Reef-building corals live only in tropical seas, where temperature, salinity and clear water allow them to develop.

There are many types of coral reef (Figure 2.35):

» Fringing reefs are those that fringe the coast of a landmass (Figures 2.36 and 2.37). Many fringing reefs grow along shores that are protected by barrier reefs and are thus characterised by organisms that are best adapted to low wave-energy conditions.

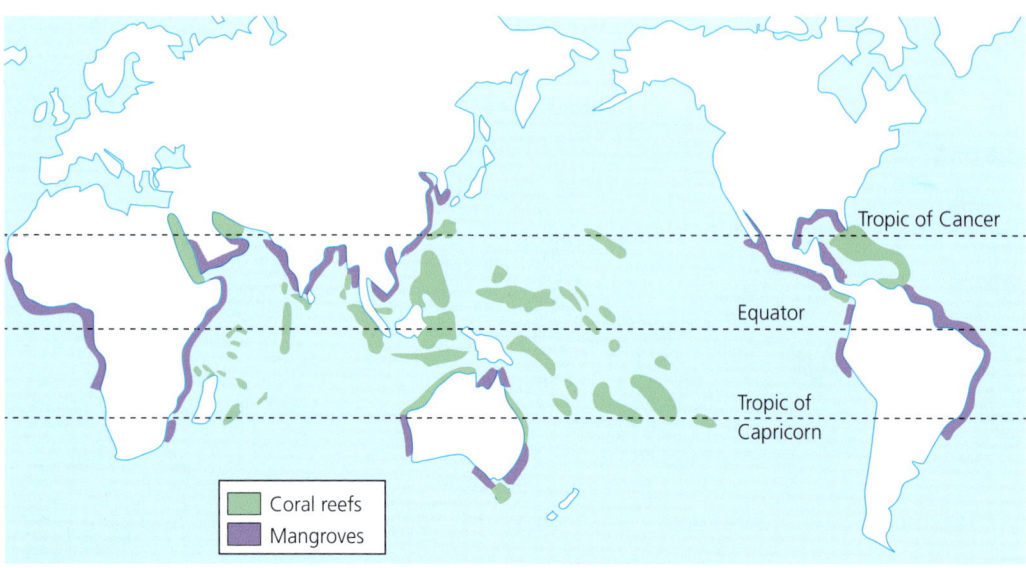

▲ **Figure 2.34** The distribution of coral reefs and mangroves

The extent to which coral reefs and mangroves can be managed

» Barrier reefs occur at a greater distance from the shore than fringing reefs and are commonly separated from it by a wide, deep lagoon. Barrier reefs tend to be broader, older and more continuous than fringing reefs. For example, the Beqa barrier reef off Fiji stretches unbroken for more than 37 km, and that off Mayotte in the Indian Ocean for around 18 km. The largest barrier reef system in the world is the Great Barrier Reef, which extends 1600 km along the east Australian coast, usually tens of kilometres offshore. Another long barrier reef is located in the Caribbean, off the coast of Belize between Mexico and Guatemala.

» Atoll reefs rise from submerged volcanic foundations. Atoll reefs are essentially indistinguishable in form and species composition from barrier reefs, except that they are confined to the flanks of submerged oceanic islands, whereas barrier reefs may also flank continents. Over 300 atolls are present in the Indo-Pacific, but only ten are found in the western Atlantic.

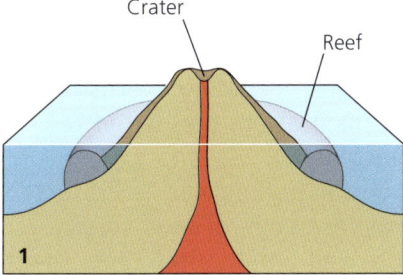
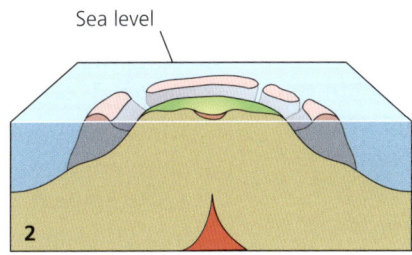
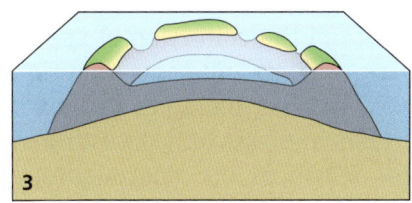

1 Rocky volcanic islet encircled by fringing coral reef

2 Reef enlarges to form barrier reef as land sinks (or sea rises)

3 Circular coral reef or atoll (with further change in level)

▲ Figure 2.35 The formation of coral reefs

▲ Figure 2.36 A fringing reef on the south coast of Antigua

▲ Figure 2.37 A fringing reef on the west coast of Antigua

Coral reefs are often described as the 'rainforests of the sea' on account of their rich biodiversity. Some coral is believed to be 2 million years old, although most is less than 10,000 years old. Coral reefs contain nearly a million species of plants and animals, and about 25 per cent of the world's sea fish breed, grow and evade predators in coral reefs. Some of the world's best coral reefs include Australia's Great Barrier Reef, many of the reefs around the Philippines and Indonesia, Tanzania and the Comoros, and the Lesser Antilles in the Caribbean.

2.3 COASTS PRESENT OPPORTUNITIES AND HAZARDS FOR PEOPLE

Coral reefs are of major biological and economic importance. Countries such as Barbados, the Seychelles and the Maldives rely on tourism based on their reefs. Florida's reefs attract tourism worth over $1 billion annually and create more than 70,000 jobs. The global value of coral reefs in terms of fisheries, tourism and coastal protection is estimated to be US$650 billion.

Threats to coral reefs

Coral reefs face many pressures, for example from the fishing industry. The industry uses dynamite to flush out fish and cyanide solution to catch live fish. Destruction of the reefs also takes many other forms – from the collection of specimens, trampling, berthing of boats, oil spills, mining and the cement industry. Indirect pressures include sedimentation from rivers and waste disposal from urban areas. Coastal development, especially for tourism, is taking its toll too. Dust storms from the Sahara have introduced bacteria into Caribbean coral, while global warming may cause coral bleaching. Bleaching occurs when high temperatures kill the algae in coral, removing their colour so the coral appears bleached. Many areas of coral in the Indian Ocean were destroyed by the 2004 tsunami.

Floating plastic pollution blocks out some sunlight, spreads diseases and cuts into coral, making it more susceptible to disease. Studies of more than 150 coral reefs in Australia, Indonesia, Malaysia and the Philippines showed that when coral came into contact with plastic, the likelihood of disease increased from 4 per cent to almost 90 per cent. In addition, coral is very vulnerable to oil pollution and runoff from fertilisers. Runoff from fertilisers or sewage can be particularly damaging, as coral reefs are adapted to low nutrient levels.

Overfishing is one of the biggest threats to coral reefs; the other is global climate change. It is believed that if atmospheric carbon dioxide levels rise to 450 ppm (they were over 420 ppm in 2024), coral reefs could die off. This would affect the lives of the up to 500 million people who depend on coral, and reduce the US$100 billion that coral provides to the human economy. Global climate change and other human activities have affected about 20 per cent of the world's coral. Global temperatures may rise by 2°C, which could lead to widespread coral bleaching, extinction of coral species, more fragile coral skeletons and greater risk of storm damage.

Sustainable techniques to protect coral reefs

To avoid permanent damage and to support people in the tropics, it is recommended that:

- the world community reduces emissions of greenhouse gases and develops plans to **sequester** carbon dioxide
- damaging human activities, including overfishing, blasting coral and sedimentation, are halted to allow coral reefs to recover
- more coral reefs are designated as Marine Protected Areas (MPAs) to act as reservoirs of biodiversity, including many remote and uninhabited reefs that are still in good condition
- the management, monitoring and enforcement of regulations is improved
- local coastal management practices are introduced
- assistance is provided to LICs and MICs
- alternative lifestyles are developed that reduce pressure on coral reefs.

> ### Activities
> 1. Identify the conditions in which coral grows.
> 2. State the difference between a fringing reef and a barrier reef.
> 3. Explain how atolls are formed.
> 4. Explain why coral reefs are so valuable.
> 5. Examine the main threats to coral reefs.

Mangroves

Mangroves are salt-tolerant forests of trees and shrubs that grow in the tidal estuaries and coastal zones of tropical areas (Figure 2.38). The muddy waters, rich in nutrients from decaying leaves and wood, are home to a great variety of sponges, worms, crustaceans, molluscs and algae. Mangroves cover about 25 per cent of the world's tropical coastline, the largest being the 570,000 ha Sundarbans in Bangladesh.

The extent to which coral reefs and mangroves can be managed

▲ Figure 2.38 Mangrove swamps

Mangroves and coral reefs are fundamentally connected ecosystems. Mangroves protect coral reefs from sedimentation from land-based sources, as well as helping to keep the water clear of particles and nutrients. Both of these functions are necessary to maintain reef health. Mangroves also provide spawning and nursery areas for many animal species that spend their adult lives on the reefs. In return, the coral reefs provide shelter for the mangroves and their inhabitants, while the calcium carbonate eroded from the reef provides sediment in which the mangroves grow.

Despite their value, many mangrove areas have been lost to rice paddies and shrimp farms. Between 2000 and 2016, 20–35 per cent of global mangrove forests were lost, mainly due to human activities but some due to erosion. Over 60 per cent of mangrove losses were due to conversion to aquaculture and agriculture. One in six species found in mangrove ecosystems is now threatened with extinction. Table 2.6 shows regional variations in the cause of mangrove losses, 2000–16. As population growth in coastal areas is set to increase, the fate of mangroves looks bleak.

Other threats to mangroves include:

Mangroves have many uses, such as providing large quantities of food, fuel, building materials and medicine. One hectare of mangrove in the Philippines can yield 400 kg of fish and 75 kg of shrimp. In addition, mangroves protect coastlines by absorbing the force of tropical cyclones and storms. They also act as natural filters, absorbing nutrients from farming and sewage disposal.

- clearance for coastal developments, for example hotels
- deforestation for wood and charcoal
- browsing (feeding on higher plants) by livestock
- overfishing in adjacent areas
- waste from nearby settlements and human activities
- pollution, for example oil spills
- changes to local landscapes, for example coastal jetties

▼ Table 2.6 Mangrove losses (per cent) by cause and region, 2000–16

Region	Approx. amount lost ('000 km²)	Loss due to settlement	Loss due to commodities	Loss due to non-productive conversion	Loss due to extreme weather events	Loss due to erosion
Northern America and the Caribbean	2.0	0	0	8	49	43
South America	0.5	0	17	7	22	54
Africa	0.3	8	7	42	23	20
Asia	2.5	1	68	7	4	20
Oceania	0.5	0	0	8	49	43

2.3 COASTS PRESENT OPPORTUNITIES AND HAZARDS FOR PEOPLE

- the control of malaria – mangroves are removed as potential breeding grounds for mosquitoes
- increased sea water temperature from power plants located along coastlines, which can damage mangrove ecosystems.

Many of the threats to mangroves come from the demands of the tourism industry and people outside the local area. There are, nevertheless, some benefits of this for local communities, such as food production and erosion control, although overall the negative impacts may outweigh the positive impacts.

Sustainable techniques to promote and manage mangroves

The Zoological Society of London (ZSL), an international conservation charity, has restored over 900 ha of abandoned fishponds in the Philippines back to mangrove forests. Over 50 per cent of mangrove forests in the Philippines had been lost. Over 1 million mangrove trees have been replanted in the Philippines since 2007. Mangroves are now being included in MPAs in Panas, Celen and Bohol in the Philippines. An eco-park featuring mangroves has been established at Katunggan It Ibajay in Panas. Indigenous women have planted over 150,000 mangrove trees on 159 ha of denuded patches of Busuanga Island since 2014. The trees have had an 80 per cent survival rate.

The WWF project 'Mangroves for community and climate aims to protect, restore and strengthen the management of approximately 2.5 million acres of mangroves in Colombia, Fiji, Madagascar and Mexico – 7 per cent of the world's mangroves.

Detailed specific example

The Coral Triangle, Southeast Asia

The richest coral reefs are in the Coral Triangle in Southeast Asia (Figure 2.39). The area includes the waters of Indonesia, Malaysia, Papua New Guinea, the Philippines, Soloman Islands and Timor-Leste. This area of 86,500 km^2 contains two-thirds of the world's coral species, approximately 600 species of reef-building corals, and more than 3000 species of reef fish (twice as many as are found anywhere else), and six of the world's seven sea turtle species.

The World Resources Institute estimates that about 60 per cent of reefs face immediate threats. In the past 50 years, about a quarter of all coral cover has died. The reefs that are in the worst shape are those off the most popular beaches, where sun cream gets into the water. In the South China Sea, island building and fishing for giant clams are changing some reefs beyond the possibility of recovery.

The problems vary spatially, hence policies must be locally adapted. The three countries with the largest numbers of people who fish on reefs are all in the Coral Triangle region: Indonesia, Papua New Guinea and the Philippines. In Indonesia and the Philippines, up to 1 million people's livelihoods depend on reefs. Trying to get all these people on board with conservation measures will prove difficult.

One approach is to give ecologically sensitive areas special status, such as that of a **marine protected area (MPA)**. In theory, this means activities that are deemed harmful, such as fishing, drilling and mining, can then be restricted or banned, with penalties for rule-breakers. The Aichi Targets, agreed in 2010 under the UN Convention on Biological Diversity, sought to have at least 17 per cent of inland water and 10 per cent of coastal and marine areas under conservation by 2020. Most countries signed up. As a result of this, about 10 per cent of the Coral Triangle marine habitats are now under effective management.

The most urgent action is regarding fishing vessels. A global register of fishing vessels would help identify wrongdoers. Moreover, simply declaring an area protected does not necessarily protect it. Around 75 per cent of global fisheries come from Asia, and 34 million people there are in engaged in capture fisheries. The industry is worth about US$20 billion annually.

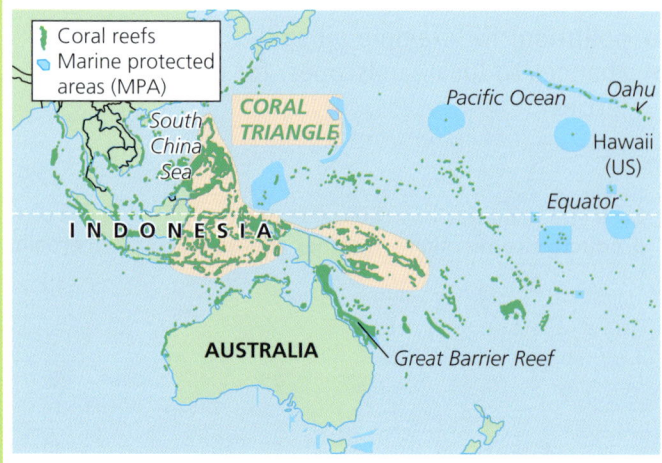

▲ Figure 2.39 Coral reefs and marine protected areas (MPAs) in the Coral Triangle

The extent to which coral reefs and mangroves can be managed

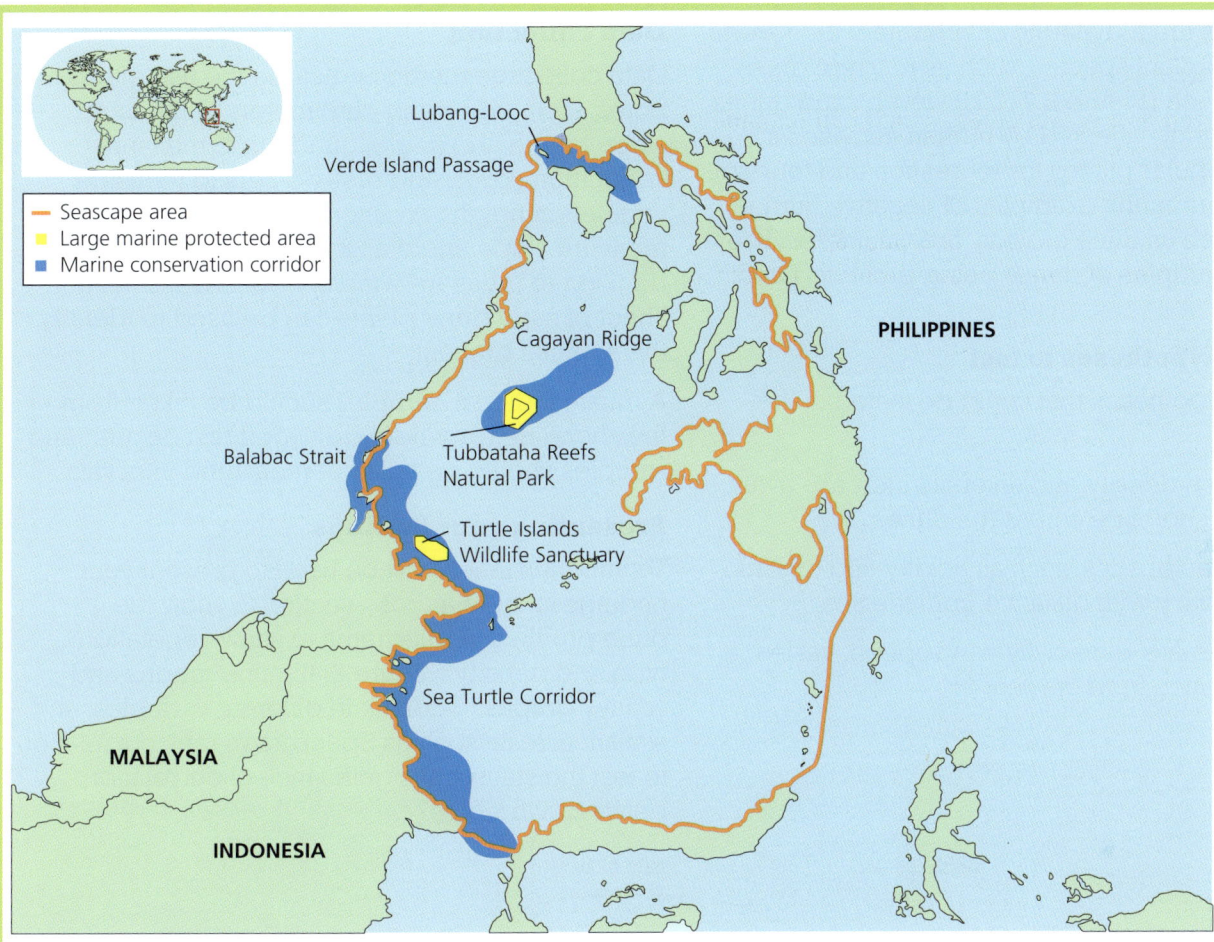

▲ Figure 2.40 The Sulu-Sulawesi Marine Ecoregion Priority Seascape

The Coral Triangle initiative has five main goals:

- The designation and effective management of priority seascapes such as the Sulu-Sulawesi Marine Ecoregion Priority Seascape (Figure 2.40), which provides a livelihood for around 40 million people, and locally marine managed areas (LMMAs) to protect biodiversity and fish stocks.
- The Ecosystem Approach to Fisheries Management (EAFM), which aims to sustainably use and conserve coastal and marine areas and their natural resources.
- To effectively manage MPAs and other marine resources.
- Adaptation methods to deal with global climate change.
- To improve the status of threatened species such as marine turtles, tuna and blue whales.

Activities

1 Explain the main threats to the Coral Triangle.
2 Describe the methods used to manage the threats to the Coral Triangle.

Fieldwork

Coastal fieldwork

There are many examples of fieldwork that can be carried out in coastal areas, for example sand dune succession, variations in sediment size and shape between the shoreline and the back of a beach, the impact of groynes on beach width, or the impacts of tourism on or development of a coastal area. In all cases, safety is paramount as coastal areas can be dangerous.

2.3 COASTS PRESENT OPPORTUNITIES AND HAZARDS FOR PEOPLE

Sand dunes are a dynamic environment for several specialised plant species, such as marram grass and sea couch (Figure 2.41). However, sand dunes are increasingly affected by human activities as they are popular places for recreation and tourism, and there are many examples of negative human impacts such as dune erosion, fires and litter. There are also examples of dunes being protected and restored.

Possible hypotheses to test

Possible hypotheses that could be tested in the fieldwork:

- Vegetation density and diversity increase away from the shoreline across the dunes.
- Soil characteristics, such as acidity and organic content, vary with distance from the sea.
- Micro-climate, especially wind speed, varies with distance from the sea.

Data collection

Data collection methods include the use of ranging poles, 20 m tapes and clinometers to measure slope length and angle; quadrats to measure vegetation cover; and metre rulers to measure vegetation height. A clinometer can be used to measure slope angle and an anemometer can be used to measure wind speed. A classification sheet of sand dune plants can be used to identify the species present.

A transect (linear sample) should be taken from the shoreline (or where driftwood first occurs) and samples should be taken at measured intervals.

Presenting your findings

Techniques that could be used to present your findings include a cross-section to show the dune profile, kite diagrams to show vegetation type and density (Figure 2.42), line graphs and scatter graphs to correlate different variables, and statistical tests such as Spearman's rank correlation to test the statistical significance of correlation between variables.

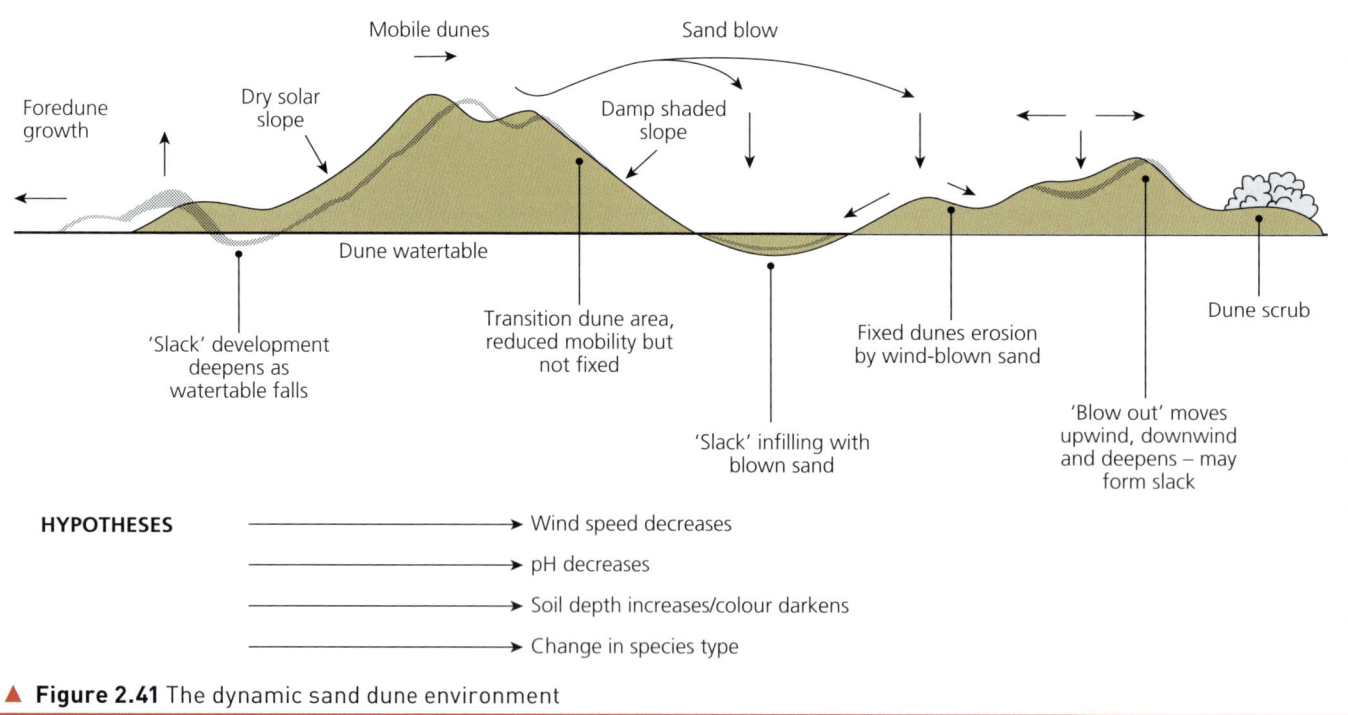

▲ Figure 2.41 The dynamic sand dune environment

The extent to which coral reefs and mangroves can be managed

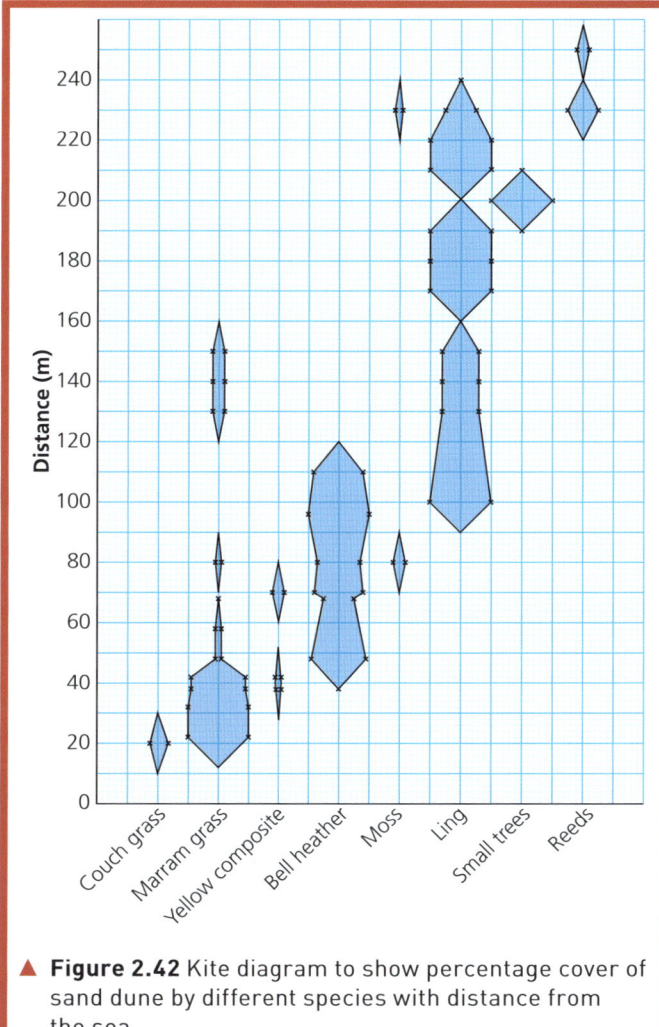

▲ **Figure 2.42** Kite diagram to show percentage cover of sand dune by different species with distance from the sea

Practice questions

1 Figure 2.43 shows a coastal landform.
  a Identify the coastal landform. [1]
  b Describe the main characteristics of the landform shown. [2]
  c Explain how this landform was formed. [3]

▲ **Figure 2.43** A coastal landform

2 a Figure 2.44 shows the global distribution of mangrove forests.
     i Describe the distribution of mangrove forests as shown on the map. [3]
     ii Explain why mangroves occur in tidal estuaries. [2]
  b Figure 2.45 shows part of a mangrove forest at low tide.
     i Explain the importance of mangroves in a coastal area that you have studied. [3]
     ii Explain one pressure on mangroves. [2]

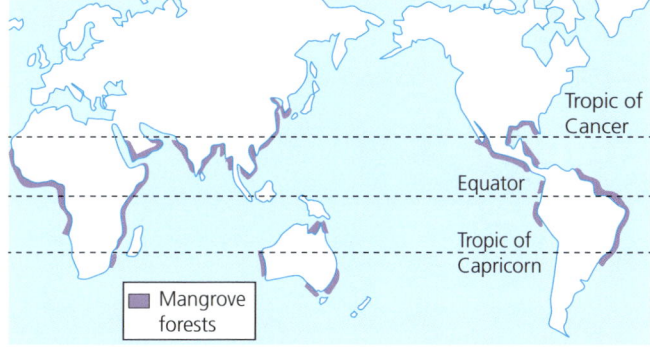

▲ **Figure 2.44** The global distribution of mangrove forests

▲ **Figure 2.45** A mangrove forest at low tide

3 Assess the opportunities and hazards of living in coastal areas. [7]

TOPIC 3

Changing ecosystems

Topics

3.1 The characteristics of the Antarctic ecosystem
3.2 The threats to the Antarctic ecosystem and how they can be managed
3.3 The characteristics of tropical rainforest ecosystems
3.4 The threats to tropical rainforest ecosystems and how they can be managed

This topic looks at:

- the characteristics of two contrasting ecosystems, Antarctica and tropical rainforests
- the threats that each faces and the strategies used to manage them.

3.1 The characteristics of the Antarctic ecosystem

This chapter will explain:
★ the location of the Antarctic ecosystem
★ the climate of Antarctica
★ the features of the Antarctic ecosystem.

The location of the Antarctic ecosystem

The Antarctic refers to everything south of the Antarctic Circle (Figure 3.1) including land, ice shelves and sea, while Antarctica refers to the landmass that is the southern polar continent. Antarctica consists of two parts, East and West Antarctica. The landmass is about 14 million km², which is about 6 million km² larger than the USA. The Antarctic ice sheet covers about 98 per cent of the continent and is on average about 1.6 km thick. Antarctica contains around 90 per cent of the world's ice and 70 per cent of its fresh water. If all the ice in Antarctica were to melt, global sea levels could rise by up to 60 m.

A large part of the West Antarctica Ice Sheet (WAIS) lies below sea level and its edges are floating ice shelves. The WAIS accounts for about 10 per cent of the total volume of the Antarctic ice sheet, about 2 million km³. If WAIS were to melt, sea levels could rise 3–4 m. The East Antarctica Ice Sheet (EAIS) rests on the Antarctica landmass; it's a thicker, larger ice sheet, and is nearly 5 km deep.

In 2024 it was reported that scientists had identified a tipping point for the loss of ice sheets in Antarctica that could cause a greater rise in sea level than previously expected. They examined how warming gets between coastal ice sheets and the ground they rest on. The warm water melts some of the ice, allowing more water to flow in. This reinforces the warming. Scientists had warned in 2022 that the Greenland ice cap was in danger of collapsing, and in 2023 they found that accelerated melting in western Antarctica was inevitable for the rest of the century.

> **Interesting note**
>
> Antarctica is unique among the continents in that it is the only one with no indigenous population. The first 'indigenous' child to be born there was an Argentinian boy, Emilio Marcos Palma, born in 1978. Since then, more than ten children have been born in Antarctica. It is inhabited by around 5000 people in summer (scientists and support staff) and about 1000 people in winter.

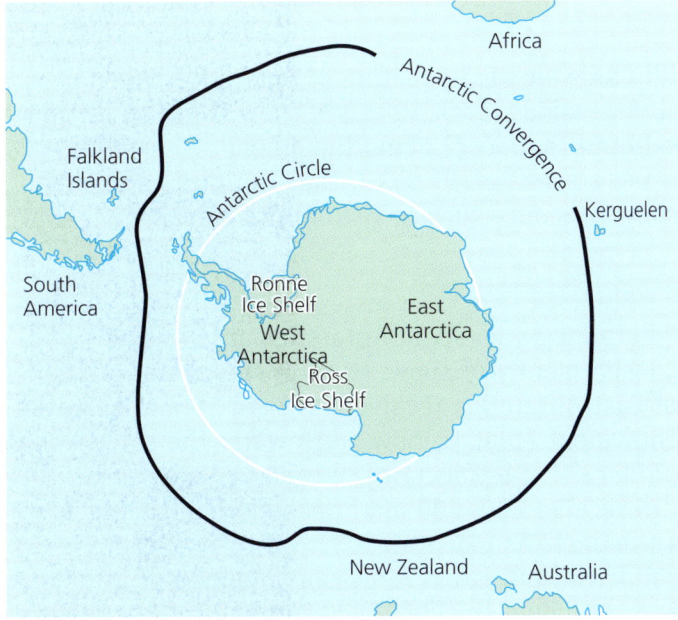

▲ **Figure 3.1** Antarctica and the Antarctic Circle

The climate of Antarctica

Temperature

The main characteristic of Antarctica's climate, in terms of both duration and intensity, is that it is cold. On ice sheets in the interior, the winters are

3.1 THE CHARACTERISTICS OF THE ANTARCTIC ECOSYSTEM

severe and the summers non-existent. However, in coastal areas, the winters are warmer and the summers cooler. There are many reasons why Antarctica is so cold:

» Due to high **air pressure**, the continent is relatively cloud-free and receives little **rainfall**. High pressure areas have descending air so there is limited uplift or **cloud** formation. This allows much of the outgoing long-wave radiation to escape from the Earth's surface, making the region very cold.
» Due to its high latitude, the low angle of the sun in relation to the ground means it receives less radiation. In the brief summer, however, there is sufficient solar radiation to melt snow and ice and to support life.
» Due to its colour, Antarctica reflects much solar radiation. This reflectivity is called **albedo**. Dark surfaces absorb more solar radiation (up to 90 per cent), whereas fresh snow may absorb as little as 10 per cent and reflect the other 90 per cent.
» There is a lack of dust and water vapour in the atmosphere. Dust and water vapour are important for trapping long-wave radiation from the Earth's surface, which heats the Earth's atmosphere. Hence, limited amounts of long-wave radiation are trapped over Antarctica to heat up the atmosphere.

Antarctica's temperatures are low throughout the year (Figure 3.2). The mean annual temperature is below freezing everywhere south of 60°S. On the continent, the mean annual temperature varies from about -60°C inland to around -10°C by the coast. There are seasonal contrasts too. On the continental interiors, temperature ranges from about -70°C in July to -25°C in January. In contrast, along the coast it ranges from -40°C in winter to -2°C in summer. The height of much of Antarctica also leads to lower temperatures. In general, temperatures fall 1°C per 100 m. Around 50 per cent of Antarctica is over 2000 m and 25 per cent over 3000 m. If those areas were at sea level, Antarctica would be 20–30°C warmer!

Winds are particularly important in Antarctica and can intensify the chilling effects of low temperatures (Table 3.1).

▼ **Table 3.1** The effect of wind speed on temperature – the wind chill effect

Air temperature (°C)	Wind speed (metres per second)				
	Calm	2.5	5	10	20
0	19	0	-7	-12	-18
-5	16	-6	-13	-19	-26
-10	13.5	-11	-19	-26	-33
-20	9	-21	-31	-40	-44
-30	4.5	-31	-43	-54	-63
-40	0	-41	-55	-68	-78

Blizzards are also a feature of Antarctica. When surface winds reach 10 m/s^{-1}, snow is picked up and saltated (bounced) along the ground. When winds reach a speed of 15 m/s^{-1}, they can pick up enough snow to reduce surface visibility to zero.

Precipitation

Precipitation levels in Antarctica are low and consist mainly of snow. Many of the central parts of East Antarctica receive less than 50 mm of water equivalent (that is, how much water the snow would produce if it fell as rain), whereas coastal areas can receive 200–600 mm. Once more, **altitude** has an impact. As moist air is forced to rise, there is condensation, cloud formation and precipitation, and maximum snowfall occurs at about 1600 m.

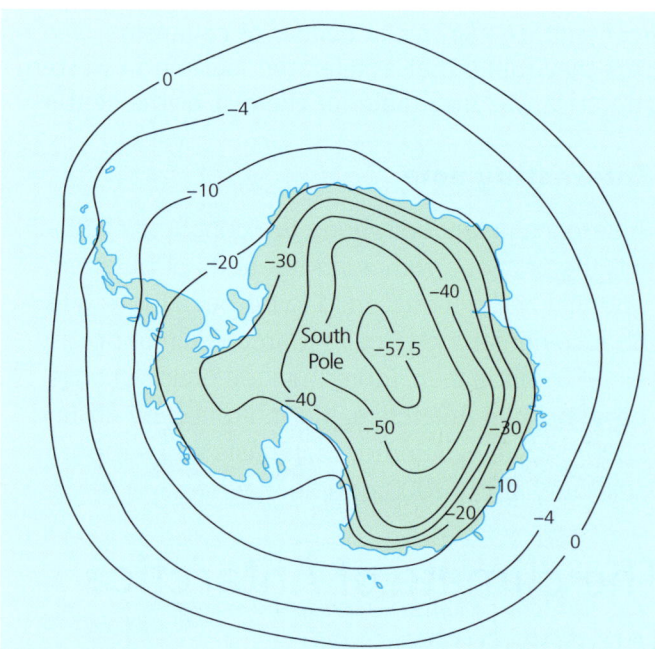

▲ **Figure 3.2** Mean annual surface temperatures in Antarctica (°C)

Activities

1. Using Table 3.1, state the wind chill when:
   a. air temperature is 0°C and wind speed is 5 m/s⁻¹
   b. air temperature is -10°C and wind speed is 10 m/s⁻¹.
2. Table 3.2 shows climatic data for McMurdo, Antarctica.

   a. Calculate the mean annual temperature and the total annual rainfall.
   b. Explain why higher temperatures occur between November and February.

▼ Table 3.2 Climatic data for McMurdo, Antarctica

	Jan	Feb	Mar	Apr	May	Jun	Jul	Aug	Sep	Oct	Nov	Dec	Annual
Temperature (°C)	-2.9	-9.5	-18.2	-20.7	-21.7	-23.0	-25.7	-26.1	-24.1	-18.9	-9.7	-3.4	
Rainfall equivalent (mm)	15.0	21.2	24.1	18.4	23.7	24.9	15.6	11.3	11.8	9.7	9.5	15.7	

3. Copy and complete the climate graph for McMurdo, Antarctica shown in Figure 3.3.

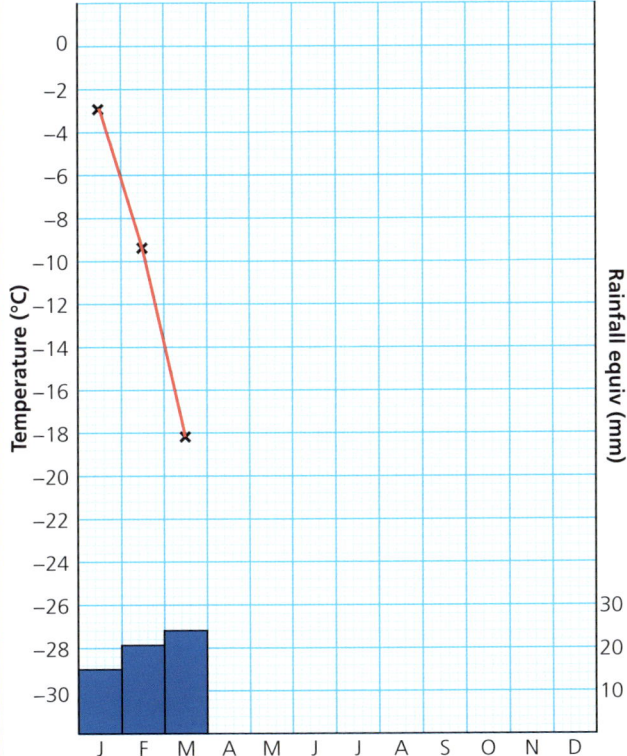

▲ Figure 3.3 Partially completed climate graph for McMurdo, Antarctica

The features of the Antarctic ecosystem

Despite the low temperatures, limited moisture, poor **soils** and lack of sunlight for six months, there are many **ecosystems** in and around Antarctica (Figure 3.4). There around 200 species of lichen, more than 100 species of mosses and liverworts, 30 species of macrofungi, and large numbers of algae. Although the growing season is short, many species exist in the milder parts of the Antarctica Peninsula. Opportunities for photosynthesis and water availability are extremely limited, and growth rates and reproduction are very low, and some species, notably glass sponges in Terra Nova Bay on the Ross Sea, can live for up to 15,000 years.

3.1 THE CHARACTERISTICS OF THE ANTARCTIC ECOSYSTEM

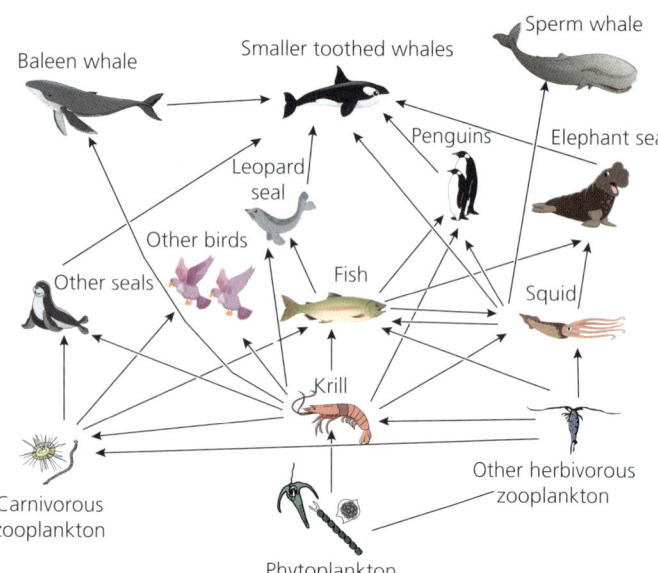

▲ Figure 3.4 Ecosystems around Antarctica

Antarctica's marine and aerial environments support an abundance of species, including penguins, seabirds, seals, whales and krill. The Southern Ocean contains some of the coldest sea water in the world. Cold water is very dense and sinks, displacing lower water to the surface. This upwelling water brings up oxygen and nutrients to the surface. This provides nutrients for surface algae, which then provide food for krill, a shrimp-like marine crustacean, which in turn provide food for fish, seals, whales and sea birds. Marine ecosystems in Antarctic waters may be relatively simple but they are, nonetheless, varied.

Krill (Figure 3.5) are about 5–6 cm long and feed on phytoplankton and zooplankton. Their biomass in Antarctic waters is estimated at around 500 million tonnes. Krill is commercially important, and between 150,000 and 200,000 krill are harvested in the Southern Ocean each year.

▲ Figure 3.5 Krill

The Southern Ocean has over 250 species of fish that can survive temperatures below freezing, such as the Antarctic icefish. The Patagonian toothfish declined by about 50 per cent between 1985 and 2020 due to overfishing. Many species of whale are seasonal visitors to the Southern Ocean and these were hunted intensively in the early parts of the twentieth century. Species such as blue, humpback, minke, killer, southern right and sperm whales feed and breed around the Antarctic. Commercial harvesting of killer whales (orcas) last occurred in 1979–80. The population of orcas may now be in excess of 80,000. Likewise, seals are found in large numbers in the Southern Ocean despite having previously been harvested. Seal species present include leopard, Ross, southern elephant and Weddell. Leopard seals can be over 3 m in length and weigh up to 500 kg. They feed on fish, penguins and other seals' pups. The Southern Ocean is home to nearly 90 per cent of benthic (deep sea) fauna.

Bird life includes terns, petrels, skuas, albatrosses and flightless penguins. Species of penguin include emperor, chinstrap, Adélie and royal. Emperor penguins are the largest – an adult male can be over 1 m tall and weigh around 40 kg. The emperor penguin is the only one to breed on the Antarctic Peninsula in winter. Albatrosses can have a wingspan of over 3.5 m. There are believed to be around 20,000 breeding pairs of albatrosses around the islands of the Southern Ocean.

> ### Activities
> 1 Study Figure 3.4. Create a food chain with seven levels in it.
> 2 Suggest why food chains in the Antarctic oceans are more developed than food chains in the Antarctic continent.

The interrelationships between abiotic and biotic factors in the Antarctic ecosystem, and how flora and fauna adapt to survive

As elsewhere, abiotic factors in Antarctica include climate (for example temperature, precipitation and wind), nutrients, sunlight, soils, rocks and permafrost. Biotic factors include the mammals, birds, fish, insects, mosses and lichens found in the oceans, on land and in the soil.

There is a close relationship between abiotic factors and biotic factors. The lack of sunlight for six months, low temperatures, limited soil development and high wind speeds make life in central Antarctica difficult. However, around coastal areas, temperatures may be higher and conditions less extreme, allowing more species to survive. Many species use Antarctica for specific purposes, such as egg laying and raising their young.

In contrast, the oceans around Antarctica have a plentiful supply of nutrients at the surface due to the upwelling ocean currents. These support a large amount of plankton and krill, which in turn support large and diverse food webs that include fish, seals, penguins and whales.

Even sea ice can support ecosystems based on single-cell species such as bacteria and phytoplankton. These then support grazers such as zooplankton and krill, which in turn can support fish and other carnivores.

Life on land in Antarctica is somewhat limited. Most of the plant life includes mosses, lichen and liverworts, which support about 60 microspecies of invertebrates such as springtails and nematodes. The largest permanent dweller on Antarctica is the Antarctic midge (2–6 mm long).

There are many species adaptations to living in and around Antarctica. Seals, whales and penguins have thick layers of blubber (insulating fatty tissue). Blubber contains red blood cells that help regulate blood flow to the animals' skin surface. During warm conditions, the blood vessels expand and blood cells are taken to the surface. However, in cold conditions, blood vessels contract, taking blood cells away from the surface to vital organs.

Penguins have different types of feathers. On the surface there are coarse, abrasion-resistant, waterproof feathers. Over the course of a year these get damaged, so penguins replace these feathers each summer during a 'catastrophic moult'. Other feathers include soft, downy feathers, which provide insulation near the penguin's skin surface.

Penguins also form huddles. Emperor penguins may form large-scale huddles during major storms. The temperatures in the centre of the huddle can reach 37 °C. Individuals cool down by moving towards the edge of the huddle and warm up by moving towards the centre.

Antarctic fur seals have two types of fur:

» Coarse, long-guard hairs, which protect the under-fur from water and abrasion.
» Dense (up to 50,000 hairs per cm^2) water-resistant insulating fur.

Antarctic seals' eyes are adapted for hunting in dim conditions. They have large pupils that dilate to let more light in. Ross seals' eyes can be up to 7 cm in diameter. Seals also have very sensitive whiskers, which can detect the movement of prey at large distances.

Antarctic icefish have developed anti-freeze proteins in their blood. They are the only adult vertebrate with no red blood cells. The proteins bind to ice crystals, reducing their growth and preventing them from joining other ice crystals that could damage the icefish with their jagged edges.

Some species of sea spiders, worms and crustaceans reach very large sizes. It is believed that the high availability of oxygen in cold polar waters allows some species to keep on growing. This is known as gigantism.

> ### Activities
> 1 Describe the relationship between abiotic factors and biotic factors in Antarctica.
> 2 Explain how Antarctic fur seals are adapted to life in the Southern Ocean.

3.2 The threats to the Antarctic ecosystem and how they can be managed

This chapter will explain:

★ the threats posed to Antarctica
★ the impacts of the destruction of the Antarctic ecosystem
★ to what extent the strategies and techniques used to manage the Antarctic ecosystem have been successful.

The threats posed to Antarctica

Climate change

Global climate change and the resultant melting ice sheets are major problems for Antarctica, in particular West Antarctica. In February 2023, Antarctic sea ice extent set a new record low (Figure 3.6), putting at risk the species that depend on stable sea ice as a breeding habitat. In 2023, Australian scientists reported that the strong ocean currents in the Southern Ocean could slow by more than 40 per cent by 2050. Moreover, the currents have already slowed by about 30 per cent since the 1990s. The reason is the melting taking place at the edge of Antarctica, which is adding more fresh water into the Southern Ocean, reducing its density. This reduces its ability to sink and to drive the deep flows that spread nutrients around the world.

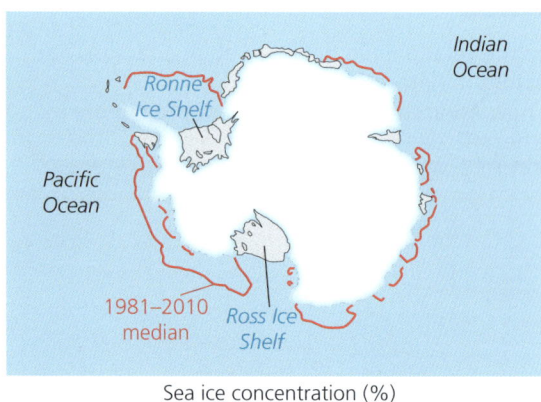

▲ **Figure 3.6** The decline of ice around Antarctica, 1981–2010 compared with 2023

This has occurred with a global warming of about 1.1°C. If warming rises by 2°C – which is predicted by some to occur by 2100 – this would lead to major changes in the ocean and to the continent. It is not just ice at the edges that is melting, the whole ice cap is losing ice. Warmer water below the West Antarctica ice sheet is causing thinning from below.

Global warming is also having an impact on emperor penguins, as they breed and raise their young on sea ice. They are especially vulnerable to the loss of sea ice, to rising temperatures and altered wind systems. It is estimated that a rise in temperature of 1.5°C could lead to a decline of 19–31 per cent in emperor penguin colonies – but if global warming continues (business as usual) to 2100, up to 80 per cent of emperor penguin colonies could become extinct.

Resource exploitation

Mining and mineral exploitation are prohibited on Antarctica. However, scientific investigations into biotic resources are increasing. Some genetic and biochemical resources have been used to develop krill oil, food-related products and anti-freeze proteins. It is estimated that there are over 200 companies and research organisations involved in biological prospecting in Antarctica.

Fishing

Fishing in Antarctic waters is allowed and is increasing as fish stocks in other seas are increasingly overfished. Fishing practices can lead to a decline in predator species, increasing unintentional catch of juvenile fish, destruction of habitat by trawlers, and the death of non-target species caught up in nets and longlines. Longline fishing involves lengthy fishing lines with up to 20,000 baited hooks, which may also capture sea birds as they fish. Overfishing has led to the decline of species such as Antarctic hake and Patagonian toothfish. Krill is endemic to Antarctic waters and many species depend on it as their main food source. However, krill is sensitive to rising ocean temperatures and it is also impacted by large-scale fishing and pollution in oceans.

Tourism

Antarctica receives upwards of 100,000 tourists per year. Transporting people to Antarctica involves emissions of 'black carbon' and other greenhouse gases by ships and aeroplanes. In addition, there is a risk of non-native insects, seeds and spores being carried in on vessels or by passengers. Most of the tourism pressure is on the Antarctic Peninsula, where 90 per cent of the tourism is concentrated.

Before the Covid-19 pandemic, some 74,000 tourists travelled to Antarctica, and around 20,000 remained on cruise ships visiting the continent. The numbers dropped to 23,000 during the pandemic but since then numbers have increased to around 100,000, with 75,000 'shore' visits.

Pollution

Pollution is an increasing problem. Microplastics are washed in from surrounding islands and seas, or they may be brought in by tourists or scientists. Sea birds often mistake plastic for food and eat it or feed it to their young. The remains of fishing equipment have been washed up on nearby islands. Seals and penguins are known to have become entangled in fishing nets, and the chances of rescue or release are slim.

> **Activities**
> 1 Explain why the emperor penguin is vulnerable to global climate change.
> 2 Explain why plastics and fishing equipment are a threat to Antarctica's ecosystems.

The impacts of the destruction of the Antarctic ecosystem

There has been continued human presence throughout the most biodiverse parts of Antarctica over the past 200 years. This has led to infrastructure developments, vegetation trampling, sealing, whaling, fishing, wildlife and soil disturbance, pollution and the introduction of alien species. However, human activity in Antarctica should be easier to regulate than elsewhere because it is restricted to science, television programmes and tourism, with few other human activities or permanent inhabitants.

Climate change is predicted to have major impacts on Antarctic ecosystems. For example:

- Rapid warming in the west Antarctic Peninsula and a change from snowfall to summer rain has led to an increase in plant communities across the Antarctic Peninsula on newly exposed land, as well as the introduction of 'alien' organisms such as grasses, flies and bacteria.
- Loss of sea ice west of the Antarctic Peninsula has led to a decline in krill and Adélie penguins. As krill is a major source of food for many marine animals and sea birds, this could lead to the decline of numbers of many species. The decline of sea ice means that species that rely on it are seeing their habitat shrink.
- Models suggest a 3°C increase in temperature by 2100, which should increase phytoplankton productivity.

Historical changes are also important – for example, the whaling industry led to the near extinction of blue whales and baleen whales in the Southern Ocean. The extinction or near extinction of any single species has a knock-on effect on other species and contributes to loss of biodiversity. This can be seen here, as with less competition for krill from whales, the number of fur seals increased dramatically in just 20 years, from a few hundred to over 20,000. The seals came onshore at Signy Island in the South Orkney Islands, and caused much vegetation destruction through trampling, and eutrophication from the large amount of excrement left on the island.

3.2 THE THREATS TO THE ANTARCTIC ECOSYSTEM AND HOW THEY CAN BE MANAGED

> **Activities**
> 1 Outline the impacts of global climate change on Antarctica's ecosystems.
> 2 Explain how changes in the whaling industry led to ecosystem destruction on Signy Island.

An evaluation of the strategies and techniques used to manage the Antarctic ecosystem

International agreements

The main strategy for managing Antarctica has been the 1959 Antarctic Treaty. This established a system for international cooperation, especially in the name of science, and the demilitarisation of Antarctica. It also prohibits all forms of mining and mineral exploitation. This enables sustainable management of Antarctica's ecosystems.

The principles and objectives of Article 9 of the 1959 Antarctic Treaty include:

» the use of Antarctica for peaceful purposes only
» the facilitation of scientific research in Antarctica
» the facilitation of international scientific cooperation in Antarctica
» the facilitation of the exercise of the right of inspection
» questions relating to the exercise of jurisdiction in Antarctica
» preservation and **conservation of living resources** in Antarctica.

The Treaty has been successful so far, partly due to the remoteness and isolation of Antarctica. Antarctica is the only continent on which there has not been a war. Its mineral resources have not been exploited, although its living natural resources such as whales, seals, fish and krill have been exploited. However, as Antarctica warms up and ice melts there could be increased pressure to develop its mineral resources.

Moreover, as global warming increases, the negative impacts on Antarctica are also likely to increase (Figure 3.7). Likewise, if the volume of tourists visiting the continent continues to increase, the impacts are likely to intensify. The same is true for pollution.

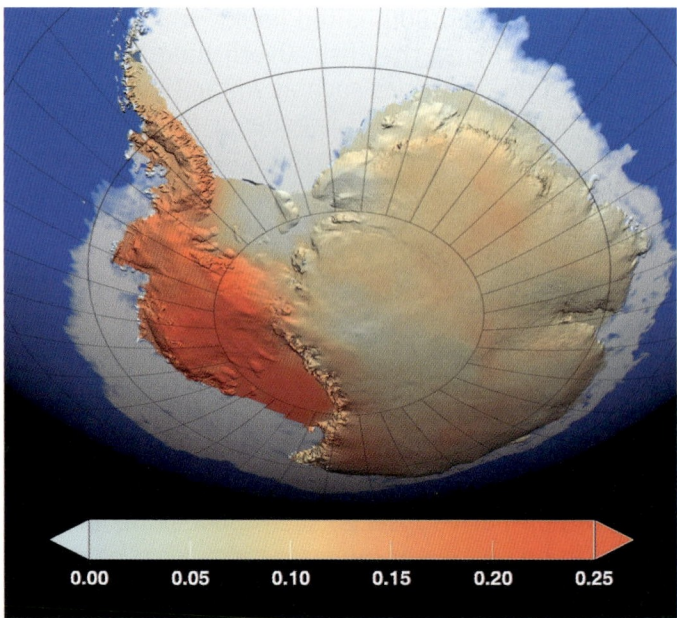

▲ **Figure 3.7** Antarctic warming since 1957 (°C)

> **Activities**
> 1 Outline the main principles and objectives of the Antarctic Treaty.
> 2 Compare the potential impacts of low- and high-global warming emissions on Antarctica.

Environmental impact surveys

An **environmental impact survey** is a detailed survey to establish the likely impacts of a new development. Before any development starts, the environmental impact survey examines and identifies environments and species that are of high **conservation** importance or at risk from the development. These surveys make it possible to analyse the potential impacts of new developments on ecosystems and introduce management strategies to protect environments at risk.

By 2018, the wharf at the Rothera Research Station off the Antarctic Peninsula (Figure 3.8) was 25 years old and beyond repair. Moreover, more modern boats were longer and deeper than the wharf would allow to dock. The 'do nothing' and '**do minimum**' approaches were considered unsafe for the berthing of vessels, so a new wharf had to be constructed.

An evaluation of the strategies and techniques used to manage the Antarctic ecosystem

![Rothera Research Station]

▲ **Figure 3.8** Rothera Research Station

▲ **Figure 3.9** The location of the Rothera Research Station, Antarctica

Some 24,000–30,000 m³ of local rock was sourced for the new wharf. Local material had to be used as the risk of introducing alien species was too high. Levels of biodiversity at Rothera were not high. Although seals were present, they did not breed in the area. The most significant environmental impacts were:

» the introduction of non-native species
» terrestrial and marine pollution from oil spills
» the removal of rock
» the removal of ice-free areas for terrestrial habitats
» the disturbance of marine mammals by underwater noise.

The wharf was started in 2018 and completed in 2020.

> ### Activities
> 1 Define the term 'environmental impact survey'.
> 2 Outline the potential impacts of new buildings on Antarctica's environment.

3.3 The characteristics of the tropical rainforest ecosystem

This chapter will explain:

★ the global distribution of tropical rainforests
★ the characteristics of the equatorial climate
★ the characteristics and structure of the tropical rainforest ecosystem.

The global distribution of tropical rainforests

Evergreen **tropical rainforests** are located in equatorial areas, largely between 10°N and 10°S (Figure 3.10). There are some areas of rainforest found outside these regions, but these tend to be more seasonal in nature. The main areas of rainforest include the Amazon rainforest in Brazil, the Congo rainforest in central Africa and the Indonesian–Malaysian rainforests of Southeast Asia. There are also many small fragments of rainforest, such as those on the island of Madagascar and in the Caribbean. Tropical rainforests everywhere are under increasing threat from human activities such as farming and logging. The result is that rainforests are disappearing and those that remain are not only smaller, but are broken up into fragments.

Interesting note

Tropical rainforests cover 6 per cent of the world's land surface but hold 50 per cent of the world's species. The Amazon rainforest alone is home to 10 per cent of the world's known species.

The characteristics of the equatorial climate

The main characteristics of an equatorial climate include:

» hot conditions – generally above 26°C – throughout the year, as the overhead sun is always high in the sky by the middle of the day
» high levels of rainfall, often over 2000 mm (Figure 3.11)
» tropical rainforests characterised by low pressure – rising air that cools and produces clouds and rains
» a lack of seasons – the temperatures are high throughout the year
» the difference between daytime and night-time temperatures (known as the diurnal range) is higher than the seasonal differences in temperature
» rainfall is mainly convectional – heating causes air to rise and then cool, and rain may fall on as many as 250 days each year
» cloud cover varies – in the morning it may be limited, but by the afternoon towering cumulonimbus clouds mark the start of the **convectional rains**
» the presence of clouds tends to reduce the amount of heat that is lost at night – hence the diurnal range is less than in hot desert areas
» the **humidity** (moisture in the atmosphere) is high, and relative humidities of 100 per cent are often reached in the late afternoon
» wind speeds within the rainforest are reduced by the large numbers of trees present
» with increasing latitude there is increased seasonality of the rainforest.

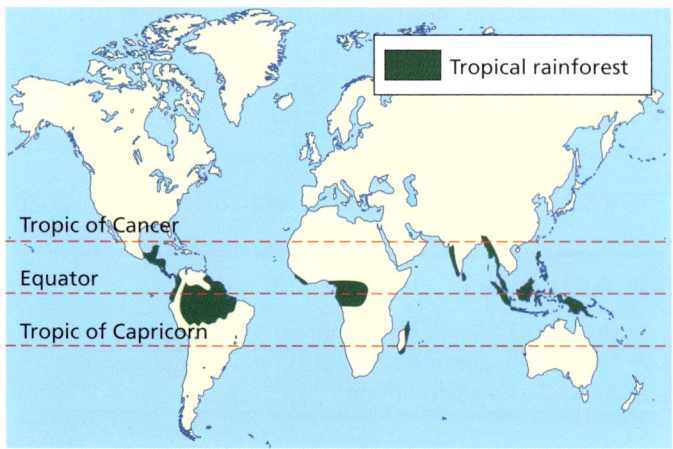

▲ Figure 3.10 World distribution of tropical rainforests

The characteristics and structure of the tropical rainforest ecosystem

The tropical rainforest ecosystem includes biotic factors (such as plants and animals) and abiotic factors (such as heat, rainfall, oxygen and minerals). These are closely interrelated (Figure 3.12).

▲ Figure 3.11 Convectional uplift causes cloud and rain on most days in a tropical rainforest, such as here in Sarawak

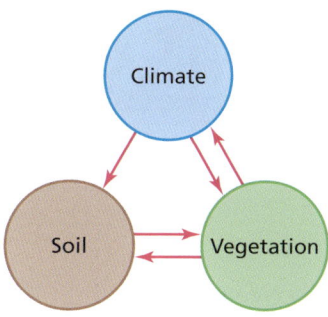

▲ Figure 3.12 The links between climate, soil and vegetation in a tropical rainforest ecosystem

The data for Manaus in Brazil (Table 3.3) show that the warmest months are September and October, with a maximum temperature of 34°C. In contrast, all of the months from December to September share the minimum temperature of 24°C. Thus the annual temperature range is 10°C.

Rainfall in Manaus is high, nearly 2100 mm annually. There is a definite wet season between November and May, whereas the months of June to October are relatively dry.

Vegetation is evergreen, enabling photosynthesis to take place all year round. This is possible due to the high temperatures and presence of water throughout the year. The vegetation is layered, and the shape of the crowns varies at each layer (Figure 3.13). Species at the top of the canopy receive most of the sunlight, whereas species that are located near the forest floor are adapted to darker conditions and generally have a darker pigment so as to photosynthesise at low light levels. There is a great variety in the number of species in a rainforest – this is known as **biodiversity**. A single hectare in a rainforest may contain as many as 300 different species. Typical rainforest species include figs, teak, mahogany and yellow woods.

> **Activities**
>
> 1 Calculate for Manaus:
>   a the mean average monthly temperature
>   b the annual precipitation.
> 2 Compare the seasonal variations in rainfall and sunshine (hours) for Manaus.

▼ Table 3.3 Climate data for Manaus, Brazil

	J	F	M	A	M	J	J	A	S	O	N	D	Av/Total
Temperature:													
Daily max (°C)	31	31	31	31	31	31	32	33	34	34	33	32	32
Daily min (°C)	24	24	24	24	24	24	24	24	24	25	25	24	24
Average monthly (°C)	28	28	28	27	28	28	28	29	29	29	29	28	
Rainfall:													
Monthly total (mm)	278	278	300	287	193	99	61	41	62	112	165	220	
Sunshine:													
Sunshine (hours)	3.9	4	3.6	3.9	5.4	6.9	7.9	8.2	7.5	6.6	5.9	4.9	5.7

3.3 THE CHARACTERISTICS OF THE TROPICAL RAINFOREST ECOSYSTEM

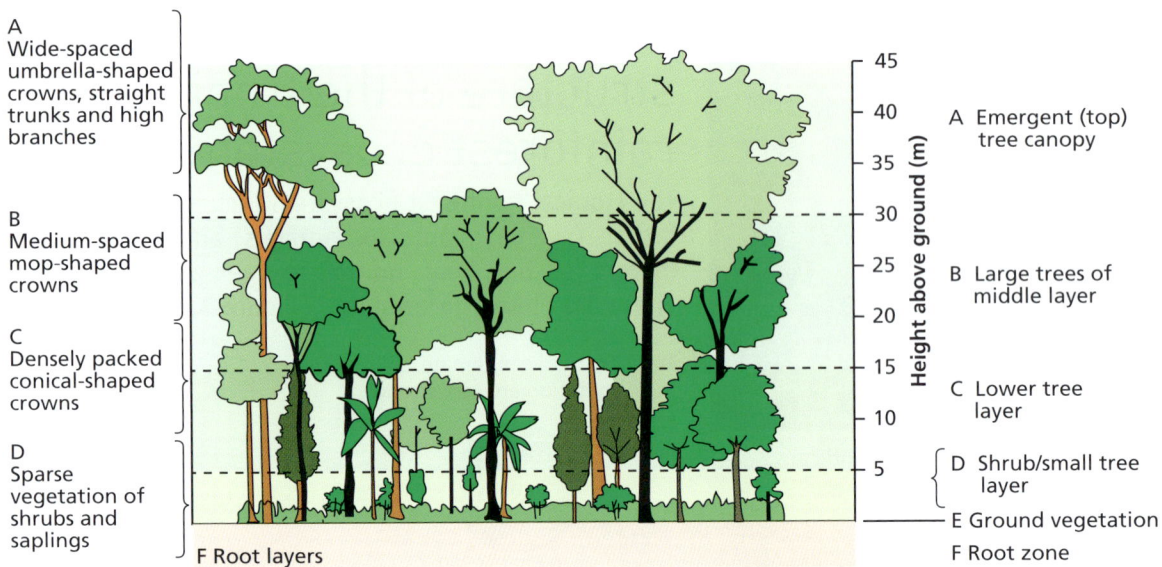

▲ Figure 3.13 Vegetation structure of the tropical rainforest

Soils

Rainforests are the most productive land-based ecosystems. Ironically, the soils of tropical rainforests are quite infertile. This is because most of the nutrients in the rainforest are contained in the biomass (living matter). Rainforest soils are typically deep due to the large amount of weathering that has taken place, and they are often red in colour due to large amounts of iron being present in the soil. Nevertheless, there are some areas in which tropical soils may be more fertile: in floodplains and in volcanic areas the soils may be enriched by flooding or by the weathering of fertile lava flows.

How flora and fauna adapt to survive

Flora

Tropical vegetation in rainforests has many adaptations. Some trees have leaves with drip-tips (Figure 3.14a), which are designed to get rid of excess moisture. In contrast, other plants have saucer-shaped leaves in order to collect water. Pitcher plants have developed an unusual means of getting their nutrients. Rather than taking nutrients from the soil, they have become carnivorous and get their nutrients from insects and small frogs that are trapped inside the pitcher (Figure 3.14b). This is one way of coping with the very infertile soils of the rainforest. Other plants are very tall. To prevent being blown over by the wind, very large trees have developed buttress roots that project out from the main trunk above the ground, which gives the plant extra leverage in the wind.

▲ Figure 3.14 Adaptations of rainforest plants: **a** drip-tip; **b** pitcher plant

Fauna

Although rainforests cover less than 6 per cent of the Earth's surface, they account for over 50 per cent of all animal species on Earth. Tropical rainforests are very biodiverse (Figure 3.15). Many species, such as orangutans, are arboreal (they live entirely in the trees) and rarely come to the ground, as they would be easy prey for large carnivores there.

Top carnivores, such as jaguars, tigers and leopards, are highly camouflaged to aid in hunting. Many other species are highly camouflaged to avoid becoming prey, such as stick insects and the Indian oakleaf butterfly. Sloths are covered with a layer of green algae that camouflages their fur in their arboreal environment. Some animals have evolved to look larger or more scary than they really are: the larvae of the lobster moth look like scorpions, but they are defenceless. Many butterflies have designs that look like large eyes on their wings, in order to confuse potential predators. Some species, such as the monarch butterfly, are poisonous, which helps deter predators. It is more than likely that there are many rainforest species that have not yet been discovered – according to the World Wide Fund for Nature, new species are being discovered in the Amazon 'every other day'. Between 2016 and 2020, over 600 new species of plants and animals were discovered in the Amazon alone.

> ### Activities
> 1 Suggest how the vegetation in Figure 3.15 is adapted to conditions in the rainforest.
> 2 Explain how animals are adapted to living in tropical rainforests.

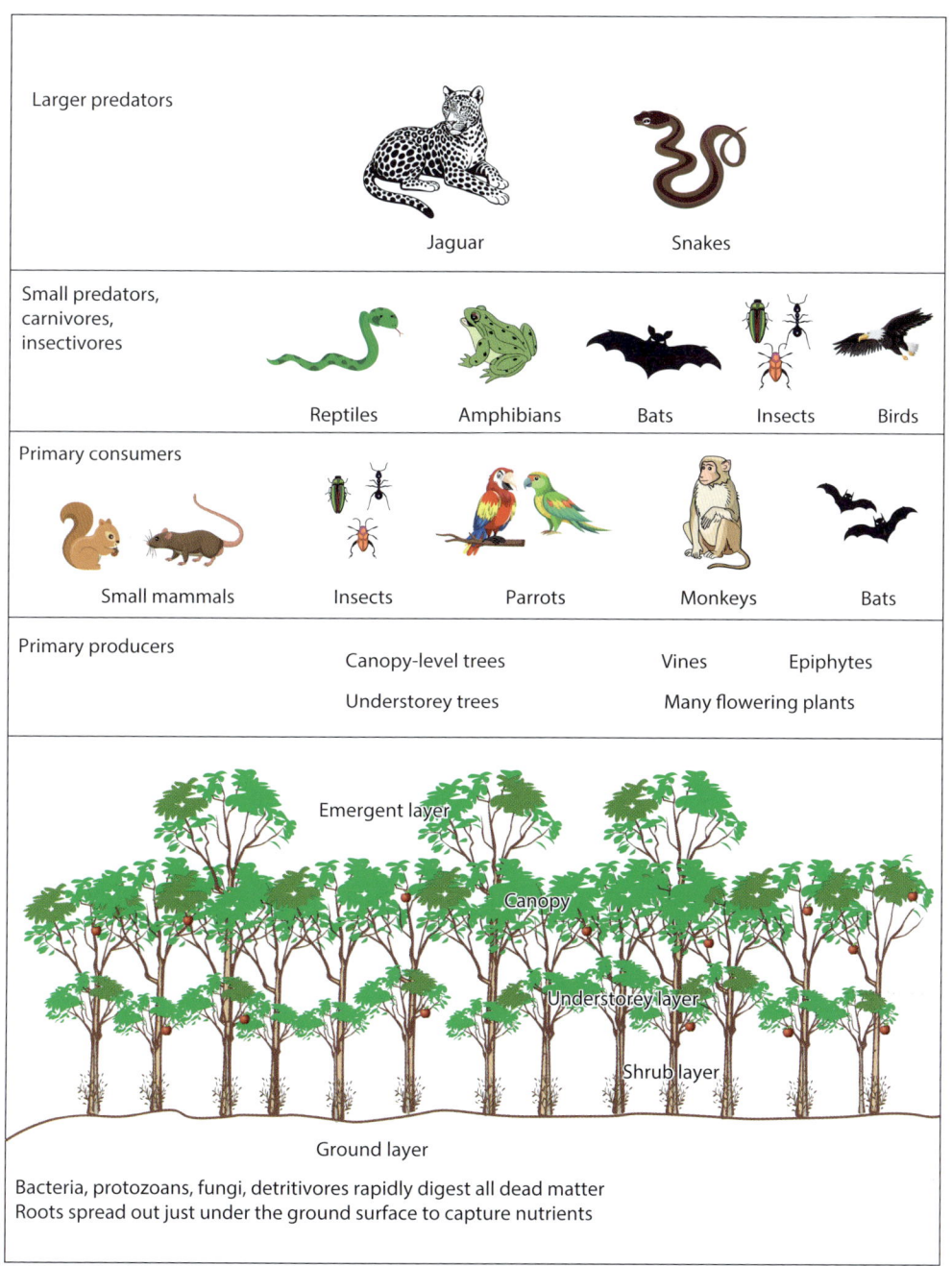

▲ Figure 3.15 A simplified food chain from the tropical rainforest

3.4 The threats to the tropical rainforest ecosystem and how they can be managed

This chapter will explain:

★ the main threats to tropical rainforest ecosystems
★ the impacts and causes of deforestation in the Brazilian Amazonian rainforest
★ to what extent the strategies and techniques used to manage tropical rainforest ecosystems have been successful
★ detailed specific example: Danum Valley Conservation Area, Borneo.

The main threats to tropical rainforest ecosystems

About 200 million people live in areas that are or were covered by tropical rainforests. These areas offer many advantages for human activities, such as farming, hydroelectric power, tourism, fishing and food supply, mineral development and forestry (Figure 3.16 and Table 3.4). Rainforests also play a vital role in regulating the world's climate, and they account for 50 per cent of the world's plants and animals. They are vital too for the protection of soil and water resources.

▲ Figure 3.16 A tropical rainforest along with shifting cultivation – rice growing in Sarawak

▼ Table 3.4 The value of tropical rainforests

Industrial uses	Ecological uses	Subsistence uses
• Charcoal	• Watershed protection	• Fuelwood and charcoal
• Saw logs	• Flood and landslide protection	• Fodder for agriculture
• Gums, resins and oils	• Soil erosion control	• Building poles
• Pulpwood	• Climate regulation, e.g. balancing levels of carbon dioxide and oxygen	• Pit sawing and saw milling
• Plywood and veneer		• Weaving materials and dyes
• Industrial chemicals		• Rearing silkworms and beekeeping
• Medicines	• Special woods and ashes	
• Genes for crops	• Fruit and nuts	
• Tourism		

However, tropical rainforests are under threat:

» The year-round growing season is very attractive for farmers, although the poor quality of the soil results in the land being farmed for only a few years before it is abandoned (Figure 3.17). Nevertheless, large-scale **plantations** occur in areas of tropical rainforest, producing crops such as palm oil, which is increasingly being used for the **biofuels** industry.
» Cattle ranching is a threat to rainforests as it replaces a multi-layered forest ecosystem, which intercepts most of the rainfall, with a single-layered grassland ecosystem, which gets trampled by cattle and can produce much overland runoff and soil erosion.
» The construction of roads and railways has two main impacts. The first is direct – rainforests are cut down to make way for roads, such as the Trans-Amazonian Highway. As a result of the increased accessibility provided by the new roads and railways, some settlers will cut down more rainforest to produce farmland.

The impacts and causes of deforestation in the Brazilian Amazonian rainforest

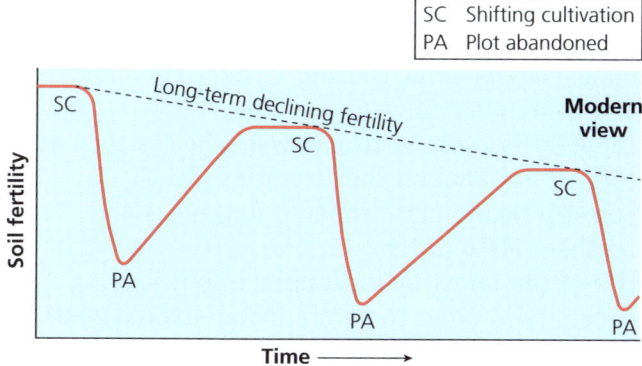

▲ **Figure 3.17** The effect of rainforest clearance on long-term soil fertility

» The second impact of the construction of roads and railways is that settlements will begin to develop along these roads and railways to provide more services for the new settlers. Each of these developments leads to a decline in the area of tropical rainforest and the services that it provides to people.
» High rainfall totals, especially in hilly areas, favour the development of hydroelectric power (HEP), such as at Batang Ai in Sarawak, Malaysia (Figure 3.19).

» Areas of rainforest have a long history of commercial logging. Tropical hardwoods, such as teak and mahogany, are prized by furniture manufacturers.
» Mineral developments, such as iron ore at Carajas in Brazil and ilmenite on the southeast coast of Madagascar, are also developed in some rainforest areas and these resources may be exploited.

The impacts and causes of deforestation in the Brazilian Amazonian rainforest

Impacts

There are a large number of effects of deforestation (Figure 3.18), including:

» disruption to the circulation and storage of nutrients
» surface erosion and compaction of soils
» increased flood levels and sediment content of rivers
» climatic change
» loss of biodiversity and animal habitats.

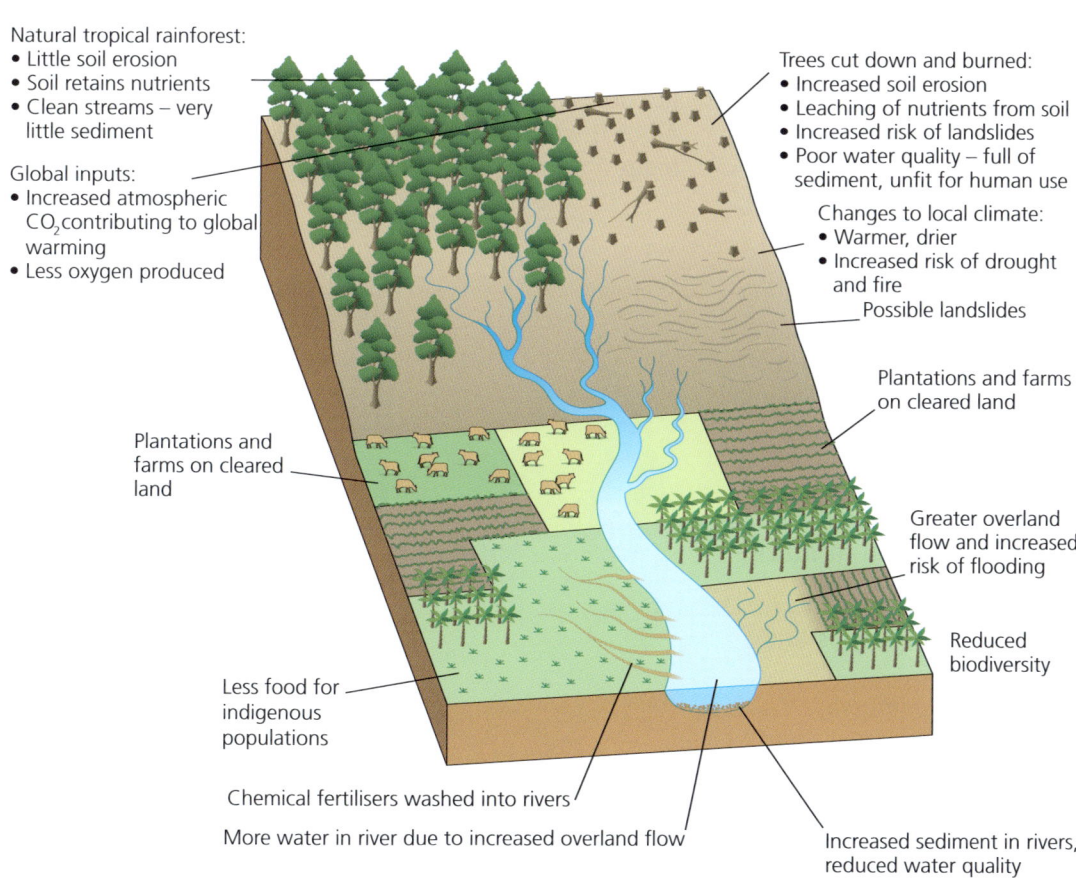

▲ **Figure 3.18** Some impacts of deforestation

3.4 THE THREATS TO THE TROPICAL RAINFOREST ECOSYSTEM AND HOW THEY CAN BE MANAGED

▲ **Figure 3.19** The rainforest at Batang Ai, affected by flooding, shifting cultivation and soil erosion

Deforestation disrupts the closed system of **nutrient cycling** within tropical rainforests. Inorganic elements are released through burning and are quickly flushed out of the system by the high intensity rains.

Soil erosion is also associated with deforestation. As a result of soil compaction, there is a decrease in infiltration, an increase in overland runoff and surface erosion.

As a result of the intense surface runoff and soil erosion, rivers have a higher **flood peak** and a shorter time lag. However, in the dry season river levels are lower, the rivers have greater turbidity (murkiness due to more sediment), an increased bed load, and carry more silt and clay in suspension.

Other changes relate to **climate**. As deforestation progresses, there is a reduction of water that is re-evaporated from the vegetation, hence the recycling of water is diminished. Evaporation rates from savanna grasslands are estimated to be only about one-third of that of the tropical rainforest. Thus, mean annual rainfall is reduced, and the seasonality of rainfall increases.

Causes

There are six main causes of deforestation in Brazil:

» Agricultural colonisation by landless migrants and speculative developers along highways and agricultural growth areas.
» Conversion of the forest to cattle pastures, especially in eastern and southeastern Para and northern Mato Grosso.
» Mining, for example the Greater Carajas Project in southeastern Amazonia, which includes a 900 km railway and extensive deforestation to provide charcoal to smelt the iron ore; another threat from mining comes from the small-scale informal gold mines, *garimpeiros*, causing localised deforestation and contaminated water supplies.
» Large-scale hydroelectric power schemes such as the Tucurui Dam on the Tocantins River.
» Forestry taking place in Para, Amazonas and northern Mato Grosso.
» Use of the forest by local populations is a necessity in order to earn a living and have an acceptable quality of life.

Deforestation in Brazil shows several trends:

» It is a recent phenomenon.
» It has partly been promoted by government policies.
» There is a wide range of causes of the deforestation.
» It includes new areas of deforestation as well as the extension of previously deforested areas.
» A major cause of the deforestation is land speculation and the granting of land titles to those who 'occupy' parts of the rainforest.

According to the news organisation Mongabay, in 2023 deforestation from conversion of forest to soy farming continued to spread through Brazil's Cerrado and Amazon rainforest. There was around 27,000 ha of deforestation and forest degradation in the Cerrado in just four months, while the Amazon saw over 30,000 ha. All of it was located near grain silos used by the seven biggest soy traders in Brazil (Figures 3.20 and 3.21).

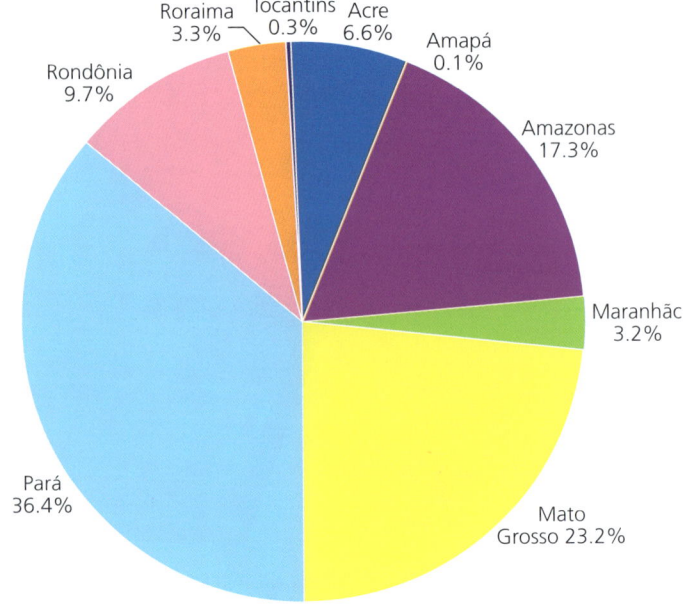

▲ **Figure 3.20** Deforestation of the Amazon by state, 2023

The strategies and techniques used to manage tropical rainforest ecosystems

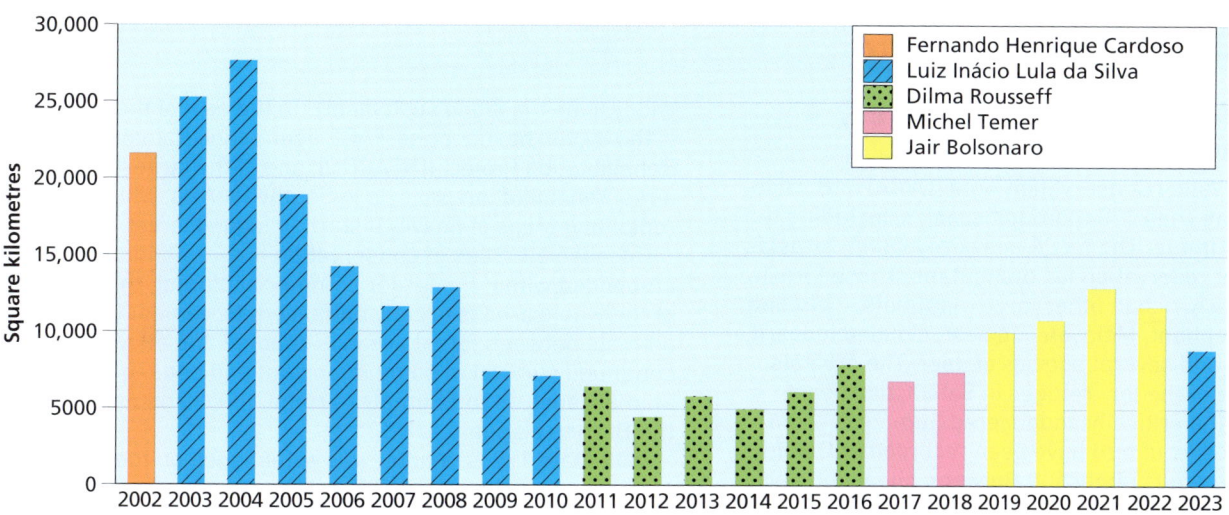

▲ **Figure 3.21** Deforestation of the Amazon under the president in office at the time, 2002–23

Figure 3.20 shows that most of the deforestation occurred in just two states, Pará and Mato Grosso. However, the long-term trend in deforestation is going down. The current decline is largely due to Brazilian President Lula, who aims to eliminate illegal deforestation by 2030. This contrasts with the significant increase in deforestation under the previous president, Jair Bolsonaro.

> ### Activities
> 1 Comment on the value of tropical rainforests for human populations.
> 2 Outline the main impacts of deforestation on the natural environment.
> 3 Explain the main causes of deforestation in Brazil.
> 4 Comment on the trends in deforestation in Brazil.

The strategies and techniques used to manage tropical rainforest ecosystems

Because tropical rainforests are important for many services, such as flood control, climate regulation, a source of genetic material, timber and food, and many others (Table 3.4, page 80), attempts have been made to protect rainforests. An example of the sustainable management of tropical rainforests is the Heart of Borneo (HoB) project. The governments of Malaysia, Indonesia and Brunei created the HoB, an area of over 20 million ha, to sustainably protect the rainforest and water resources for the welfare of present and future generations. It protects forest cover, manages protected areas and maintains biodiversity and environmental services in productive activities such as farming, forestry, fishing and ecotourism. Nevertheless, many areas of rainforest are still being cut down, in part due to population growth and the need to find more food supplies, but also due to people who want to cut down rainforests to get at mineral resources below ground.

Elsewhere in Borneo, over two-thirds of Borneo's dipterocarps (tall, straight trees providing high quality timber) are threatened with extinction. Up to 350 species of mammals, birds, reptiles and fish are at risk of extinction, for example clouded leopards, orangutans and Sumatran rhinoceroses. Other impacts include habitat losses, carbon emissions, fires (used to clear forests), land degradation and decreased water quality – palm oil plantations often use chemical fertilisers and pesticides which can pollute rivers and groundwater. Land drainage increases soil erosion. There are also social issues as the palm oil industry often removes indigenous people from their land.

3.4 THE THREATS TO THE TROPICAL RAINFOREST ECOSYSTEM AND HOW THEY CAN BE MANAGED

> **Detailed specific example**

Danum Valley Conservation Area, Malaysian Borneo

The Danum Valley Conservation Area (DVCA) in Borneo contains more than 120 mammal species, including ten species of primate. The DVCA and surrounding forest is an important reservation for orangutans. These forests are particularly rich in other large mammals, including the Asian elephant, Malayan sun bear, clouded leopard, bearded pig and several species of deer. The area also provides one of the last refuges in Sabah, northeast Borneo, for the critically endangered Sumatran rhino. Over 340 species of bird have been recorded at Danum, including the argus pheasant, Bulwer's pheasant, and seven species of pitta bird.

The DVCA covers 43,800 ha, comprising almost entirely lowland dipterocarp forest (dipterocarps are valuable hardwood trees). It is the largest expanse of pristine forest of this type remaining in Sabah (Figure 3.22).

Until the late 1980s, the area was under threat from commercial logging. The establishment of a long-term research programme between Yayasan Sabah and the Royal Society in the UK created local awareness of the conservation value of the area and provided important scientific information about the forest and what happens to it when it is disturbed through logging. Danum Valley is controlled by a management committee containing all the relevant local institutions – the wildlife, forestry and commercial sectors are all represented. This allows Danum Valley to be managed sustainably. To the east of the DVCA is the 30,000 ha Innoprise-FACE Foundation Rainforest Rehabilitation Project (INFAPRO), one of the largest forest rehabilitation projects in Southeast Asia, which is replanting areas of heavily disturbed logged forest.

Because all areas of conservation and replanting are embedded within the larger commercial forest, the value of the whole area is greatly enhanced. Movement of animals between forest areas is enabled and allows the continued survival of some important and endangered animals such as the Sumatran rhino, the orangutan and the Borneo elephant.

In the late 1990s, a hotel was established on the northeastern edge of the DVCA. It has established flourishing ecotourism in the area and exposed this unique forest to a wider range of visitors than was previously possible. As well as raising revenue for the local area, it has raised the international profile of the region as an important centre for conservation and research. This is a very good example of the sustainable development of an area of tropical rainforest.

Activities

1 What was the main threat to the Danum Valley before the late 1980s?
2 Why is the DVCA important for the conservation of species?

▲ Figure 3.22 The Danum Valley Conservation Area

Practice questions

1 Figure 3.23 shows a food web for the sea around Antarctica.

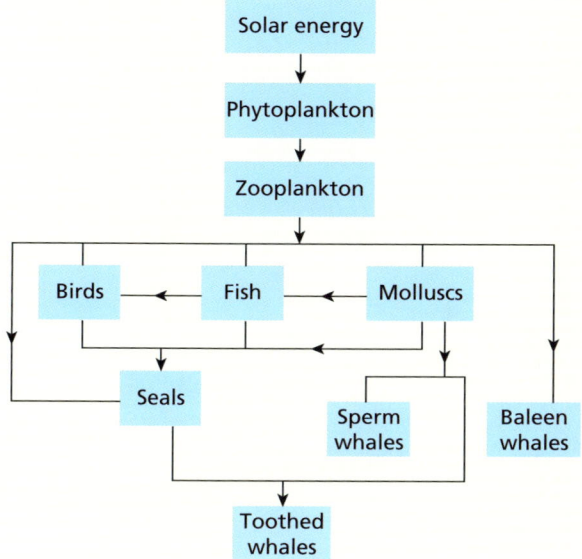

▲ **Figure 3.23** A food web for the sea around Antarctica

a i Identify the main source of energy for this ecosystem. [1]
  ii Identify the organism that feeds on phytoplankton. [1]
  iii State the organism that feeds on birds. [1]
  iv State the top carnivore in the Antarctic marine ecosystem. [1]
b Explain why Antarctica's ecosystems are largely marine-based rather than land-based. [3]

2 View Figure 3.16 on page 80 which shows an area of tropical rainforest.
  a Identify how this area of rainforest is being used. [1]
  b Compare the characteristics of the vegetation in the rainforest with that of the cleared area. [2]
  c Explain two reasons why tropical rainforests are being destroyed by human activities. [4]
  d Explain the consequences of the deforestation of tropical rainforests. [4]

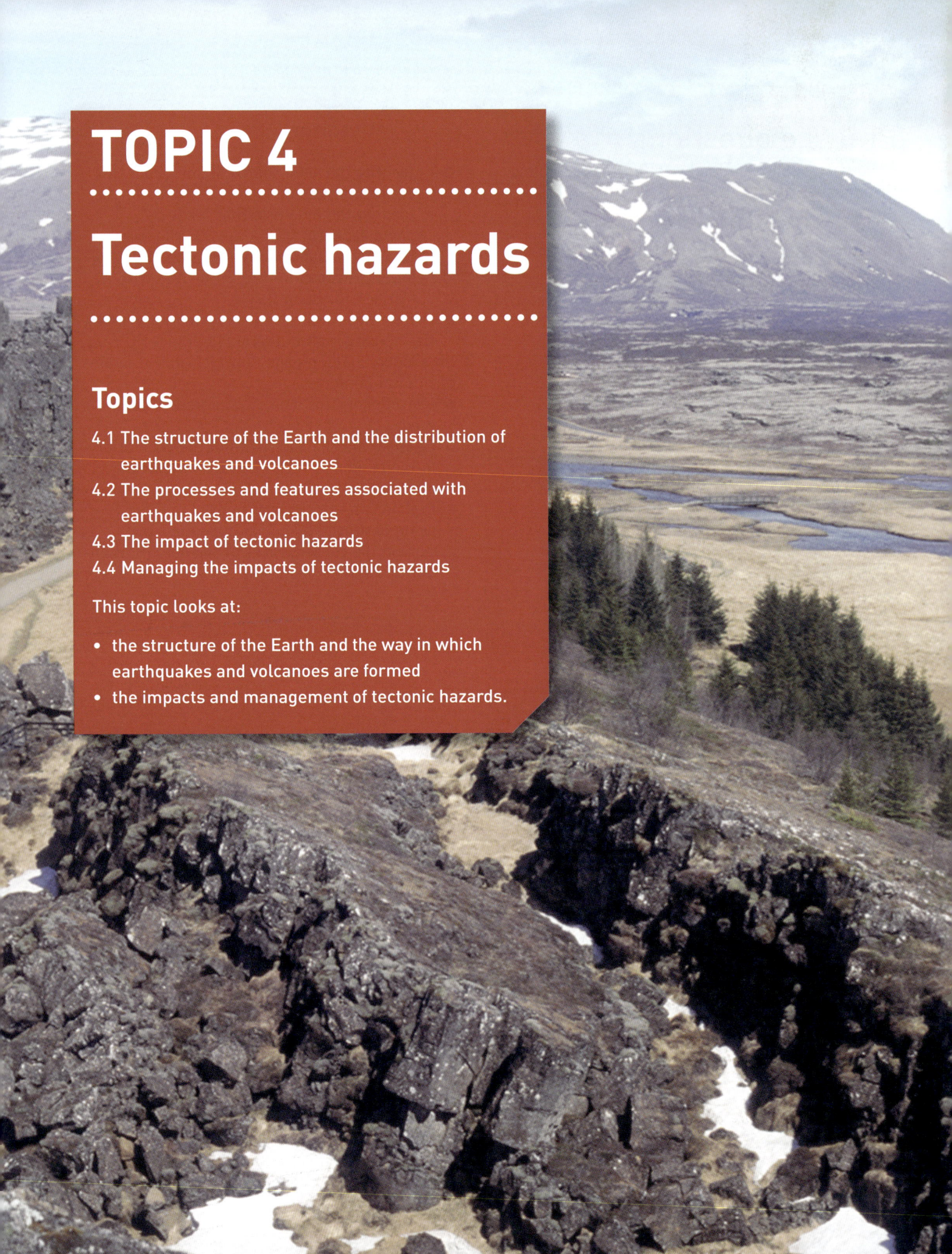

TOPIC 4

Tectonic hazards

Topics

4.1 The structure of the Earth and the distribution of earthquakes and volcanoes
4.2 The processes and features associated with earthquakes and volcanoes
4.3 The impact of tectonic hazards
4.4 Managing the impacts of tectonic hazards

This topic looks at:

- the structure of the Earth and the way in which earthquakes and volcanoes are formed
- the impacts and management of tectonic hazards.

4.1 The structure of the Earth and the distribution of earthquakes and volcanoes

This chapter will explain:

★ the characteristics of the layers of the Earth, namely the inner core, outer core, mantle, crust and lithosphere
★ the main tectonic plates and how they move
★ the main types of plate boundary and how they relate to earthquakes and volcanoes.

The characteristics of the layers of the Earth

There are four main layers within the Earth (Figure 4.1):

» The core is divided into two parts:
  • The inner core is solid. It is five times denser than surface rocks.
  • The outer core is semi-molten.
» The mantle is semi-molten and about 2900 km thick.
» The crust is solid and its depth varies between 10 km and 70 km. It is divided into two main types:
  • Continental crust: Mostly formed of granite and less dense than the oceanic crust.
  • Oceanic crust: Mostly formed of basalt. As it is denser than the continental crust, it plunges beneath it when they come together.
» The lithosphere is the solid outer part of the Earth and includes the brittle upper part of the mantle and the crust.

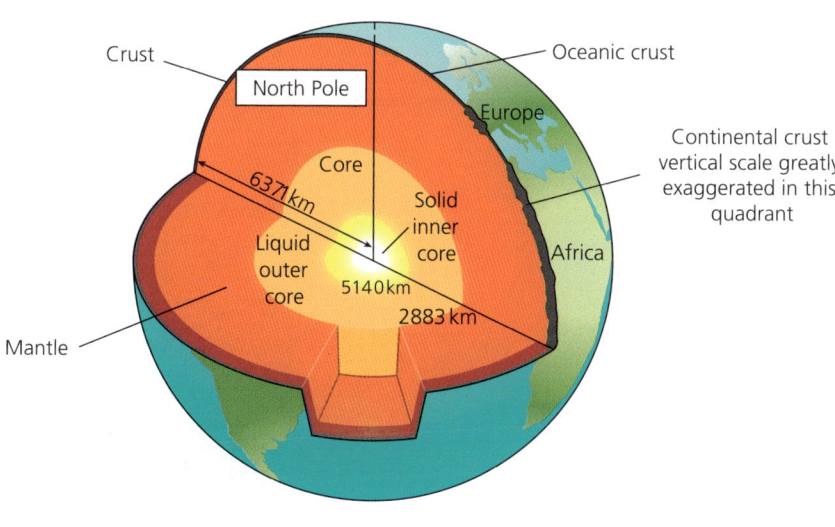
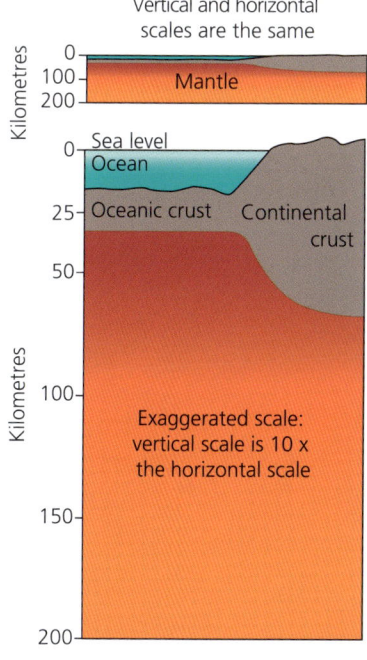

▲ Figure 4.1 The structure of the Earth

4.1 THE STRUCTURE OF THE EARTH AND THE DISTRIBUTION OF EARTHQUAKES AND VOLCANOES

> **Activities**
> 1 Compare the oceanic crust with the continental crust.
> 2 Compare the inner core with the outer core.

Tectonic plates and their movement

Figure 4.2 shows the location of tectonic plates and their direction of movement relative to one another. When the plates collide, they produce tectonic activity and new landforms. There are seven large plates (five of which carry continents) and a number of smaller plates:

» The main plates are the Pacific, Indo-Australian, Antarctic, North American, South American, African and Eurasian Plates.
» Smaller plates include the Caribbean, Iranian, Arabian and Juan de Fuca Plates.

Tectonic plates are constantly moving. Two main mechanisms for plate movement have been identified:

» Convection currents
» Gravitational sliding.

Convection currents

Convection currents are partly responsible for the movement of tectonic plates. Heat produced by radioactive decay from the Earth's core causes rock (**magma**) to rise towards the crust and spread (Figure 4.3). The magma cools and sinks. This process is repeated in a cycle and is responsible for the movement of crustal plates.

Gravitational sliding

It is now believed that gravity is also important in driving tectonic plate movement. There are two ways in which gravitational sliding operates (Figure 4.3):

» Ridge push: At mid-ocean ridges, new oceanic crust is pushed apart by rising magma from the mantle. As the mid-ocean ridge is higher than the ocean floor, gravity causes the plates to move downwards.
» Slab pull: Gravity acting upon the colder and denser subducting (plunging) plate causes it to sink into the mantle and pull the remaining plate behind it.

> **Activities**
> 1 Explain the difference between slab pull and ridge push.
> 2 Identify one oceanic plate, one continental plate and one small plate.

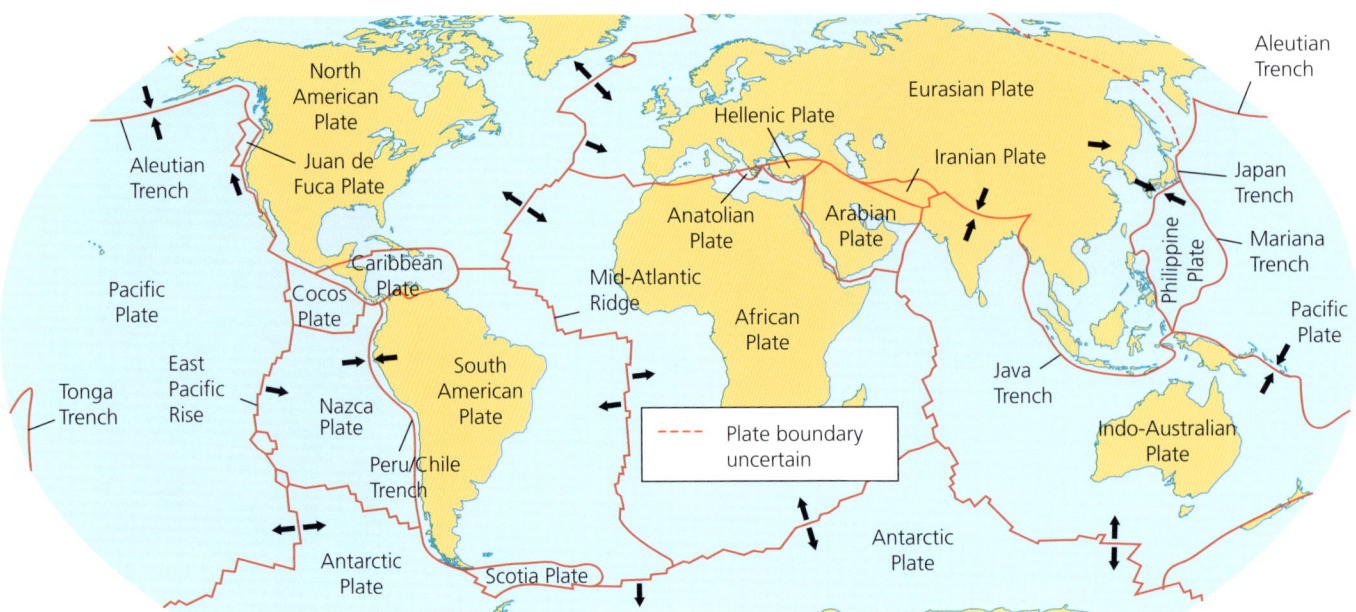

▲ Figure 4.2 The world's main tectonic plates

Types of plate boundaries

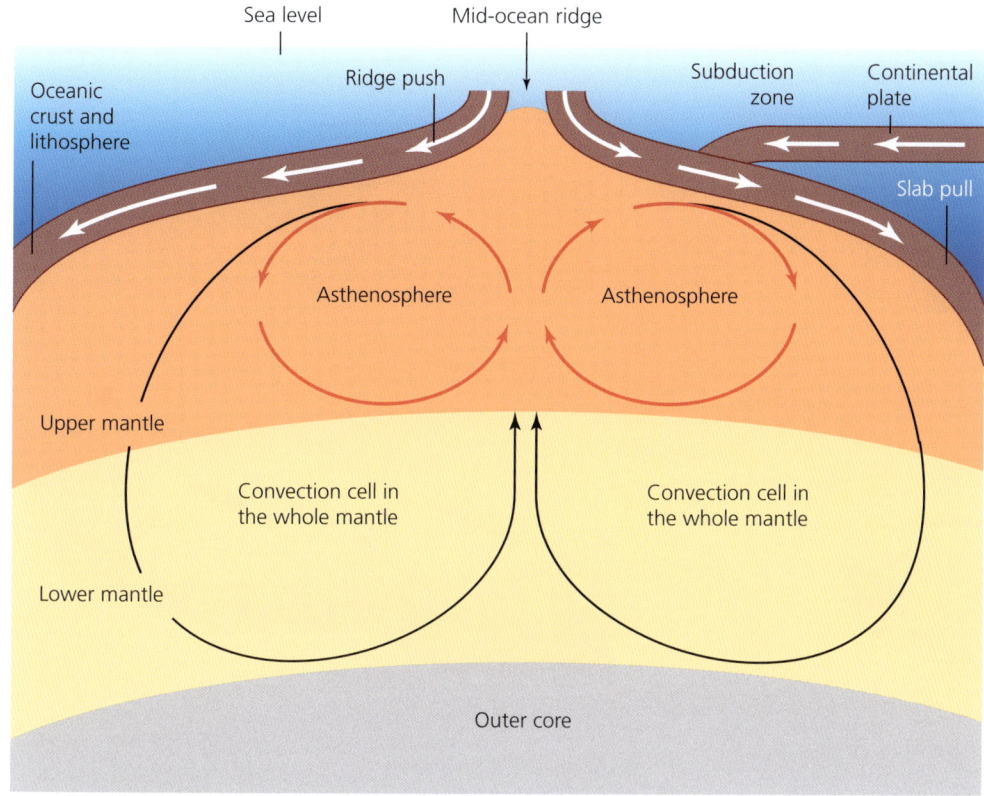

▲ **Figure 4.3** Convection currents, ridge push and slab pull

Types of plate boundaries

There are a number of different types of plate boundaries (Table 4.1 and Figure 4.4). These include:

» **divergent/constructive boundaries**, in which new oceanic crust is created
» **convergent/destructive boundaries**, in which older crust is destroyed
» **convergent/collision boundaries**, where plates are folded and crumpled
» **conservative/transform boundaries**, where plates slip past each other, causing earthquakes to occur.

Different plate boundaries are associated with different tectonic activities: volcanic eruptions, folding and earthquake activity (Figure 4.5).

▼ **Table 4.1** The types of plate boundary

Divergent/ constructive boundary	• Two plates move apart from each other, causing sea floor spreading. • New oceanic crust is formed, creating mid-ocean ridges. • Volcanic activity is common, e.g. the Mid-Atlantic Ridge (Europe is moving away from North America).
Convergent/ destructive boundary	• The oceanic crust moves towards the continental crust and sinks beneath it due to its greater density. • Deep-sea trenches and island arcs are formed. • The continental crust is folded into fold mountains. • Volcanic activity is common, e.g. the Nazca Plate is sinking under the South American Plate.
Convergent/ collision boundary	• Two continental crusts collide. • As neither can sink, they are folded up into fold mountains, e.g. the Indian Plate collided with the Eurasian Plate to form the Himalayas.
Conservative/ transform boundary	• Two plates slip sideways past each other, but land is neither destroyed nor created, e.g. the San Andreas Fault in California.

4.1 THE STRUCTURE OF THE EARTH AND THE DISTRIBUTION OF EARTHQUAKES AND VOLCANOES

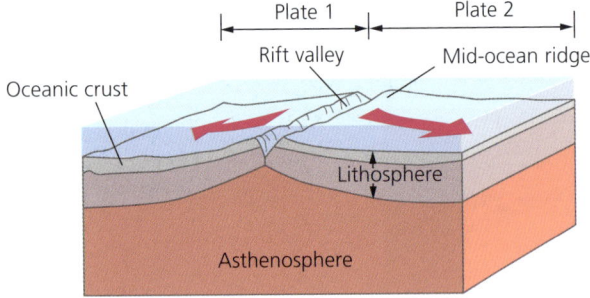

a Divergent/constructive boundary

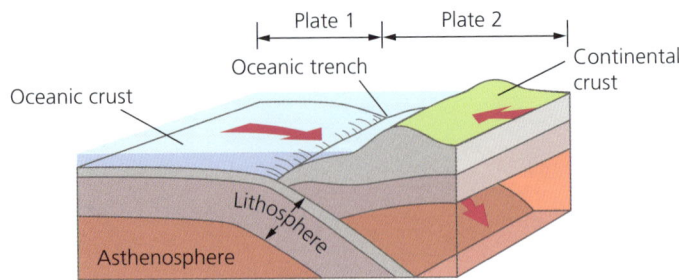

b Convergent/destructive boundary

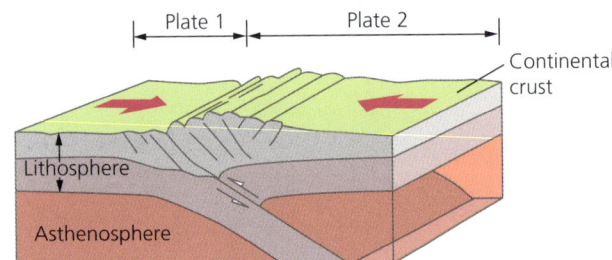

c Convergent/collision boundary

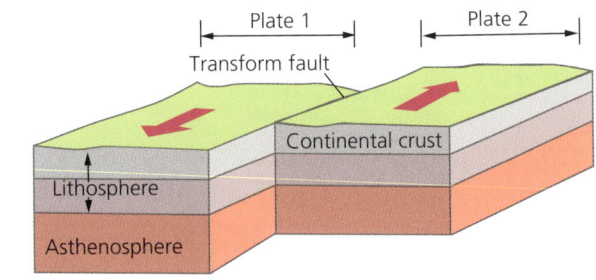

d Conservative/transform boundary

▲ **Figure 4.4** The types of plate boundary

▲ **Figure 4.5** Tectonic activities **a** A folded landscape, Himalaya foothills **b** Thingvellir Rift Valley, Iceland **c** Volcanic eruption of Soufrière, Montserrat, with the former capital city Plymouth in the foreground **d** Tourists standing by the boiling mud springs, Soufriere, St Lucia

Activities

1. Name the type of plate boundary located:
   a off the west coast of central America
   b in the south Atlantic Ocean
   c where the Indian Plate meets the Eurasian Plate.
2. Using Figure 4.5 a–c, identify one location of:
   a a convergent/collision boundary
   b a divergent/constructive plate boundary
   c a convergent/destructive plate boundary.

The location of earthquakes

Earthquakes are located very unevenly around the world (Figure 4.6). About 500,000 earthquakes are detected each year by sensitive instruments. They are associated with all types of plate boundaries. Most of the world's earthquakes occur in linear chains along plate boundaries, such as along the west coast of South America. Some earthquakes appear in areas away from plate boundaries, such as in the Midwest of the USA. These earthquakes could still be related to plate movement, as the North American Plate is moving westwards.

Some earthquakes are the result of human activity. The building of large dams and deep reservoirs increases pressure on the ground. Mining removes underground rocks and minerals, which may cause collapse or subsidence of the overlying materials. Testing of nuclear weapons underground has been known to trigger earthquakes, too.

The location of volcanoes

There are over 1300 active volcanoes in the world, many of them under the ocean (see Figure 4.6). Three-quarters of the world's active volcanoes are located in the 'Pacific Ring of Fire', the area around the Pacific Ocean. Good examples include Mount Pinatubo (Philippines), Krakatoa (Indonesia) and Popocatépetl (Mexico). These volcanoes are related to plate boundaries, notably convergent/destructive plate boundaries (for example Mount St. Helens in the USA and Soufrière in Montserrat in the Caribbean, see Figure 4.8) and divergent/constructive boundaries (for example Eldfell volcano on Heimaey, Iceland). The continuing eruption of Soufrière in Montserrat occurs at the boundary of the North American and Caribbean Plates. Some volcanoes, such as Mauna Loa and Kîlauea in Hawaii, and Teide on Tenerife, are located over hotspots. These are isolated plumes of rising magma that have burned through the crust to create active volcanoes (Figure 4.7).

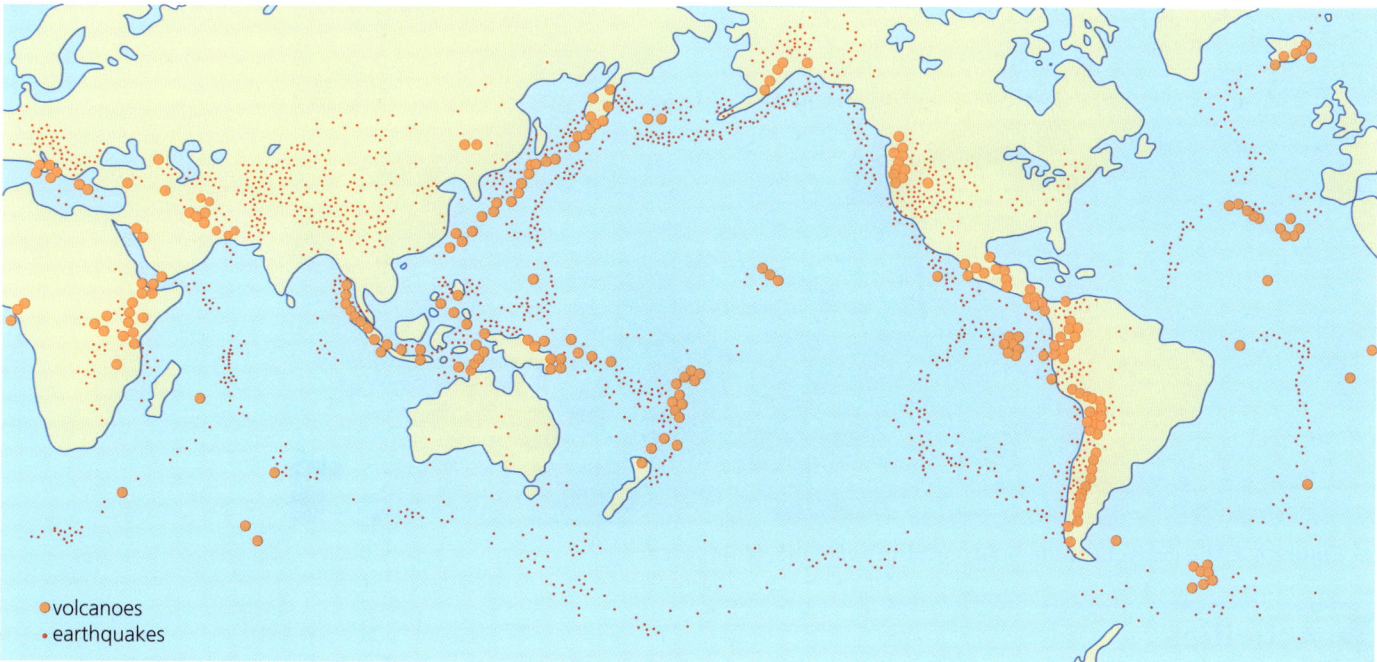

▲ Figure 4.6 World distribution of volcanoes and earthquakes

4.1 THE STRUCTURE OF THE EARTH AND THE DISTRIBUTION OF EARTHQUAKES AND VOLCANOES

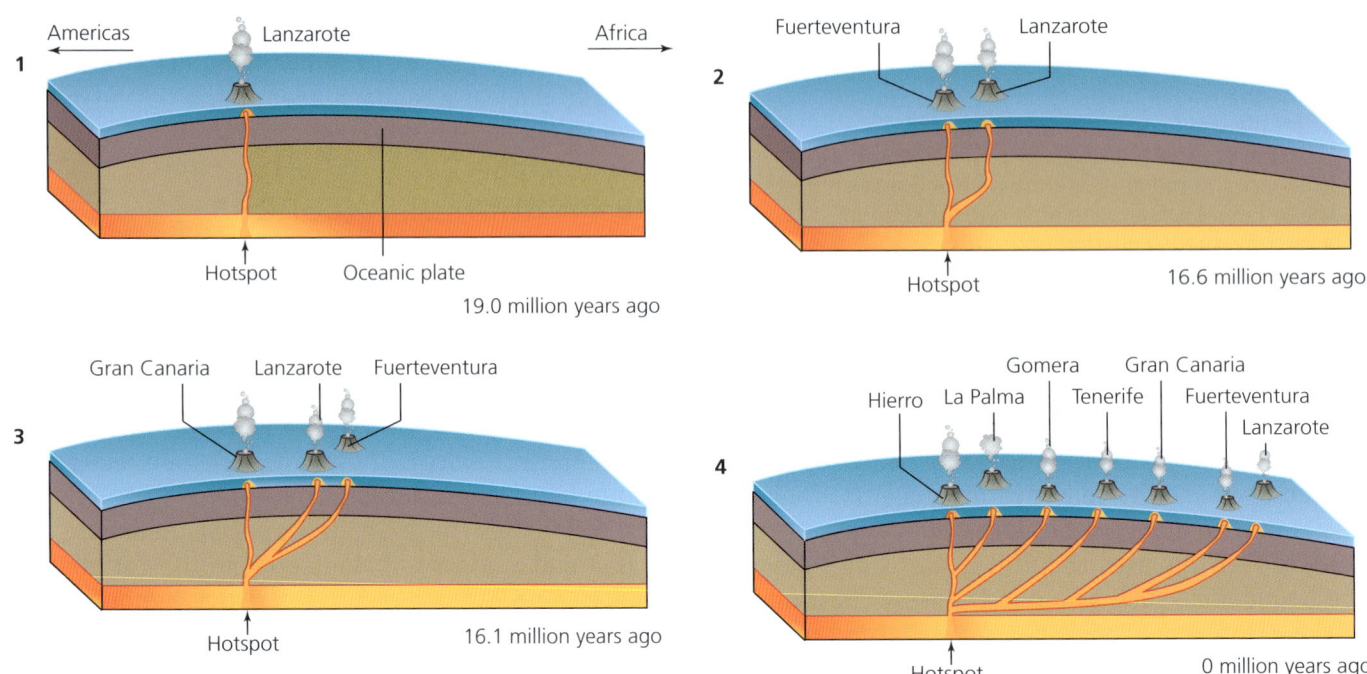

▲ **Figure 4.7** Volcanic activity associated with hotspots

▲ **Figure 4.8** The impacts of the Soufrière Hills volcano, Montserrat

> ### Activities
> 1. Describe the global distribution of earthquakes as shown on Figure 4.6 (page 91).
> 2. Explain why some volcanoes occur in locations away from plate boundaries.

4.2 The processes and features associated with earthquakes and volcanoes

This chapter will explain:

★ the processes at plate boundaries that cause earthquakes and volcanoes
★ the main characteristics of earthquakes
★ the main types of volcanoes
★ how volcanoes are classified
★ the main features of volcanoes
★ the hazards associated with volcanoes

The processes at plate boundaries that cause earthquakes and volcanoes

At divergent/constructive plate boundaries two plates move apart from each other, causing **sea floor spreading**. New oceanic crust is formed, creating mid-ocean ridges and pushing the plates apart; volcanic activity is common. Less viscous, runny basaltic lava forms shield mountains. A narrow belt of shallow focus, low-magnitude earthquakes is formed. An example is the Mid-Atlantic Ridge (Europe is moving away from Northern America).

At convergent/destructive plate boundaries the oceanic crust moves towards the continental crust and is subducted (sinks) beneath it due to its greater density, pulling the rest of the plate behind it. Deep-sea trenches and island arcs are formed. Volcanic activity is common. More acidic, explosive lava results in cone volcanoes. **Pyroclastic flows** and **lahars** may occur (Figure 4.9). The plate movement compresses and folds the seafloor rocks/sediments, forming fold mountains. Earthquakes are a mix of shallow-, intermediate- and deep-focus and can be high magnitude. An example is the Nazca Plate sinking under the South American plate.

At a convergent/collision boundary two continental crusts collide. As neither can be subducted (sink), the crust and sediments become crumpled up into fold mountains. There is no volcanic activity due to the lack of magma. Earthquakes are shallow focus but powerful due to the friction as the plates move. A good example is the Indian Plate colliding with the Eurasian Plate to form the Himalayas. Prior to this, the Tethys Sea (continental crust) had collided into the Eurasian Plate. However, once it had fully subducted, the Indian Plate that it was dragging behind it collided with the Eurasian Plate to form a collision boundary.

At a conservative/transform plate boundary, two plates slip sideways past each other, but land is neither destroyed nor created. There is no volcanic activity due to the lack of magma. Earthquakes are shallow focus. Stress can build up close to the surface as plates 'get stuck' and this is released as earthquakes. The San Andreas Fault in California is a good example of a conservative/transform plate boundary, where the Pacific Plate is slipping past the North American Plate.

▲ **Figure 4.9** The impact of lahars at Plymouth, Montserrat

4.2 THE PROCESSES AND FEATURES ASSOCIATED WITH EARTHQUAKES AND VOLCANOES

> **Activities**
> 1 State the types of plate boundaries where:
>   a volcanic activity occurs
>   b earthquakes occur.
> 2 Identify the type of location where volcanoes occur that is not a plate boundary.

The main characteristics of earthquakes

Earthquakes involve sudden, violent shaking of the Earth's surface. They occur after a build-up of pressure causes rocks and other materials to give way. Most of this pressure occurs at plate boundaries, when one plate is moving against another. Earthquakes are associated with all types of plate boundary.

The **focus** refers to the place beneath the ground where the earthquake takes place. **Deep-focus earthquakes** are associated with **subduction zones**. **Shallow-focus earthquakes** are generally located along divergent/constructive boundaries and along conservative/transform boundaries (Figure 4.10). The **epicentre** is the point on the ground surface immediately above the focus.

Some earthquakes are caused by human activity, including nuclear testing, building large dams, drilling for oil/natural gas (fracking) and coal mining.

Seismic waves are the 'shock waves' generated by the release of energy following an earthquake. There are different types of seismic waves:

» **Body waves** spread out from the focus of the earthquake.
» **Surface waves** spread out from the epicentre. Surface waves are slower than body waves.

Body waves include **primary waves** (P-waves) and **secondary waves** (S-waves). Primary waves are compressional waves and are very fast (4–7 km/second). They can pass though liquids. In contrast, secondary waves are slower (2–5 km/second) and cannot pass through liquids.

Surface waves are produced when the energy from an earthquake reaches the Earth's surface and causes a rolling or swaying motion. They are the slowest type of wave, but they cause the most damage.

Scientists use these waves to determine the exact location of the epicentre of the earthquake.

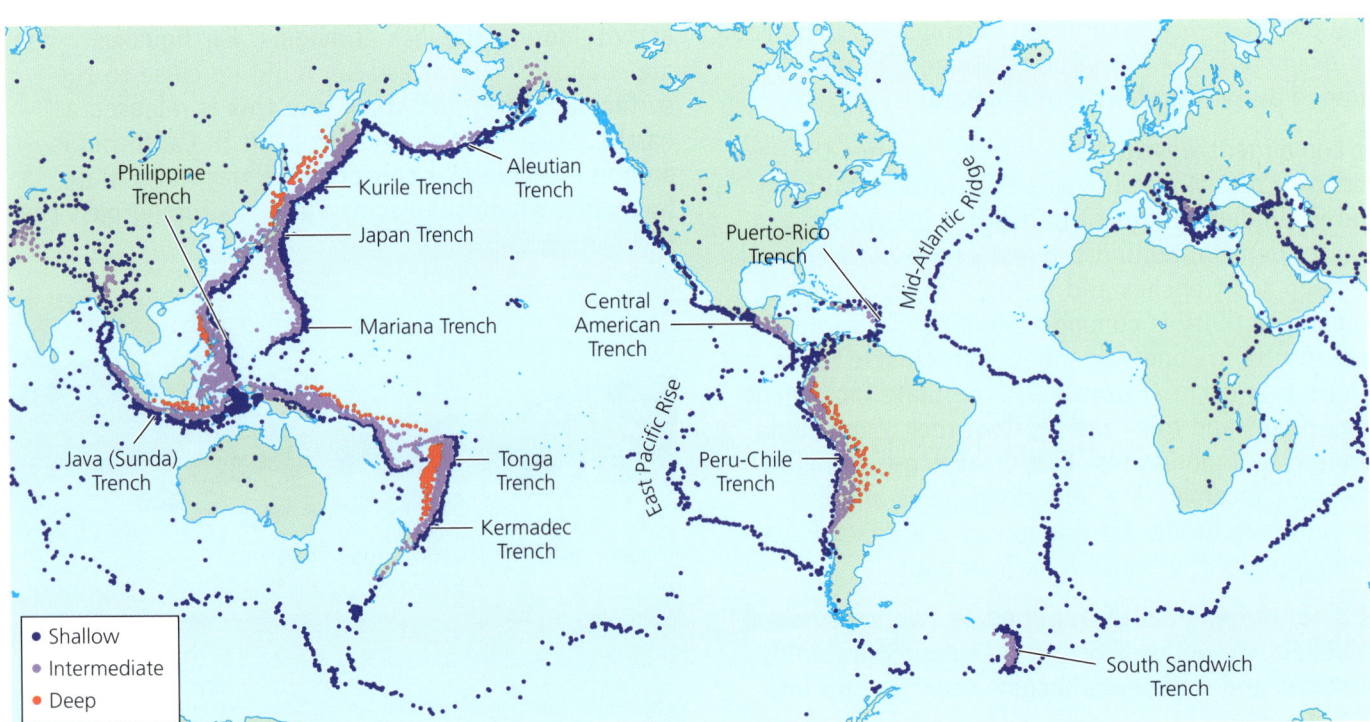

▲ **Figure 4.10** The distribution of shallow-, intermediate- and deep-focus earthquakes, 1900–2017

The main types of volcanoes

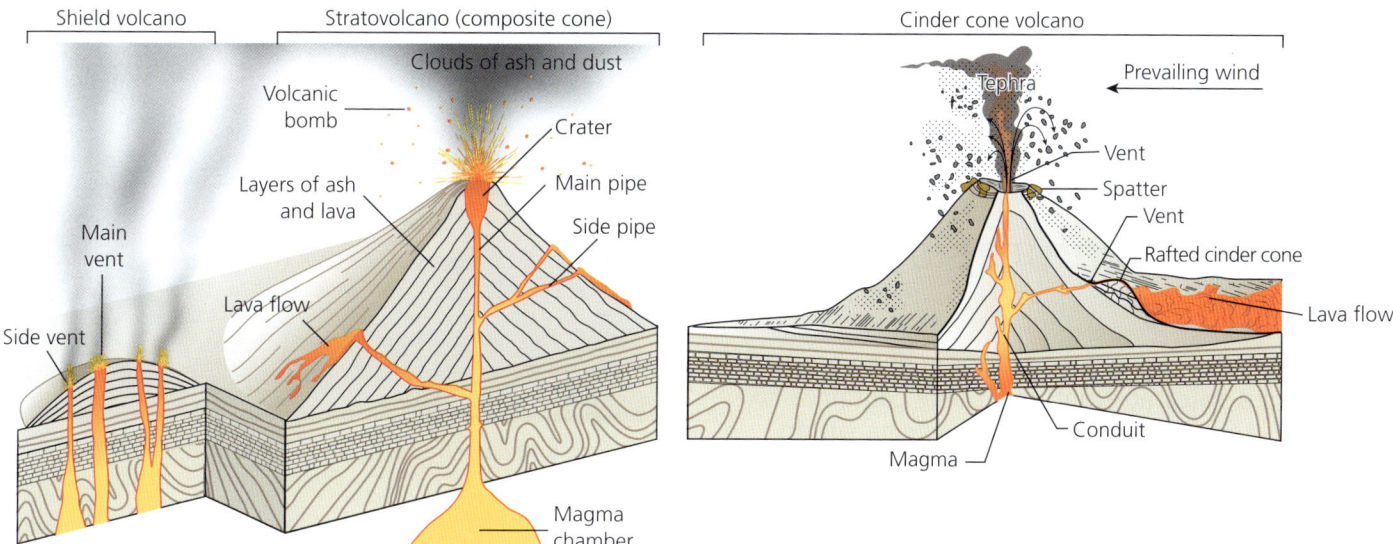

▲ **Figure 4.11** Shield, stratovolcano (composite cone) and cinder cone volcanoes

> ### Activities
> 1. Describe the distribution of shallow- and deep-focus earthquakes.
> 2. Distinguish between body waves and surface waves.

The main types of volcanoes

The shape of a volcano depends on the type of lava it contains.

- **Shield volcanoes** are gently sloping and produced by very hot, runny basaltic lava, for example Mauna Loa in Hawaii. They can cover a very wide area. They are made up of many layers of lava from repeated eruptions.
- **Cinder cone volcanoes** are steep-sided, conical hills produced by thicker, less fluid material. They are made up of loose, pyroclastic fragments built up around a volcanic vent. Most cinder cones have a crater at the summit. Parícutin in Mexico was formed by ash and cinders building up into a symmetrical cone as a result of eruptions in the 1940s.
- **Stratovolcanoes** (also known as **composite cone volcanoes**) are built up of many layers of hardened lava and tephra (volcanic ash, cinders, pumice and bombs), for example Mount Nyiragongo in Democratic Republic of the Congo and Mount Etna in Sicily. A stratovolcano is one that has had more than one eruption, and later eruptions may change the shape of the volcano. For example, Vesuvius was altered by the eruption in AD 79, forming a new version of itself, although some parts of the previous volcano survived the eruption. The shape of the volcano also depends on the amount of change there has been since the last volcanic eruption.

Cone volcanoes are generally associated with convergent/destructive plate boundaries, whereas shield volcanoes are characteristic of divergent/constructive boundaries and hotspots (areas of weakness in the middle of a plate).

> ### Activities
> 1. Distinguish between shield and cone volcanoes.
> 2. Identify the types of plate boundaries associated with shield and cone volcanoes.

The classification of volcanoes

Active volcanoes are those that have erupted in recent times, such as Grindavik in Iceland, Kîlauea in Hawaii and Whakaari/White Island in New Zealand, which all erupted in 2023–24 and could erupt again.

Dormant volcanoes are volcanoes that have not erupted for many centuries but may erupt again, such as Mount Rainier in the USA.

4.2 THE PROCESSES AND FEATURES ASSOCIATED WITH EARTHQUAKES AND VOLCANOES

Extinct volcanoes are not expected to erupt again. Kilimanjaro in Tanzania is an excellent example of an extinct volcano. The Le Puys region of France is an area of extinct volcanoes, which continue to influence settlements and tourism.

The main features of volcanoes

A **crater** is the depression at the top of a volcano following a volcanic eruption. It may contain a lake. A **vent** is the channel that allows magma within the volcano to reach the surface in a volcanic eruption. Magma is the molten material from the Earth's interior. The **magma chamber** refers to the reservoir of magma located deep inside the volcano. A **secondary cone** is a small volcano found on the side of the main volcano. For example, Mount Etna has several secondary cones (Figure 4.12).

▲ **Figure 4.13** Lava flows, Grindavik, Iceland, 2023–24

flows in Grindavik, Iceland (Figure 4.13) caused much disruption and led to the death of one person.

Tephra refers to small fragments of rock ejected into the air by an erupting volcano. Most tephra falls back on to the slopes of the volcano, so enlarging it. But billions of smaller and lighter pieces less than 2 mm in diameter (ash) can be carried hundreds of kilometres by winds. **Ash falls** are potentially very hazardous and can disrupt human activities across thousands of square kilometres.

Ash fallout to the ground can cause significant disruption and damage to buildings, transportation, water and wastewater, power supply, communications equipment, agriculture and primary production. This can lead to potentially substantial societal impacts and costs, even at thicknesses of only a few millimetres. Additionally, when ingested, the fine-grained ash can cause health impacts to humans and animals.

▲ **Figure 4.12** Secondary cones on Mount Etna

The hazards associated with volcanoes

Lava flows are streams of molten rock that come from an erupting vent. They can be erupted from an explosive or nonexplosive event. The speed at which lava flows move across the ground depends on many factors, such as the type of lava erupted, the steepness of the ground, whether the lava flows as a broad feature or in a confined channel, and the rate of lava production at the vent. In 2023–24, lava

Volcanic ash clouds are a major hazard to aviation. The 2010 eruption of Eyjafjallajökull in Iceland had a major impact on air travel. Over 300 airports in Europe were closed between 15 and 21 April 2010. More than 100,000 flights were cancelled, affecting around 7 million passengers and causing about US$1.7 billion in lost revenue for the aviation industry.

A lahar is a fast-flowing mudflow made up of volcanic ash and water. The water may come from surface water (for example, from the lake in the volcano's crater), rain or melting snow. Lahars flow down valleys and material is deposited over low-lying areas. Following the eruption of Mount Pinatubo in 1991, a tropical cyclone brought heavy rain and caused a number of lahars to form.

The hazards associated with volcanoes

▲ **Figure 4.14** Pyroclastic flow on the Soufrière Hills volcano, Montserrat – a cone-shaped volcano formed by the subduction of the North American and South American Plates beneath the Caribbean Plate

Pyroclastic flows (Figure 4.14) are thick clouds of volcanic ash, pumice (volcanic rock) and gas that flow at high speed from a volcano into the valleys below. The temperature inside the pyroclastic flow can reach 700°C and flows can move at speeds of over 500 km/h. When they cool down, they form a layer of ash across the landscape that solidifies into pumice.

A **volcanic block** is a fragment of rock, over 64 mm in diameter, that is ejected in a solid condition from a volcano during an explosive eruption. Blocks are formed from material that was created during previous eruptions. In contrast, a volcanic bomb is a mass of partially molten rock (tephra) formed when a volcano ejects viscous fragments of lava during an eruption.

Volcanic gases may escape through vents in the ground. The most common volcanic gases include carbon dioxide, sulphur dioxide, hydrogen sulphide and carbon monoxide. Some of these gases are irritating or poisonous, or cause breathing problems. The release of sulphur dioxide may cause acid rain to form. The escape of carbon dioxide from Lake Nyos in Cameroon in 1986, possibly due to a small volcanic eruption, led to the death of over 1700 people and about 3500 livestock. This happened quickly so there was no warning and it was an infrequent event, but it was a high-magnitude event with a major impact. Fortunately, the gases did not spread to other areas.

Interesting notes

- The greatest volcanic eruption in recorded history was Tambora in Indonesia in 1815. Some 50–80 km^3 of material was blasted into the atmosphere.
- In 1883 the explosion of Krakatoa was heard from as far as 4776 km away.
- The world's largest active volcano is Mauna Loa in Hawaii, which is 120 km long and over 100 km wide.

Activities

1. Distinguish between a volcano's crater and its chamber.
2. Explain three hazards related to volcanic activity.

4.3 The impact of tectonic hazards

This chapter will explain:

★ the reasons why people live in areas at risk from earthquakes and volcanic eruptions
★ the impacts of earthquakes and volcanic eruptions
★ how the magnitude of tectonic events is measured.

The reasons why people live in areas at risk from earthquakes and volcanic eruptions

All natural environments provide opportunities and challenges for human activities. Some countries, such as Iceland and the Philippines, were created by volcanic activity. People may choose to live in volcanic areas because they are useful:

» Volcanic soils are rich, deep and fertile, and allow intensive agriculture to take place.
» Volcanic areas are important for tourism.
» Some volcanic areas are seen by people as being symbolic and are part of the national identity, such as Mount Fuji in Japan.

There are several ways of looking at people's vulnerability:

» One view is that people choose to live in hazardous environments because they understand the environment. In this situation, people choose to live in an area because they feel there are potential advantages to living there and that these outweigh the risks. These advantages include:
  • fertile soils, which increase food production
  • geothermal energy and geothermal spas
  • minerals that can be used in the construction industry
  • volcanoes, which attract tourists
  • employment opportunities in tourism.
» Another view is that some people live in hazardous environments because they have very little choice over where they live, as they cannot afford to move. Many communities have lived in these areas for generations.
» Many people may not be aware of the potential risks associated with the hazard. The perception of the risk of the hazard may be low due to:
  • the infrequent occurrence of hazardous events
  • better monitoring, warnings, predictions and building design.

▶ Activity

Suggest contrasting reasons why some people live in volcanically active areas.

The impacts of earthquakes

Hazards can be divided into two types – primary and secondary. Primary hazards are the direct hazards associated with natural events, while secondary hazards are the indirect hazards associated with natural events. Table 4.2 shows the main hazards and impacts associated with earthquake activity.

▼ Table 4.2 Hazards and impacts associated with earthquakes

Primary hazards	Secondary hazards	Impacts
• Ground shaking • Surface faulting	• Ground failure and soil liquefaction • Landslides and rockfalls • Debris flows and mudflows • Tsunamis (Figure 4.15)	• Total or partial destruction of building structures • Interruption of water supplies • Breakage of sewage disposal systems • Loss of public utilities such as electricity and gas • Floods from collapsed dams • Release of hazardous materials • Fires • Spread of chronic illness

The impacts of volcanic eruptions

▲ Figure 4.15 The 2004 tsunami caused widespread damage around the countries bordering the Indian Ocean

> **Activity**
>
> Distinguish between primary and secondary hazards, giving an example of each.

The impacts of volcanic eruptions

Likewise, there are several hazards related to volcanic activity (Table 4.3).

▼ Table 4.3 The hazards and impacts associated with volcanic activity

Primary hazards	Secondary hazards	Impacts
• Pyroclastic flows • Volcanic bombs (projectiles) • Lava flows • Ash fallout • Volcanic gases • Lahars (mudflows) (see Figure 4.9 on page 93) • Earthquakes	• Atmospheric ash fallout • Landslides • Tsunamis • Acid rainfall	• Destruction of settlements • Loss of life • Loss of farmland and forests • Destruction of infrastructure – roads, airstrips and port facilities • Disruption of communications

Measuring the magnitude of tectonic events

The magnitude of an earthquake refers to its strength, and it remains unchanged with distance from the epicentre. The intensity describes the shaking caused by an earthquake at a given location, and this decreases with distance from the epicentre. The magnitude of earthquakes is measured using different scales.

The Richter Scale

In 1935, Charles Richter of the California Institute of Technology developed the **Richter Scale** to measure the magnitude (strength or force) of earthquakes (Table 4.4). The strength or force of an earthquake is measured on a seismometer and shown on a seismograph (Figure 4.16). The Richter Scale is logarithmic. This means that an earthquake of 6.0 is ten times greater than one of 5.0, and 100 times greater than one of 4.0.

▼ Table 4.4 The annual frequency of occurrence of earthquakes of different magnitudes, based on observations since 1900

Descriptor	Magnitude (Richter Scale)	Average number/year	Hazard potential
Great	≥8	1	Total destruction, high loss of life
Major	7.0–7.9	18	Serious building damage, major loss of life
Strong	6.0–6.9	120	Large losses, especially in urban areas
Moderate	5.0–5.9	800	Significant losses in populated areas
Light	4.0–4.9	6200	Usually felt, some structural damage
Minor	3.0–3.9	49,000	Typically felt, but usually little damage
Very minor	<3.0	9000/day	Not felt, but recorded

▲ Figure 4.16 A seismograph reading from Soufrière Hills, Montserrat

4.3 THE IMPACT OF TECTONIC HAZARDS

The Mercalli Scale

In contrast, the **Mercalli Scale** (Table 4.5) relates ground movement to things that you would notice happening around you. It is a subjective scale and lacks accuracy/reliability about the magnitude of the earthquake event. However, its main advantage is that it allows ordinary eyewitnesses to provide information on the strength of an earthquake. For example, a weak earthquake (II on the scale) might be felt by only a few people at rest in the upper floors of a building, whereas a violent earthquake (IX on the scale) is likely to cause considerable damage to buildings, with partial collapse.

▼ Table 4.5 The Mercalli Scale

Strength	Observations
I	• Rarely felt
II	• Felt by people who are not moving, especially on the upper floors of buildings
	• Hanging objects may swing
III	• The effects are noticeable indoors, especially upstairs
	• The vibration is like that experienced when a truck passes
IV	• Many people feel it indoors; a few outside
	• Some are awakened at night
	• Crockery and doors are disturbed, and standing cars rock
V	• Felt by nearly everyone – most people are awakened
	• Some windows are broken, plaster becomes cracked and unstable objects topple
	• Trees may sway and pendulum clocks stop
VI	• Felt by everyone – many are frightened
	• Some heavy furniture moves, plaster falls
	• Structural damage is usually quite slight
VII	• Everyone runs outdoors
	• Noticed by people driving cars
	• Poorly designed buildings are damaged
VIII	• Damage to ordinary buildings – many collapse
	• Well-designed buildings survive but suffer slight damage
	• Heavy furniture is overturned and chimneys fall
IX	• Damage occurs even to buildings that have been well designed
	• Many buildings are moved from their foundations
	• Ground cracks and pipes break
X	• Most masonry structures are destroyed; wooden ones may survive
	• Railway tracks bend and water slops over banks
	• Landslides and sand movements occur
XI	• No masonry structure remains standing; bridges are destroyed
	• Large cracks occur in the ground
XII	• Total damage
	• Waves are seen on the surface of the ground and objects are thrown into the air

The Moment Magnitude Scale

Scientists are increasingly using the **Moment Magnitude Scale** (M), which measures the amount of energy released in an earthquake. For every increase of 1 on the M scale, the amount of energy released increases by more than 32. Every increase of 0.2 represents a doubling of the energy released. Although this may be a more reliable and scientific measurement, a subjective scale may be enough for those who are affected by an earthquake, and the media still generally refer to the Richter Scale.

The Volcanic Explosivity Index

The strength of a volcano is measured by the **Volcanic Explosivity Index** (VEI) (Figure 4.17). This is based on the amount of material ejected in the explosion, the height of the cloud it caused, and the amount of damage that results. Any explosion above level 5 is considered to be very large and violent.

A **supervolcano** is a volcano of VEI 8. The scale is logarithmic, so VEI 8 is ten times more powerful than VEI 7, 100 times more powerful than VEI 6, and 1000 times more powerful than VEI 5 (Mount Pinatubo, 1991). The last VEI 8 was 74,000 years ago (Mount Toba, Indonesia).

Supervolcanoes tend to be much larger than 'normal' volcanoes – the Yellowstone magma chamber, for example, is over 50 km wide. The likely impacts of a VEI 8 eruption include:

» almost complete loss of life within about 1000 km of the eruption
» destruction of all crops and livestock, leading to a global famine
» economic and social devastation.

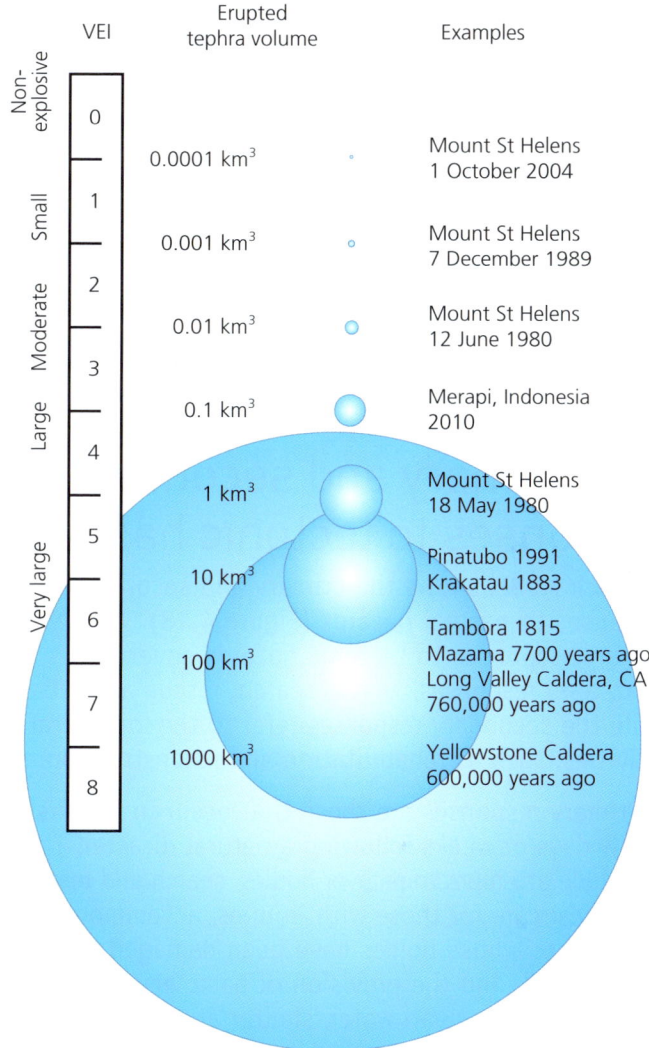

▲ **Figure 4.17** The Volcanic Explosivity Index

Activities

1. Distinguish between earthquake magnitude and intensity.
2. The Richter Scale is logarithmic. How much stronger is an earthquake of 7.0 on the Richter Scale compared with one of 5.0?

4.4 Managing the impacts of tectonic hazards

This chapter will explain:

★ the primary and secondary responses to tectonic hazards
★ the strategies and techniques used to manage the impacts of tectonic events
★ detailed specific example: Tiburon Peninsula earthquake, Haiti, 2021
★ detailed specific example: Mount Etna, Sicily, Italy.

The primary and secondary responses to tectonic hazards

The Park Model shows the primary and secondary responses to natural hazards (Figure 4.18). The primary responses to tectonic hazards cover the short-term responses, that is search, rescue and relief, while the secondary responses refer to the long-term activities, such as rehabilitation and reconstruction.

» Primary responses are those that occur in the first few days or weeks following a natural disaster. They include attempts to recover bodies, search and rescue, medical care for those with injuries, accommodation for those whose homes have been damaged or destroyed, and the provision of basic amenities such as food, clean water, shelter and healthcare.
» Secondary responses usually occur after the first few weeks or months and can continue for years. They include the rebuilding of housing, education, and improving infrastructure. In many LICs, reconstruction may take decades.

> ### Activities
> 1. Distinguish between primary and secondary responses.
> 2. Identify whether the following are primary or secondary responses:
>   a. Rebuilding homes
>   b. Providing tents
>   c. Providing toilets
>   d. Rebuilding bridges.

Managing the impacts of tectonic events

The impacts of volcanic eruptions can be reduced in several ways:

» Spraying lava flows with water to cool them down, helping them to solidify. This was successfully carried out in Heimaey, Iceland. The islanders used sea water to spray on to the lava flow and managed to divert it away from reaching the town. In contrast, some residents of Goma in the Democratic Republic of the Congo tried to spray water on to lava flows from the

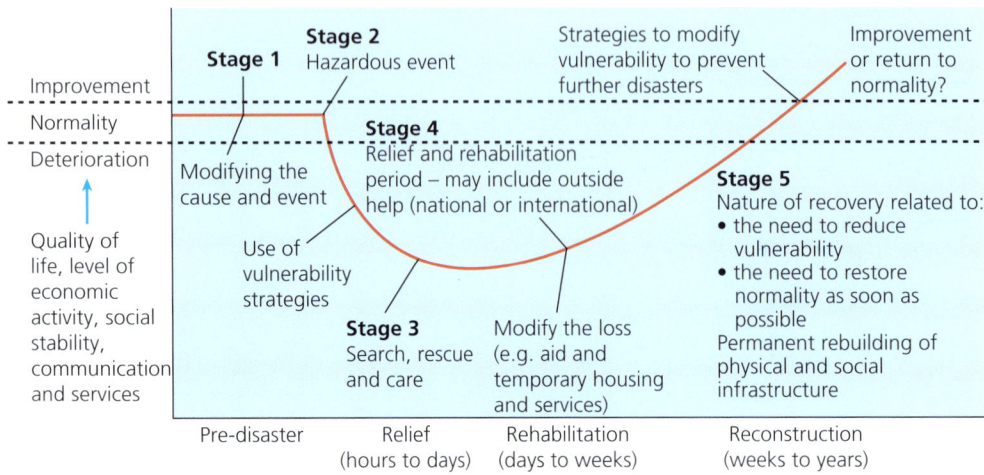

▲ Figure 4.18 The primary and secondary responses to hazards – the Park Model

Nyiragongo volcano, but this proved unsuccessful as they had neither enough water nor the infrastructure to deal with the advancing lava flows.
- Digging diversion channels to divert lava flows away from settlements: This has been used successfully on Mount Etna, Sicily, although the lava needs to be diverted away from human habitation and infrastructure.
- Adding 'cold' boulders to a lava flow in an attempt to cool the lava and stop it moving.

However, if the eruption is a pyroclastic flow, there is little that can be done to prevent the impacts other than evacuation.

Predicting volcanoes

The main methods of predicting volcanoes include:

- using seismometers to record swarms of tiny earthquakes that occur as the magma rises
- using chemical sensors to measure increased sulphur levels
- using lasers to detect the physical swelling of the volcano
- measuring small-scale uplift or subsidence, changes in rock stress and changes in radon gas concentration
- using ultrasound to monitor low-frequency waves in the magma resulting from the surge of gas and molten rock, as happened with Mount Pinatubo, El Chichón and Mount St. Helens.

Overall, relatively few people are killed by volcanic eruptions. This may be because most of the world's active volcanoes are monitored, especially in areas with high population density. Nevertheless, volcanoes can erupt without warning. If Vesuvius in Italy were to erupt again, millions of people could be affected. In addition, high-magnitude events, such as a VEI 8 supervolcano, will occur again at some point, and there is little that can be done to reduce some of the global impacts of such an event.

> ### Activities
> 1 Outline two ways of controlling lava flows.
> 2 Outline two ways in which volcanoes can be predicted.

Preparing for earthquakes

The main ways of preparing for and managing earthquakes include:

- better forecasting, warning, evacuation and emergency procedures
- improving building design and safe houses
- land-use planning for building location.

Forecasting and warning

There are a number of ways of predicting and monitoring earthquakes. These include:

- measuring crustal movement – the small-scale movement of plates
- recording changes in electrical conductivity
- noting strange and unusual animal behaviour, for example among fish (such as carp) or toads
- checking historical evidence – there are possibly trends in the timing of earthquakes in some regions.

Evacuation and emergency procedures

Loss of life can be minimised by ensuring that the population is educated in emergency evacuation procedures. For example, schools in Japan practise earthquake drills, and there is disaster education in many schools, colleges and states in the USA.

The US Centers for Disease Control and Prevention (CDC) provides information on preparing for hazards, for example 'drop, cover, and hold on' in the case of an earthquake. In an earthquake, people should get outside if possible. If trapped indoors, they should shelter under a sturdy object such as a kitchen table. When the shaking has stopped, they should get outside and move away from the building.

Building design

Buildings can be designed to cope with the shock waves that occur in an earthquake. For example, single-storey buildings are more able to absorb shock waves than multi-storey buildings as the potential for swaying is reduced. Also, tall buildings can be built with a 'soft storey' at the bottom, such as a car park on raised pillars. This soft storey may collapse in an earthquake so that the upper floors sink down on to it, which cushions the impact.

Building reinforcement strategies include building on foundations that are laid deep into the underlying bedrock, and steel frames that can withstand shaking can also be used.

In an earthquake, high-rise buildings respond slowly, and shock waves are increased as they move up the building. If buildings are too close together, vibrations may be amplified between buildings and increase the damage. Therefore maximising space between buildings can help minimise this amplification.

4.4 MANAGING THE IMPACTS OF TECTONIC HAZARDS

Safe houses

Poverty, corruption and poor governance mean that buildings in many countries are not earthquake-proof. Billions of people live in houses that cannot withstand shaking. Yet safer ones can be built cheaply, using straw, adobe (cheap, sun-dried brick) or old tyres, by applying a few general principles (Figure 4.19).

In rich cities in fault zones, the added expense of making buildings earthquake-resistant has become a fact of life. Concrete walls are reinforced with steel, for instance, and a few buildings even rest on elaborate shock absorbers. Strict building codes were credited with saving thousands of lives when a magnitude 8.8 earthquake hit Chile in February 2010. But in less developed countries, like Haiti, conventional earthquake engineering is often unaffordable, even though there are some cheap solutions.

In Peru in 1970 an earthquake killed more than 70,000 people, many of whom died when their houses crumbled around them. Heavy, brittle walls of traditional adobe cracked instantly when the ground started to move. Existing adobe walls can be reinforced with a strong plastic mesh installed under plaster. During an earthquake, these walls crack but do not collapse, allowing occupants to escape. Plastic mesh could also work as a reinforcement for concrete walls in Haiti and elsewhere.

Researchers in India have successfully tested a concrete house reinforced with bamboo. A model house for Indonesia rests on ground-motion dampers – old tyres filled with bags of sand. Such a house might be only a third as strong as one built on more sophisticated shock absorbers, but it would also cost much less – and so is more likely to get built in Indonesia. In northern Pakistan, straw is available. Traditional houses are built of stone and mud, but straw is far more resilient, and warmer in winter.

Building location

Land-use planning is another way of reducing earthquake risk. Densely populated areas and important services, such as hospitals and fire services, should not be built close to known fault lines or areas prone to

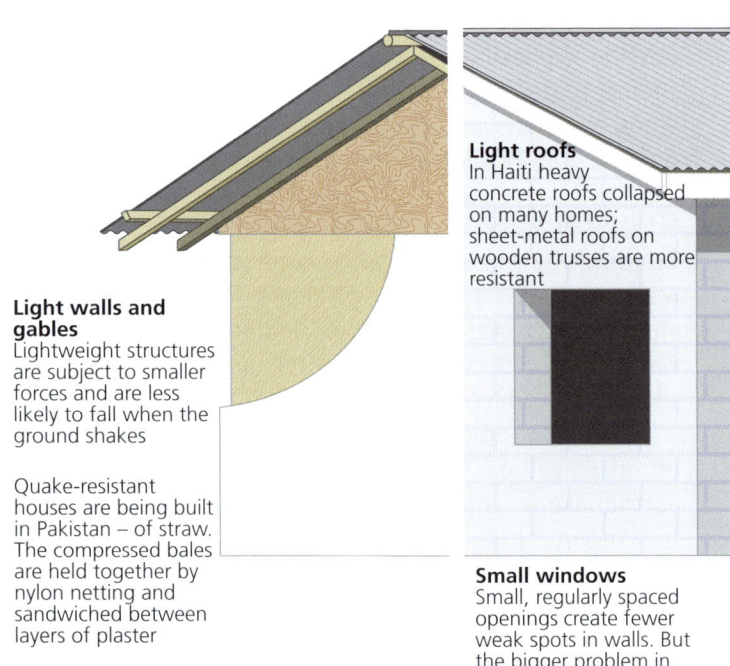

▲ Figure 4.19 Safe-house design

landslides. Areas with weak rocks, faulted (broken) rock and soils should, ideally, be avoided for buildings. In California, since 1972, no building can be built within 17 m of a surface fault line, although this would not offer much protection from an earthquake.

Ultimately, although it is possible to suggest areas where most earthquakes are likely to occur, it is not possible to predict with certainty exactly where they will occur, when they will occur or their magnitude. However, taking precautions may help reduce their impact.

Certain areas are very much at risk from earthquake damage – notably areas with weak rocks, faulted rocks and soft soils. Many oil and water pipelines in tectonically active areas are built on rollers so that they can move with an earthquake rather than fracture (Figure 4.20).

> ### Activities
> 1 Identify ways in which earthquakes can be predicted.
> 2 Explain the term 'safe house'. Briefly explain how houses can be made 'safe'.

▲ **Figure 4.20** A pipeline on rollers to allow movement during an earthquake

Detailed specific example

Tiburon Peninsula earthquake, Haiti, 2021

Haiti is located on the Gonâve Microplate, a small strip of the Earth's crust that is squeezed between the North American and Caribbean Plates. This makes it vulnerable to earthquakes. In August 2021, a 7.2 magnitude earthquake struck the country's Tiburon Peninsula. It was a shallow-focus earthquake (10 km deep) and the epicentre was about 150 km west of the capital Port-au-Prince (Figure 4.21). The earthquake was caused by the eastward movement of the Caribbean plate relative to the North American plate.

The earthquake caused around 2250 deaths, more than 12,750 injuries and damage estimated at US$1.5–1.7 billion. There were over 900 aftershocks, two of which were in 2022, killing a further two people. The earthquake was followed by the impact of Hurricane Grace the next day, which triggered thousands of landslides. At least 137,500 buildings were damaged or destroyed. Les Cayes, Haiti's third-largest city, was badly affected, with over 37,000 homes destroyed and 46,000 others damaged, see Figure 4.22. More than 50 medical facilities were damaged and six destroyed, and over 1000 schools were damaged or destroyed.

Shelter was the greatest long-term need. Emergency services were required given the needs of the population – including food, water, medicines and Covid-19 vaccines. Rebuilding is ongoing, including to infrastructure (communications, transport, WASH (water, sanitation and hygiene) and electrical supplies).

Six months after the earthquake, Haiti began to look at long-term recovery and reconstruction. It needed around US$2 billion to rebuild social services and housing, health and education infrastructure, and food supply. After that, aid funds were spent on agriculture, industry and key infrastructure. Haiti's financial situation meant it was reliant on international aid for its recovery. It is the lowest-income country in the Western Hemisphere and is included by the UN on its list of the 45 least developed countries in the world, the only country from the Caribbean.

However, many donor countries were facing their own financial problems or were less able or unwilling to provide aid. In addition, Haiti was competing with other countries for funding, notably Afghanistan, Ethiopia, Ukraine, and Turkey and Syria (following the earthquake in 2023).

Research using remote sensing of the damage to buildings found that those with deep foundations and timber roofs fared best during the earthquake. The research, undertaken by international scientists with the aid of local volunteers who took notes and uploaded photographs, has helped ensure that new buildings will be more earthquake-proof for the future.

4.4 MANAGING THE IMPACTS OF TECTONIC HAZARDS

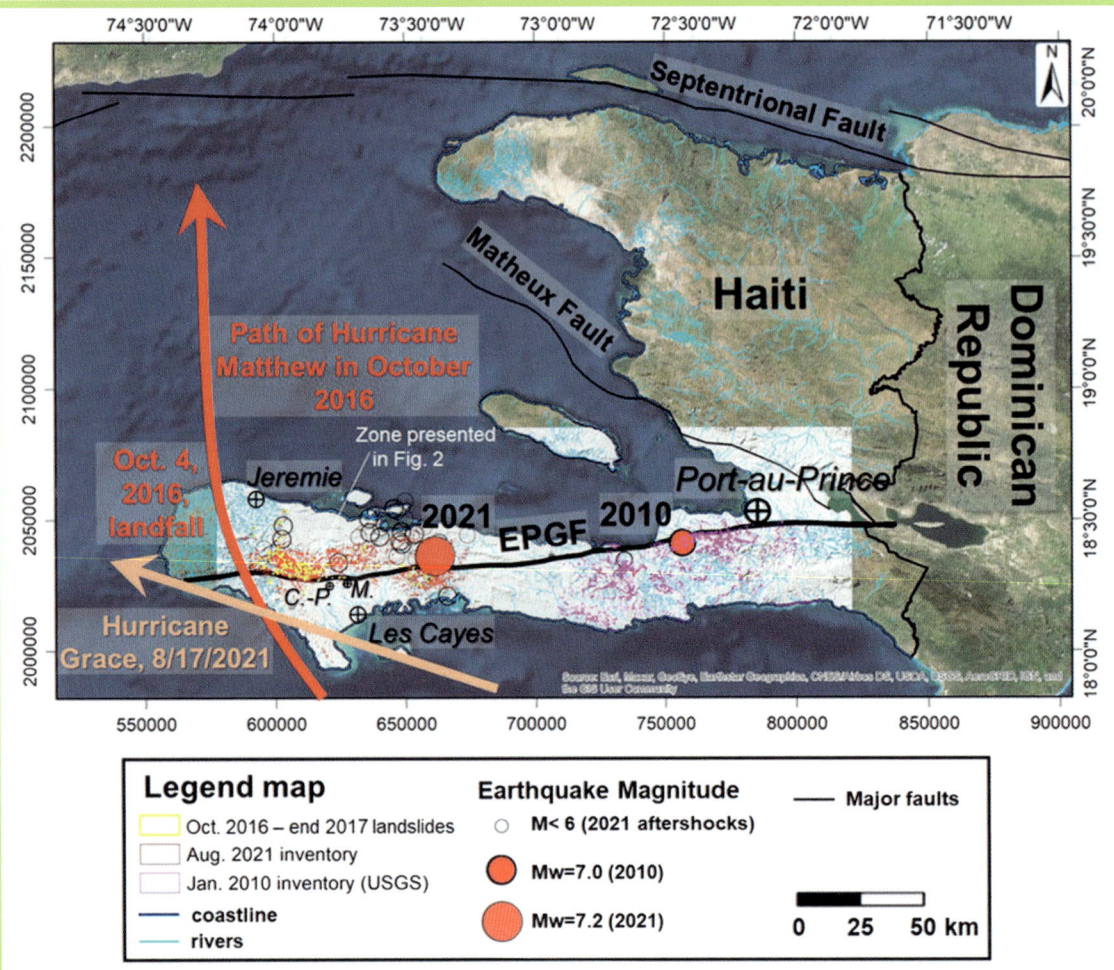

▲ **Figure 4.21** The location of the Tiburon Peninsula earthquake and the track of Hurricane Grace

According to the World Bank, the 2021 Disaster Risk Management and Reconstruction (DRM) Project for Haiti focused on supporting Haiti to improve its disaster-response capacity, and enhancing the resilience of critical transport infrastructure.

Key activities and systems developed under the project included risk assessments, communications and early warning systems, and improved road sections. In response to the August 2021 earthquake, the government was able to:

- mobilise Municipal Civil Protection Committees
- conduct a rapid and sound assessment of damages and losses
- achieve greater coordination among groups and individuals working in the DRM system.

Activities

1 Explain why earthquakes occur in Haiti.
2 Suggest how government actions after the 2021 earthquake helped manage the recovery process.

▲ **Figure 4.22** Search and rescue can be physically demanding as well as extremely dangerous

Managing the impacts of tectonic events

Detailed specific example

Mount Etna, Sicily, Italy

Mount Etna is an active stratovolcano on the east coast of the island of Sicily, Italy. It is located on the convergent plate margin of the African and Eurasian Plates (Figure 4.23). It is one of the world's most active volcanoes. In 2021 it increased in height by approximately 30 m, to around 3369 m. In historic times, only 77 deaths have been caused by its eruptions, the last two being in 1987. Nevertheless, Etna is a very active volcano.

There are different types of eruption at Etna. Most occur at one of the five craters near the summit (Figure 4.24). There are about 300 side vents, ranging in size from large craters to small holes. Ironically, the highly explosive summit eruptions are less hazardous than the flank eruptions, as these can occur close to inhabited areas. Since 2000 there have been at least four periods of flank eruptions and six periods of summit eruptions.

Volcanic activity has occurred at Etna for around 500,000 years. It evolved in three main phases from a shield volcano caused by submarine fissure eruptions to a stratovolcano caused by explosive activity. Over time, volcanic activity has moved from the southeast to the northwest.

▲ **Figure 4.24** The craters near the summit of Etna

Around 8000 years ago, a massive landslide caused a large depression on the side of Etna – the Valley of the Ox – and triggered a large tsunami. Major eruptions have caused the summit to collapse many times, forming calderas. A collapse 2000 years ago formed the Piano Caldera, although this has largely been filled in by subsequent lava flows.

In 1971 lava flows buried the Etna Observatory and destroyed the Etna cable car. The 1991–93 eruptions threatened the town of Zafferana, but diversion of the lava flow saved the town. Initially the lava flow overtopped the barriers, so engineers used explosives near the source of the lava flow to divert it away from Zafferana.

Between 1995 and 2001 there was an increase in volcanic activity. In 2001 an eruption on the south side of Etna occurred during the tourist season and damaged the cable car station (Figure 4.25). In 2002–03, a large eruption destroyed the tourist station Piano Provenzana. Some of the footage of this eruption was used in the film *Star Wars: Episode III – Return of the Sith*.

Some eruptions continue for a very long time. One in 2008, for instance, lasted for 417 days. Eruptions in 2011 and 2012 led to the closure of Catania Airport several times. Starting in 2021, a series of explosive eruptions affected surrounding settlements, with rocks and volcanic ash falling as far away as Catania. Etna erupted 11 times in just three weeks.

The Global Volcanism Program reports that since 1955 Etna has experienced 17 VEI 1 eruptions, 24 VEI 2 eruptions and seven VEI 3 eruptions.

▲ **Figure 4.23** The location of Mount Etna and tectonic plates

4.4 MANAGING THE IMPACTS OF TECTONIC HAZARDS

▲ Figure 4.25 Damage to tourist infrastructure by lava flows at Etna, Sapienza Refuge ski area

There have been many responses to the threat of volcanic activity on Mount Etna. In the 1970s the government took photos of Etna temperature readings in order to monitor change. During an eruption in 1992 they built an earth barrier that was 200 m long and 20 m high to prevent lava from reaching Zafferana. It contained the lava flow for three months. US marines used explosives to blast a hole in the lava, and then dropped blocks of concrete into it to cool it down. During an eruption in 2002 evacuations occurred, schools were closed and a state of emergency was declared. The Institute of Vulcanology monitors radon gas, uses GPS to examine any changes in size or shape of the volcano, and uses very sensitive seismometers to detect earth tremors. Following the eruptions of 2002–3 the government pledged financial aid to farmers and those working in the tourism industry.

Despite the relatively low number of fatalities, Etna has proved disruptive to many local residents and to the tourism sector. Methods used to control the impacts of Mount Etna have included constructing earthen dams to divert lava, bombing the lava flows, cooling lava using sea water, blocking lava tunnels to reduce the lava flow and monitoring air quality (for sulphur dioxide) and ground deformation to make predictions about likely activity.

> ### Activities
> 1 Explain the impacts of volcanic activity on Mount Etna.
> 2 Analyse ways in which the threats posed by Mount Etna have been managed.

Practice questions

1 Figure 4.26 shows the distribution of shallow-, intermediate- and deep-focus earthquakes.

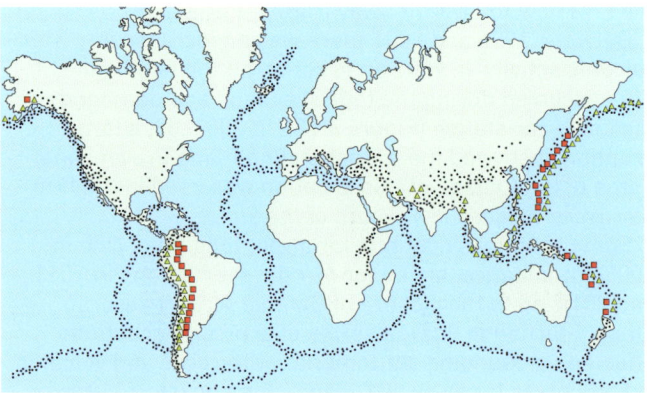

Depth of focus
- Shallow (0–70 km)
- Intermediate (70–300 km)
- Deep (300–700 km)

▲ Figure 4.26 The distribution of shallow-, intermediate- and deep-focus earthquakes

  a Describe the distribution of shallow-focus and deep-focus earthquakes. [4]
  b Suggest reasons to explain the distribution of shallow-focus and deep-focus earthquakes. [4]

2 Figure 4.27 shows the summit of Mount Etna in Sicily.

▲ Figure 4.27 The summit of Mount Etna

  a Identify **two** features of the volcano shown on the photograph. [2]
  b Describe **two** hazards related to volcanoes. [4]
  c Explain **two** reasons why people live in areas at risk from volcanic activity. [4]

3 Assess the extent to which the impacts of earthquakes can be managed. [7]

TOPIC 5

Climate change

Topics

5.1 The natural and human causes of climate change
5.2 The impacts of climate change at a range of geographic scales
5.3 The responses to climate change

This topic looks at:

- the natural and human causes of climate change
- the impacts of and responses to climate change.

5.1 The natural and human causes of climate change

This chapter will explain:

★ the evidence for climate change
★ the factors that influence natural climate change
★ how human influence affects climate change.

The evidence for climate change

There is a range of evidence for climate change, including temperature data, ice cores, extent of sea ice and the records provided in historic writings and paintings.

Global temperature data

Average annual temperatures have increased by approximately 1.1°C since pre-industrial times. Figure 5.1 shows global average surface temperatures 1880–2023 compared with the average for the twentieth century. The upward trend in mean temperature is clear.

Ice cores

Ice core analysis shows that carbon dioxide (CO_2) levels, which had been steady for the previous 10,000 years at around 260–280 ppm, began to increase in the nineteenth century (Figure 5.2) and have now reached approximately 420 ppm, an increase of about 50 per cent in around 150 years. CO_2 is increasing at a rate of about 15 ppm every six years, while methane (CH_4) has doubled in the last 200 years.

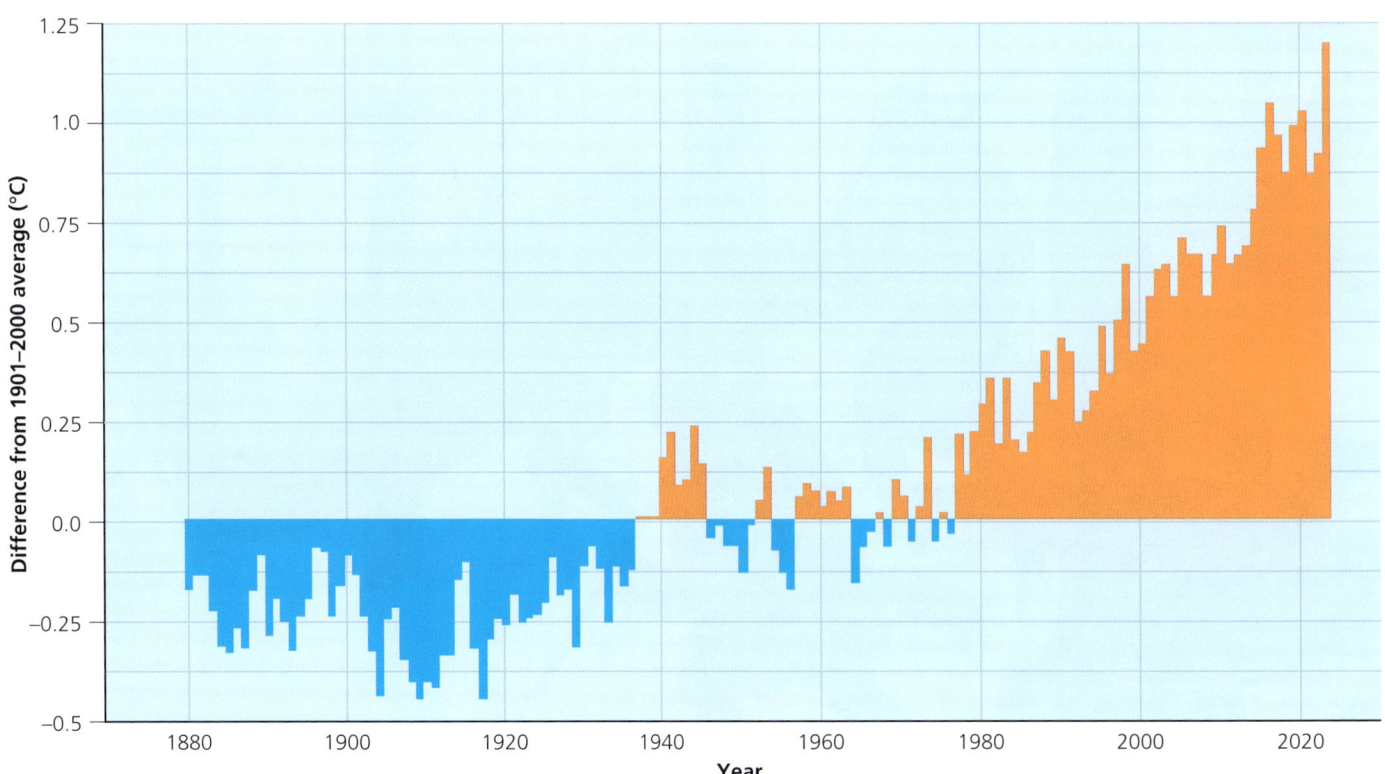

▲ Figure 5.1 Global average surface temperature 1880–2023 compared with the average for the twentieth century

The evidence for climate change

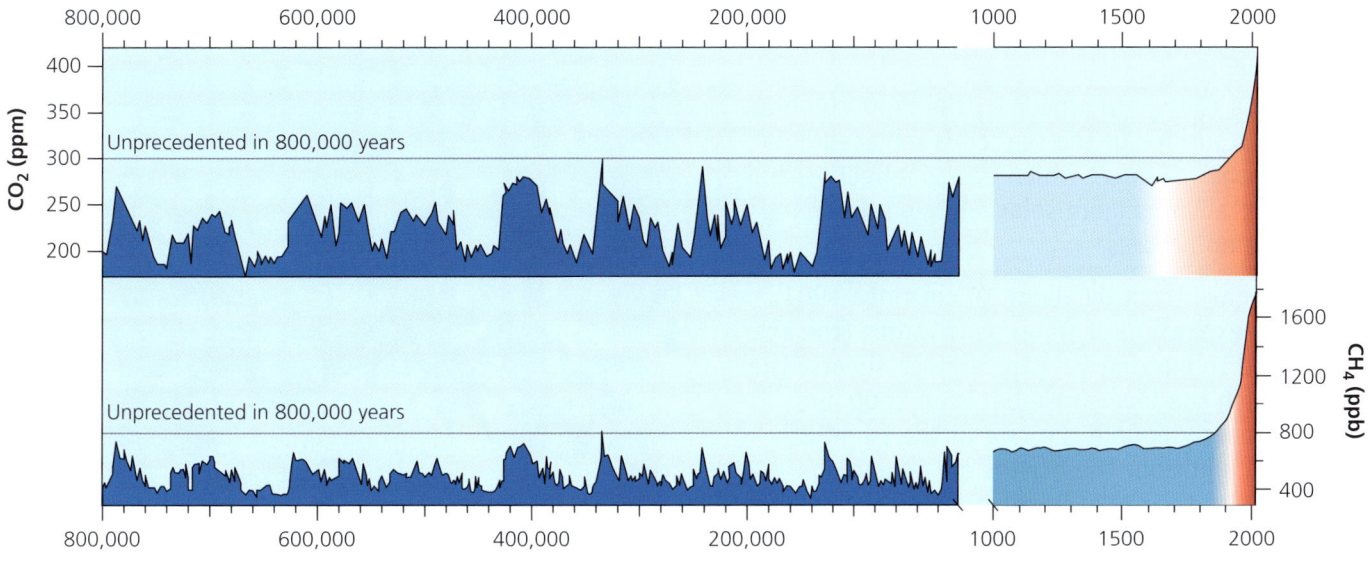

▲ **Figure 5.2** Changes in atmospheric carbon dioxide (CO_2) and methane (CH_4) over the last 800,000 years, as recorded in ice cores and atmospheric sampling

Interesting note

The history of climate change science is believed to have started with Eunice Newton Foote, an American scientist, inventor and women's rights campaigner. In 1856, she published a paper demonstrating the effects of CO_2 on temperature, in which she concluded: 'An atmosphere of that gas would give our Earth a higher temperature'.

Sea ice

Ice sheets, such as the West Antarctica ice sheet, are shrinking and thinning. Many of the world's glaciers are retreating and many smaller ones have disappeared. Snow cover in the Northern Hemisphere has decreased over the last 50 years and snow is melting earlier.

Levels of sea ice have been declining since the 1850s and the rate of this has accelerated. In each of the maps in Figure 5.3, the minimum amount of sea ice was towards the end of the period.

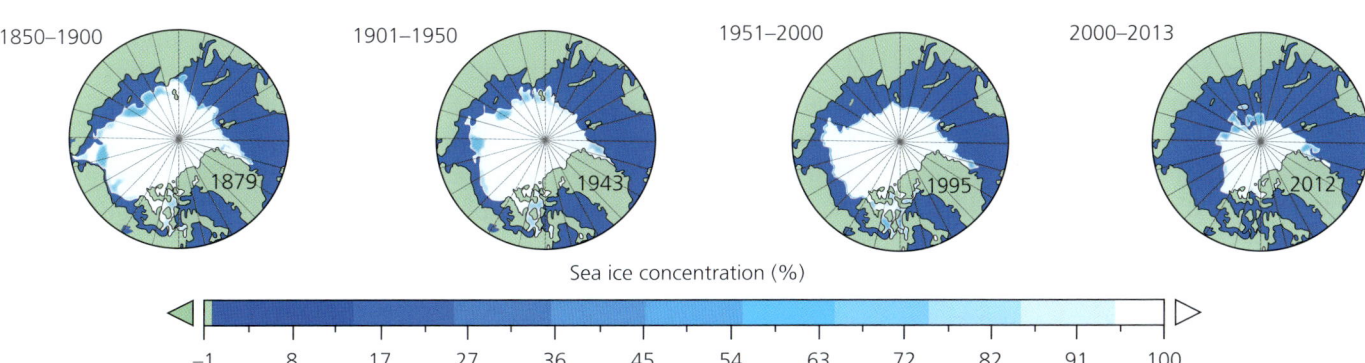

▲ **Figure 5.3** Sea ice cover maps for the annual minimum in September, for the periods 1850–1900, 1901–1950, 1951–2000 and 2001–2013

5.1 THE NATURAL AND HUMAN CAUSES OF CLIMATE CHANGE

The decline of sea ice leads to two changes:

1 Sea levels rise.
2 There is a change in albedo as the light-coloured reflective surface of the sea ice is replaced by a darker ocean surface, which is less reflective and therefore absorbs more solar energy, reinforcing the heating process.

Sea levels have risen by about 16 cm since 1901, due to the expansion of water as it warms (the **steric effect**) and to melting ice caps and ice sheets. The rate at which sea levels rose between 2000 and 2020 was nearly double the rate of the twentieth century. In addition, the oceans are becoming more acidic as they take in some 20–30 per cent of total anthropogenic (human-made) carbon dioxide emissions. Since the Industrial Revolution, oceanic acidity has increased by 30 per cent.

> **Interesting note**
>
> The decline in snow and ice cover is having a major impact on species adapted to cold conditions, for example polar bears and Arctic hares. Many species may need to move to higher latitudes or altitudes, but the options for those already inhabiting high-altitude or high-latitude environments are limited. The same is true in oceans – some lower-latitude species such as tuna, cod and Alaska pollock are moving into higher-latitude oceans.

Historic writing and paintings

Some historic writings and paintings give clues about past climate. The first eight years of Charles Dickens' life (born 1812) had white Christmases (when snow falls on 25 December), and he refers to white Christmases in several of his books, including *The Pickwick Papers* (1836) and *A Christmas Carol* (1843).

Pieter Bruegel the Elder painted *The Hunters in the Snow* (1565) (Figure 5.4) following the harsh European winters of 1564 and 1565. Some people have wondered if *The Hunters* was a record of the start of the Little Ice Age.

We can conclude from these and other writings and art that there were colder periods during the past, and that these conditions were very different from the climate change that we are now experiencing.

▲ **Figure 5.4** *The Hunters in the Snow*, by Pieter Bruegel the Elder

> **Activities**
>
> 1 Suggest how the painting *The Hunters in the Snow* by Pieter Bruegel the Elder and the writings of Charles Dickens suggest change in climate in Europe from the sixteenth century to the mid-nineteenth century to the present.
> 2 Describe the changes in Arctic sea ice between the periods 1850–1900 and 2000–13 shown in Figure 5.3 (page 111).

The natural causes of climate change

Orbital changes

The Quaternary Period refers to the most recent 2.6 million years of the Earth's history, up to and including the present day. The Quaternary has been characterised by frequent fluctuations of warming and cooling of the Earth, the growth and retreat of major ice sheets and ice caps, rises and falls of sea levels, the extinction of large mammals and the evolution of humans.

The Quaternary is a very short part of Earth's geological history. It is split into two parts, the Pleistocene Era and the Holocene Era:

» The **Pleistocene** refers to the geological era that lasted from about 2.6 million years ago to about 11,700 years ago. It was characterised by the growth and decline of ice sheets and ice caps.
» The **Holocene** refers to the most recent 11,700 years, when the Earth's climate has been relatively warm and the major ice sheets have retreated.

Nevertheless, the Earth is still in an ice age, as evidenced by the ice in Antarctica and Greenland. However, we are in a warm phase, known as an **interglacial period**. A cold phase, when ice grows, is referred to as a **glacial period**. Now, approximately 10 per cent of the Earth's surface is covered by ice. During the glacial periods, about 30 per cent of the Earth's surface is covered by ice.

Milankovitch cycles

One of the most important theories to help explain the natural causes of climate change in the Quaternary Period is that put forward by Milutin Milankovitch. The Milankovitch cycle theory (Figure 5.5) states that the amount of solar energy reaching the Earth varies with changes in the Earth's orbit, its tilt and its 'wobble' – that is, the direction of its rotation. The Earth's orbit varies over a timescale of about 100,000 years. When the Earth is further from the sun, it receives less energy. Also, when the tilt is greater, the seasons are longer. The 'wobble' determines which hemisphere – Northern or Southern – is facing the sun. This varies over a 19,000–23,000-year timescale.

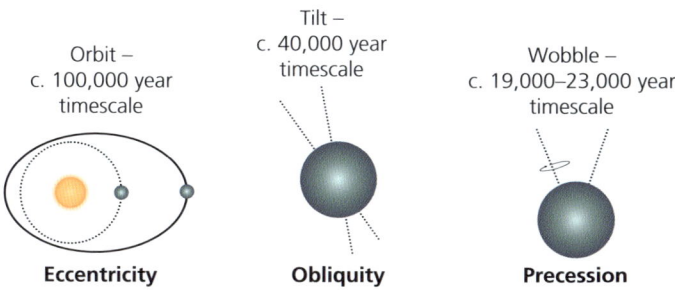

▲ **Figure 5.5** The Milankovitch cycles

At the start of the Quaternary Period, the continents were largely in the positions that they are today. Since then, however, the Earth has wobbled in its rotation around the sun. These 'wobbles' have been partly responsible for the growth and decline of glaciations. By around 800,000 years ago a pattern had emerged: cold glacial phases that lasted up to 100,000 years, followed by a warmer interglacial period lasting about 10,000 to 15,000 years.

Sunspots

There is evidence that the sun has an 11-year cycle in which it brightens and dims – although the output only varies by around 0.1%. This change may have a very small effect on surface temperatures. Between 1880 and 1955, solar radiation gradually increased. However, since 1955 it has been gradually decreasing. So, although global temperatures have increased, due to global warming, solar output has decreased.

Volcanic activity

Other natural causes of climate change include volcanic eruptions (Figure 5.6). Those more likely to cause changes to climate are large eruptions, especially in tropical areas. The resulting atmospheric dust is also believed to block or reflect incoming solar energy, thereby leading to lower temperatures on Earth. Low latitudes receive a greater amount of solar energy than high latitudes, which is why volcanic eruptions in low latitudes have a greater impact on climate change – they block out a higher amount of solar radiation than high-latitude eruptions.

According to the United Nations' Intergovernmental Panel on Climate Change (IPCC), long- and short-term variations in solar activity play only a very small role in altering the Earth's climate.

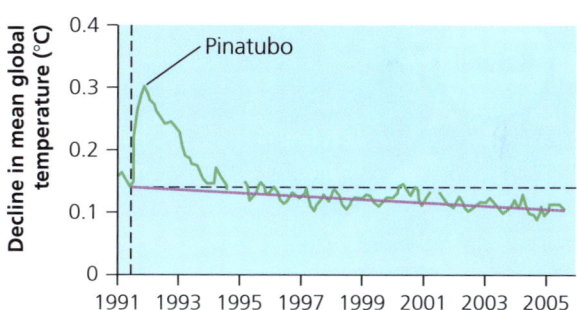

▲ **Figure 5.6** The impact of the 1991 Mount Pinatubo eruption (Philippines) on temperature between 1991 and 2005: the purple line shows the global decline in sun-blocking aerosols following the eruption

> **Activities**
> 1 Describe what happens to the Earth during:
>   a the 100,000-year cycle
>   b the 41,000-year cycle.
> 2 Explain how the eruption of Mount Pinatubo affected global climate.

The human influences on climate change

The **greenhouse effect** is the process by which certain gases (greenhouse gases) allow short-wave radiation to pass through the atmosphere but trap a

5.1 THE NATURAL AND HUMAN CAUSES OF CLIMATE CHANGE

proportion of the outgoing long-wave radiation from the Earth. This traps heat within the atmosphere like a greenhouse and leads to a warming of the atmosphere. It a natural process and vital for life on Earth.

The most common greenhouse gas is water vapour, which accounts for about 50 per cent of the natural greenhouse. However, the gases that account for human causes of climate change are carbon dioxide (CO_2), methane (CH_4) and chlorofluorocarbons (CFCs).

The enhanced greenhouse effect

Atmospheric levels of carbon dioxide rose from around 315 parts per million (ppm) in 1950 to over 420 ppm by 2020, and they are predicted to rise to 600 ppm by 2050. This rise is due to issues such as burning **fossil fuels** (coal, oil and natural gas) and land-use changes to make way for agriculture, settlements and infrastructure. Deforestation reduces the store of carbon on the Earth (that is, in the trees) and carbon is released into the atmosphere when the trees are burned, adding to the amount of atmospheric carbon dioxide.

Agriculture is a major cause of climate change through processes including land-use change (deforestation, for example), methane emissions from cattle and rice fields, and the use of machinery in farming and in the transport of goods to market.

The **enhanced greenhouse effect** (EGHE) (Figures 5.7 and 5.8) is known by many terms, such as global climate change, climate crisis and global warming. As a result of the EGHE, temperatures have been rising since around 1850.

> ### Activities
> 1. Identify three greenhouse gases.
> 2. Explain how greenhouse gases affect temperature on Earth.

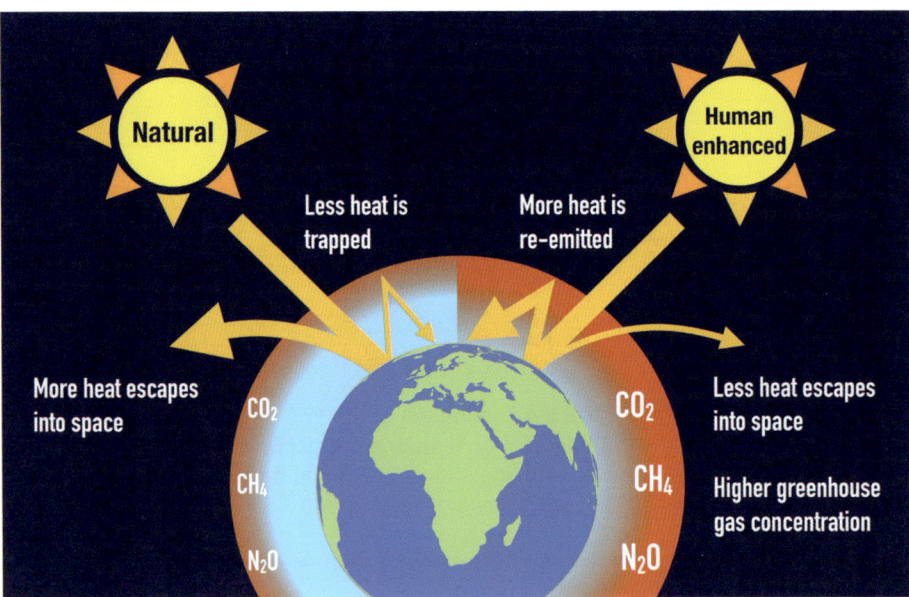

▲ **Figure 5.7** The greenhouse effect and the enhanced greenhouse effect

▲ **Figure 5.8** Some impacts of global warming – sunbathing on glaciers and irrigation needed to play tennis

5.2 The impacts of climate change at a range of geographic scales

This chapter will explain the present-day and predicted future impacts of climate change:

★ the changes to global temperature and weather patterns
★ the changes to sea levels
★ future predictions.

The changes to global temperature and weather patterns

Figure 5.9 shows the change in global surface temperatures between 1880 and 2020. Between 1880 and 1940, annual temperatures were below the 1880–2020 long-term average. It was coldest (about 0.4°C colder) between about 1900 and 1910. Temperatures generally fluctuated above and below the long-term average between 1940 and 1980. However, since 1980 temperatures have always been higher than the long-term average, reaching approximately 1.0°C hotter than the long-term average around 2015. It is predicted to be about 1.1–1.2°C above the long-term average in 2025.

According to a 2021 report by the IPCC (the United Nations body that assesses the science for climate change), the enhanced greenhouse effect should be seen as a 'code red' for humanity. It warns of increasingly extreme heatwaves, droughts, flooding and high temperatures over the next few decades. It also warns of an impending **tipping point**. It points out that:

» since 1970, global surface temperatures have risen faster than in any other 50-year period over the last 2000 years
» global surface temperatures are about 1.1–1.2°C higher than in 1850–90
» the years 2017–24 were the warmest on record since 1850
» the rate of sea level rise has increased more than three-fold since 1971 – over 260 million people are at risk from rising sea levels and this could rise to 400 million people by 2100
» human influence is more than 90 per cent likely to be the cause of the global retreat of glaciers and the decrease of Arctic sea ice
» hot extremes have become more frequent and intense since the 1950s, for example there were droughts in the Amazon in 2005, 2010, 2015 and 2020, whereas cold extremes have become less frequent and less intense.

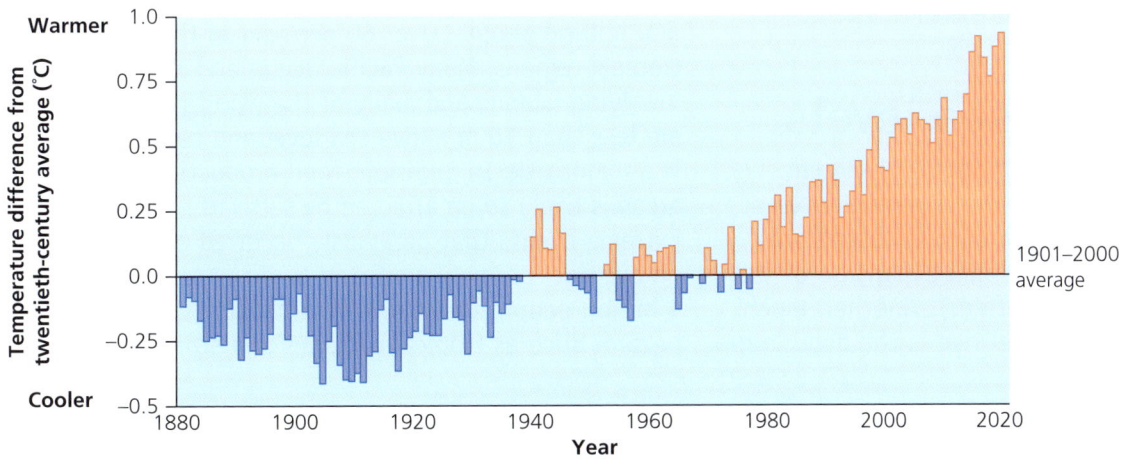

▲ **Figure 5.9** Annual global surface temperature change for land and ocean

5.2 THE IMPACTS OF CLIMATE CHANGE AT A RANGE OF GEOGRAPHIC SCALES

The report also states that the enhanced greenhouse effect is much worse than previously thought, and that 40 per cent of the world's population are vulnerable to climate change. The report claims that climate extremes are going beyond humanity's and species' ability to cope. Between 2000 and 2020 there was a fifteen-fold increase in the number of deaths due to storms, floods and droughts, especially in emerging countries.

Extremes of weather are becoming more frequent and severe. The heatwave that killed over 70,000 people in Europe in 2003 was said to be a one-in-a-thousand-year event – but now it is said to be a one-in-ten-year event. In 2024, around 1300 pilgrims who made the journey to Mecca died due to heat exhaustion and dehydration. Temperatures there reached 50°C in the shade.

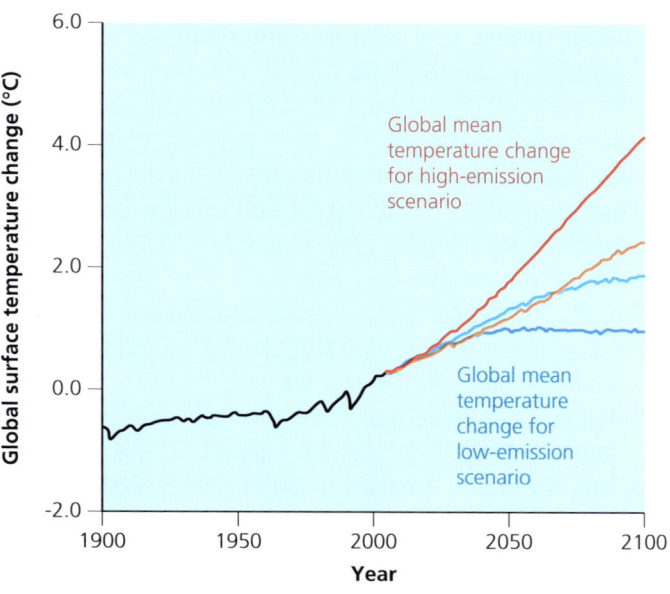

▲ Figure 5.10 Predicted temperature changes 2000–2100 under high- and low-emissions scenarios

The changes to sea levels

Global mean sea level has risen about 21–24 cm since 1880. As already stated, the rising water level is mostly due to a combination of meltwater from glaciers and ice sheets and thermal expansion of sea water as it warms (the steric effect). Another, much smaller, contributor to sea level rise is the decline in the amount of water on land – in aquifers, lakes and reservoirs, rivers and soil moisture. This shift of liquid water from land to ocean is largely due to people depleting groundwater.

The global mean sea level in the ocean rose by 3.6 mm per year from 2006 to 2015, which was 2.5 times the average rate of 1.4 mm per year throughout most of the twentieth century. By 2100, global mean sea level is likely to rise by at least 0.3 m above 2000 levels (Figure 5.11). In the United States, about 30 per cent of the population lives in areas where sea level plays a role in flooding, shoreline erosion and hazards from storms.

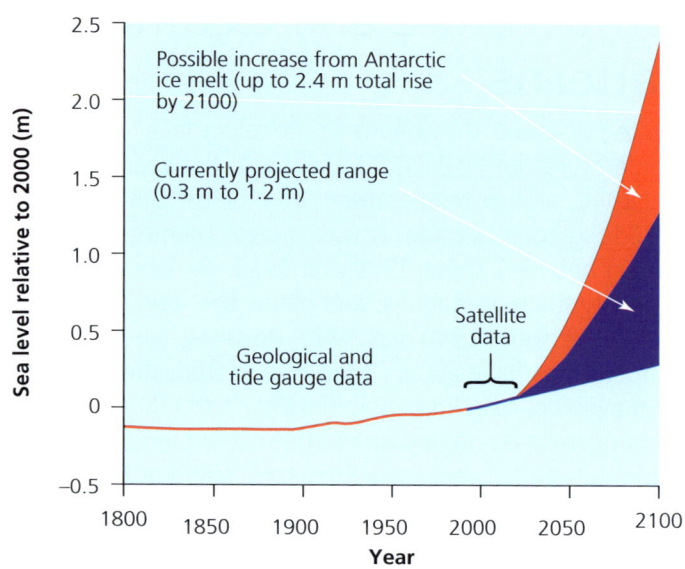

▲ Figure 5.11 Global mean sea level rise, 1800–2100 (projected)

Over 260 million people are at risk from rising sea levels, and by the year 2100 this could rise to over 400 million people. Over 60 per cent of these live in tropical regions. Many Pacific Ocean and Indian Ocean low-lying islands and many of the world's megacities are less than 10 m above sea level and coastal flooding is a major risk. In addition, food production in coastal areas is likely to be affected by saltwater intrusion. Extreme events that previously occurred once a century on average, are now predicted to occur annually. Up to 1 billion people are at risk from coastal-specific climate hazards such as tropical cyclones and storm surges.

Future predictions

Greenhouse gas concentrations in the atmosphere will continue to increase unless annual emissions decrease substantially. Increased concentrations are expected to raise average temperatures, alter patterns and amounts of precipitation, reduce ice and snow cover, increase the frequency, intensity and/or duration of extreme events, increase threats to human health, increase the acidity of oceans and cause ecosystems to shift their locations.

Precipitation is likely to increase, especially in tropical and high-latitude regions. The strength of the winds associated with tropical storms is likely to increase, although some scientists believe the storms will be more frequent but not necessarily stronger.

Sea ice and snow cover in the Northern Hemisphere has decreased since about 1970. Over the next century, these trends will continue.

Food production

Food production is likely to be affected by rising global temperatures. The decline in water resources may make it difficult for many farmers to continue the type of farming they currently practise. They may have to change the type of farming or the crops they grow, or they may even be forced out of farming altogether. Places in the Middle East, such as in Egypt, Dubai and Saudi Arabia, are likely to experience extreme heatwaves and farmlands may become desertified. In parts of Africa, especially Southern Africa, cereal crop productivity will be reduced as 30 per cent of the area that is used for maize production and 50 per cent of the area that is used for beans is likely to go out of production due to the enhanced greenhouse effect.

▶ Activities

1 Compare the greenhouse effect and the enhanced greenhouse effect.
2 Using Figure 5.11 (page 116), explain why there are variations in the predicted sea level rise by 2100.

5.3 The responses to climate change

This chapter will explain:

★ the strategies used to manage the impacts of climate change
★ how successful mitigation and adaptation strategies and techniques have been in managing the impacts of climate change
★ detailed specific example: the impacts of, and responses to, climate change in Bangladesh.

The strategies used to manage climate change – national and international agreements

The major international agreements to manage climate change include the Paris Agreement, the Kyoto Protocol and the United Nations Framework Convention of Climate Change (UNFCCC).

The **UNFCCC** is the United Nations' (UN) process to negotiate an agreement to limit global climate change. To achieve this, it proposes to limit the amount of greenhouse gases in the atmosphere. It was signed in 1992 at the UN Conference on the Environment (the 'Earth Summit') in Rio de Janeiro. Its main objective is to stabilise greenhouse gases in the atmosphere. By 2022, the UNFCCC had 198 members ('Parties'), who meet annually at COP (Conference of the Parties) meetings.

The **Kyoto Protocol**, signed in 1997, became effective in 2005. In 2020 there were 192 member countries. The Kyoto Protocol aims to reduce global climate change by decreasing atmospheric greenhouse concentrations to a level that would prevent dangerous human interference. The Protocol had differentiated responsibilities, recognising that HICs were historically responsible for most greenhouse gas emissions. The Protocol's first commitment period was 2008–12. Although many HICs reduced their emissions, total global emissions increased from 1990 to 2010. A second agreement existed from 2012 to 2020. Some 37 countries had binding targets, including the EU (28 countries at that time), Australia and Ukraine. Some countries withdrew from the Protocol, including Canada and Russia.

The **Paris Agreement** (2015) is an international treaty on climate change, agreed by 196 countries. It is a separate agreement under the UNFCCC. The long-term goal of the Paris Agreement is to keep global climate change to less than 2°C above pre-industrial levels, and ideally to less than 1.5°C above pre-industrial levels. To achieve this, greenhouse gas emissions need to be reduced and net zero (that is, when emissions of greenhouse gases are balanced by removal of greenhouse gases from the atmosphere) reached by about 2050. The Paris Agreement has been criticised by some for not being strict enough. In 2017, the USA, under President Trump, announced that it would pull out of the Paris Agreement. It withdrew in 2020, but rejoined in 2021 under President Biden. The Paris Agreement does not cover emissions from international aviation or shipping.

Despite these agreements, greenhouse gases in the atmosphere continue to rise, as do global temperatures.

An evaluation of mitigation and adaptation strategies

Climate change **adaptation** refers to methods of learning to live with climate change, whereas **mitigation** refers to measures to prevent or limit global climate change (Figure 5.12). Adaptation is easier but less effective. Mitigation would be better in the long term, but is more difficult to achieve.

Adaptation includes:

» building higher flood walls
» using more air conditioning systems to cool living/work spaces
» ensuring availability of vaccines for diseases new to a region, such as malaria

An evaluation of mitigation and adaptation strategies

- building flood shelters
- growing crops that are adapted to drought or heat.

Mitigation refers to attempts to reduce global climate change, including:

- burning less fossil fuel
- cutting down fewer trees
- planting more vegetation
- using sustainable transport
- conserving energy
- using renewable energy sources
- eating less red meat and dairy products.

The disadvantages of mitigation include cost, for example the cost of developing nuclear power or building HEP schemes, and the difficulty of encouraging changes in people's behaviour, such as using less energy, using public transport, cycling or walking to work, college or school. The less mitigation that society achieves today, the more it will need to adapt in the future.

Adaptation may be easier, but it does not get rid of the problem. Adaptation will become progressively less effective if there is no mitigation, as global climate change will become more extreme over time. However, using clean energy sources, such as solar and wind power, will reduce the use of fossil fuels.

Other measures to reduce global climate change include insulating homes, switching off electricity sources when not in use, changing from a meat- and dairy-based diet to a vegetarian diet, reducing consumption of goods and using fewer plastics.

> ### Activities
> 1 Suggest reasons why international agreements to manage the impact of climate change have been unsuccessful.
> 2 Explain how green infrastructure and water- and energy-conservation can be examples of adaptation and mitigation.

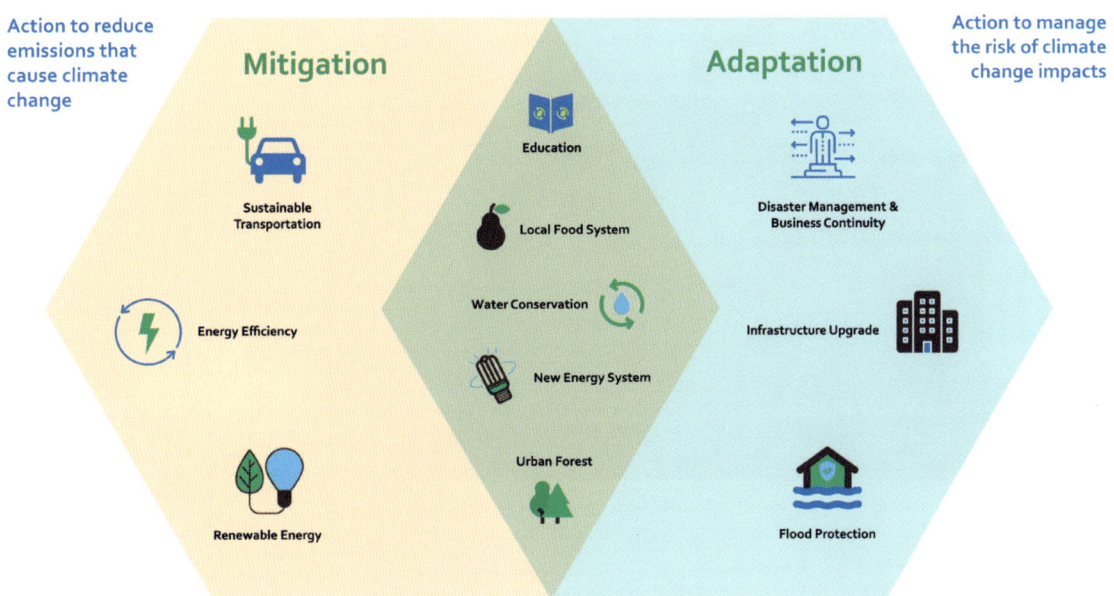

▲ **Figure 5.12** Adaptation and mitigation strategies for climate change

5.3 THE RESPONSES TO CLIMATE CHANGE

> **Detailed specific example**

The impacts of, and responses to, climate change in Bangladesh

The impacts of climate change

Bangladesh accounts for less than 1 per cent of global greenhouse gas emissions, but it is one of the most vulnerable countries to climate devastation. This is known as a climate injustice.

Between 2000 and 2019 Bangladesh experienced economic loss of US$3.7 billion and 185 extreme weather events. Around 90 million Bangladeshis, over 50 per cent of the population, have a 'high' exposure to flooding. Two-thirds of Bangladesh is less than 5 m above sea level (Figure 5.13) and around one-third of the population live in coastal areas. A 50 cm rise in sea level could cause Bangladesh to lose 11 per cent of its land and up to 18 million people would be forced to leave their homes. A sea level rise of 2 m could displace up to 50 million people. Rising sea levels also cause salinisation, which is when salt accumulates in soil and severely reduces the soil's fertility.

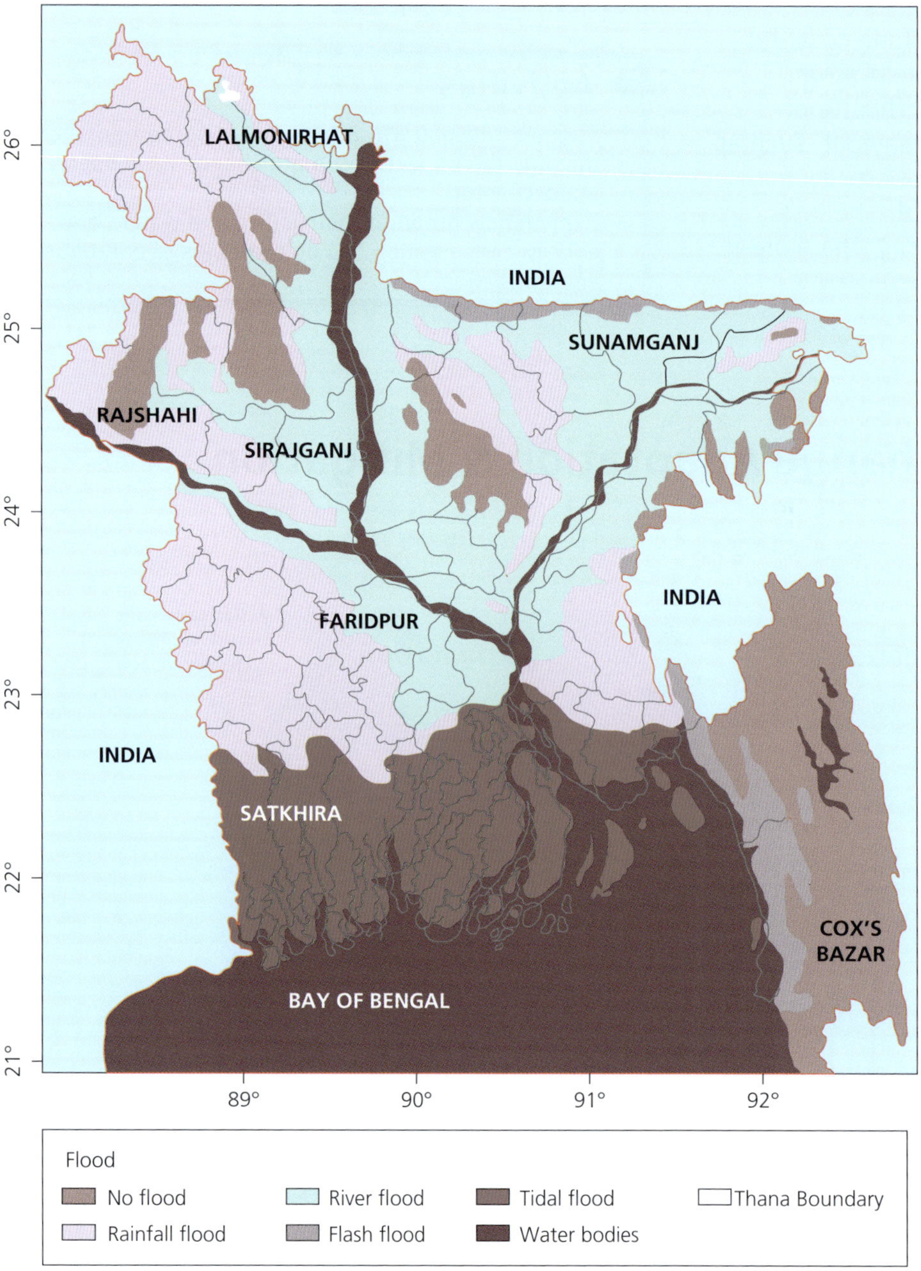

▲ Figure 5.13 Flood risk in Bangladesh

An evaluation of mitigation and adaptation strategies

Extreme events are becoming more frequent in Bangladesh. Up to 50 per cent of people living in Bangladesh's informal settlements are believed to be there due to flooding elsewhere. Dhaka, the capital, now has a population density of up to 47,500 people/km^2.

Tropical cyclones are becoming more intense. For example, in 2007 Cyclone Sidr brought winds of 240 km/h and killed over 3400 people. In 2009, Cyclone Aila led to the deaths of 200 people and left about 200,000 people homeless. Cyclone Amphan in 2020 was the strongest cyclone in Bangladesh's history. It led to the deaths of 128 people and caused US$14 billion worth of damage in Bangladesh and India.

The responses to climate change

To deal with such events, Bangladesh has developed a Climate Change Strategy Action Plan (BCCSAP). This was first produced in 2009, with five adaptation plans and one for mitigation. The Nationally Determined Contributions (NDCs) of 2021 planned:

- to reduce greenhouse gas emissions by >27 million tonnes of carbon dioxide equivalent (Mt CO_2e) (6.75 per cent) using Bangladesh's own resources by the end of 2030
- to reduce greenhouse gas emissions by >60 Mt CO_2e (>15 per cent) using international support by the end of 2030
- to reduce greenhouse gas emissions by around 90 Mt CO_2e (21.5 per cent) using combined resources by the end of 2030.

Strategies and techniques to manage climate change

Adaptation projects have included the building of embankments, over 4500 cyclone shelters and 300 flood shelters, and the excavation/re-excavation of canals.

Mitigation schemes have included the planting of millions of trees, including mangrove forests. As a result, emissions are expected to have been reduced by over 2 Mt CO_2e by 2025.

Bangladesh is pursuing a low carbon future with greater emphasis on renewable energy and increased energy efficiency. Farming methods used to adapt to the changing climate include:

- early-harvest, rapid-developing varieties, for example rice that develops in 100–20 days (traditional varieties require 140–50 days)
- drought-resistant early varieties
- salt-tolerant rice varieties
- flooding-tolerant crop varieties
- stress-tolerant crop varieties.

Bangladesh joined the Kyoto Protocol in 2013 and signed the Paris Agreement in 2016.

> ### Activities
> 1 Describe the distribution and type of flood risk in Bangladesh, as shown in Figure 5.13 (page 120).
> 2 Distinguish between adaptation and mitigation strategies that have been used in Bangladesh.

5.3 THE RESPONSES TO CLIMATE CHANGE

Practice questions

1 Figure 5.14 shows annual global surface temperature change for land and ocean, compared to the 1901–2000 average.
   a Estimate the temperature of the year that was furthest below the 1901–2000 average. [1]
   b Estimate the temperature of the year that was furthest above the 1901–2000 average. [1]
   c Identify the decades when temperatures were lowest. [1]
   d Identify the decade when temperatures were highest. [1]
   e Estimate the increase in temperature between the decades with the lowest temperatures and the decade with the highest temperatures. [2]
2 Distinguish between the greenhouse effect and global warming. [4]
3 a Define climate change mitigation and climate change adaptation. [2]
   b Evaluate techniques of climate change mitigation and adaptation. [7]

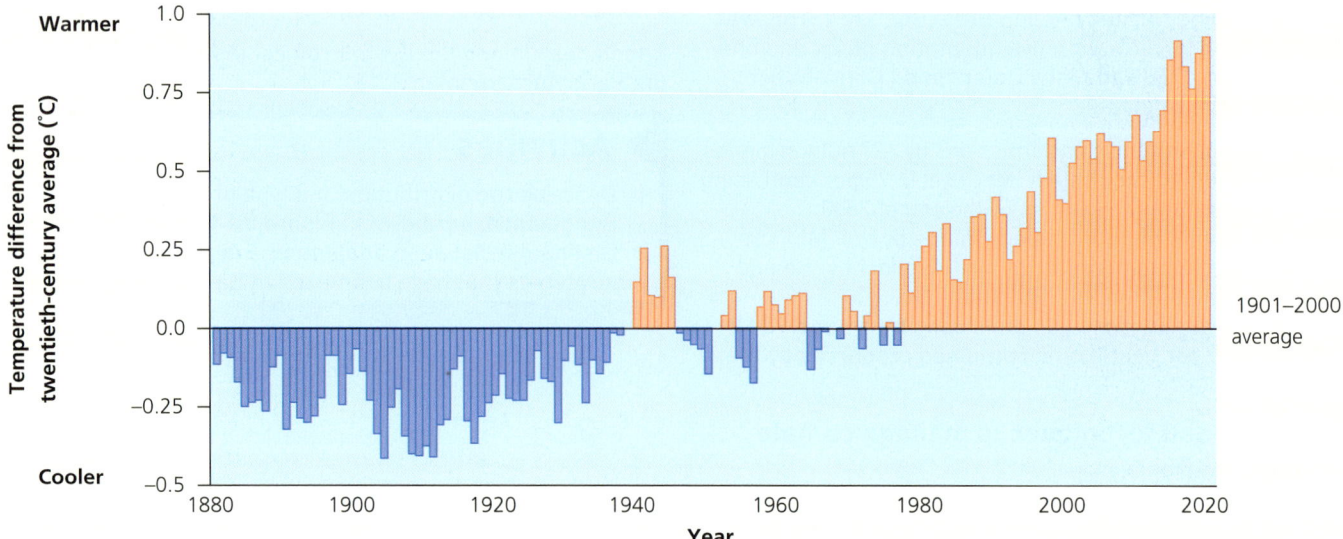

▲ Figure 5.14 Annual global surface temperature change for land and ocean, compared to the 1901–2000 average

TOPIC 6
Changing populations

Topics

6.1 Populations grow and decline
6.2 Population structures change over time
6.3 The causes and impacts of international migration

This topic looks at:

- the changes that occur to populations over time
- what causes international migration and the impacts of this.

6.1 Populations grow and decline

This chapter will explain:

★ the patterns and trends in global population growth and the rapid increase in the world's population
★ the reasons for the growth and decline of a country's population
★ the impacts of pro-natalist and anti-natalist policies on birth rates
★ the strengths and limitations of the demographic transition model.

The patterns and trends in global population growth: the rapid increase in the world's population

During most of the period since humankind first evolved global population was very small, reaching perhaps some 125,000 people a million years ago.

» It has been estimated that 10,000 years ago, when people first began to domesticate animals and cultivate crops, world population was no more than 5 million. Known as the Neolithic Revolution, this period of economic change significantly altered the relationship between people and their environments.
» As a result of technological advances, the **carrying capacity** of populated areas improved and population increased. By 3500 BCE global population had reached 30 million, and by 0 CE this had risen to about 250 million (Figure 6.1).
» **Demographers** estimate that world population reached 500 million by about 1650. From this time, population grew at an increased rate (Figure 6.2).

▲ Figure 6.1 The Colosseum, Rome – construction began around 70 CE, when the world's total population was about 250 million

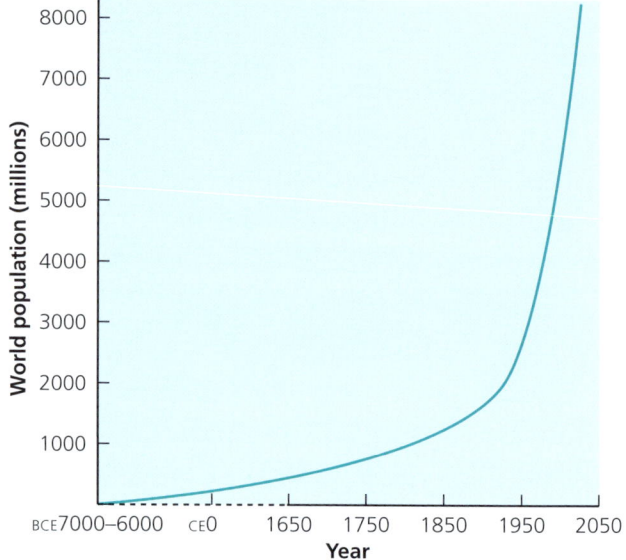

▲ Figure 6.2 World population growth

By about 1800, global population had doubled to reach 1 billion. Table 6.1 shows the time taken for each subsequent billion to be reached. It took less than 50 years for world population to double from 4 billion in 1974 to 8 billion in 2022. Global population is projected to surpass 9 billion around 2038 and 10 billion by 2060. These figures are based on the UN's medium projection variant. As with all **population projections**, there is a degree of uncertainty because so many different factors will affect future population growth.

▼ Table 6.1 World population growth by each billion

Population	Year	Number of years to add each billion
1 billion	1804	All of human history
2 billion	1927	123
3 billion	1960	33
4 billion	1974	14
5 billion	1987	13

▼ Table 6.1 (Continued)

Population	Year	Number of years to add each billion
6 billion	1998	11
7 billion	2010	12
8 billion	2022	12
9 billion	2038	16
10 billion	2060	22

The rapid growth in human population in recent centuries has been due to significant advances in:

» public health and medicine, including greater access to clean drinking water, improvements in sanitation and advances in disease control
» nutrition and general living standards.

These massively important improvements in human welfare have lowered considerably the risk of dying, particularly among children (**child mortality rate**). Thus, increasing numbers of people have survived to reproductive age, alongside the gradual increase in human lifespan.

Recent demographic change

Fastest in Africa; slowest in Europe

Figure 6.3 shows population growth and projected population growth by world region from 1950 to 2100, while Table 6.2 gives the percentage share of world population by world region for 1980, 2010, 2023 and 2050.

» Asia dominates global population, with 59.1 per cent of the world's population in 2023.
» Africa significantly increased its share of global population between 1980 and 2023 and overtook Europe as the second most populous world region in 1996.
» Together, Asia and Africa accounted for 77.2 per cent of the world's population in 2023.

Asia and Africa are projected to drive much of the world's future population growth, with the fastest rate of growth still being in Africa. This continent is projected to have more than a quarter of the world's population by 2050. By that year, it is expected that just over 80 per cent of the world's population will live in either Asia or Africa.

Europe's share of world population has been in decline for a long time, falling from 22.0 per cent in 1950 to 9.2 per cent in 2023. It is projected to decline further, to 7.3 per cent in 2050. Europe's total population, in absolute terms, began to decline in 2021. Northern America is the other world region where the population share has declined significantly, but at a slower rate than that of Europe. However, in contrast to Europe, there is no indication that Northern America's population is ever going to decline in absolute terms in the time period illustrated by Figure 6.3.

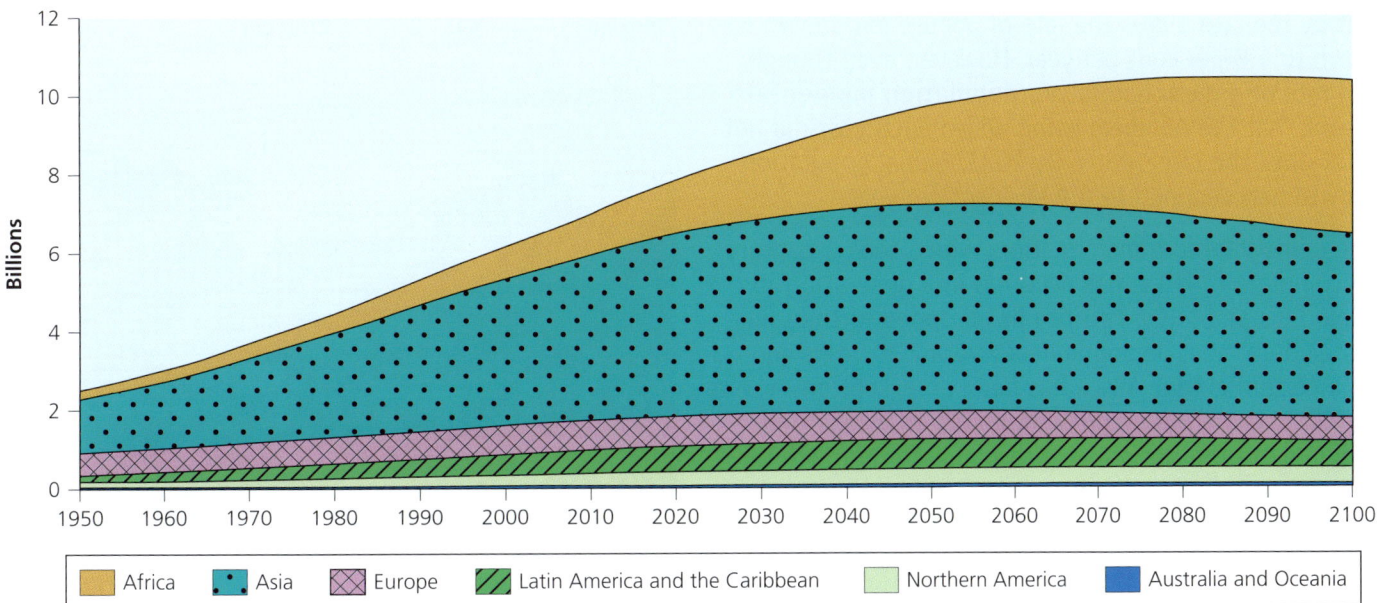

▲ Figure 6.3 Population change by world region, 1950–2100 (projected)

6.1 POPULATIONS GROW AND DECLINE

Table 6.2 Global population distribution by world region

Region	1980	2010	2023	2050
Africa	10.8	15.1	18.1	25.6
Asia	59.3	60.4	59.1	54.5
Latin America/Caribbean	8.2	8.5	8.3	7.7
North America	5.6	4.9	4.7	4.3
Europe	15.6	10.6	9.2	7.3
Australia/Oceania	0.5	0.5	0.6	0.6

Population momentum

World population data for the period 1950–2023 show that the rate of population growth was much higher in LICs and MICs than in HICs. However, it is only since the mid-twentieth century that population growth in LICs and MICs has overtaken that in HICs. HICs had their period of high population growth in the nineteenth and early twentieth centuries, while for LICs and MICs high population growth has occurred since about 1950. Countries with high levels of fertility tend to be those with relatively low incomes per capita. Over time the growth of the world's population has become increasingly concentrated among the world's lowest-income countries.

The highest ever global population growth rate was reached in the early to mid-1960s, when population growth in LICs and MICs peaked at 2.4 per cent per year. At this time the term 'population explosion' was widely used to describe this rapid population growth, but by the late 1990s the rate of population growth was down to 1.8 per cent per year. However, even though the rate of growth had fallen, **population momentum** meant that the numbers being added each year did not peak until the late 1980s. By 2023, global population growth had declined to 0.9 per cent per year.

The demographic transformation, which took a century to complete in HICs, has occurred in a generation in some LICs and MICs. Fertility has dropped further and faster than most demographers foresaw 30 years ago. To a certain extent Africa has been an exception, where in just over half of countries families of at least four children are the norm and with population growth of 2.4 per cent (2023). However, population growth rates in Africa have fallen significantly in recent decades, following the trend in other world regions.

The present to 2050

Between now and 2050, almost all of the increase in numbers of children and adults under 65 will occur in low- and lower-middle-income countries. In high- and upper-middle-income countries, the population under 65 will decline, with future growth occurring entirely among the population aged 65 years or over.

Table 6.3 shows the ten most populous countries in the world for 2023 and 2050. In mid-2023, India (Figure 6.4) overtook China as the world's most populous country. Both nations now have populations over 1.4 billion, meaning each represents nearly 18 per cent of the world's population.

Table 6.3 The world's ten largest countries in population, 2023 and 2050

Country	2023 population (millions)	Country	2050 population (millions)
India	1428.6	India	1670.5
China	1411.3	China	1303.5
USA	335.0	USA	383.6
Indonesia	278.7	Nigeria	377.5
Pakistan	240.5	Pakistan	367.8
Nigeria	223.8	Indonesia	328.9
Brazil	204.0	Brazil	230.8
Bangladesh	173.5	Bangladesh	223.5
Russia	146.9	DR Congo	217.5
Mexico	131.0	Ethiopia	214.8

▲ Figure 6.4 Traffic congestion in Jaipur, northwest India

> **Interesting note**
>
> The Population Reference Bureau estimates that throughout the history of human population, about 108 billion people have lived on Earth. This means that about 6.5 per cent of all people ever born are alive today.

> **Activities**
>
> 1 With the help of Figure 6.2 (page 124) and Table 6.1 (page 124), briefly describe the growth of the human population over time.
> 2 Define the terms:
>   a Carrying capacity
>   b Demographer.
> 3 Study Table 6.2 (page 126). Describe the changes and projected changes in global population distribution by world region.

The reasons for the growth and decline of a country's population

The **birth rate** is defined as the number of births per thousand population in a year. If the birth rate of a country is 17/1000 (17 per 1000), this means that, on average, for every 1000 people in this country, 17 births will occur in a year. The **death rate** is the number of deaths per thousand population in a year. If the death rate for the same country is 8/1000, it means that, on average, for every 1000 people, eight deaths will occur in a year.

The difference between the birth rate and the death rate is called the **rate of natural change**. If it is positive, it is termed natural increase. If it is negative, it is known as natural decrease. In the case given above, there is a natural increase of 9/1000 (17/1000 − 8/1000). The rate of natural increase may also be shown as a percentage, so in this example 9/1000 is equivalent to 0.9 per cent. This is the current rate (2023) of natural increase for the world as a whole – look at the birth and death rates given in Table 6.4, which shows how much birth and death rates vary by world region.

▼ **Table 6.4** Birth and death rates by world region, 2023

Region	Population (millions)	Birth rate (per 1000)	Death rate (per 1000)
World	8009	17	8
Africa	1453	32	8
Asia	4739	14	7
Latin America/Caribbean	652	14	8
North America	375	11	10
Europe	744	9	12
Oceania	45	15	7

Natural change and net migration

Population change in a country is affected by:

» natural change – the difference between births and deaths
» **net migration** – the balance between immigration and emigration.

The **immigration rate** is the number of immigrants per 1000 population entering a receiving country in a year. The **emigration rate** is the number of emigrants per 1000 population leaving a country of origin in a year.

The dividing zigzag red line on Figure 6.5 indicates that the relative contributions of natural change and net migration can vary over time within a particular country, as well as between countries at any one point in time. The model is a simple graphical alternative to the popular equation:

$$P = (B - D) +/- M$$

Where P = population, B = births, D = deaths, and M = migration

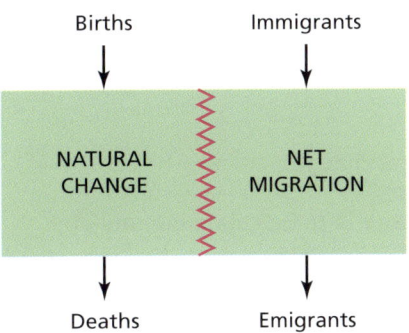

▲ **Figure 6.5** Input–output model of population change

6.1 POPULATIONS GROW AND DECLINE

Figure 6.6 shows some simple demographic calculations for the imaginary island of Pacifica.

Population at beginning of year: 5000

Population change during the year:
Births: 150   Deaths: 60
Immigrants: 20   Emigrants: 10

Rates of change based on data above
Birth rate: 30/1000   Death rate: 12/1000
Rate of natural change: +18/1000
Immigration rate: 4/1000   Emigration rate: 2/1000
Rate of net migration: +2/1000

Total population at end of the year
= 5100 (natural change of 90 + 10 for net migration)

▲ **Figure 6.6** Demographic calculations for the island of Pacifica

For most countries natural change is a more important factor in population change than net migration, but there are exceptions:

» Since the 1990s, positive net migration has been the primary source of population growth in HICs.
» Migration is projected to be the only driver of population growth in HICs after 2020.
» By 2050 it is expected that the population of HICs will begin to decline, as net migration will no longer be sufficient to compensate for the excess of deaths over births.

Sixty per cent of the increase in the UK's population between 2001 and 2020 was due to the direct contribution of net migration.

> ### Activities
> 1 Define:
>   a The birth rate
>   b The death rate
>   c The rate of natural change.
> 2 What is net migration?
> 3 Look at Table 6.4 (page 127). Calculate the rate of natural change for each region.
> 4 Look at Figure 6.6. Imagine that the population at the beginning of the year was 4000 rather than 5000. Calculate the rates of change for this new starting population.

Fertility
Definitions and global variations

The birth rate and the death rate are the most basic measures of fertility and mortality. Because these measures are so basic, demographers refer to them as the 'crude birth rate' and the 'crude death rate' because they take no account of gender or age.

Fertility varies widely around the world. It can also change significantly within a country or world region over time. For more accurate measures of fertility than the birth rate, the **fertility rate** and the **total fertility rate** are used. The crucial factor in fertility is the percentage of young women of reproductive age. Another important measure is **replacement level fertility**.

» The fertility rate is the number of live births per 1000 women aged 15–44 years in a given year.
» The total fertility rate (TFR) is the average number of children born to a woman in a country during their lifetime.
» Replacement level fertility is the level at which each generation has just enough children to replace themselves in a population. A TFR of 2.12 is generally considered to be replacement level fertility.

Table 6.5 shows how much the total fertility rate varies by world region. In terms of individual countries, the total fertility rate (mid-2023) varied from a high of 6.7 in Niger and 6.4 in the Central African Republic, to a low of 0.8 in South Korea and 0.7 in China, Macau SAR and Hong Kong SAR.

▼ **Table 6.5** The total fertility rate by world region, 2023

Region	Total fertility rate
World	2.2
Africa	4.3
Asia	1.9
Latin America/Caribbean	1.9
North America	1.6
Oceania	2.1
Europe	1.4

The factors affecting levels of fertility

The factors affecting fertility can be grouped into four main categories (Table 6.6).

» The most important factor is the number/proportion of women in a population in the reproduction age range, particularly in the younger part of this range.
» Education can be a major influence on fertility (Figures 6.7 and 6.8). Education can provide increased knowledge of contraception and family planning, greater social awareness, more opportunity for employment and a wider choice of action generally. There is a very strong correlation between the increasing proportion of girls enrolled in secondary education and falling rates of fertility (Figure 6.9).
» Lower population growth is genrally associated with higher economic growth. Economic growth allows greater spending on health, housing, nutrition and education, which is important in lowering mortality and in turn reducing fertility.

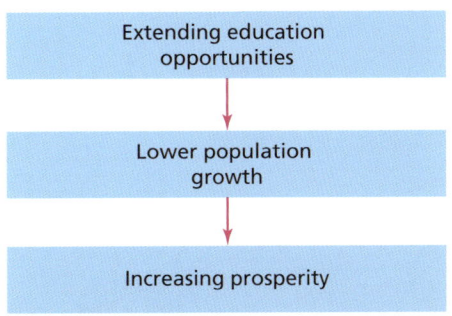

▲ Figure 6.7 Education and development

▲ Figure 6.8 A school in Indonesia – this country sees education as essential for its future development

▼ Table 6.6 The factors affecting levels of fertility

Demographic	<ul><li>Female age is the most important factor affecting fertility.</li><li>When infant and child mortality rates are high, parents often have more children to compensate for these expected deaths.</li><li>The average age of marriage impacts levels of fertility (Figure 6.10).</li></ul>
Social/cultural	<ul><li>In some societies tradition demands high rates of reproduction.</li><li>Education, especially female literacy, is a key factor in lowering fertility.</li><li>In some countries religion is an important, although a generally declining, influence on fertility in terms of being comfortable with artificial contraception.</li><li>In most countries, fertility rates are higher in rural areas than in urban areas.</li><li>In countries/regions where extended family networks are close by, grandparents are much more likely to help with childcare.</li><li>In most countries, population policies have aimed to reduce fertility by investing in birth control programmes.</li><li>Lifestyle factors (smoking, drinking alcohol, obesity, etc.) can influence reproductive health.</li></ul>
Economic	<ul><li>In rural areas of some low-income countries, children are seen as an economic asset because of the work they do and also because of the support they are expected to give their parents in old age.</li><li>In high-income countries the general perception is reversed, with the cost of the child-dependency years a major factor in the decision to begin or extend a family.</li><li>The lack of affordable housing and other 'cost of living' issues have been cited as a major reason for low fertility rates in recent years in a number of countries.</li></ul>
Political	<ul><li>At various times in the past individual countries have encouraged larger families, regarding a higher population as important for national status and power.</li><li>The duration of internal and international conflicts can also have a huge impact on fertility.</li></ul>

6.1 POPULATIONS GROW AND DECLINE

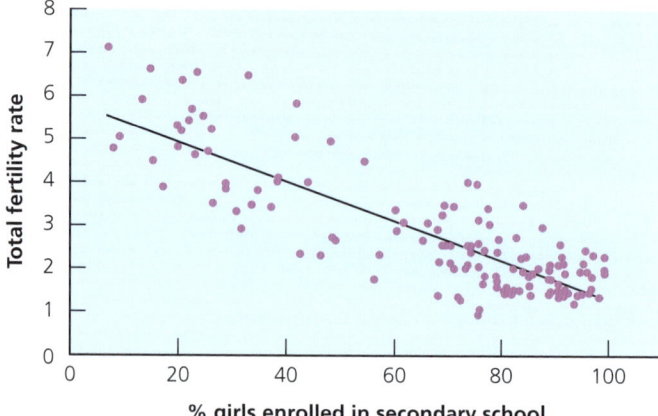

▲ **Figure 6.9** An international comparison between female secondary school enrolment and total fertility rates

▲ **Figure 6.10** A wedding party in Tokyo, Japan – the Japanese government is trying to encourage higher fertility

Fertility decline

The global peak population has been continually revised downwards in recent decades. This is in sharp contrast to warnings in the 1960s and 1970s of a 'population explosion'. The main reason for this slowdown in population growth is that fertility levels in all parts of the world have fallen faster than was previously expected. In the second half of the 1960s, after a quarter-century of increasing growth, the rate of population growth began to slow down. Since then, many LICs and MICs have seen the quickest falls in fertility ever known and thus the earlier population projections did not materialise. Between 1970 and 2020, fertility declined in every country in the world:

» Globally, fertility has decreased from an average of five births per woman in 1950 to 2.2 births per women in 2023.
» Global fertility is projected to fall to 2.1 births per woman by 2050.

» With global fertility decline, a growing number of countries have reached or fallen below replacement level fertility. In 2023, 103 countries and territories had total fertility rates below 2.1.

The movement to below replacement level fertility is undoubtedly one of the most dramatic social changes in history.

> ### Activities
> 1 Define the total fertility rate.
> 2 Describe the differences in the total fertility rate by world region.
> 3 Study Table 6.6 (page 129). Discuss the influence on fertility of one factor from each of the four categories.
> 4 What is the evidence that global fertility is in decline?

The impacts of pro-natalist and anti-natalist policies on birth rates

In countries where governments have developed policies relating to fertility, these policies may be:

» **pro-natalist** – encouraging higher fertility
» **anti-natalist** – encouraging lower fertility.

Since the establishment of modern population policies from the 1950s, most countries that have tried to control fertility have sought to lower it. It began in earnest with India in the early 1950s. Because its family planning programme was perceived to be working, it was not long before many other developing countries followed India's policy of government investment to reduce fertility. The most extreme anti-natalist policy ever introduced commenced in China in 1979. This was known as the one-child policy.

As fertility has fallen rapidly in the vast majority of countries, it may seem obvious that family planning policies have been successful. However, a counter-argument is that, as countries develop with improvements in female literacy and other development indicators, fertility decline is at least partly due to the personal choices made by large numbers of women. With economic development, the 'real cost' of raising children becomes more obvious and the benefits of directing income in other directions become more attractive for many people. A reasonable assertion is that both factors have proceeded hand in hand and that the government policies have speeded up a process that would have taken maybe decades more to reach where it is now.

With the continuous fall in global fertility, in recent decades an increasing number of countries have sought to slow or halt fertility decline. Their concerns have centred round:

- the socioeconomic implications of their ageing populations
- the long-term prospect of absolute population decline.

Pro-natalist countries include Japan, Canada, South Korea, Singapore and many countries in Europe. France's relatively high fertility in European terms can be partly explained by its long-term active family policy, adapted in the 1980s to accommodate the entry of women into the labour force. France has taken steps to encourage fertility on a number of occasions over the last 70 years. Measures to encourage couples to have two or more children include:

- longer maternity and paternity leave
- higher child benefits
- preferential treatment in the allocation of government housing.

Overall, France has tried to reduce the economic cost to parents of having children. French economists argue that although higher fertility means more expenditure on child-care facilities and education, in the longer term it gives the country a more sustainable age structure.

In recent decades, the Chinese government has become increasingly concerned about the adverse effects of its one-child policy. These include:

- demographic ageing
- an unbalanced sex ratio
- a generation of 'spoiled' only children
- a social divide as an increasing number of wealthy people 'bought their way round' the legislation.

With growing concern about these issues, the Chinese government relaxed the rules from 2016 to allow couples to have two children. In 2021, China again amended its Population and Family Planning Law to permit couples to have three children. However, the recent efforts by the Chinese government to encourage higher fertility seem to have failed. The birth rate, which was 12.2/1000 in 2016, has declined year-on-year, to 10.9/1000 in 2023. This seems to be due to a combination of two main factors:

- New established social norms initiated by the one-child policy.
- The conditions that many Chinese people of reproductive age see as being adverse to starting or extending a family, such as the perceived high cost of raising children in cities.

Attempting to reverse declining fertility is not easy and requires careful planning and well-targeted expenditure. A number of studies into the issue of low fertility have stressed the importance of removing the obstacles to marriage and facilitating access to decent housing. The financial concerns of young people have been an important factor in the rising average age at marriage and the increasing age of parents at the birth of first and subsequent children.

> ### Activities
> 1. Explain the difference between pro-natalist and anti-natalist policies.
> 2. How has France tried to encourage a higher birth rate?
> 3. Why did China change from an anti-natalist policy to a pro-natalist policy?

The demographic transition model

Demographic transition is the change from a mainly rural, agricultural society with high fertility and mortality rates, to a largely urbanised society with much lower fertility and mortality rates. The **demographic transition model** (Figure 6.11) helps explain how and why the characteristics of a country's population change over time. A model is a simplification of reality, but it helps us understand the most important aspects of a process.

6.1 POPULATIONS GROW AND DECLINE

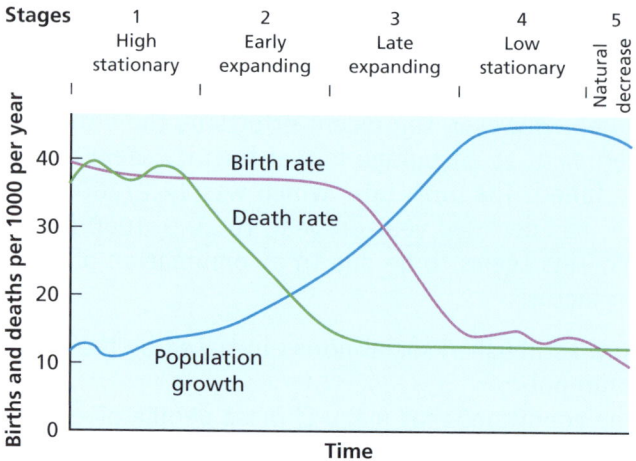

▲ Figure 6.11 The demographic transition model

» **The high stationary stage (stage 1):** The birth rate is high and stable, while the death rate is high and fluctuating due to the sporadic incidence of famine, disease and war. Population growth is very slow and there may be periods of considerable decline. Infant mortality is high and life expectancy is low (Figure 6.12). A high proportion of the population is under the age of 15. Society is pre-industrial, with most people living in rural areas and dependent on subsistence agriculture.

▲ Figure 6.12 A graveyard dating from the eighteenth century in the UK – inscriptions show that life expectancy at that time was very low

» **The early expanding stage (stage 2):** The death rate declines significantly. The birth rate remains at its previous level as **social norms** governing fertility take time to change. As the gap between the birth and death rates widens, the rate of natural change increases to a peak at the end of this stage. Infant mortality falls and life expectancy increases.

The proportion of the population under 15 increases. The decline in mortality is mainly due to better nutrition, improved public health (particularly in terms of clean water supply and efficient sewage systems) and medical advance. Considerable rural-to-urban migration occurs during this stage.

» **The late expanding stage (stage 3):** After a period of time, social norms adjust to the lower level of mortality and the birth rate begins to decline. The birth rate declines due to:
- better education about, and improved access to, contraception
- declining infant mortality due to advances in healthcare, nutrition and public health
- increasing gender equality and employment opportunities for women
- changing cultural expectations about family size.

Urbanisation generally slows and average age increases. **Life expectancy at birth** continues to increase and infant mortality to decrease. Countries in this stage usually experience lower death rates than nations in the final stage due to their relatively young population structures (Figure 6.13).

▲ Figure 6.13 Children helping their parents milk goats in the Gobi region of southern Mongolia

» **The low stationary stage (stage 4):** Both birth and death rates are low. The former is generally slightly higher, fluctuating somewhat due to changing economic conditions. Population growth is slow. Death rates rise slightly as the average age of the population increases. However, life expectancy still improves as age-specific mortality rates continue to fall.

- The natural decrease stage (stage 5): In an increasing number of countries the birth rate has fallen below the death rate, which results in natural decrease. In the absence of net migration inflows, these populations decline.

No country as a whole retains the characteristics of stage 1. This stage only applies to the most remote societies on Earth, such as isolated tribes in New Guinea and the Amazon basin who have little or no contact with the outside world. LICs tend to be in stage 2, with some moving into stage 3. Most MICs are in stage 3, with some in stage 4. All HICs are now in stages 4 or 5. Stage 5, natural decrease, is mainly confined to Eastern and Southern Europe and parts of East Asia. Examples include Italy, Germany, Bulgaria, Belarus, Ukraine, Japan and South Korea.

Criticisms of the demographic transition model

Critics of the model see it as too Eurocentric, as it was based on the experience of Western Europe. They argue that many developing countries may not follow the sequence set out in the model. It has also been criticised for its failure to take into account changes due to migration.

The model presumes that all countries will eventually pass through all stages of the transition, just as most HICs have done. Because these countries have achieved economic success and enjoy generally high standards of living, completion of the demographic transition has come to be associated with socioeconomic progress, which many LICs in particular might struggle to achieve.

There are a number of important differences in the way that developing countries have undergone population change compared to the experiences of most developed countries before them:

- In developing countries, birth rates in stages 1 and 2 were generally higher than they were in developed countries when they were in these early stages.
- The death rate in developing countries fell much more steeply and for different reasons. For example, the rapid introduction of modern medicine, particularly in the form of inoculation against major diseases, has had a huge impact on lowering mortality.
- Some developing countries had much larger base populations, thus the impact of high growth in stage 2 and the early part of stage 3 has been far greater in terms of absolute population growth.
- For those developing countries in stage 3, the fall in fertility has also been steeper. This has been due mainly to the relatively widespread availability of reliable modern contraception.
- The relationship between population change and economic development has been much more tenuous in some developing countries.

Activities

1. What is a geographical model (such as the demographic transition model)?
2. Explain the reasons for falling mortality in stage 2 of demographic transition.
3. Why does it take some time before the fall in fertility follows the fall in mortality?
4. What are the characteristics of stage 5?
5. Discuss the merits and limitations of the demographic transition model.

6.2 Population structures change over time

This chapter will explain:

★ the factors influencing population structures
★ the issues of youthful populations
★ the issues of demographic ageing
★ detailed specific example: renewed concerns about population size in India in the twenty-first century.

The factors influencing population structures

The most studied aspects of **population structure** are age and sex. Other aspects of population structure that can also be studied include race, language, religion and social/occupational group. The two fundamental influences on population structure are natural increase and net migration. These factors have already been examined in detail in Section 6.1 (page 127).

Age and sex structure are usually illustrated by the use of **population pyramids** (Figure 6.16). Pyramids can be used to show either absolute or relative data. Absolute data shows the figures in thousands or millions, while relative data shows the numbers in percentages. The latter is most frequently used as it allows for easier comparison of countries with different population sizes. Each bar represents a five-year age group. The male population is represented to the left of the vertical axis, and females to the right. Figure 6.15 provides some useful tips for understanding population pyramids.

▲ Figure 6.14 Elderly people are making up an increasing proportion of the population in many countries

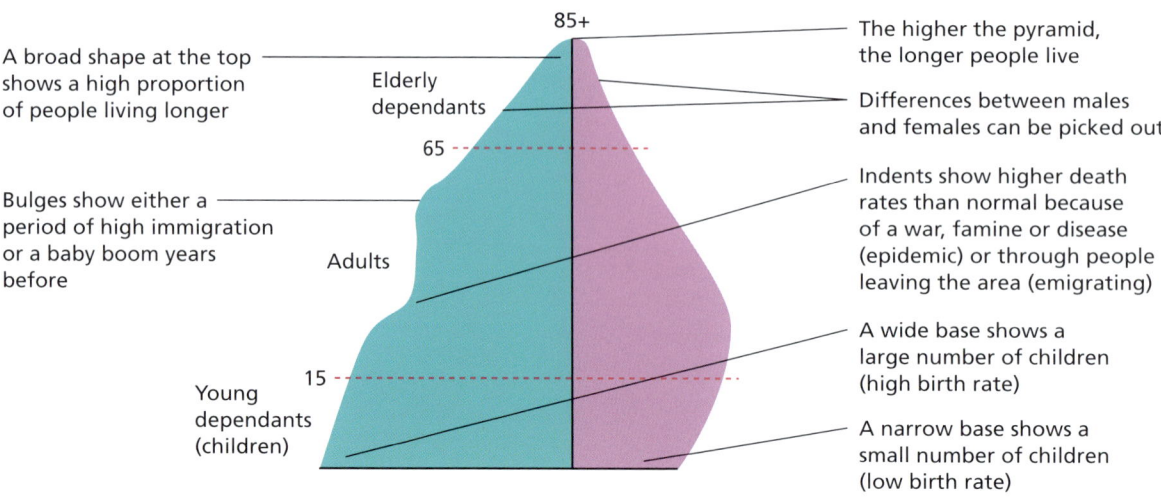

▲ Figure 6.15 An annotated population pyramid

The factors influencing population structures

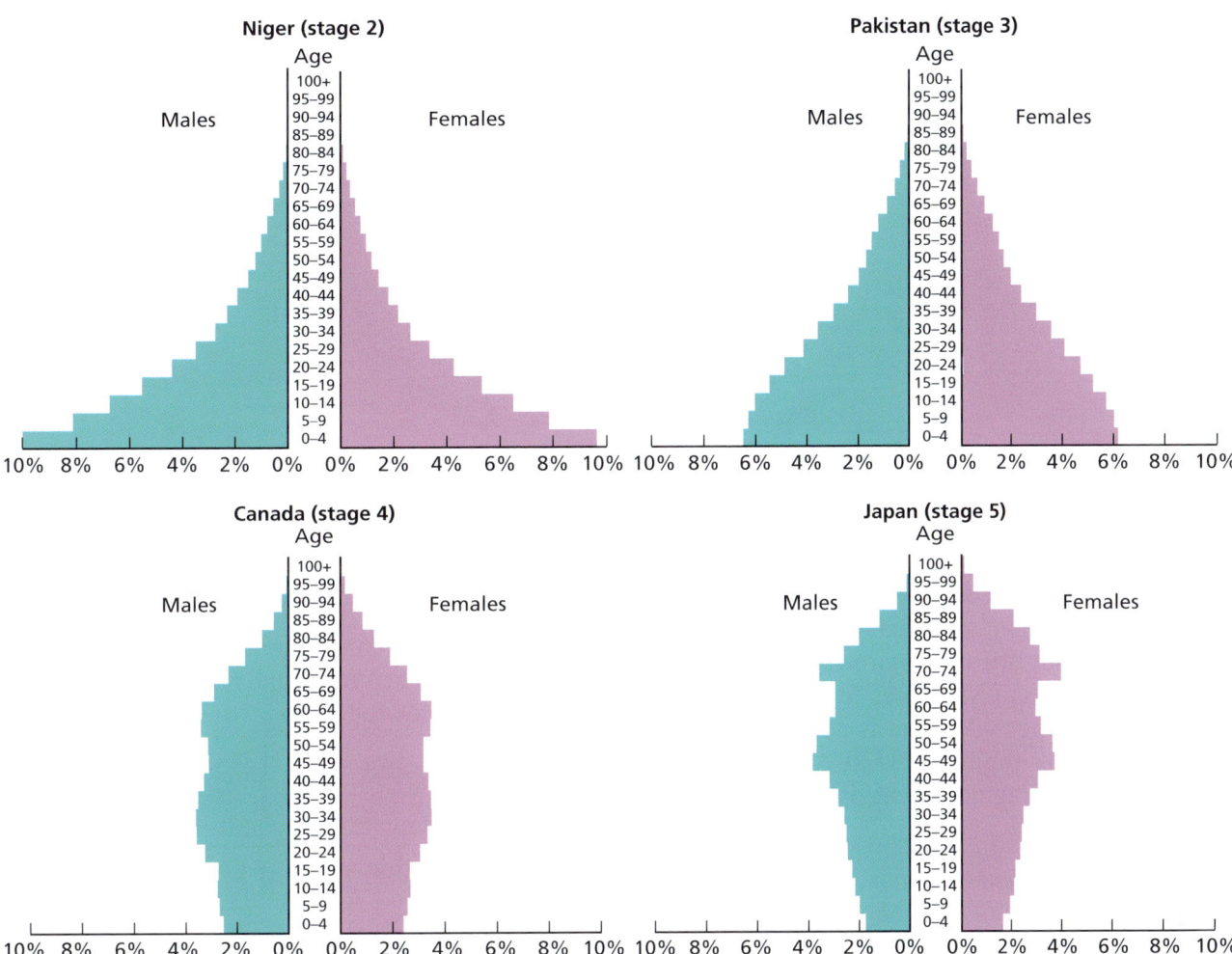

▲ **Figure 6.16** Four population pyramids, 2022

Population pyramids change significantly in shape as a country progresses through demographic transition. The different timing and pace of fertility decline around the world has produced distinct contrasts in age profiles. The four pyramids in Figure 6.16 relate to the demographic transition model, with Niger in stage 2, Pakistan in stage 3, Canada in stage 4 and Japan in stage 5.

» Niger (stage 2): The wide base reflects one of the highest birth rates in the world. The marked decrease in width of each successive bar indicates relatively high mortality and limited life expectancy. Youth dependency is extremely high, but aged dependency is very low.
» Pakistan (stage 3): The base of this pyramid is narrower than that of Niger, reflecting a considerable fall in the birth rate after decades of government-promoted birth control programmes. The percentage population under 15 years of age is significantly lower than that of Niger. The share of the population aged 65 and over is greater, indicating a higher life expectancy compared to Niger.
» Canada (stage 4): A lower birth rate compared to Pakistan is illustrated by a much narrower base to the pyramid. The greater uniformity of the pyramid with increasing age indicates a further decline in mortality and greater life expectancy compared to Pakistan. Aged dependency in Canada is considerably higher than in Pakistan and Niger.
» Japan (stage 5): The distinctly inverted base reflects the lowest birth rate of all four countries. The width of the rest of the pyramid is a consequence of the lowest mortality and highest life expectancy of all four countries. The birth rate is lower than the death rate.

6.2 POPULATION STRUCTURES CHANGE OVER TIME

▼ Table 6.7 GNI per capita and demographic data for Niger, Pakistan, Canada and Japan

Indicator	Niger	Pakistan	Canada	Japan
Birth rate	45 per 1000	28 per 1000	10 per 1000	7 per 1000
Death rate	8 per 1000	7 per 1000	8 per 1000	12 per 1000
Natural change	3.7%	2.1%	0.2%	-0.5%
Infant mortality rate	40 per 1000	52 per 1000	4.5 per 1000	1.7 per 1000
Life expectancy	62	66	82	85
Population <15	49%	37%	16%	12%
Population 65+	2%	4%	19%	29%
GNI per capita (PPP)	$1330	$5800	$51,690	$44,570
Urban population	17%	37%	82%	92%

Table 6.7 compares these four countries for a range of important indicators. The difference in GNI per capita between Niger and Pakistan is consistent with the general difference between countries in stages 2 and 3 of demographic transition, although many countries in stage 3 have a higher GNI per capita than Pakistan. The much higher GNI figures for Canada and Japan are also to be expected. There is no significant difference in income per capita between countries in stages 4 and 5 of demographic transition.

The **dependency ratio** is an important measure used in the analysis of population pyramids. Dependants are viewed as people who are too young or too old to work. The dependency ratio is the relationship between the working or economically active population and the non-working population. The formula for calculating the dependency ratio is:

$$\text{Dependency ratio} = \frac{\text{\% population aged 0--14} + \text{\% population aged 65 and over}}{\text{\% population aged 15--64}} \times 100$$

A dependency ratio of 80 means that for every 100 people in the economically active population, there are 80 people dependent on them. In this example, the dependency ratio is related to the base number of 100. However, sometimes it is expressed in relation to the base number of 1. In this example, the dependency ratio would be 0.8.

A low dependency ratio is a significant economic advantage as there will be a large number of people in the economically active age range compared to a combination of the young dependent age group and the elderly dependent age group. In LICs, children form the great majority of the dependent population. In contrast, in HICs and MICs there is a more equal balance between young and elderly dependents.

The dependency ratio can be subdivided into the **youth dependency ratio** and the **elderly dependency ratio** as follows:

$$\text{Youth dependency ratio} = \frac{\text{\% population aged 0--14}}{\text{\% population aged 15--64}} \times 100$$

$$\text{Elderly dependency ratio} = \frac{\text{\% population aged 65 and over}}{\text{\% population aged 15--64}} \times 100$$

The dependency ratio is important because the economically active population will in general contribute more to the economy in income tax, value-added tax (VAT) and corporation tax (tax on a business). In contrast, the dependent population tend to be bigger recipients of government funding, particularly for education, healthcare and public pensions. An increase in the dependency ratio can cause significant financial problems for governments if they do not have the financial reserves to cope with such a change. Net migration can have a significant influence on the dependency ratio of a country or region. For example, a high rate of immigration of young adults and children into a HIC may:

» reduce the decline in the population under 15
» increase the number of people in the economically active population.

The **demographic dividend** is a term used by economists and others to refer to the process through which the changing age structure of a country can assist its economic growth. The general sequence leading to the demographic dividend begins with a considerable fall in fertility in a developing country that previously had a high fertility rate. This leads to a reduction in youth dependency and an increase in the proportion of the population who are of working age. In a stable economic environment, living standards should improve in terms of income per capita. Economists sometimes refer to the demographic dividend as a 'window of opportunity' for economic growth.

The highest demographic dividend will be in Africa, where the economically active population will increase the most. Although economic growth is due to a range of factors, the concept of the demographic dividend accounts for much of the variation in the past economic performances of different countries.

Population structure: differences within countries

In countries where there is strong rural-to-urban migration, the population structures of the areas affected can be significantly changed. These differences show up clearly on population pyramids. Out-migration from rural areas is age-selective, with single young adults and young adults with children dominating this process. Thus, population pyramids for rural areas affected by out-migration will indicate fewer people than expected in these age groups.

In contrast, population pyramids for urban areas that attract migrants will show age-selective in-migration, with substantially more young adults than expected. Such migrations may also be sex-selective. If this is the case, it should be apparent on the population pyramids.

Sex structure

The sex ratio is the number of males per 100 females in a population. The 'natural' sex ratio at birth is around 105 boys per 100 girls. Male births consistently exceed female births due to a combination of biological, social, cultural, economic and technological factors. For example:

» In many countries, more couples decide to complete their family on the birth of a boy than on the birth of a girl.

» Gender selection through sex determination is more common in some countries than others.

After birth, the gender gap generally begins to narrow until eventually females outnumber males. This is because male mortality is higher at every age than female mortality, and because women tend to live longer than men. The narrowing of the gender gap happens most rapidly in the lowest-income countries, where infant mortality is markedly higher among males than females. Here, the gap may be closed in less than a year.

In 2020, the proportion of women in the world's population was just under 50 per cent, with most countries having a female share between 49 and 51 per cent. However, there are some notable outliers:

» In South and East Asia, particularly China and India, there are significantly more males than females, due mainly to large differences in the sex ratio at birth.
» Several countries in the Middle East have a large 'excess' of males. In the UAE it is almost four to one, and in Oman it is about three to one. In both countries, the main reason is a large male migrant population.
» In many Eastern European countries there are far more females than males because of large differences in life expectancy.

> ### Activities
> 1 Study Figure 6.16 (page 135). Compare the population structures of Niger and Japan, making appropriate reference to youth dependency and elderly dependency.
> 2 What do you understand by the term 'demographic dividend'?
> 3 a Define the term 'sex ratio'.
>   b State two ways in which the 'natural' sex ratio can be distorted.

The issues of youthful populations

Different patterns of population growth can bring both benefits and problems to the countries concerned. This is particularly the case when countries have a very high percentage of either young people or old people in their populations.

Rapid population growth results in a large young dependent population. The global average proportion

6.2 POPULATION STRUCTURES CHANGE OVER TIME

of the population under the age of 15 is 25 per cent. Table 6.8 shows the significant variation by world region around this figure.

▼ **Table 6.8** Variation by world region for the percentage of the population under the age of 15 and aged 65 and over

Region	Percentage of population under 15	Percentage of population 65 and over
World	25	10
High-income countries	16	19
Middle-income countries	25	9
Low-income countries	42	3

The 2023 World Population Data Sheet shows:

- LICs have more than two and a half times the population under 15 that HICs have.
- Twenty-nine countries had 40 per cent or more of their populations under 15. All of these countries were in Africa, with the one exception of Afghanistan.

Countries with large youthful populations have to allocate considerable resources to look after them. Young people require resources for health, education, food, water and housing above all. The money required to cover such needs may mean there is little left to invest in agriculture, industry and other aspects of the economy. A national government may see this as too large a demand on the country's resources and as a result may introduce family planning policies to reduce the birth rate. However, individual parents may have a different view, in that they see a large family as valuable in terms of the work children can do on the land. Alongside this, people in lower-income countries often have to rely on their children in old age because of the lack of state welfare benefits.

As a large young population moves up the age ladder, it will become a large working population when it enters the economically active age group. This will become an advantage if a country can attract investment and create jobs. Such a situation can create an upward spiral of economic growth. However, without investment and job creation, the unemployment rate will be high. This may cause many young adults to emigrate because of the lack of opportunities in their own country. Eventually the large number of people in this age group will reach old age, increasing elderly dependency.

The Gambia is a small LIC in Western Africa. The 2023 World Population Data Sheet gave its GNI per capita as US$ 2470. Although the country's total fertility rate fell from 6.1 in 1970 to 4.4 in 2023, the rate of natural increase is 27/1000. Forty-three per cent of the population are under 15, with only 2 per cent aged 65 years and over. The country's family planning programme has had only limited success. The Gambia's young and fast-growing population is placing big demands on its resources.

- Many parents in The Gambia struggle to provide basic housing for their families. There is huge overcrowding and a lack of sanitation, with many children sharing the same bed.
- Rates of unemployment and underemployment are high and wages are low.
- Many schools (Figure 6.17) operate a two-shift system, with one group of students attending in the morning and a different group attending in the afternoon.
- Another sign of population pressure is the large number of trees being chopped down for firewood. As a result, desertification is increasing at a rapid rate.

▲ **Figure 6.17** A school in The Gambia

The issues of demographic ageing

Ageing of population (demographic ageing) is now regarded by many experts as the most formidable population challenge facing the world. The current rate of population ageing is unprecedented in human history. Never before have so many people, in both absolute and relative terms, reached age 65 and over. Global life expectancy increased from 34 years in 1913 to 72 years in 2022, and is expected to continue on this long-term trajectory.

The issues of demographic ageing

Demographic ageing is a rise in the median age of a population. It occurs when:

» fertility declines
» life expectancy increases
» larger cohorts of a population progress to older age groups.

Table 6.8 (page 138) shows how the income regions of the world compare to the global average of 10 per cent of the population aged 65 years or over.

» HICs have more than six times as many of the population aged 65 and over as LICs have.
» In terms of individual countries, in 2023 Monaco topped the global league table both with a median age of 55.4 years and with 36 per cent of its population aged 65 years and over.
» For countries of a more substantial size Japan is the 'oldest', with a median age of 48.6 years and 29 per cent of its population aged 65 years and over.
» In contrast, a group of nine countries had a low of just 2 per cent of their populations aged 65 years and over in 2023 – UAE, Qatar, Chad, Burundi, Uganda, Zambia, The Gambia, Mali and Niger.

Demographic ageing puts healthcare systems and public pensions, and indeed government budgets in general, under increasing pressure. The fastest-growing segment of the population is the so called 'oldest-old': those who are 85 years old or more. It is this age group who are most likely to need expensive residential care.

Some countries have made relatively good pension provision by investing wisely over a long period of time. However, others have more or less adopted a pay-as-you-go system as the elderly dependent population rises. It is this latter group of countries that will be faced with the biggest problems in the future. At present, very few countries are generous in looking after their elderly. Poverty among the elderly is a considerable problem, but technological advances might provide a solution by improving living standards for everyone. If not, other less popular solutions, such as increased taxation, will have to be examined. Figures for the European Union (Figure 6.18) show:

» a decline in the working age population from 2020; by 2070 it will be approximately 85 per cent of the size it was in 2020
» a decline of the total population from about 2040
» an increase in the old-age dependency ratio (right-hand scale).

However, some demographers argue that there needs to be a certain rethinking of age and ageing, with older people adopting healthier and more adventurous lifestyles than people of the same age only a few generations ago. Sayings such as '60 is the new 50' have become fairly commonplace. A few years ago the United Nations established the UN Decade of Healthy

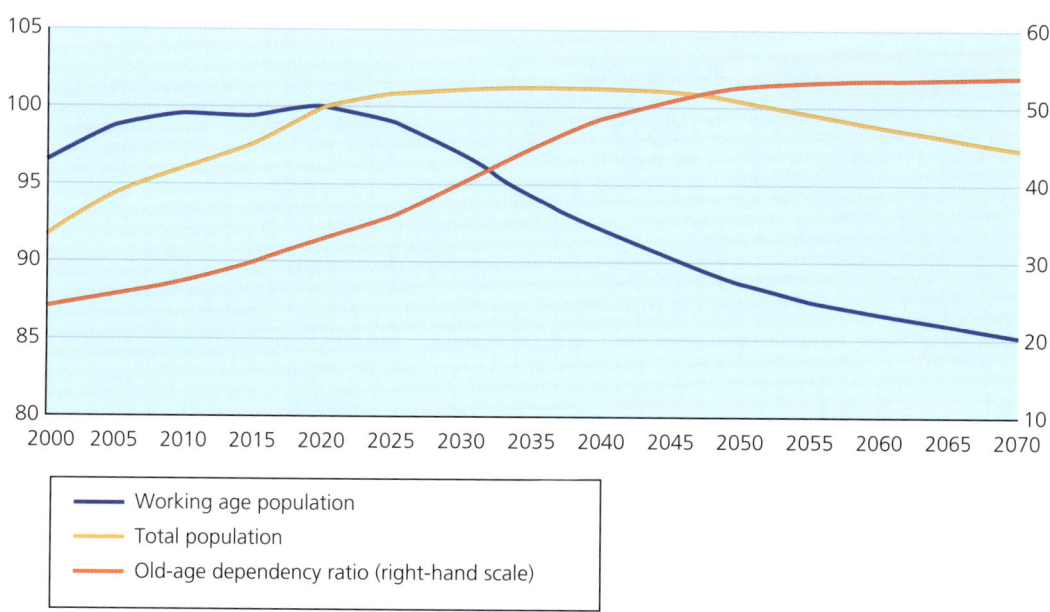

▲ **Figure 6.18** Working age population, total population and old-age dependency ratio for the EU, 2000–70 (left-hand scale: indices, 2020 = 100; right-hand scale: old-age dependency ratio as percentages of the 15–64 age group)

6.2 POPULATION STRUCTURES CHANGE OVER TIME

Ageing (2021–30). A major theme of this initiative is **'demographic preparedness'** – the idea that there is time to enact policies and encourage behaviours to limit the potential adverse effects of the demographic changes that do occur. An important behavioural change centres on increasing physical activity in all age groups, so that people are not only healthier in the earlier years of life, but also enter their advanced years in better physical and mental condition.

It is argued that we should think not just of chronological age, but also of prospective age – the remaining years of life expectancy people have. It is easy to underestimate the positive aspects of ageing, with many older people making a big contribution to childcare by looking after their grandchildren, while a significant number of older people work as volunteers in numerous capacities.

➡ Detailed specific example

Renewed concerns about population size in India in the twenty-first century

In 1950, India's population had increased to over 350 million. This was about 100 million more than the population had been in 1920, only thirty years earlier. There was a general perception in the Indian government that the country was overpopulated and that this was hindering economic development. Rural and urban birth control clinics rapidly increased in number. Two years later, India became the first country to introduce a policy designed to reduce fertility and to aid development, with a government-backed family planning programme.

- Financial and other incentives were offered in some states for those participating in programmes, especially sterilisation.
- Abortion was legalised in 1972.
- In 1978, the minimum age of marriage was increased to 18 years for females and 21 years for males.

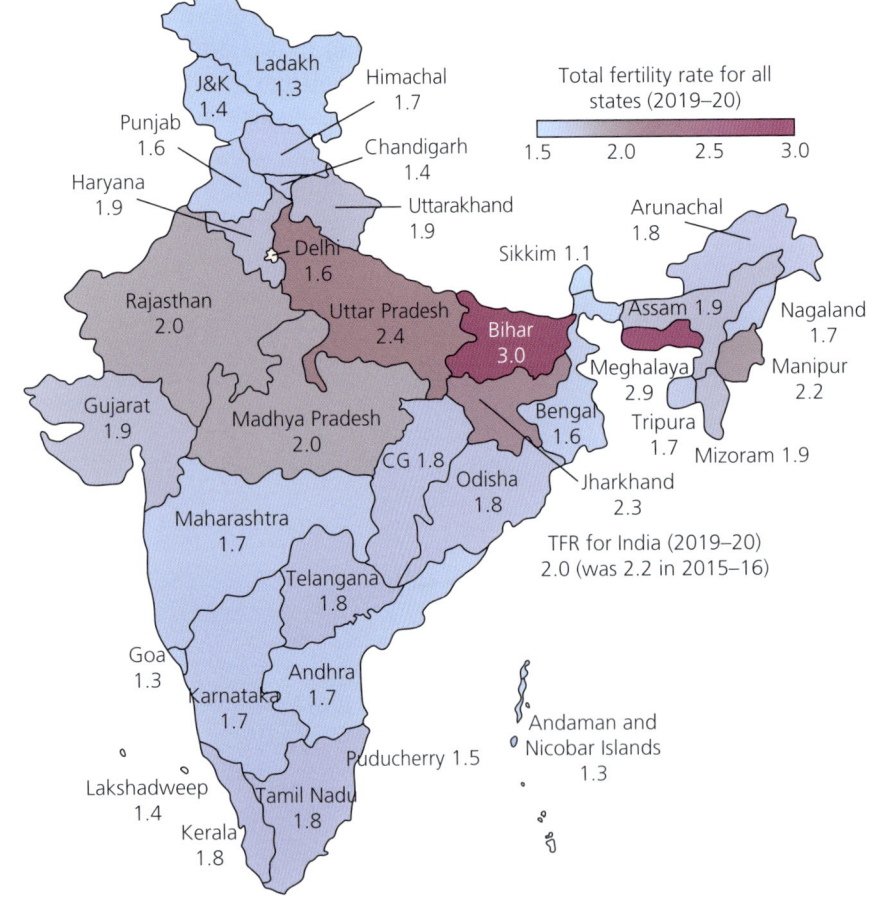

▲ **Figure 6.19** Total fertility rate by state in India, 2019–20

The issues of demographic ageing

In the mid-1970s the sterilisation campaign became increasingly coercive, reaching a peak of 8.3 million operations a year. International condemnation of such a coercive approach led India to return to its former approach of providing reproductive health and family planning services in an attempt to establish a stable population. Under India's federal political structure, state governments increasingly set their own priorities with regard to population policy.

India's population growth rate began to decline from 1981 and by 2020 it reached replacement level fertility, a significant point in the country's demographic transition. However, the national average figure covers wide regional differences (Figure 6.19). Table 6.9 shows the decline in India's birth rate from 1960 to 2023.

▼ **Table 6.9** India's falling birth rate, 1960–2023

Year	Total population (million)	Birth rate (per 1000)
1960	451	42.5
1970	555	39.5
1980	699	36.2
1990	873	31.8
2000	1057	27.0
2010	1234	21.4
2023	1429	18.0

▲ **Figure 6.20** Substandard housing in a small town in northern India

In very recent years, there has been a renewed high level of debate about the size of India's population. In 2019, the country's Prime Minister Narendra Modi stated that the large size of India's population was hindering its development, stating 'There is a need to have greater discussion and awareness on population explosion'. Some leading politicians have argued for the introduction of a national two-child policy, but opponents view such a policy as a return to coercion.

India now has the world's largest population at 1.43 billion (2023). Under the UN's 'medium variant' projection, its population will surpass 1.5 billion by the end of the 2020s. India's population will continue to grow slowly until it peaks at 1.7 billion in 2064. The country has a working age population of about 500 million and it is still expanding. Some observers see this situation as a demographic dividend, but others see it as potentially an unsustainable drain on the country's resources. Greater investment in education and training will be vital for a positive outcome. Despite declining fertility, population momentum will see the total population increase for another four decades before it stabilises. The median age in India is 28 years, compared to 39 years in China.

An important body of opinion stresses the need to focus on the high-fertility regions of the country (Figure 6.19). The Indian government has identified 146 high-fertility districts. Most are in the northern states of Bihar, Rajasthan, Jharkhand, Madhya Pradesh and Uttar Pradesh. All these states have well below average development scores in education and health. New initiatives have focused on increasing supplies of contraceptives and on family planning campaigns. Bihar, with a 2021 population of 128 million, is the most densely populated state in India. The state's population grew by over 18 per cent between 2011 and 2020, the highest growth in India. The fertility rate has resulted in a high young dependent population. Bihar has lagged behind India in general in socioeconomic development. A considerable proportion of the population live below the poverty line.

In March 2023, the Indian government stated that the minimum age of marriage for women would increase from 18 years of age to 21, putting it on a par with the age for men. At the time of writing, the proposal was progressing through parliament. The new legislation will come into effect two years after it is agreed. A major Indian media organisation has termed the forthcoming policy 'a momentous decision'.

Southern states such as Kerala and Tamil Nadu have adopted an approach to fertility more related to overall development. As levels of education, health and GDP have improved, fertility rates have declined significantly. Some academics have stressed the importance of India utilising its **gender dividend**. This is the increase in economic growth that can result from greater investment in women and girls, including promoting secondary and tertiary education and female workforce participation. According to the World Bank, just 23 per cent of Indian women perform paid work, compared to 63 per cent in China.

▶ **Activities**

1 Study Table 6.9. Discuss the reasons behind the steady decline of India's birth rate.
2 Why have some academics stressed the importance of India utilising its gender dividend?

6.3 The causes and impacts of international migration

This chapter will explain:

★ the types of migrant
★ the causes of migration
★ the impacts of migration on countries of origin and destination and on migrants themselves
★ the different approaches to managing international migration
★ detailed specific example: international migration from Mexico to the USA.

International **migration** has been a major process in shaping the world as it is today. Its impact has been economic, social, cultural, political and environmental. Few people now go through life without changing residence several times. How many people in your class have lived in different parts of your country and/or in other countries?

remained relatively stable as a proportion of the world's population.

> Migration is a complex issue. As such, it is one that can be exacerbated by misinformation and politicisation to alarming degrees.
>
> World Migration Report, 2022

▲ **Figure 6.21** The border crossing between Uzbekistan and Turkmenistan in Central Asia

Migration is defined as the movement of people across a specified boundary, national or international, to establish a new permanent place of residence. The United Nations defines 'permanent' as meaning a change of residence lasting more than one year. Movements with a timescale of less than one year are termed 'circulatory movements'.

Figure 6.22 shows the cumulative migrant stock of each country, using data from the UN Department of Economic and Social Affairs. Currently, about 3.6 per cent of the world's population is living outside the country of their birth. This amounted to almost 281 million people in 2020. The international migrant population globally has increased in size, but it has

The types of migrants

International migrants can be grouped into the following broad categories:

» Economic migrants: People who migrate with a view to being employed.
» **Refugees**: People who have been forced to flee their homes because of war, violence or persecution. People who obtain refugee status are given protections under international laws and conventions.
» Asylum seekers: People who are seeking international protection from dangers in their home country, but whose claim for refugee status has yet to be determined.

In the latter part of the twentieth century and the beginning of the twenty-first century, some of the world's most violent and protracted conflicts have been in the developing world, particularly in Africa and Asia. These troubles have led to numerous population movements of a significant scale. Not all have crossed international frontiers to merit the term refugee movements, but have instead involved **internal displacement**. This is a major global problem and is showing little sign of slowing down. Table 6.10 shows that a total of 89.4 million people were living in displacement worldwide at the end of 2020, up from 84.8 million the previous year.

The types of migrants

Total number of international immigrants, 2020

The total number of people living in a given country that were born in another country. This measures the cumulative migrant stock.

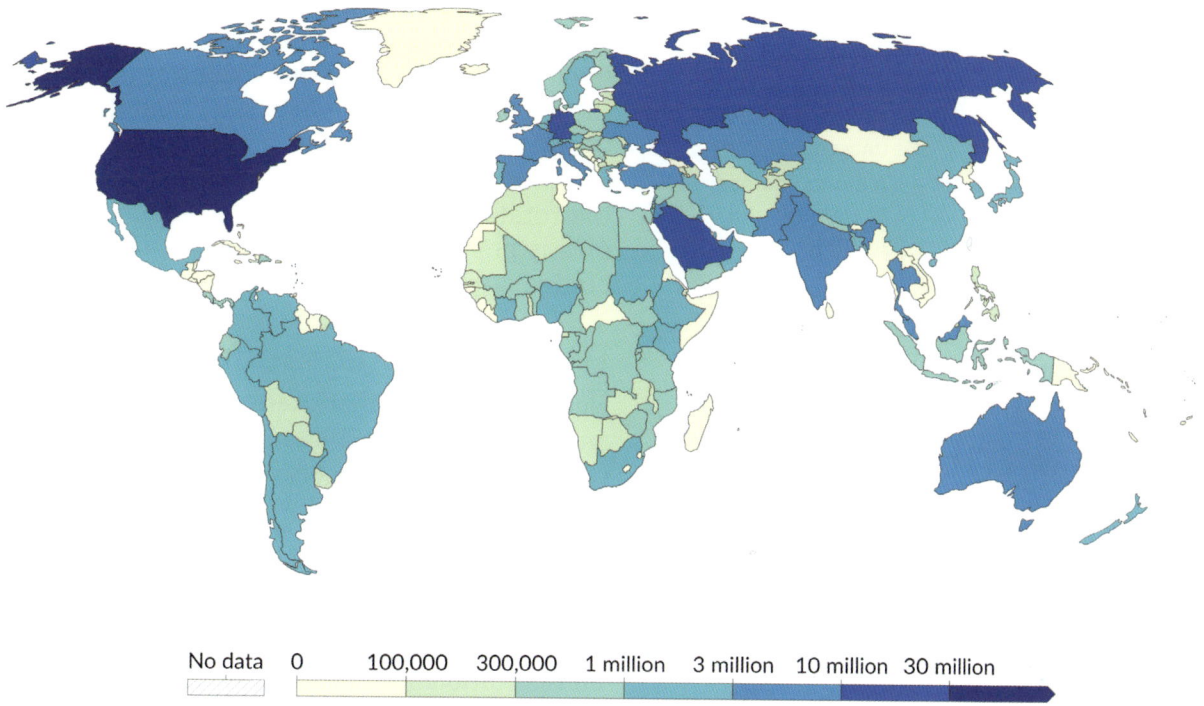

▲ **Figure 6.22** The total number of international immigrants by country, 2020

▼ **Table 6.10** The number of displaced persons globally, according to UN category

Displaced persons	Number in 2020
Refugees	26.4 million
Asylum seekers	4.1 million
Internally displaced persons (IDPs)	55 million

Refugees

By the end of 2020 the global refugee population was 26.4 million, the highest on record.

» The long, ongoing conflict in the Syrian Arab Republic accounted for 6.7 million of the global total of refugees. The impact of this conflict has been huge. Back in 2010, Syria was a **country of origin** for fewer than 30,000 refugees and asylum seekers, while it was also the third-largest host country (**country of destination**), with more than 1 million refugees, mainly from Iraq.

» As well as Syria, other major source countries for refugees were Afghanistan (2.6 million) and South Sudan (2.2 million). Refugees from these three countries, along with those from Myanmar and the Democratic Republic of the Congo, accounted for over half of the world's refugee population.

» Globally, 38 per cent of refugees were under 18 years of age.

» Turkey is the largest host country, with over 3.6 million refugees, mainly from Syria.

» Seventy-three per cent of refugees are hosted in neighbouring countries. For example, Lebanon hosts a large number of refugees from Syria.

Many refugees are housed at very high densities in huge refugee camps. Apart from immediate problems concerning overcrowding, sanitation and the disposal of waste, long-term environmental damage may result from deforestation and soil degradation. The high cost of refugee camps is financed by the host countries and by international charities.

6.3 THE CAUSES AND IMPACTS OF INTERNATIONAL MIGRATION

The World Development Report (2023) highlights the differences between the movements of economic migrants and refugees (Table 6.11).

▼ **Table 6.11** The differences between the movements of economic migrants and refugees

Factor	Economic migrants	Refugees
Age group	Primarily working age adults	A high proportion of children, some unaccompanied
Origin	A wide range of countries	A limited number of countries (increasingly so)
Destination and distance	Countries where they believe there is a demand for their skills, often regardless of distance	Prioritising safety and security over employment. Tend to move to safe countries bordering their country of origin
Pace	Most economic migrants spend a considerable time planning and preparing to move	Refugee movements are characterised by their suddenness and rapid pace

The causes of migration

Forced and voluntary migration

A basic distinction is between voluntary and forced migration (Figure 6.23). **Voluntary migration** is where the individual or household has a free choice about whether or not to move. **Forced migration** occurs when the individual or household has little or no choice but to move. This may be due to environmental or human factors.

Arguably the worst form of forced migration is human trafficking. This involves illegally transporting people from one country or area to another, usually for the purpose of forced labour (modern slavery) or sexual exploitation. Because of the very nature of this practice, high quality data is difficult to obtain, but an estimate published by the International Labour Organization in 2017 puts the total at about 25 million people. About a fifth of this population have crossed an international border.

Barriers to migration

Figure 6.23 shows that there are barriers to migration. In earlier times, the physical barriers (Figure 6.24) and dangers of the journey and the monetary costs involved were major obstacles. Examples of dangers include crossing seas and oceans or traversing deserts and mountain ranges, when the means of transport were very rudimentary compared to now. The relatively low real cost of modern transportation and the associated high level of safety have considerably reduced these barriers. However, physical geography can still prove to be a formidable barrier for undocumented migrants. Between 2014 and 2020 the Mediterranean Sea saw the highest number of deaths along a migration corridor, where over 21,000 people lost their lives.

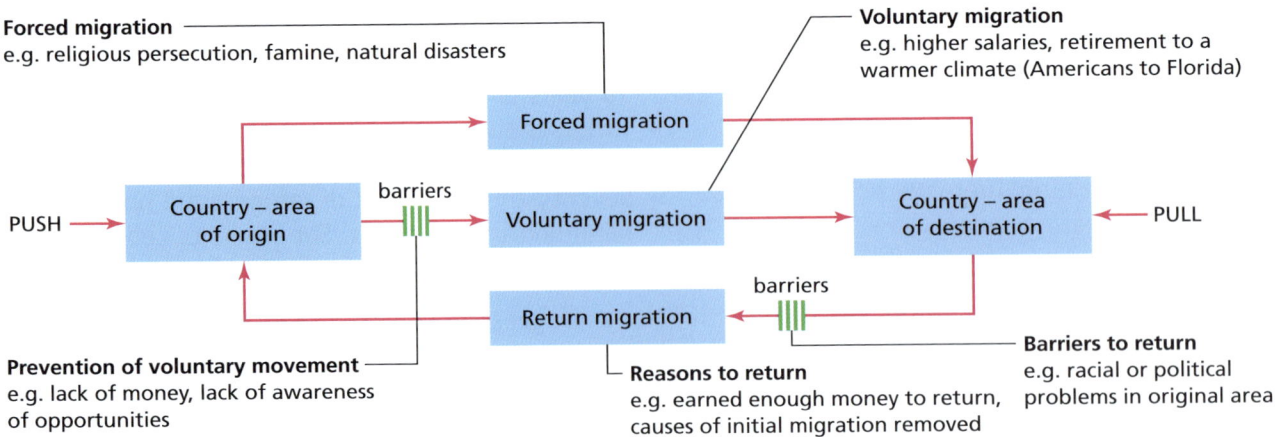

▲ **Figure 6.23** Voluntary and forced migration

The causes of migration

▲ **Figure 6.24** Iguacu Falls, Brazil: the physical environment is much less of a barrier to migration than it once was

In the modern world, the main barriers to international migration are the legal restrictions that countries place on migration. Most countries now have very strict rules on immigration, and some countries restrict emigration. Both North Korea and Taliban-controlled Afghanistan place strong restrictions on residents leaving these countries. Wherever they are going, a range of documentation is required for potential migrants to gain entry into another country. Obtaining a visa or a work permit can be a lengthy process.

Demographer E.S. Lee (1966) produced a series of principles of migration in an attempt to bring together all aspects of migration theory at that time. Of particular note was his origins, destination and intervening obstacles model, which emphasised the role of push and pull factors. Here he suggested four classes of factors that influence the decision to migrate:

1 Those associated with the place of origin.
2 Those associated with the place of destination.
3 Intervening obstacles that lie between the places of origin and destination.
4 A variety of personal factors that moderate factors 1, 2 and 3.

Lee stressed the point that the factors in favour of migration would generally have to considerably outweigh those against, because people are naturally reluctant to uproot themselves from established communities.

Push and pull factors

Push and pull factors encourage people to migrate. **Push factors** are the observations that are negative about the area in which the individual is presently living, while **pull factors** are the perceived better conditions in the place to which the migrant wishes to go. The greater the perceived difference between origin and destination, the more likely a person is to attempt to migrate. Push and pull factors can be divided into the categories shown in Table 6.12.

▼ **Table 6.12** Examples of push and pull factors

Factor group	Push	Pull
Economic	• Unemployment and underemployment • Low-paid and insecure jobs, often in the informal sector • Fragile economies susceptible to high inflation and other economic shocks	• Better employment prospects, with greater job security and higher wages • Employment more heavily weighted towards the formal sector • Higher overall standard of living
Social	• Poor housing conditions, with overcrowding and lack of adequate sanitation • Regions lacking electricity • Limited social systems – health, education, benefits, pensions and so on • Desire to find work abroad to provide for family members • Inadequate policing and justice systems, with high crime rates	• Better housing conditions, with connections to utility systems • More comprehensive social systems, providing greater security for migrants and their families • Family reunification – joining family members already at the destination • A high level of governance and personal security
Political	• Both international and internal conflicts (such as Myanmar or Syria) have caused many people to flee from conflict zones	• The hope of living in more stable political environments or where there is less corruption

6.3 THE CAUSES AND IMPACTS OF INTERNATIONAL MIGRATION

▼ Table 6.12 (Continued)

Factor group	Push	Pull
Environmental	• The insecurity of fragile environments (food, water and energy) • Regions prone to natural disasters • Some world regions are particularly susceptible to climate change • High levels of pollution (air, water, land) that increase a range of health problems	• Stricter environmental legislation to combat environmental problems • Higher environmental quality in both urban and rural areas • People retiring to warmer climates

Demographers view the push/pull discussion as a useful starting point, but as lacking in detailed explanation. For example:

» Research shows that an individual's social networks, which are neither traditional push nor pull, are a very strong influence on the decision-making of potential migrants.
» Migration decisions for an individual are often made collectively, at the household level.
» Demographers now prefer the notion of 'determinants', with a measurable causal influence. This is the objective of trying to be as specific as possible in identifying the causes of migration, for example employment opportunities in the ICT sector in a destination country, and quantifying the number of migrants involved if credible data is available.

Economic migration

Much international migration is economic – that is, people seeking employment in other countries. The rise in labour-related migration has included both temporary and permanent workers. It has been across all employment categories. Of the major economies, only Japan has not had a significant influx of migrant workers. This is because Japanese governments over a long period of time have been reluctant to accept large-scale immigration, particularly for low-skilled labour. While the inflow of skilled labour remains the priority for developed countries, some countries also welcome less skilled workers, particularly to work in agriculture, construction, care for elderly people, and other business and household services.

Although many migrants rely on family contacts and migrant networks, others may have little choice but to use a company that arranges jobs for people. These companies will try to match a potential migrant to a job in a more affluent country. For example, workers in Bangladesh can pay up to US$2000 for a job in Saudi Arabia.

However, people also move for other reasons as well as economic migration:

» Older people retiring to warmer climates, both within the country in which they already live and abroad. A major international example is from northern European countries such as Sweden, the UK and Germany to southern European nations like Spain, Portugal, Greece and Italy. A significant internal migration for retirement purposes is from the northeastern states of the USA to the warmer southern states such as Florida and California. A high level of migration for retirement purposes can have a big impact on the age structure of receiving areas.
» Medical professionals are in demand internationally, and patients migrate for health reasons. Many countries struggle to recruit enough medical professionals from their own medical schools. They may look abroad to bridge the gap between domestic demand and supply. This is a controversial issue as it may mean attracting staff from lower-income countries, leaving those countries without enough doctors and nurses to meet their own demand.
» Air pollution has increased in importance as a push factor in migration, because of the medical problems it can cause, particularly for children. Parents, eager to protect a child's health, may look to move to a safer environment even if this impacts their standard of living.

> ### Activities
> 1. Define migration.
> 2. What is the difference between voluntary migration and forced migration?
> 3. Discuss three significant push factors in migration.
> 4. Suggest how the barriers to migration have changed over time.
> 5. State the difference between a refugee and:
>   a an internally displaced person
>   b an asylum seeker.
> 6. What is economic migration?

The impacts of migration

Table 6.13 is a summary of the possible impacts of international migration. Many of these factors are also relevant to internal migration. Because migration can be such an emotive issue, you may not agree with all of these statements and you may consider that some important factors have been omitted. The spatial impact of migration has spread, with migrants leaving and entering an increasing number of countries. While many traditional migration streams have remained strong, significant new streams have developed. The great majority of migrants from HICs go to other affluent nations. Migrants from MICS and LICs go in roughly even numbers to HICs and to other LICs and MICs that have stronger economies than their own.

The impact of migration on countries of origin

Remittances

Recent international migration reports have stressed the sharp rise in the number of people migrating to the world's richest countries for work. Such movement is outpacing family-related and humanitarian movements in many countries. The value of **remittances** these migrants send back has increased considerably over the past 25 years.

» International remittances totalled US$702 billion in 2020, of which US$540 billion went to LICs and MICs.
» Some economists argue that remittances are the most effective source of financing in MICs and LICs. Although foreign direct investment is larger, it varies with global economic fluctuations.

▼ **Table 6.13** The impacts of international migration

	Impact on countries of origin	Impact on countries of destination	Impact on migrants themselves
Positive	• Remittances are a major source of income in some countries • Emigration can ease the levels of unemployment and underemployment • Reduces pressure on health and education services and on housing • Return migrants can bring new skills, ideas and money into a community	• Increase in the pool of available labour may reduce the cost of labour to businesses and help reduce inflation • Migrants may bring important skills to their destination • Increasing cultural diversity can enrich receiving communities • An influx of young migrants can reduce the rate of population ageing	• Wages are higher than in the country of origin • There is a wider choice of job opportunities • There is a greater opportunity to develop new skills • They have the ability to support family members in the country of origin through remittances • Some migrants have the opportunity to learn a new language
Negative	• Loss of young adult workers who may have vital skills, e.g. doctors, nurses, teachers, engineers (the 'brain-drain' effect) • An ageing population may be left in communities with a large outflow of (young) migrants • Agricultural output may suffer if the labour force falls below a certain level • Migrants returning on a temporary or permanent basis may question traditional values, causing divisions in the community	• Migrants may be perceived as taking jobs from people in the long-established population • Increased pressure on housing stock and on services such as health and education • A significant change in the ethnic balance of a country or region may cause tension • A larger population can have a negative impact on the environment	• The financial cost of migration can be high • Migration means separation from family and friends in the country of origin • There may be problems settling into a new culture (assimilation) • Migrants can be exploited by unscrupulous employers • Some migrations, particularly those that are illegal, can involve hazardous journeys

6.3 THE CAUSES AND IMPACTS OF INTERNATIONAL MIGRATION

▲ **Figure 6.25** MoneyGram is an electronic money transfer system, by which remittances may be sent

» Remittances exceed considerably the amount of official aid received by MICs and LICs.
» Migration advocates stress that these revenue flows help reduce poverty, encourage investment, and reduce the impact of global recession, when private capital flows decrease.

The major sources of remittances are the USA, Western Europe and the Persian Gulf. The top destinations for remittances in 2020 were India, China, Mexico and the Philippines (Figure 6.26). Dependency on remittances is usually measured as the ratio of remittances to GDP. In 2020 there were 29 countries (out of 177 countries reported) that had a remittance-to-GDP ratio above 10 per cent. However, there are concerns that a heavy reliance on remittances can:

» create a culture of dependency, potentially reducing participation in the labour force
» make a country vulnerable to significant changes in the level of remittances.

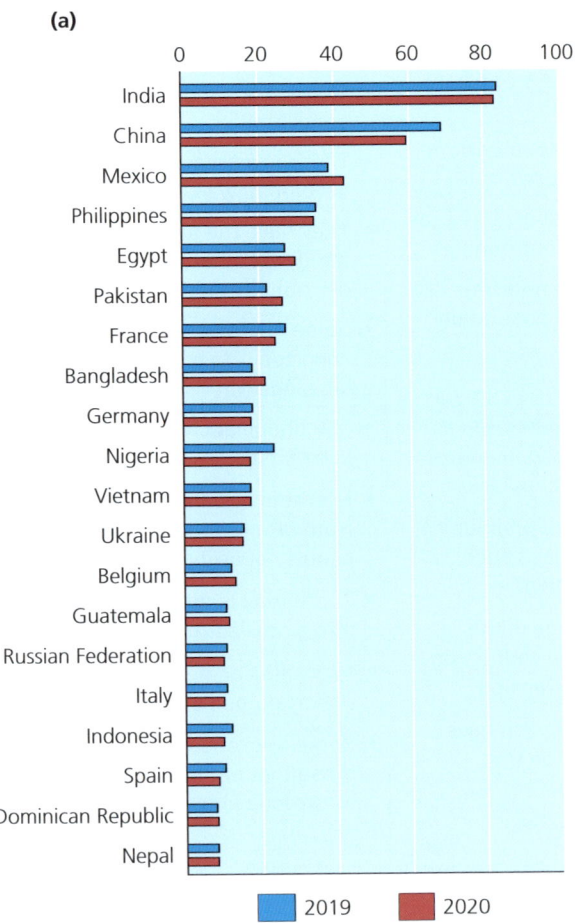

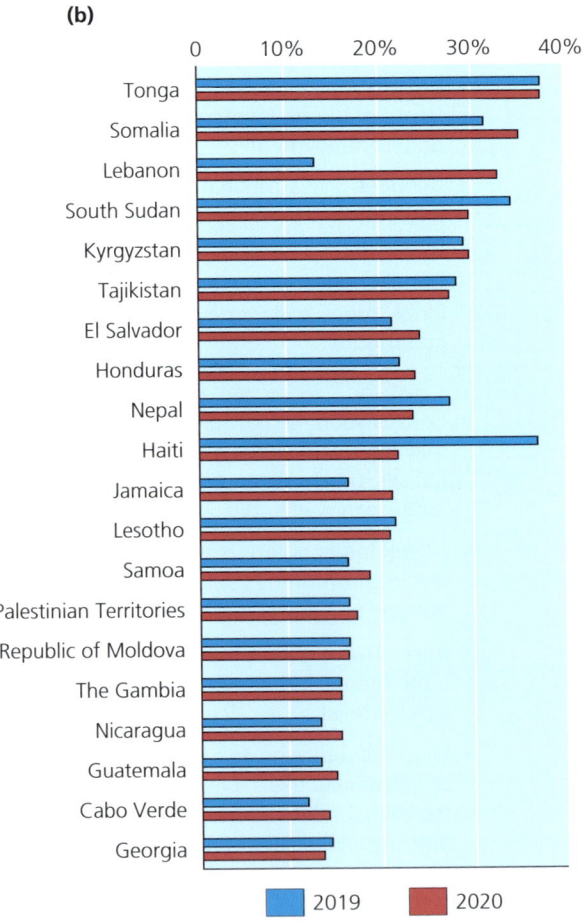

▲ **Figure 6.26** The top 20 recipient countries/territories of international remittances, by total in $US billion (a) and share of GDP (b), 2019–20

Other potential advantages of emigration

Other possible advantages of emigration include reducing population pressure on resources such as food and water, and lowering levels of unemployment and underemployment, which are all major problems in many LICs. Emigration can also reduce pressure on the housing stock and on key services such as health and education, which are areas of heavy expenditure for LICs. The links that migrants maintain with their home communities can also help to develop new skills and technologies, particularly when migrants return home permanently and establish new businesses.

Diaspora

The word '**diaspora**' describes the dispersal of a people, originally belonging to one country, around the world. Significant diaspora populations from many countries of origin have been established in many receiving countries. For example, the 20 million people who make up the Indian diaspora are scattered over 135 countries. The Indian diaspora (Figure 6.27) is comprised of people of Indian birth and Indian ancestry. Because of the inclusion of the latter, some estimates put the Indian diaspora as large as 30 million. The largest group is in the USA.

▲ Figure 6.27 Southall, the centre of London's Indian community

Potential disadvantages

In terms of disadvantages, the loss of workers with important skills – the so called 'brain-drain' effect – is of concern in many countries of origin. In extreme cases in some communities there may not be enough people to continue to farm effectively. Population structure can be adversely affected if migration is very gender-selective, and the loss of many young people can advance the ageing of the population.

The impact of migration on countries of destination

Any increase in the labour force is generally welcomed by businesses, particularly if migrants have skills that are in short supply. For example, health services in many HICs rely heavily on foreign nurses and doctors.

» Greater competition in the labour force tends to limit wage rises, which helps keep inflation low. Low inflation is an important factor in economic stability.
» Many people value an increase in cultural diversity, as it can enrich communities.
» An influx of young migrants can help to reduce the rate of population ageing and lower the dependency ratio. This has a positive financial impact on countries.

The negative impact of immigration is more contentious. A significant influx of migrants can put pressure on the available housing stock, causing overcrowding and pushing up prices. Similarly, it will increase demand on health, education and other services to varying degrees. Trade unions often voice concerns over migration levels if they feel that their members are losing out on employment prospects because of an increase in the level of competition for work. In areas where there has been a significant change in the ethnic balance, tension between ethnic groups may increase.

Each receiving country has its own sources of immigrants. This is the result of historical, economic and geographical relationships. Earlier generations of migrants form networks that help new arrivals to overcome legal and other obstacles. The distribution of immigrants in receiving countries is uneven, with significant concentrations in economic core regions (Figure 6.28).

▲ Figure 6.28 Chinatown, San Francisco – a major Chinese community in an American city

6.3 THE CAUSES AND IMPACTS OF INTERNATIONAL MIGRATION

The impact of migration on migrants themselves

The prospect of higher wages, a wider choice of job opportunities and the chance to develop new skills are all important attractions to potential migrants. For many, this will enable them to provide financial support for their families in the country of origin. Other benefits may also become apparent, such as developing language skills. Eventually, obtaining citizenship in their country of destination may allow migrants to bring family members from their country of origin to join them.

On the debit side, the financial cost of migration may be considerable, but this may be shared among family members in anticipation of the remittances to follow. For many migrants, by far the greatest cost is separation from family and friends in their home country. The 'culture shock' of settling in a new country may be lessened if migrants secure housing and employment in an established migrant community in their new country. The risk of exploitation is a real concern for many migrants, and receiving countries vary in how effectively they challenge such practices.

The World Development Report (2023) identified four major challenges and barriers that most migrants face:

- » Uncertainty: Dealing with unexpected and uncertain outcomes, including the possibility of unemployment, social isolation, psychological stress and harm during transit.
- » Unfamiliarity: Monetary and non-monetary costs associated with the language, social norms and culture of their destination society.
- » Job search: Qualifications, skills and experience gained in one country may not transfer easily to another country. Migrants often have no option but to 'downgrade' to lower-skill employment.
- » Financing: The upfront costs of migration, such as travel and relocation, visas, processing and payments to intermediaries can be high. Such costs can vary widely across migration corridors.

> **Activities**
> 1. State two positive and two negative impacts of international migration on countries of origin.
> 2. Give three benefits of international migration for countries of destination.
> 3. Present a brief analysis of Figure 6.26 (page 148).
> 4. With reference to an example, explain the term 'diaspora'.
> 5. Discuss the four major challenges and barriers to migration identified in the World Development Report, 2023.

The different approaches to managing international migration

International migration is a major global issue. Today, few countries favour a large influx of migrants, for a variety of issues. Within individual countries, attitudes to immigration can vary considerably. Attitudes tend to harden when economies are going through difficult periods and become more relaxed when economic conditions are good. Countries sometimes make incoming international migration more difficult for political and other reasons, and then find themselves short of labour in key economic sectors. Such a situation may be due not just to harsher barriers put up to deter incoming migrants, but also because existing migrants feel unwelcome and return home or migrate to another country.

The International Organization for Migration (IOM) is the leading international organisation for migration. It works to:

- » advance understanding of migration issues
- » assist in the challenges of migration management.

The IOM works with other international organisations to try to broker disputes between countries about population movements, particularly with regard to refugees and illegal immigration. In general, developing countries want their populations to have greater migration access to developed countries. In contrast, developed countries have tended to tighten controls on immigration.

The different approaches to managing international migration

International migration is a politically sensitive issue in many countries. There are clear examples in Europe and Africa:

» Within the EU, there is a contrast between the more welcoming approach of some countries to migrants escaping from conflict in the Middle East, Afghanistan and parts of Africa, compared to the harder line taken by countries such as Poland and Hungary, which openly state their concerns about **cultural identity**.
» Environmental degradation and conflicts in parts of Africa have led to significant movements of people, which have placed severe pressures on already poor host countries.

Countries have an unequal role in establishing migration policies. Most origin countries have little influence on regulating migration. In contrast, destination countries define and regulate who crosses their borders, who is legally allowed to stay, and with what rights. Labour economics and international law provide the two main concepts (Figure 6.29) for understanding migration patterns and the design of appropriate migration policies:

» Labour economics focuses on the match between migrants' skills and the labour requirements of destination countries.
» Under international law, migrants' motives determine destination countries' obligations. Those who meet the criteria for refugee status are entitled to international protection under the 1951 Refugee Convention.

Figure 6.29 provides a useful framework for understanding and managing international migration. On the diagram, the 'problem' area is the lower-left quadrant. These are 'distressed migrants' who neither qualify as refugees nor are a strong employment match at their destination. The migratory routes of this group of people are often irregular and unsafe, posing considerable challenges for destination countries. In addition, where migrants stand in the

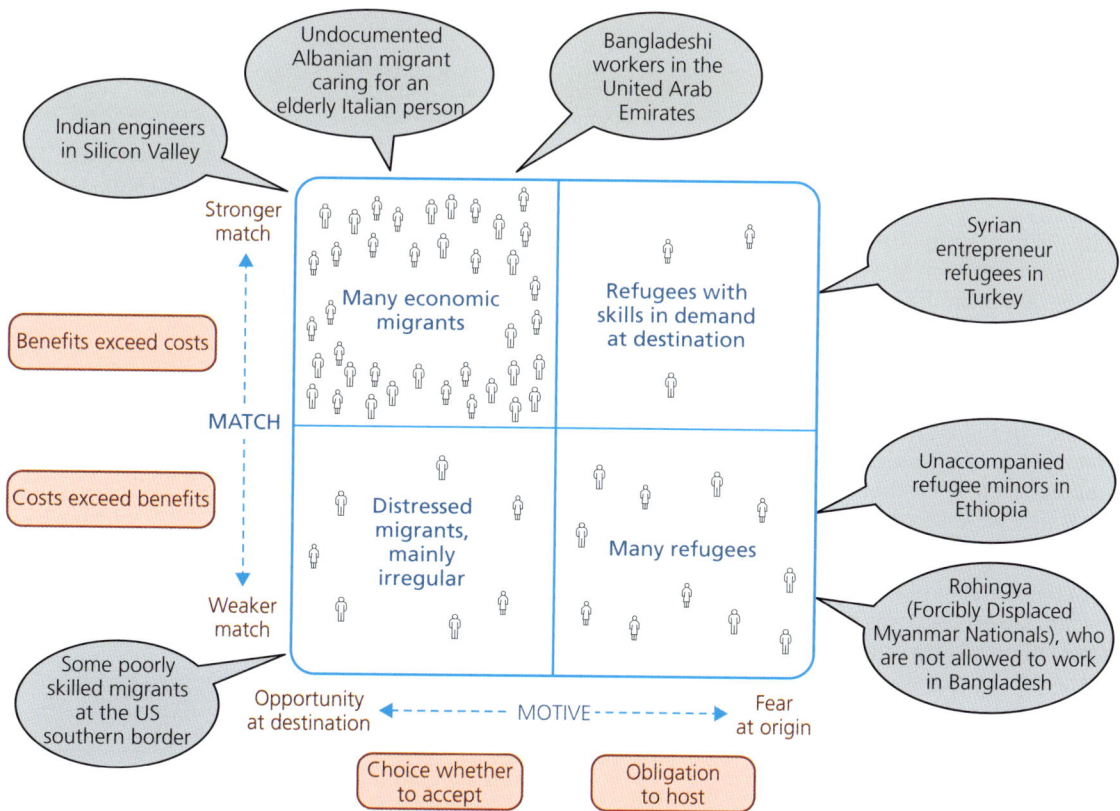

▲ **Figure 6.29** The Match and Motive Matrix: 'Match' determines the net gains of receiving migrants; 'Motive' determines their international protection needs

6.3 THE CAUSES AND IMPACTS OF INTERNATIONAL MIGRATION

Match and Motive Matrix depends, at least in part, on whether the migrant has the right to work at the level of his/her qualifications. For example, a migrant who is a qualified midwife in their country of origin might not satisfy the qualification criteria in a country of destination.

The World Development Report (2023) states that government policies should aim to both:

» maximise the development gains of migration – for the migrants, the origin societies and the destination societies
» provide refugees with adequate international protection.

A range of organisations have stressed the importance of increased cooperation between origin countries, transit countries (where relevant) and destination countries. For example, over the past decade there has been improved dialogue between the USA (a destination country) and a number of countries in Latin America that are either countries of origin or transit countries (or both).

Australia's offshoring policy

Australia is the only country in the world to enforce a policy of mandatory detention and offshore processing for all asylum seekers who arrive without a valid visa. This policy is particularly aimed at people arriving in Australian territorial waters by boat. Currently there is a single offshore detention centre on the Pacific island nation of Nauru. Although this policy has been heavily criticised internationally, Australian governments have defended it as necessary to deter people traffickers who exploit desperate asylum seekers. The UK announced a similar policy in 2022, with offshore processing in Rwanda. Due to numerous legal challenges, no asylum seekers had been sent to Rwanda at the time of writing in 2024.

> ### Activities
> 1 Name the leading international organisation for migration.
> 2 Write a brief analysis of Figure 6.29 (page 151).

Detailed specific example

International migration from Mexico to the USA

One of the largest international migration streams in the world over the past 50 years has been from Mexico to the USA. This migration has formed the largest immigrant community in the world. It has been primarily a **labour migration**, and the result of the following push and pull factors:

- much higher average incomes in the USA
- lower unemployment rates in the USA
- faster growth of the labour force in Mexico, with significantly higher population growth than in the USA
- higher quality of life in the USA, according to most indicators.

Most migration from Mexico has taken place since the 1970s. However, previous surges occurred in the 1920s and 1950s, when the US government allowed the recruitment of Mexican workers as **guest workers**. As Figure 6.30 shows, persistent **mass migration** between the two countries did not take hold until the late twentieth century. However, between 2010 and 2021 there was a decline of about 1 million (9 per cent) in the US–Mexican population. This has been attributed to two factors:

- Increased immigration enforcement by the US authorities (Figure 6.31)
- A strengthening of the Mexican economy.

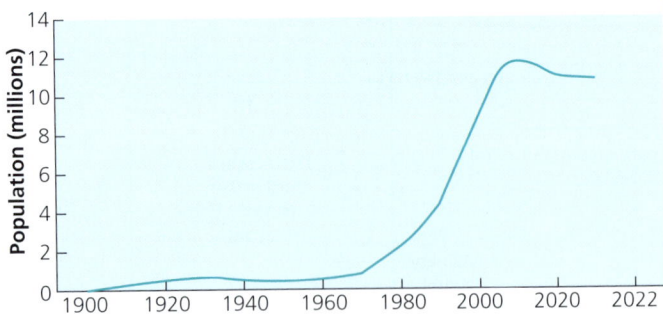

▲ **Figure 6.30** Changes in the Mexican-born population in the USA, 1900–2022

The different approaches to managing international migration

▲ **Figure 6.31** Fencing along the US border with Mexico, and the US Border Patrol

Mexicans accounted for about 24 per cent (10.7 million) of the 45.3 million foreign-born residents in the USA in 2021. However, for several years beginning in 2013, Mexico had been overtaken by India and China as the top country of origin for new immigrants to the USA. However, in 2021 Mexico was back in first place. The USA is the destination country for an astonishing 97 per cent of all Mexican emigrants, and 8 per cent of all people born in Mexico were living in the USA in 2020. Canada and Spain are the next most popular destinations for Mexican emigrants.

There is a very strong concentration of the US–Mexican population in particular states. For the 2015–19 period, most immigrants from Mexico lived in California (36 per cent), Texas (22 per cent), Illinois (6 per cent) and Arizona (5 per cent). The next six most populous states – Florida, Washington, Georgia, North Carolina, Colorado and Nevada – accounted for an additional 13 per cent of the Mexican-born population. The US cities with the largest number of Mexicans were the greater metropolitan areas of Los Angeles, Chicago, Dallas and Houston.

Within urban areas, the Mexican population is concentrated in areas of poor housing and low average income. In more peripheral areas, the Mexican population is largely found in low-cost housing areas where proximity to farm employment is an important factor. The main reasons for this spatial distribution are:

- proximity to the US southern border with Mexico
- the location of demand for immigrant farm workers
- urban areas where the Mexican community is long-established.

Table 6.14 shows that a far higher proportion of the Mexican-born population in the USA are in the economically active age group compared to the US-born population. Additionally, like other immigrant groups, Mexicans are more likely to be employed in low-paid unskilled and semi-skilled jobs that are unpopular with the US-born population. These are the so called '3D' jobs – that is, jobs that are 'dirty,

dangerous and dull/difficult'. The Mexican-born population is heavily represented in agriculture, construction, hospitality and catering, and in household services.

▼ **Table 6.14** The age distribution of the US population by origin, 2021

Age group	Mexican immigrants	US-born population
Under 18	3 per cent	25 per cent
18–64	85 per cent	59 per cent
65 and over	12 per cent	17 per cent

In the USA the Federation for American Immigration Reform (FAIR) has opposed large-scale immigration from Mexico and elsewhere, arguing that it:

- undermines the employment opportunities of low-skilled US workers
- has adverse environmental effects because of the increased population
- threatens established US cultural values.

Those opposed to FAIR see its actions as uncharitable and arguably racist. Such individuals and groups highlight the advantages that Mexican and other migrant groups have brought to the country. They stress:

- the long-term economic benefits of immigration to the US economy
- the historical role the USA has played in welcoming immigrants from difficult circumstances abroad.

Donald Trump was president of the USA 2017–21. While in office he promised to build a new 'wall' in an attempt to make illegal migration from Mexico much more difficult. Much of this illegal migration is not from Mexico itself, but from a range of countries in Latin America for which Mexico is their potential crossing point into the USA. By the time President Trump left office in 2021, new border infrastructure had been constructed only along a limited part of the border. The border and how it is monitored and enforced is a controversial issue in both the USA and Mexico.

The impact of migration on Mexico and on Mexican migrants

Sustained large-scale migration has had a range of impacts on Mexico, some of them clear and others debatable. Significant impacts include the high value of remittances, which totalled over US$63 billion in 2023, more than any other country except India in that year. Remittances as a percentage of Mexico's GDP grew from 2 per cent in 2010 to 3.8 per cent in 2020, and to an estimated 4.5 per cent in 2023.

The Mexican economy in the two opening decades of this century (2000s and 2010s) has been more stable than in the 1980s and 1990s, when the country faced a number of serious economic crises.

6.3 THE CAUSES AND IMPACTS OF INTERNATIONAL MIGRATION

There is little doubt that in the 1980s and 1990s (and earlier), migration to the USA:

- reduced unemployment pressure, as migrants tend to leave areas where unemployment is particularly high
- lowered pressure on the housing stock and public services
- supported the finances of many households and communities through remittances.

On the negative side, emigration has:

- changed the population structure of many communities due to the emigration of large numbers of young adults
- witnessed the loss of skilled and enterprising people
- depleted the agricultural workforce in areas where emigration has been dominated by young men, with women and children often taking the place of now-absent men in agriculture.

In recent decades, two important changes occurred to impact the strength of migration from Mexico. The significant loss of jobs in the USA in the 'Great Recession' of 2007–09 occurred in industries in which immigrants tend to be heavily represented. This encouraged some migrants to move back to Mexico. Secondly, the aftermath of the recession also made the USA a less attractive proposition to many potential migrants. On the Mexican side of the border, declining fertility had slowed the growth of the working age population. To an extent, smaller average family size reduced the need for migration as a means of family financial support.

An important characteristic of migration, which can be viewed as either positive or negative according to the observer's personal views, is the return of migrants to Mexico with changed values and attitudes. The new values and attitudes of returning migrants often clash with traditional views, particularly in more isolated rural communities.

In terms of individual migrants, the most successful movements tended to be from communities in Mexico where a **migrant culture** had become established and where strong diaspora communities had been built up in particular destinations in the USA. Mexican culture has had a sustained impact on many areas of the USA, particularly urban areas close to the border. As a result, many Mexican migrants find reassuring similarities between the two countries.

These is no doubt that the Mexican population of the USA has undergone a process of assimilation over time. There are three facets to assimilation:

- Economic
- Social
- Political.

Assimilation tends to occur in the order presented above, with economic assimilation occurring first. While many migrants from Mexico would be in the low-skilled category, their children and grandchildren usually aspire to, and gain, higher qualifications and skills. Such economic mobility results in greater social contact with the mainstream population. Eventually, more people from migrant populations get involved in politics and the migrant community gains better political representation.

The migration status of an individual migrant is of great importance to the choices that individual can make. Because of its very nature, it is difficult for the authorities to be precise about illegal immigration. In 2019, the Pew Research Center estimated that 77 per cent of US immigrants were 'Documented' and 23 per cent 'Undocumented'. Nearly half of all immigrants have become US citizens.

Migration and sustainability

The literature on the relationship between migration and sustainability often stresses that these issues are frequently managed separately by government at all scales. The potential impact of migration on sustainability increases with the level of migration. The impacts are environmental, economic and social, and occur in the countries of both origin and destination. In recent decades, immigration into the USA has been at very high levels. However, Mexico has also been affected by large numbers of migrants moving through the country from further south in Latin America. The huge scale of this movement has resulted in increasing cooperation between the US and Mexican governments, and also between other governments in Latin America.

The US Strategy to Address the Root Causes of Migration in Central America aims to establish more prosperous, safe and democratic conditions, so that potential migrants are not compelled to leave their home countries. However, what the USA can realistically do at scale to limit the desire of people in a range of Latin American countries to travel north remains to be seen.

In recent years more attention has been directed to immigration governance in the USA. There is a growing recognition that the contribution of migrants to US society tends to increase with integration. For example, access to better education gives migrants greater opportunity to realise their potential.

> ### Activities
> 1 Describe the changes in the Mexican-born population in the USA illustrated by Figure 6.30 (page 152).
> 2 Outline the main factors that have encouraged migration from Mexico to the USA.
> 3 Present a brief analysis of Table 6.14 (page 153).
> 4 Why has the Mexican-born population in the USA fallen in recent times?

The different approaches to managing international migration

Practice questions

1 Figure 6.32 shows global birth rates in 2023.

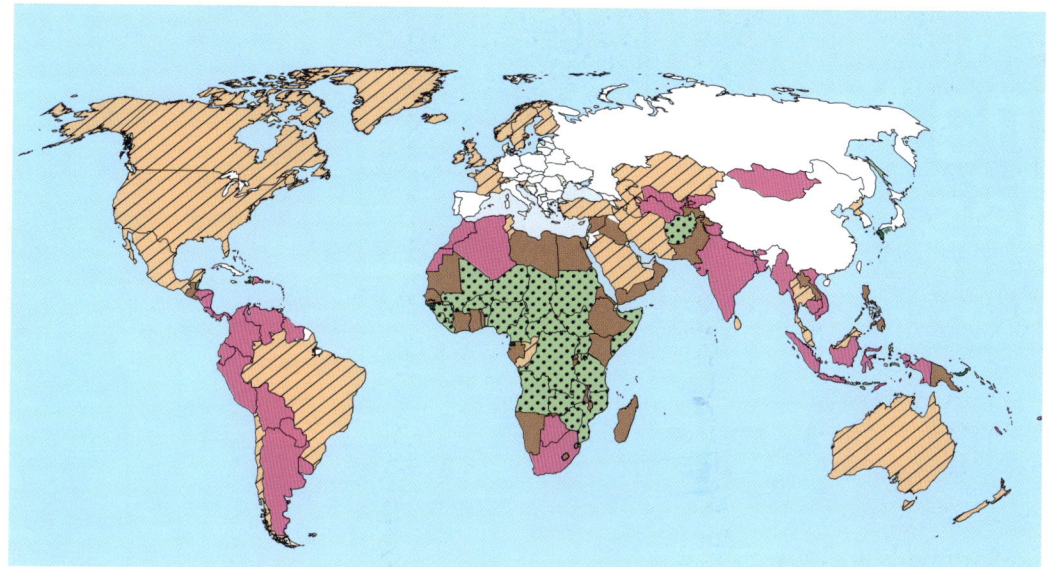
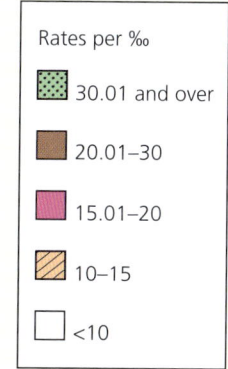

▲ **Figure 6.32** Global birth rates, 2023

  a Define the term 'birth rate'. [1]
  b Describe the pattern of global birth rates in 2023. [3]
  c Suggest reasons for the high birth rates in your answer to (b). [3]

2 Figure 6.33 shows the global pattern of death rates in 2023.

  a Define the term 'death rate'. [1]
  b Describe the global pattern of death rates in 2023. [3]
  c Suggest reasons for the high death rates in your answer to (b). [3]

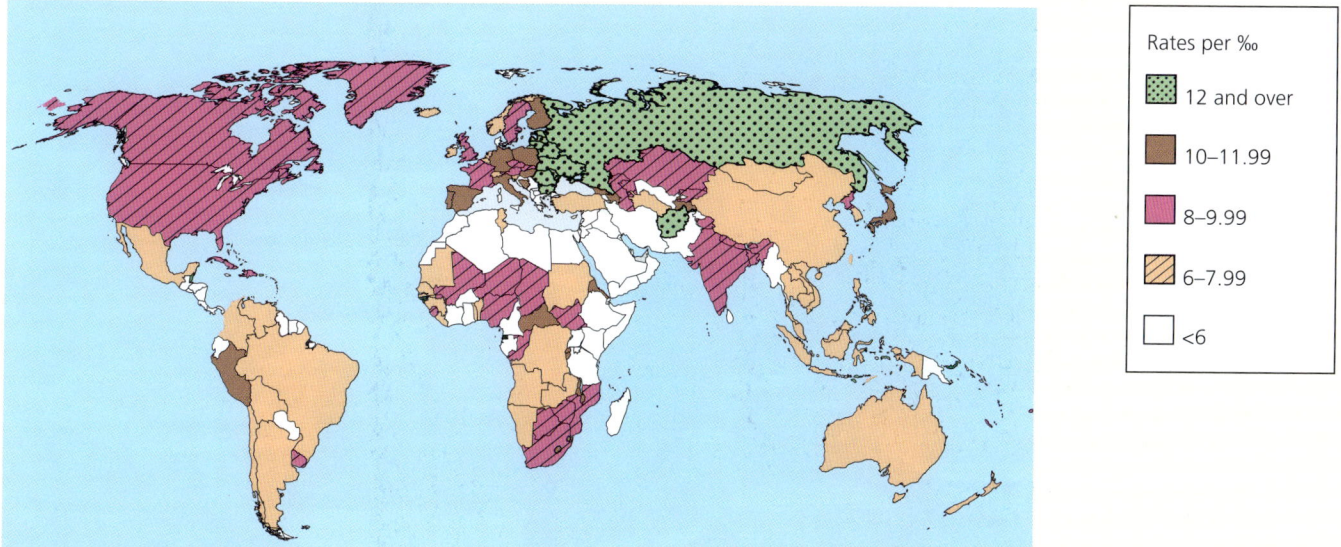

▲ **Figure 6.33** The global pattern of death rates, 2023

3 a Define the term 'population structure'. [1]
  b Explain how rural-to-urban migration causes changes in population structure. [3]

4 Assess the advantages and disadvantages of ageing populations. [7]

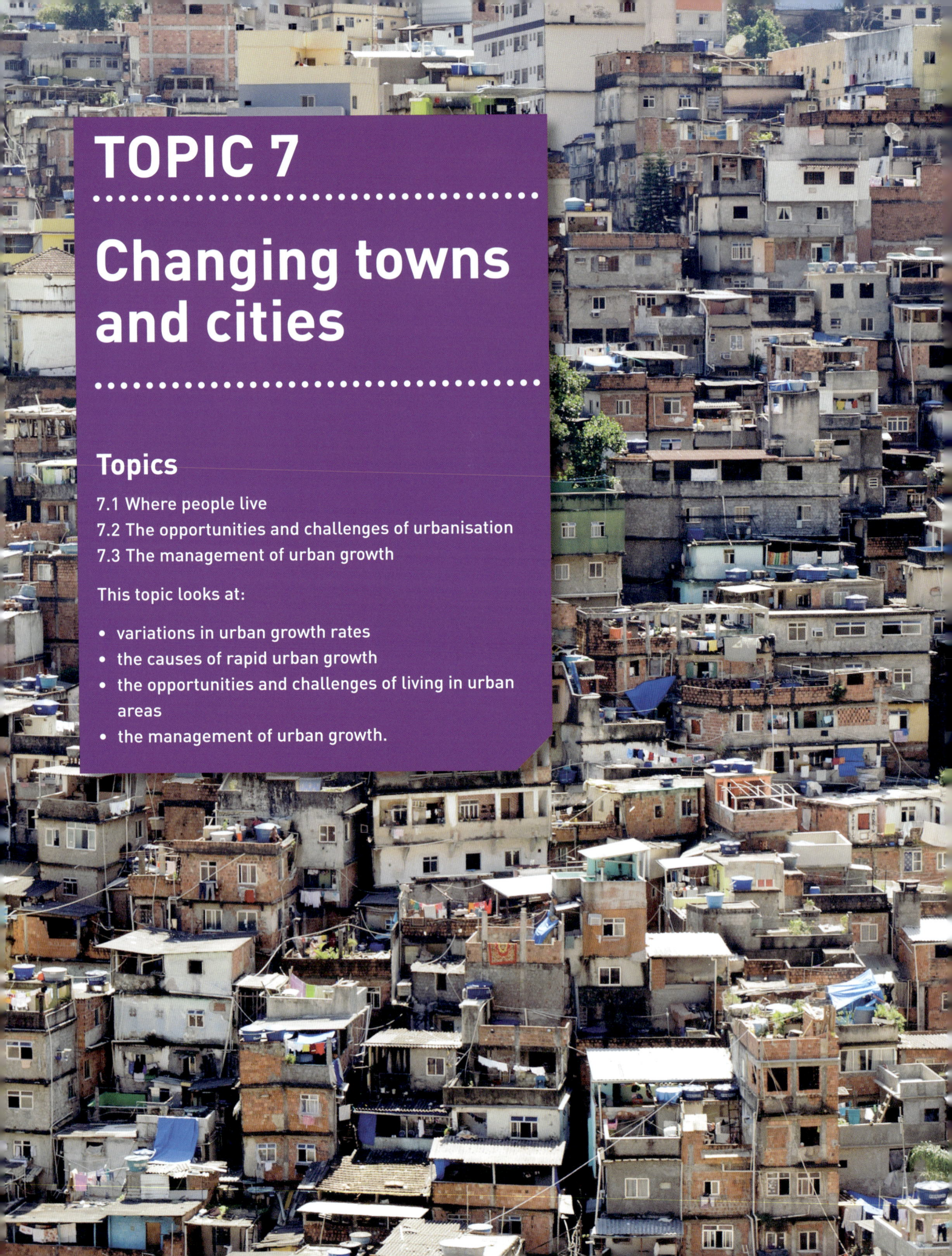

TOPIC 7
Changing towns and cities

Topics

7.1 Where people live
7.2 The opportunities and challenges of urbanisation
7.3 The management of urban growth

This topic looks at:

- variations in urban growth rates
- the causes of rapid urban growth
- the opportunities and challenges of living in urban areas
- the management of urban growth.

7.1 Where people live

This chapter will explain:

★ the reasons for variations in urbanisation
★ the causes of rapid urban grown in LICs.

Urbanisation is the process by which an increasing proportion of a country's population comes to live in towns and cities. It may include rural-to-urban migration and the reclassification of rural settlements as they are engulfed by expanding cities. It is one of the most significant geographical phenomena of the twentieth and twenty-first centuries. In contrast, **urban growth** refers to the physical growth (expansion) of the urban area and/or the growth in the number of people living in urban areas. Urban populations are those living in areas classified as urban. However, definitions vary by country. Most urban populations have a large size, specific urban characteristics such as a CBD, and economic activities that are mainly services and manufacturing and administrative functions.

Figure 7.1 shows that the **least low-income countries** (LLICs) have the lowest rate of urbanisation, and that levels of urbanisation generally increase as countries become more developed (Figure 7.2). However, for some HICs levels of urbanisation may decrease as some people prefer to live in smaller settlements and rural areas due to commuting/working from home and the perceived benefits of living in smaller settlements. This trend increased in some countries, including the UK, during the Covid-19 pandemic as larger settlements were seen to present more risk of the spread of the disease.

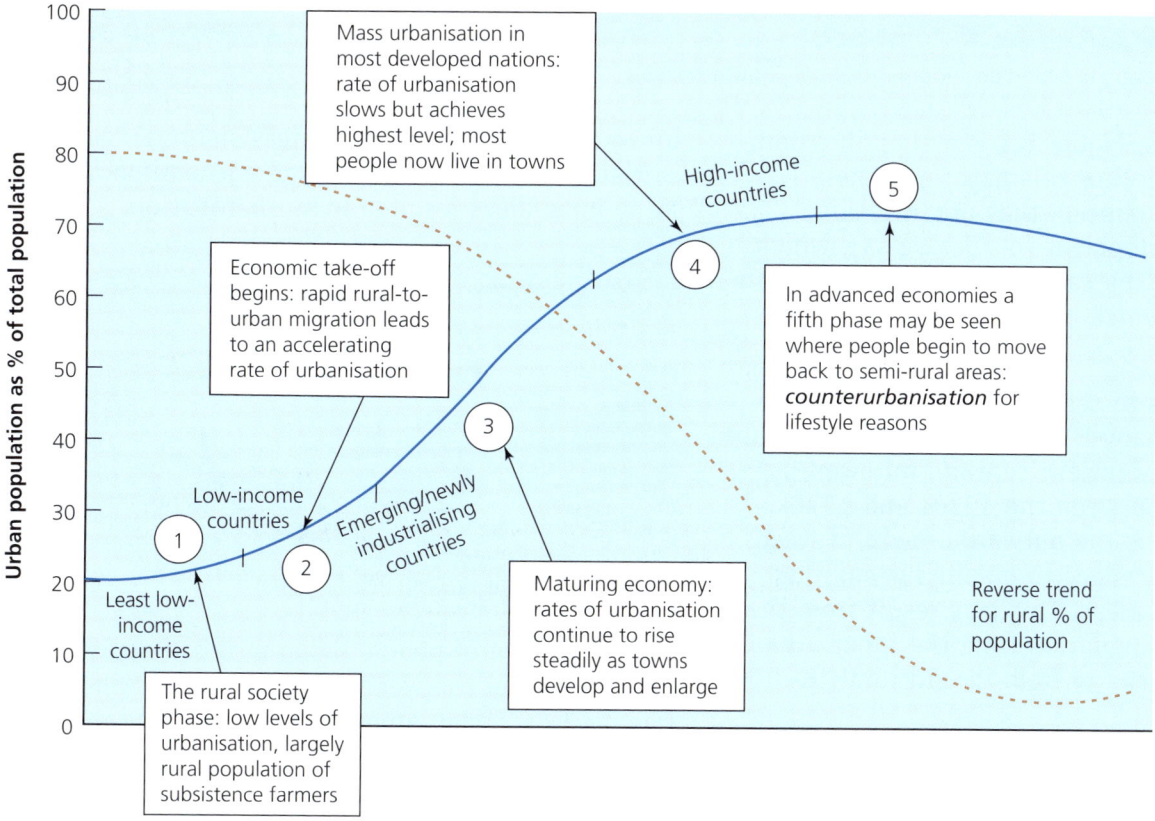

▲ Figure 7.1 The process of urbanisation

7.1 WHERE PEOPLE LIVE

▲ Figure 7.2 A high-density urban development, Seoul, South Korea

As with all models, Figure 7.1 is a simplification. Not all countries follow this path, and there is an underlying assumption that urbanisation accompanies development (or vice versa), which is not necessarily the case, and that urbanisation is a good thing, which again is not necessarily the case.

It is also possible to see a cycle of urbanisation (Figure 7.3). In many HICs, urbanisation dominated during the early twentieth century, followed by suburbanisation between the 1930s and 1970s. **Suburbanisation** is the outward growth of towns and cities at their edges to engulf surrounding villages and rural areas. This may result from the outmigration of population from the inner urban areas to the **suburbs** or from inward rural-to-urban movement.

From the 1970s, counterurbanisation became more widespread. **Counterurbanisation** is a process involving the movement of population from large urban areas to smaller urban areas and rural settlements. From the 1990s the regeneration of many inner areas resulted in **reurbanisation**.

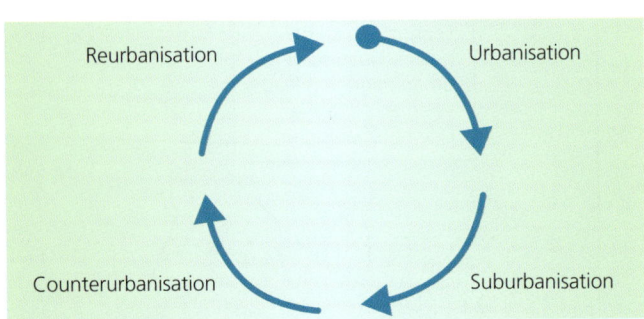

▲ Figure 7.3 The cycle of urbanisation

The proportion of people living in urban areas varies spatially (between places) and over time (Figure 7.4). Some regions are experiencing rapid urbanisation, whereas others are seeing rates of urbanisation slow down.

The reasons for variations in urbanisation

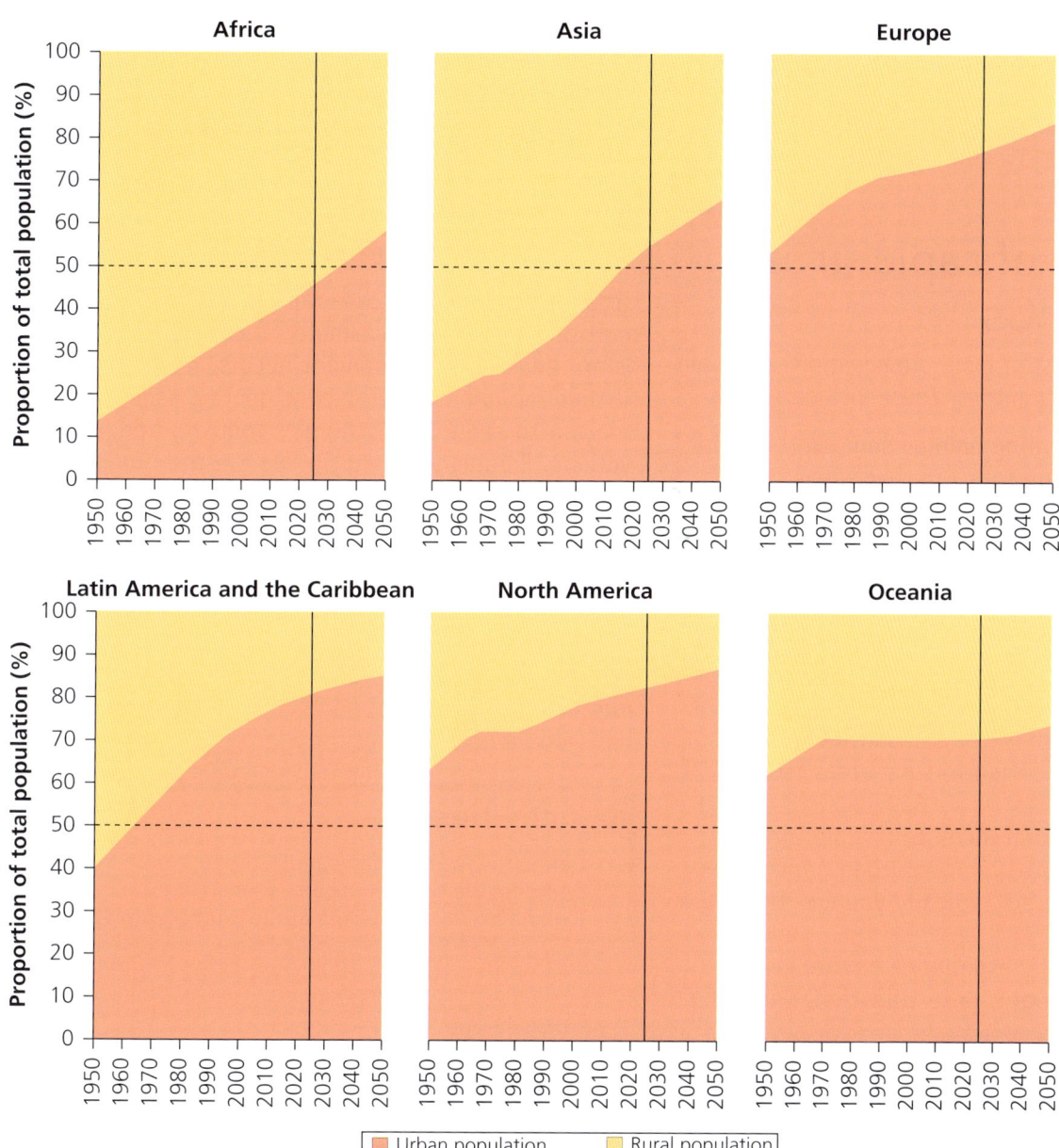

▲ **Figure 7.4** Changes in urban and rural populations by region, 1950–2050 (projected)

The reasons for variations in urbanisation

In the least low-income countries, rates of urbanisation are relatively low. Countries with very low levels of urbanisation include Malawi and Uganda (16 per cent), Papua New Guinea (13 per cent) and Burundi (12 per cent). These are largely agricultural economies with many subsistence farmers.

As countries begin to develop, there is rapid rural-to-urban migration as many people leave rural areas in search of better-paying jobs in urban areas. Countries at this stage include Kenya (with about 25 per cent of the population living in urban areas) and Afghanistan (about 27 per cent).

In maturing economies, the rate of urbanisation continues to increase quite rapidly at first as towns and cities enlarge. India's urban population increased from 25 per cent in 1980 to 36 per cent in 2023, whereas China's urban population increased from 20 per cent in 1980 to 64 per cent in 2023.

7.1 WHERE PEOPLE LIVE

The causes of rapid urban growth in LICs

In the final stage (stage 5 in Figure 7.1 on page 157), large urban areas may experience a decline in population while smaller towns and rural areas experience an increase. For example, Japan's urban population peaked at 115 million in 2018 and fell to 114 million by 2021.

The causes of rapid urban growth in LICs

There are many reasons for rapid urbanisation, especially in LICs and emerging countries. These include:

- the prospects of finding employment, better paid jobs and more secure jobs in urban areas (this is an economic pull factor)
- better provision of education and health facilities in urban areas (social pull factor)
- fewer economic opportunities in rural areas – farming and related work can be low paid, insecure and increasingly subject to climate extremes and natural hazards (economic and physical push factors)
- poor access to clean water and sanitation, healthcare and education in rural areas (social push factors)
- higher rates of natural increase (higher birth rates than death rates) among urban populations (urban populations are typically younger than rural populations)
- some rural areas may experience environmental stress (such as drought, desertification, declining soil fertility) and so people move away to urban areas where they are less dependent on the physical environment
- many urban areas, especially large urban areas, may be considered to be a 'melting pot' for different populations and people may feel that they can integrate in urban areas, irrespective of their nationality, culture and political views.

Urbanisation may increase rapidly due to the multiplier effect. For example, if a new activity such as an industrial development occurs, it can trigger further economic development nearby, especially in relation to services for the new industry and/or its workers. Employment multipliers are the number and type of new jobs created related to the new economic activity. Income multipliers refer to the value of the activities created.

Consequently, rural-to-urban migration may be large-scale, leading to rural depopulation and the growth of population in urban areas. The majority of those who migrate are young adults, especially those with skills and a good education.

> ### Activities
> 1. Define the term 'urbanisation'.
> 2. Describe the cycle of urbanisation.
> 3. Explain briefly the causes of rapid urban growth.

▲ **Figure 7.5** There is a concentration of employment opportunities in central districts of large urban areas, as here in Pudong, Shanghai

7.2 The opportunities and challenges of urbanisation

This chapter will explain:

★ the opportunities of urban living
★ the challenges of urban living
★ the impacts of urban sprawl on the rural-to-urban fringe.

The opportunities of urban living

In 2018, 55 per cent of the world's population lived in urban areas. By 2050, this could reach 68 per cent. This is because urban areas offer many opportunities, as they attract many people to live there. For example, many large urban areas offer numerous cultural facilities such as museums, art galleries, sporting venues, historic buildings and places of worship. There are also more cultures and ethnicities represented in urban areas. Diversity and inclusiveness are more likely in an urban area due to the varied backgrounds of its residents.

▲ **Figure 7.6** Multiculturalism is a characteristic of large urban areas

Many urban areas provide a greater number of well-built homes, and different types of homes including houses, flats and apartments. The range of services is greater in urban areas, including educational, healthcare, retail, as well as leisure and entertainment facilities such as cinemas, theatres, restaurants and clubs. Many of these are very accessible to urban residents due to the public transport that is available. Many cities are walkable – the so called '15-minute' cities, with high population densities and good access to services, facilities and employment.

For many, it is the number and range of jobs that are available in urban areas that is the biggest attraction. Most migrants move to large urban areas because they believe they can earn more than in a rural area or small urban area. Urban areas are centres of business, commerce and industry, and up to 80 per cent of global GDP is generated by cities.

For example, in New York City, the 'Big Apple', there are many financial jobs associated with Wall Street, political/administrative jobs associated with the United Nations, jobs in entertainment associated with Broadway and Times Square, and many others. It also has some of the world's top universities such as Columbia University, New York University and Pace University. New York has an advanced public transport system. The subway system has over 1000 km of track, the Metropolitan Transportation Authority runs a fleet of over 6000 buses, and there is a very popular ferry system, the Staten Island Ferry. In terms of entertainment, many residents and tourists are attracted by what is on offer on Broadway, Times Square, Chinatown and Little Italy. New York also offers much open space, such as Central Park, Prospect Park and Coney Island. It is also associated with sports such as baseball and tennis. The New York Yankees are internationally known and the US Tennis Open is held at Flushing Meadow in the Bronx.

7.2 THE OPPORTUNITIES AND CHALLENGES OF URBANISATION

The challenges of urban living

On the other hand, there are challenges associated with urban living.

Inequality

Inequalities in income are a feature of many urban areas (Figure 7.7). Urban areas contain many wealthy residents, such as people working in financial services, lawyers, doctors and businesspeople, and also low-paid workers such as cleaners, caterers and informal workers, as well as those who are out of work. This leads to large income inequalities.

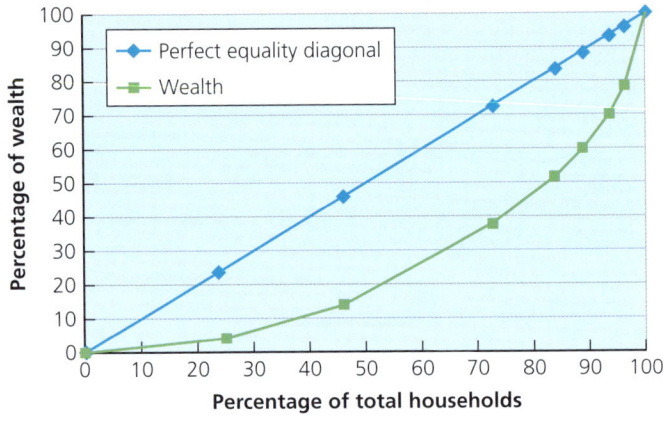

▲ **Figure 7.7** Income inequality in Rio de Janeiro, shown on a Lorenz curve

Service provision

Service provision in urban areas is highly variable. For example, Woodstock is an inner suburb of Cape Town, South Africa, whereas Blikkiesdorp is a relocation camp made up of about 1600 one-roomed tin structures (Figure 7.8). Woodstock experienced urban renewal and gentrification (the movement into the area of young professional workers) in the 1970s and 1980s, and this led to the growth of service provision such as fashionable restaurants, ICT and other businesses and offices (Figure 7.9), whereas Blikkiesdorp is a lower-income area and has not attracted much service provision (Figure 7.10).

Housing

There is a lack of sufficient quality housing in all countries. There are at least four aspects to the management of housing stock:

» Quality of housing – with adequate water, sanitation and space

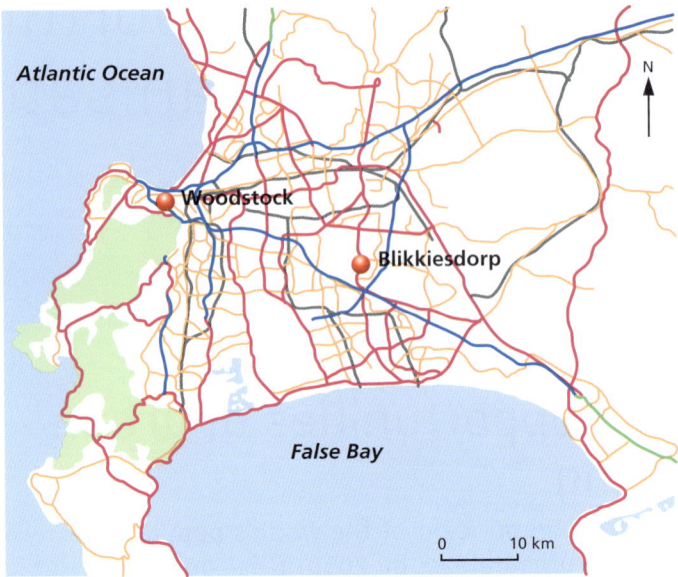

▲ **Figure 7.8** The location of Woodstock and Blikkiesdorp, Cape Town

▲ **Figure 7.9** Services in Woodstock, Cape Town

▲ **Figure 7.10** Housing in Blikkiesdorp, Cape Town

The challenges of urban living

- Quantity of housing – having enough housing to meet demand
- Availability and affordability of housing
- Housing tenure (e.g. ownership or rental).

Increased demand for housing is caused by:

- people moving to urban areas in search of work and a higher standard of living
- longer life expectancy
- young people leaving home earlier
- more families breaking up and moving into separate homes
- more people preferring modern housing with good facilities.

The cost of housing and living is higher in urban areas than in rural areas. Urban areas may be overcrowded, and people may feel more stress and pressure in their daily lives, for example commuting to work. Air quality can be low due to emissions from factories, vehicles and public transport networks. There is also much less open space in urban areas compared with rural areas. Transport can be very congested, leading to longer travel times to get to places such as hospitals.

High competition for space and resources leads to smaller houses and gardens in large urban areas. There is a shortage of affordable housing in many cities. In some LICs, parts of the urban population make their own homes (informal housing). Up to 1 billion people live in informal settlements, which may be very overcrowded (Figure 7.11) and lacking in basic facilities, such as sanitation, clean water and electricity. Up to 50 per cent of displaced people live in urban areas, placing stress on these areas to provide for all their residents.

Employment

Employment is another challenge in urban areas. This may occur because the number of people seeking work exceeds the number of jobs that are available. There are a number of options available:

- The government can create more jobs through public works programmes, for example building infrastructure that benefits the urban area.
- People can find jobs themselves in the informal economy.
- People can job-share.
- People may work part-time.

Transport

Transport problems in urban areas include congestion and poor air quality. Congestion can delay journeys and make people late for school, work or appointments. It increases fuel consumption and adds to vehicular emissions of greenhouse gases and other

▲ **Figure 7.11** High-density housing, Petionville, Port-au-Prince, Haiti – the brightly coloured homes on the left of the photo were painted as part of a government project to improve impoverished areas

7.2 THE OPPORTUNITIES AND CHALLENGES OF URBANISATION

pollutants. It may lead to frustration and 'road rage' and it can have a negative impact on people's health through stress and poor air quality. In 2013 the World Health Organization announced that air pollution can cause cancer. Diesel fumes are carcinogenic, and air pollution has a causal link with respiratory diseases.

Waste management

Waste management is a major issue for large urban areas. For example, London produces about 7 million tonnes of rubbish per year from homes, public buildings and businesses. Some of this is disposed of in landfill sites, some is burned to generate electricity, and some is recycled. On average most households produce around 1000 kg of waste yearly.

About half of London's waste is recycled and most of the rest goes to landfill, but many landfill sites are likely to be full by 2026. London plans to recycle 65 per cent of its waste by 2030. The city produces 1.5 to 1.75 million tonnes of food waste annually, valued at about £2.5 billion. Much of this waste goes to landfill but some is incinerated to create energy from waste.

Unplanned settlements

Unplanned settlements in LICs may be bulldozed without warning. However, since authorities are unable to offer alternative housing, they often leave the unplanned settlements alone. Some say that these unplanned settlements are the solution to the housing shortage in LICs, as less wealthy people have built their homes and do not require housing to be supplied by the government.

> ### Activities
> 1 Outline the opportunities of living in urban environments.
> 2 Identify the problems of waste management in urban environments.
> 3 Suggest reasons why housing can be an issue in urban areas.

Favela Barrio Project

The Favela Barrio Project (Favela Neighbourhood Project) began in Rio de Janeiro in 1994. It aimed to recognise favelas (informal settlements) as neighbourhoods of the city in their own right and to provide the residents with essential services (Figure 7.12). Around 120 medium-sized favelas (those with 500–2500 households) were chosen to be part of the project.

The first phase of the project addressed the built environment, aiming to provide:

» paved and formally named roads
» water supply pipes and sewerage/drainage systems
» crèches, leisure facilities and sports areas
» relocation for families who were then living in high-risk areas, such as areas subject to frequent landslides
» channelled rivers to stop them from changing course.

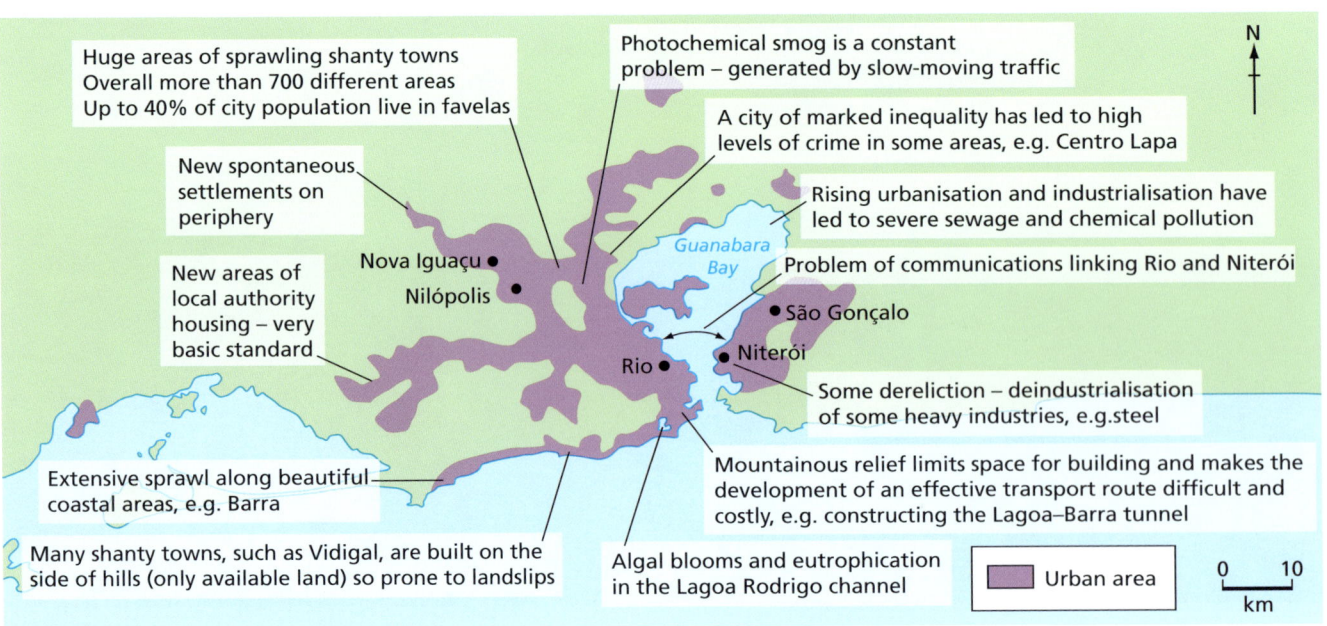

▲ Figure 7.12 Informal housing in Rio de Janeiro

The second phase of the project aimed to bring the favela dwellers into mainstream society and keep them away from crime. This is being done by:

- generating employment, for example by creating cooperatives of dressmakers, cleaners and construction workers, and helping them to establish themselves in the labour market
- improving education and providing relevant courses, such as ICT
- giving residents access to credit so they can buy construction materials and improve their homes.

The project has been used as a model of its type. The government is also helping people to become homeowners.

A number of developments have taken place to try to improve the quality of the education system. *Amigos da Escola* (school friends) encourages people from the community to volunteer their skills to improve opportunities offered by their local schools. *Bolsa Escola* (school grants) gives monthly financial incentives to low-income families to keep their children at school.

Rocinha is a central favela with a population of about 200,000 inhabitants (Figure 7.13). Over 90 per cent of the buildings are now constructed from brick and have electricity, running water and sewerage systems. Rocinha has its own newspapers and radio station. There are food and clothes shops, video rental shops, bars, a travel agent and, at one stage, a McDonald's fast-food outlet. Nearly all of this was achieved by the local community and their initiatives, often without government support. However, the lack of security of tenure means that many companies and/or political organisations do not invest in informal settlements as any improvements could be discarded.

▲ **Figure 7.13** Rocinha, Rio de Janeiro

Many of these improvements and developments are the result of Rocinha's location, close to wealthy areas such as São Conrado and Copacabana. Many of the residents work in these wealthy areas surrounding Rocinha and, although monthly incomes are low, they are not as low as elsewhere in the city and in Brazil as a whole. These regular incomes have allowed improvements to be carried out by the residents themselves. A distinction has been made between settlements known as 'slums of hope' (the more central informal settlements, such as Rocinha, that are closer to the city centre and where access to employment is therefore better), and those known as 'slums of despair' (the informal settlements on the outskirts of the city, such as Caxias, where access to employment is more limited).

Ways in which people can adapt to living in urban areas (if they have the choice and resources to do so) include:

- by living near public transport routes
- by shopping locally
- by utilising public spaces like parks and facilities such as libraries
- by creating communities.

> ### Activities
> 1. Identify three measures or pieces of evidence of inequality within Rio de Janeiro.
> 2. Suggest why there is a housing shortage in Rio de Janeiro.
> 3. Outline the opportunities of living in squatter settlements.
> 4. Distinguish between settlements that are called 'slums of hope' and those that are called 'slums of despair'.

The impacts of urban sprawl on the rural-to-urban fringe/surrounding areas

Urban sprawl is uncontrolled, outward growth at the edges of urban areas. Positive aspects include greater land availability and cheaper land, so that companies can expand and larger houses with bigger gardens can be built. However, negative impacts include the erosion of agricultural land, less food production, increased commuting, a decline in air quality, degradation of local ecosystems, a decline in local biodiversity and a reduction of ecosystem services such as climate

7.2 THE OPPORTUNITIES AND CHALLENGES OF URBANISATION

▼ Table 7.1 Issues in the urban fringe

Land use	Positive aspects	Negative aspects
Agriculture	• Some farms may diversify into alternative activities, such as 'pick your own food'	• Farms on the urban fringe often suffer litter, trespass and vandalism • Some land is left derelict in the hope that planning permission for development will be granted
New developments	• Some developments are well-sited and landscaped, such as business and science parks	• Some developments, such as out-of-town shopping areas, may cause traffic congestion and pollution • Many businesses are unregulated, for example scrap metal processing/storage
Urban services	• Some can be seen as attractive compared with dense built-up areas, e.g. cemeteries have lots of open space and flowers, and reservoirs may have many water birds	• Sewage works, markets, car boot sales and landfill sites can be unattractive and polluting
Transport infrastructure	• Cycleways can improve access and promote new development	• Motorways destroy countryside, especially near junctions
Recreation and sport	• Country parks, sports fields and golf courses can include conservation	• Car racing and scrambling/climbing erode ecosystems and create localised litter and pollution
Landscape and nature conservation	• Conservation areas may be included at the edge of the city	• There may be degraded land, for example land affected by dumping or 'fly-tipping'

regulation and flood control (Table 7.1). It also leads to land-use conflicts as many different groups compete for the land.

Recent developments in the rural-to-urban fringe include the construction of roads, especially peripheral ring roads, housing developments, out-of-town shopping centres (Figure 7.14), science parks and industrial estates.

▲ Figure 7.14 An out-of-town supermarket, with free parking and good accessibility for workers and customers

The impacts of urban sprawl on the rural-to-urban fringe/surrounding areas

Many businesses **decentralise** to the rural-to-urban fringe because of good levels of accessibility for road transport, space to offer free parking and a clean, attractive environment. According to the World Bank, between 2016 and 2030 the physical growth of urban areas is likely to be about 1.2 million km².

Many transport schemes, such as much of the M25 in the UK, are located in the rural-to-urban fringe. Park-and-ride schemes (in which commuters park their car in a car park and then take a bus to the city) are located in the urban fringe. Housing developments also take advantage of greenfield sites (previously undeveloped sites) in the rural-to-urban fringe, but they contribute to urban sprawl.

Urban sprawl has become a feature of many cities in the USA. Some geographers predict that the whole area between Phoenix and Tucson in Arizona could become one enormous urban sprawl by 2100. Phoenix is 70 per cent larger in area than New York City, but has only 20 per cent of the population of New York City (Figure 7.15). Phoenix has been described as the 'world's least sustainable city', although many of its suburbs on the rural-to-urban fringe, such as Anthem, a planned community, are extremely popular as places to live.

The car industry and the oil industry have enabled settlements to spread out along super-corridors and highways. One downside has been the enhanced stress on society, with increases in road rage and congestion and declining air quality.

Some countries have introduced green belts (also known as green wedges, greenways or green spaces) to restrict urban sprawl and protect the natural and semi-natural environment. These areas improve air quality, reduce the risk of flooding and enable people (with cars) to access the countryside.

> **Activities**
> 1 Explain reasons for the growth of developments in the rural-to-urban fringe.
> 2 Suggest how developments in the rural-to-urban fringe impact the natural environment.

▲ **Figure 7.15** Residential urban sprawl west of Phoenix, Arizona

> **Interesting note**
>
> The first green belt was established by the Prophet Muhammad PBUH in the seventh century, around Medina in what is now Saudi Arabia.

7.3 The management of urban growth

This chapter will explain:

★ the strategies and techniques used to manage urban growth
★ the extent to which these strategies and techniques are successful
★ how the management of urban areas can be sustainable
★ detailed specific example: urban change in Shanghai.

Several issues in urban areas need to be managed, including housing, transport, industry, retail, leisure, social issues such as migration and inequality, and environmental issues such as air pollution and access to clean water. Many strategies and techniques can be used.

Managing housing

Access to affordable, high-quality housing is one of the biggest issues in urban areas. This is due in part to in-migration, but also to more people living longer and households splitting up. Potential solutions include building more social housing that is affordable for more people. However, developers make greater profits from building more expensive housing for richer residents. To deal with more single people in urban areas, more flats and apartments could be built.

Renewal of older buildings (gentrification) allows some buildings to be re-used rather than demolished. However, it may force poorer residents out of a neighbourhood. In some areas, old industrial buildings can be converted into residential buildings. For example, Lucy's Iron Foundry in Oxford was converted to residential flats (Figure 7.16). This can be a sustainable use of an old, derelict building.

In many LICs, squatter settlements (low-income housing) could be given legal status/security of tenure, which would remove the threat of them being demolished by the local or national government. In many HICs, such as the USA and the UK, dealing with the increased demand for housing in large urban areas has led to the decentralisation of people away from larger urban areas to smaller urban areas, including new towns and expanded towns.

▲ Figure 7.16 The conversion of Lucy's Iron Foundry from an industrial to a residential use

Managing transport

Managing transport can be a major issue in many urban areas. Transport can lead to congestion, poor air quality and, in some cases, road rage. However, there are many ways to reduce the number of vehicles on the roads in urban areas and many examples of sustainable forms of transport management. These include:

» encouraging the use of public transport
» the creation of low-emission zones
» car-pooling schemes (such as in BedZED in London)
» park-and-ride schemes (such as in Oxford, UK)
» congestion charge schemes
» vehicle exclusion zones

- use of vehicle-restriction practices (e.g. only allowing vehicles with an odd number plate on certain days of the week and with an even number plate on the other days, as used in Mexico City, Athens and Singapore)
- the reintroduction of trams (e.g. between Wimbledon and Croydon in London)
- permit-only parking
- pedestrianised shopping streets (such as Rose Street, Edinburgh)
- cycle lanes (such as in Copenhagen)
- transport calming (e.g. speed bumps)
- speed limits.

Singapore opened a Mass Rapid Transit (MRT) system in 1987. This has proved very successful and has enabled population and industry to move away from central areas but still be able to reach the city centre quickly. There are six MRT lines, over 130 stations and about 200 km of rail line. The MRT system and the bus network meet at Integrated Transport Hubs. The government has also tried to reduce the number of private vehicles on the road by introducing a system that allows different vehicles to use the roads only on certain days of the week.

Managing industry

Large-scale industrial polluters can be moved, for example the steel works in Beijing was relocated away from the city prior to the 2008 Olympic Games. In many HICs there has been a large-scale decline of industry in urban areas, notably the car industry in Detroit. The change from manufacturing industry to service industries tends to have a positive impact on the natural environment. Many science parks are landscaped, with green spaces and water features.

A more recent development, which accelerated during the Covid-19 pandemic, was the number of people working from home. This trend was more noticeable in HICs than LICs due to the greater coverage of, and access to, ICT. It reduces the number of people commuting to work and, arguably, is more sustainable.

Managing retail

Issues relating to retailing include the higher cost of living in many large urban areas, the development of more out-of-town retailing, which benefits car owners, and the decline of many more central retail units due to the developments on the urban fringe. This relates mainly to HICs that have a large proportion of car-owning households. Many retailing premises have closed in recent years, in part due to the growth of online retailing and in part due to the Covid-19 pandemic. Retailing in cities is very diverse. At one end, there are formal shopping areas, such as in city centres and retail parks. At the other end, much retailing is informal and carried out by individuals who sell goods by the roadside. Some markets may be permanent, whereas others might occur at a location once a week or once a month. Some 'retailing' may occur at car-boot sales, where individuals or households sell some of their own goods to other people.

Managing leisure

In some urban areas, the amount of open space available for recreation and sport is limited. This is true in LICs and in some HICs. Many schools have limited areas for sports to take place. However, in areas of low-density urban development, such as in the USA and Australia, there is more open space available for leisure facilities. Some forms of leisure can be very large-scale, such as national sports events or professional sports teams, and these may generate large numbers of supporters to watch the sporting events. In many cases, this may lead to congestion, litter, and noise and light pollution. However, such events may provide great enjoyment, much employment and social/community activities for local residents.

Many leisure activities are very sustainable and have limited environmental impact, for example walking, jogging and open-water swimming. Some forms of leisure may be very productive, such as the provision of allotments so that people can produce food. Some food may also be grown in people's gardens, for example by having fruit trees.

Managing social issues

Population issues in large urban areas include exclusionary zoning, in which low-income people of selected racial groups are prevented from living in certain areas. This is also called 'red-lining'. Urban areas may be considered as 'melting pots' for migrants or may lead to anti-migrant feelings. Urban poverty is another urban issue. This could be addressed by creating more jobs and building partnerships between

7.3 THE MANAGEMENT OF URBAN GROWTH

higher and lower-income areas, for example by working together to improve informal settlements or expanding housing for those in poverty. However, for many urban areas this is too costly or not a priority.

Managing environmental issues

Air pollution in urban areas can be an issue due to the concentration of industries, factories and vehicles. In addition, some households burn fuelwood, rubbish and plastic indoors, or use paraffin or low-grade coal. Potential solutions include planting more trees, having fewer vehicles on the road and using renewable energy sources rather than coal or fuelwood.

In Warsaw, Poland, the burning of coal was banned in 2023 in order to reduce air pollution. Pollution controls need to be established and enforced. In India, Prime Minister Modi has pledged to reduce the country's carbon emissions by 45 per cent by 2030. However, coal accounts for about 50 per cent of energy in India. In urban areas, such as Mumbai and New Delhi, authorities have taken to washing some roads and building sites to improve air quality. New Delhi is also affected by the burning of wheat and rice in winter. India's National Clean Air Programme was introduced in 2019. Initiatives include promoting renewable energy, developing electric vehicles and encouraging households to use liquefied petroleum gas for cooking.

Water in many urban areas may be scarce due to the large number of people present and/or climate change, as was seen in Cape Town's water shortage ('Day Zero'), which peaked in 2017–18. To overcome this, more water could be collected in water butts, greywater could be used for watering gardens and/or impermeable surfaces could be changed to permeable ones so that water does not run off. Programmes to increase the amount of urban forests or green roofs (Figure 7.17) could store more water in urban areas, and there could be more water conservation in homes, schools and industries.

▲ **Figure 7.17** Green roofs in Singapore

To improve urban environments, the number of vehicles could be restricted. New open spaces could be created, such as the Queen Elizabeth Park in East London for the London 2012 Olympic Games. Other ways of creating open spaces include green belts, green wedges and green spaces.

> ### Activities
> 1 Evaluate two strategies used to manage urban growth.
> 2 Compare the strategies to manage either housing or transport in urban areas in LICs and HICs.

➡️ Detailed specific example

Urban change in Shanghai

Urban growth in Shanghai

For around 700 years Shanghai has been one of Asia's major ports, and it has had a varied history. It thrived until 1949, when China closed itself off from trade with the West. This changed Shanghai from an international centre of production and trade to an inward-looking city. During the 1970s, China began slowly reopening its economy to the world, and Shanghai was designated one of 14 open cities. The Shanghai Economic Zone was established in 1983, and in the early 1990s an ambitious major programme of redevelopment was started, especially in the eastern hinterland around Pudong (Figure 7.18). The neighbouring cities of Suzhou and Wuxi have slowly merged with Shanghai to create one continuous megalopolis. Since economic reforms began in China in 1978, around 200 million people have moved from rural areas to urban areas. This may be the largest population movement in human history.

Site and situation

Shanghai developed on a flat, low-lying alluvial plain on the banks of the Yangtze River (Figure 7.19). It is located at the confluence of the Huangpu and Suzhou Rivers – so it has an excellent location for shipping and trade. From 1844, British, French, American and Japanese traders owned land in Shanghai. By 1920 it was China's largest and most important city, but the foreign influence declined after the end of the Second World War and following the Communist Revolution in 1949.

Managing environmental issues

▲ **Figure 7.18** Housing in Pudong, Shanghai

▲ **Figure 7.19** The situation of Shanghai in China

Economic change

Between 1949 and 1976, political influences such as the Great Leap Forward (1958–60) and the Cultural Revolution (1966–76) focused attention away from rural areas, foreign influence and capitalist development. During this period 1 million people returned to the countryside.

However, in 1979 the first generation of Special Economic Zones was created. Although Shanghai was not one of them, it benefited from relaxed housing restrictions such as the subdivision and subletting of housing.

In 1984, Shanghai was declared open to foreign development. For much of the next 20 years, Shanghai's economic growth rate was over 12 per cent per annum. In 1990, a new CBD was created in the Pudong area (Figure 7.20). Banks, stock exchanges and insurance companies moved in. By 2000, over 3000 skyscrapers had been built, including the Shanghai Financial Centre.

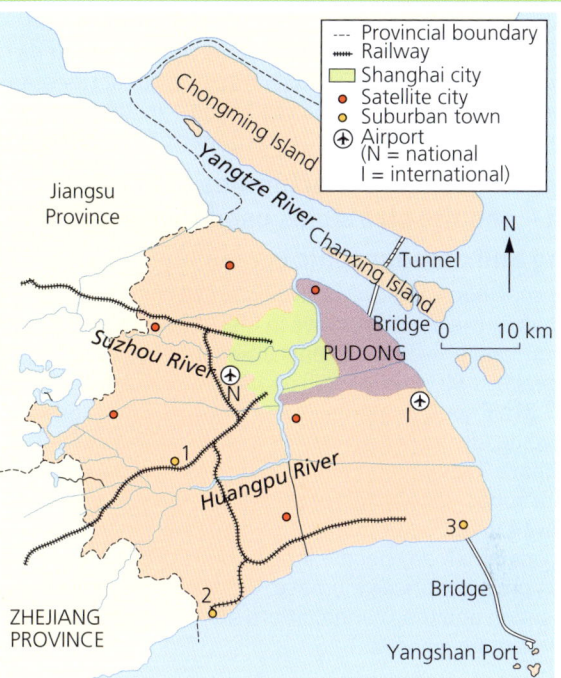

▲ **Figure 7.20** Land use in Shanghai

Having established a strong industrial base, by 1990 the city was well placed to take advantage of the new opportunities offered by globalisation. It became a major centre for export manufacturing (based on automobiles, biotechnology, chemicals and steel), and its service industry sector (trade, finance, real estate, tourism and e-commerce) helped to diversify its economy. Between 1990 and 2000, Shanghai began to re-emerge as a world city. Foreign investment was attracted. Over half of the world's top 500 transnational companies and 57 of the largest industrial enterprises set up in Shanghai, contributing to an annual regional growth rate of over 20 per cent – more than twice the national average. In 2009, Shanghai was ranked the seventh-largest city in the world with a population of 15 million. In 2024 it had grown to almost 25 million.

Since 1990, the city's manufacturing sector has steadily contracted, losing over 1 million jobs, while the business services, finance and real estate sectors have expanded. Rising demand for highly skilled labour has led to further in-migration, resulting in an increasing disparity in wealth. Shanghai's experience does lend support to the general hypothesis that world city status inevitably leads to a widening gap between lower and higher incomes.

In 2006, Yangshan Port was opened to accommodate larger ships that could enter the Huangpu and Suzhou Rivers. Yangshan was built on an island about 40 km southeast of Shanghai and connected by a 35 km bridge. It is now the world's largest port. In 2023, it handled over 49 million containers, compared with 39 million in Singapore, the world's second-largest port.

Shanghai is a city-state. It is part of the Yangtze River Delta Economic Zone, one of the fastest-growing urban areas in the world, containing 16 megacities, including

7.3 THE MANAGEMENT OF URBAN GROWTH

Shanghai. The region has a population of 75 million and earns 25 per cent of China's GNP – and 50 per cent of its foreign direct investment. The city has been described as the largest construction site in the world – in 2020, 4000 buildings with over 24 storeys were under construction.

The challenges and opportunities of urban growth – and the strategies used to manage it

Housing and demographic issues

Housing shortages and overcrowding problems are acute. Almost half of the population live in less than 5 per cent of the total land area, and in central Shanghai population density reaches 40,000–160,000 people per square kilometre (Figure 7.21). Population pressure is caused by in-migration, overcrowding, disparities in wealth and the social insecurity of Shanghai's 'floating population'. From the 1990s, whole neighbourhoods were demolished. Over 2 million residents were moved to the outer suburbs to live in better-quality accommodation. Many people are unaware of their property rights. As property prices increase, they are given insufficient compensation and they cannot afford alternative housing in their old neighbourhoods.

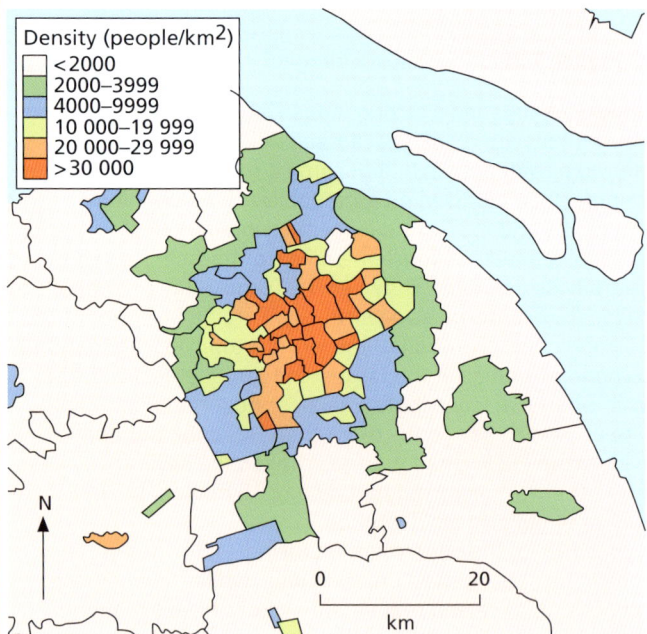

▲ **Figure 7.21** Population density in Shanghai

The Shanghai government has established a series of important policies to address these problems:

- A combination of widespread family planning and medical care, which has controlled fertility levels among the young immigrant population
- Compulsory work permits
- Educational initiatives to improve immigrant job opportunities.

These programmes have reduced population density in the heart of the city and increased it in the suburban satellite cities such as Songjiang. This has been a successful strategy, although it means that many who have been moved out under the decentralisation policy must travel back to the centre for work.

Economic growth has attracted an increasing number of Chinese living overseas and foreign migrants to live in Shanghai. Many of these live in luxurious, gated apartments.

Changes in transport infrastructure in Shanghai

As a result of population growth and the expansion of the built-up area of Shanghai, there is the need to develop transport systems. Shanghai's integrated transport system is developing a transport infrastructure that will deal with increased traffic volumes. The integrated transport system focuses on two ports, two highways and three transport networks.

Shanghai has completed the north port area of the Yangshan Port. It is the busiest port in the world, with an annual cargo of nearly 700 million tonnes. Shanghai has two international airports and four airport terminals. It receives over 70 million passengers each year. Pudong International Airport is the world's third largest in terms of cargo.

Shanghai has over 12,000 km of roads, including nearly 800 km of expressways. The target for planners has been referred to as '15, 30 and 60', meaning that motorists in suburban areas can reach an expressway within 15 minutes, then travel to the city centre in 30 minutes, and the time needed to travel between any two suburban areas is less than 60 minutes.

The Urban Expressway Network is over 4000 km long, with over 3000 km of road in the city centre and nearly 200 km of expressway. There are some 23 crossing facilities for the Huangpu River, including four bridges and 12 tunnels.

Shanghai's railway has nearly 400 km of line. Journey times have been shortened – for example, journey time fell from 90 minutes to 30 minutes to Hangzhou and from 180 to just over 70 minutes to Nanjing. The urban rail network has developed in less than 20 years. There are 13 Metro lines, which together carry over 5 million passengers daily. Approximately 25 per cent of the city centre is covered by railway stations, serving 40 per cent of the city centre's population. In addition, there are over 1000 bus lines and 17,000 buses.

The practical targets of Shanghai's transport strategy are that:

- public transport between any two parts of the city centre should take less than 60 minutes
- residents in suburban areas should be able to access the Metro with just one bus ride
- there should be direct bus links between the new towns and the urban centre.

Access to water and sanitation

The Huangpu River and the Yangtze River are the main surface water sources for Shanghai's water supply. Parts of the upper reaches of the Huangpu River pass through suburban Shanghai, with its intensive farming activities, while the lower course passes through urban areas, with intensive industrial activities.

Managing environmental issues

In urban areas throughout China, the proportion of people with access to piped water rose from 40 per cent in 1990 to nearly 99.4 per cent in 2022. By 2020 nearly 90 per cent of Shanghai's wastewater was treated. Over 70 per cent of households have access to sewerage services.

Although water is abundant in Shanghai, there are high levels of water stress. Increasing demand, pollution and saltwater intrusion are threatening the city's water security. For example, agricultural practices have led to fertilisers and insecticides getting into the urban water system. The relatively low capacity of the sewage treatment system has led to some industrial and residential waste being discharged directly into rivers. Due to organic pollution, the Huangpu River has lower water quality than the Yangtze River. In addition, saltwater intrusion is a seasonal problem in the winter and early spring dry season.

To combat the need for more freshwater, Shanghai built the Qingcaosha Reservoir, and this has been designed to provide water for up to 68 days (the theoretical maximum extent of saltwater intrusion in Shanghai).

Waste treatment

As Shanghai increases in size, it is producing more rubbish, and the sites for landfills are filling up. Increasingly, Shanghai is turning to incineration, and generating electricity at 'waste to energy' plants, such as the one at Hangzhou. The demand for such incinerators is increasing at a much more rapid pace than elsewhere in the world.

In the past, most of Shanghai's rubbish ended up in landfill sites or in unregulated heaps on the edge of the city. At such sites, there was much informal recycling of materials, but there was also contamination of the land and groundwater by methane. Plastics were carried into the rivers, and from there into the ocean.

Shanghai currently produces the most household rubbish in China – 22,000 tonnes/day. China's largest landfill, Laogang, is located near the city by the coast. However, there is now an incinerator next to the landfill and this takes in around 3000 tonnes a day. This is much more sustainable. The waste is burned at high temperatures (850°C or higher) to destroy toxins. Water is heated to produce steam, which in turn drives the turbines to generate electricity. The new incinerators at Laogang will burn 9000 tonnes a day. The aim is to increase the amount of waste incinerated from 35 per cent of domestic waste to 75 per cent.

Shanghai's Master Plan

According to Xie and Zhang (2020), the Shanghai Master Plan (introduced in 2000) focused too much on westernisation and economic growth (for which it has been very successful) and not enough on human development. They consider that Shanghai's main aim in the 2017–35 Master Plan is to be an 'excellent global city', but that growth and development have led to increased social disparities.

The UN had previously described Shanghai as the model city for sustainable development, with its 2011 'Better city, better life' programme promising balance, growth and an open and multicultural society. However, developments have been dominated by Western-dominated practices, notably a concentration on economic development. This has reinforced gaps in social and environmental justice. China's income inequality is high – the top 10 per cent of households account for 30 per cent of the wealth, while the lowest 10 per cent account for 3 per cent of the wealth. Inequality in Shanghai is also marked – in 2019, the land values in the old, central areas of Jingan, Xuhai and Huangpu for residential flats were US$8500 per m^2, whereas neighbouring districts that were once crowded with factories, such as Zhaibei, Yangpu and Putuo, were described as hubs for 'poor wretches'.

The Master Plan hopes to cap Shanghai's population at 25 million. This will be achieved by pushing excess people to marginal satellite towns. However, many of these have limited infrastructure and are under-developed with respect to schools and hospitals. Most of the employment remains in the city centre. Moving lower-income people to the edge of Shanghai leads to higher costs and more time for commuting, thereby reinforcing inequalities. The Master Plan also proposes the creation of more green spaces. For example, in 2023 the Yangshupu Power Plant became a public green space (Figure 7.22). This is a good example of sustainable development – using an old building premises for the greater good of the people.

▲ **Figure 7.22** In 2023 the Yangshupu Power Plant became a public green space

7.3 THE MANAGEMENT OF URBAN GROWTH

More than 50 per cent of people living and working in Shanghai do not have a Shanghai Hukou (registration) and thus they cannot draw on social support mechanisms. Many non-Shanghai residents are forced to live in satellite towns where social services are inferior, rather than in the city centre. In addition, Shanghai wants to attract foreign talent. The city has a lower proportion of foreign skilled workers than London, New York and Los Angeles, but it wants to be seen as internationally competitive (Xie and Zhang, 2020).

Activities

1. Compare the challenges and opportunities brought about by urban growth in Shanghai.
2. Assess the strengths and weaknesses of strategies to manage urban growth.

Fieldwork

Urban fieldwork

The central business district (CBD) is the commercial and economic core of a city. It is the heart of the city and is the area that is most accessible by public transport. It has a number of characteristics:

- Multi-storey development: High land values force buildings to grow upwards, so the floor space of the CBD is much greater than the ground space.
- Concentration of retailing: Accessibility attracts shops with a high **range** and **threshold characteristics**. Department stores are located in CBDs; specialist shops occupy less accessible areas.
- Offices are concentrated in the CBD: Centrality favours office development.
- Vertical zoning is apparent: Shops occupy ground floors while offices occupy upper floors.
- Few people live in the CBD: A few luxury flats exist and there may be some artisans or university accommodation.
- Pedestrian flows and traffic restrictions are greatest in the CBD: Pedestrianisation has reduced areas for cars since the 1960s.

Possible hypotheses to test

Possible hypotheses that could be tested in the fieldwork include:

- The CBD is the most accessible area of the city for public transport.
- Building heights are highest in the CBD.
- Land use is mainly commercial – little land is used for residential or manufacturing purposes in the CBD.
- Pedestrian flows are greatest in the CBD.
- There are more parking restrictions in the CBD than elsewhere in a city.
- Vertical zoning is apparent.

Data collection

You could:

- compare different parts of the CBD, for example the core and frame, or compare the CBD with a suburban area, or take a transect from the centre of the CBD towards the edge of a small town/city
- use a base map of the city centre to mark off land use – make sure that you have a key to include different categories, such as commercial (high order), commercial (low order), finance (for example banks), restaurants, etc.
- record pedestrian flows at selected locations, counting the number of pedestrians passing you at 2-minute intervals
- analyse bus timetables to show the number and frequency of buses over several hours
- compare ground floor usage with that of upper floors to show vertical zoning.

Along with these quantitative data forms you should take photographs on your phone to annotate later and provide qualitative data about the nature of the environment.

Presenting your findings

To present your findings you could use land-use maps, annotated photos, flowcharts of the number of buses, or located pie charts/bar charts to show pedestrian counts.

Managing environmental issues

Practice questions

1 Figure 7.23 shows changes in urban and rural populations by region between 1950 and 2050 (predicted from 2025).
   a State the continent with the largest number of urban dwellers in 2025. [1]
   b State the decade when 50 per cent of the population in Latin America and the Caribbean lived in urban areas. [1]
   c Describe the trend in levels of urbanisation in Europe between 1950 and 2050 (predicted from 2025). [3]

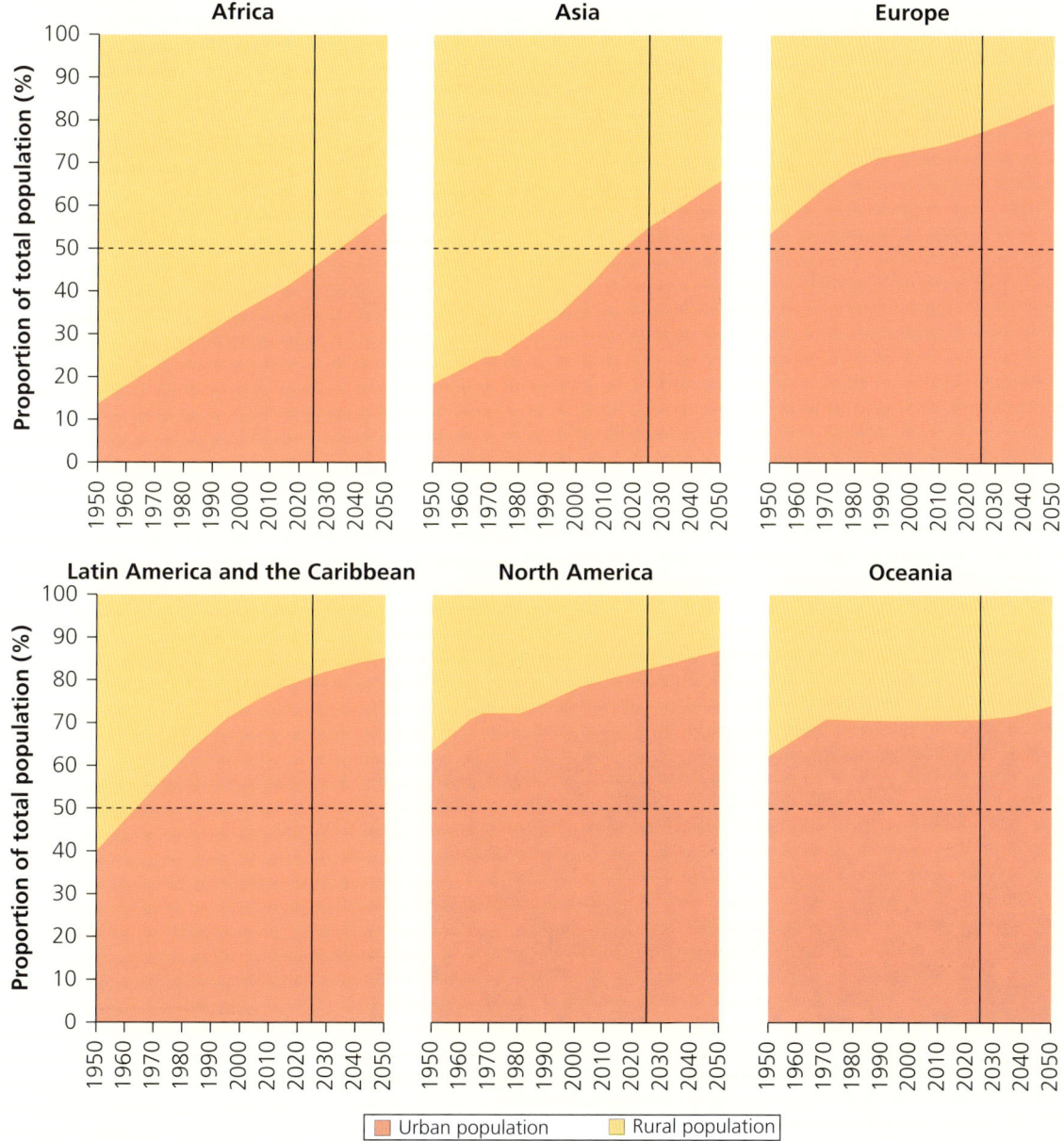

▲ Figure 7.23 Changes in urban and rural populations by region, 1950–2050 (projected)

7.3 THE MANAGEMENT OF URBAN GROWTH

2 Figure 7.24 shows seasonal variations in air quality in Delhi. Explain the seasonal variations in air quality in Delhi. [5]

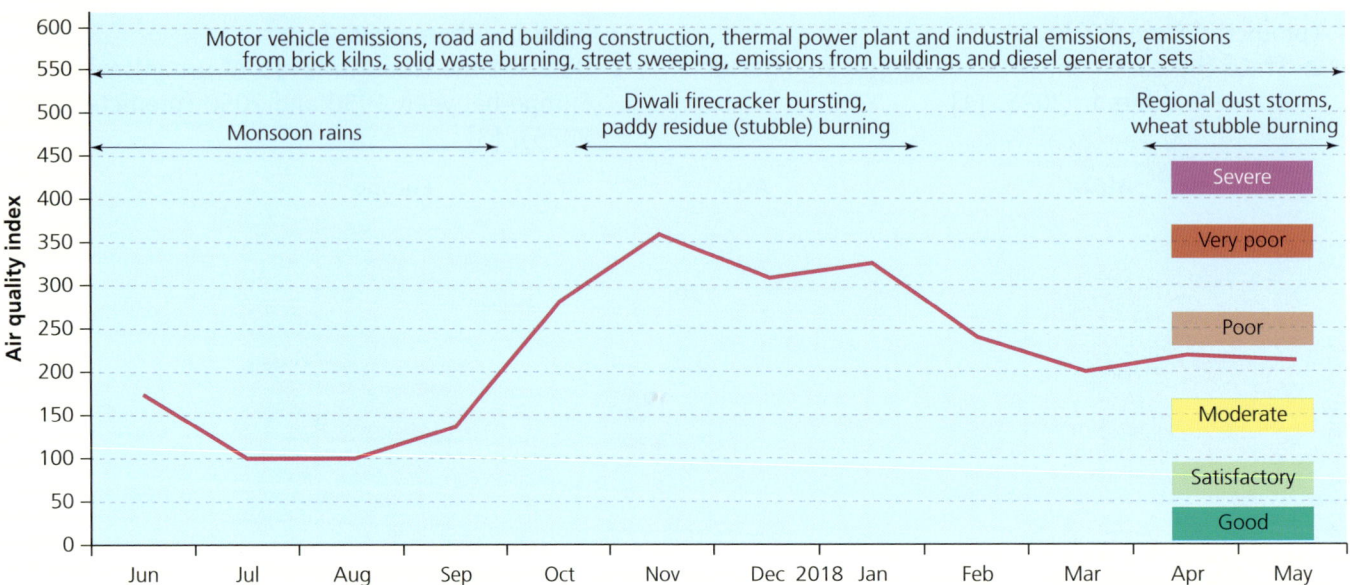

▲ Figure 7.24 Seasonal variations in air quality in Delhi

3 Figure 7.25 shows a development on the rural-to-urban fringe.
   a Define the term 'rural-to-urban fringe'. [1]
   b Suggest the advantages of the rural-to-urban fringe for new developments. [3]
   c Assess the likely impacts of new developments in the rural-to-urban fringe on the environment. [4]
4 For a named example, assess the challenges and opportunities of urban living. [7]

▲ Figure 7.25 A development on the rural-to-urban fringe

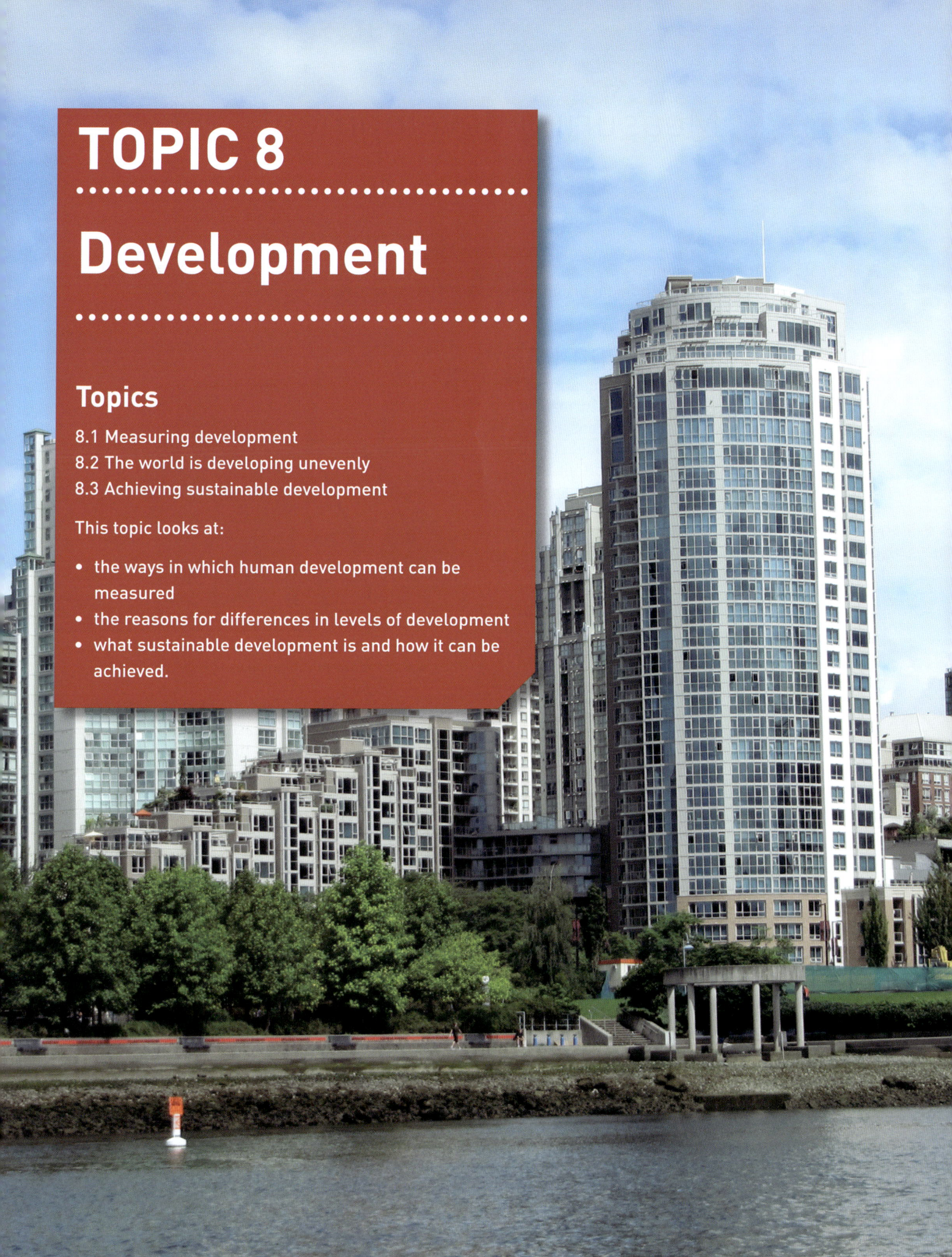

TOPIC 8
Development

Topics

8.1 Measuring development
8.2 The world is developing unevenly
8.3 Achieving sustainable development

This topic looks at:

- the ways in which human development can be measured
- the reasons for differences in levels of development
- what sustainable development is and how it can be achieved.

8.1 Measuring development

This chapter will explain:

★ indicators of development
★ the Human Development Index: a broader measure of development.

Indicators of development

Development, or improvement in the quality of life, is a wide-ranging concept. It includes increasing wealth, but also involves other important aspects of our lives (Figure 8.1). Many people would consider a good health service to be more important than wealth. People who live in countries that are not democracies, where freedom of speech cannot be taken for granted, often envy those who do live in democratic countries, though many African countries look to China as a model of success.

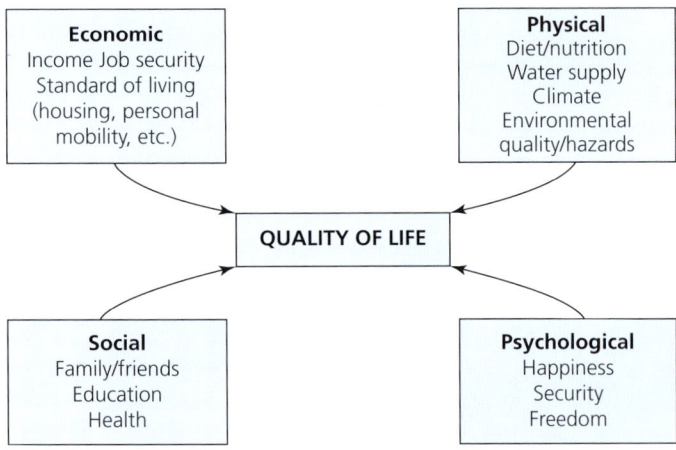

▲ Figure 8.1 Factors comprising the quality of life

Development occurs when there are improvements to the individual factors making up the quality of life and standard of living (Figure 8.2). The more factors that are improved in terms of people's wellbeing, the greater the rate of overall progress. For example, development occurs in a low-income country when:

» local food supply improves due to investment in farm machinery and fertilisers
» the electricity grid extends outwards from the main urban areas to rural areas
» levels of literacy improve throughout the country
» a new road or railway improves the accessibility of a remote province.

▲ Figure 8.2 An open-pit toilet in the Gobi Desert: in some parts of the world sanitation can be very basic

Thus, development is a process of change for the better. However, in some cases it can proceed in a way that adversely affects the most impoverished in a society. Development is not always a continuous process as it can be interrupted by major national and global events such as war, famine, disease and economic recession. The Covid-19 pandemic had a massive negative impact on development, and set back progress in LICs by many years.

The **United Nations Development Programme (UNDP)** is the UN's development agency working to eradicate poverty. It plays a crucial role in helping countries achieve the Sustainable Development Goals. The UNDP has highlighted a number of important issues concerning development:

» Development is about creating opportunities and giving people the freedom to make choices.
» Sustainability is vital for long-term development.
» Development must balance social, economic and environmental sustainability.

Table 8.1 looks at four categories of development, giving examples of how countries can improve these aspects of the quality of life of their citizens. It is not an exclusive list and you may well be able to think of other examples within each category.

▼ Table 8.1 Categories of development and examples

Category	Examples
Social	• Achieving higher levels of literacy • Reducing infant and maternal mortality • Improving life expectancy • Reducing the gender gap • Improving the quality of housing and other aspects of soft infrastructure
Economic	• Raising productivity • Increasing incomes for all levels of society • Improving hard infrastructure (roads, railways, airports) • Achieving a better balance between economic sectors • Increasing trade with other countries • Raising per capita energy consumption • Increasing level of car ownership
Environmental	• Reducing CO_2 and other greenhouse gas emissions: cutting lower atmosphere air pollutants • Improving water quality in rivers and other water bodies • Tackling land pollution and other issues such as deforestation • Conserving biodiversity
Political	• Improving democratic institutions • Implementing measures to reduce corruption • Keeping to new international human rights obligations

Economic indicators of development

There are three widely used economic indicators of development:

» Gross domestic product (GDP) is the total value of goods and services produced within a country's borders, regardless of the nationality of those involved.
» Gross national product (GNP) focuses on the income generated by a country's residents, regardless of location.
» Gross national income (GNI) provides the most comprehensive picture by incorporating both GDP and GNP.

The World Bank, the Human Development Index and the annual World Population Data Sheet published by the Population Reference Bureau all use GNI to illustrate the development differences between countries. For all three of these economic indicators, both absolute and relative data is available. Relative data, for example **GNI per capita**, tends to be the most widely used. GNI per capita (see Table 8.2) is calculated by dividing the total GNI of a country for a given year by the number of people in that country. Per capita data allows comparisons to be made between countries that have big differences in their population size. For example, the total GNI of China is greater than that of France, but GNI per capita is much higher in France.

However, such data do not take into account the differences in the cost of living between countries. For example, a dollar buys much more in Bangladesh than it does in the USA. To account for this, the GNI per capita at **purchasing power parity (PPP)** can be calculated.

While GNI per capita is a useful measure, it does not tell us anything about:

» wealth distribution within a country – in some countries the gap between highest and lowest incomes is much greater than in others
» the ways in which governments invest the money at their disposal – a number of factors can affect the efficiency with which governments operate, with a high level of **corruption** being a significant problem in some countries.

Economic growth (GNI growth) refers to an increase in real GNI (percentage in a year). 'Real' means after inflation is taken into account. Countries that experience high rates of inflation often struggle to cope with rising prices as average incomes lag behind. A high rate of inflation can adversely affect economic confidence within a country itself and also impact investment coming into the country from abroad.

> ### Activities
> 1 What is development?
> 2 Give two examples of development in a low-income country.
> 3 The UNDP stresses the importance of a balance between which three types of development?
> 4 What is the reason for calculating GNI data at purchasing power parity?

8.1 MEASURING DEVELOPMENT

▼ Table 8.2 Key development indicators by world region

Region	Life expectancy at birth (years)	Infant mortality rate (per 1000)	GNI per capita (US$)	Per capita kilocalorie supply from all foods per day (kcal)
World	72	29	25,100	2959
Africa	63	44	5645	2573
North America	77	5	75,470	3587
Latin America/Caribbean	74	13	18,412	3108
Asia	74	24	16,349	2927
Europe	78	4	47,960	3456
Oceania	79	15	43,904	3087

Other indicators of development

Many individual indicators other than those above, which are clearly economic, have been used to highlight contrasts in development. Some of the most widely used are the literacy rate, life expectancy at birth, infant mortality, calorie intake and number of people per doctor. Development varies not only between countries – it can also vary significantly within countries. For instance, the number of people per doctor is likely to be higher in rural areas. Most of the measures that can be used to study the contrasts between countries can also be used to look at regional variations within countries.

Literacy

Education is the key to socioeconomic development. It can be defined as the process of acquiring knowledge, understanding and skills. Education has always been regarded as a very important indicator of development and it has figured prominently in aggregate/composite measures of development. Adult literacy (the ability to read and write) is one of the main ways in which differences in educational standards between countries can be shown. In 2020, the global **adult literacy rate** was estimated at 86 per cent. For developed nations it is at least 96 per cent, but for LICs the average adult literacy rate is only 65 per cent. A low adult literacy rate is a great obstacle to development. Poverty and illiteracy go hand in hand. Many of the countries with the lowest literacy rates are in Sub-Saharan Africa.

Of the roughly 780 million adults worldwide who cannot read and write, nearly two-thirds are female. Improving female literacy is essential because so many aspects of development depend on it. For example, there is a very strong relationship between levels of female literacy and infant mortality rates. People who are literate are much more able to access medical and other information that will help them achieve a higher quality of life compared with those who are illiterate.

Life expectancy at birth

Life expectancy at birth is to a large extent the end result of all the factors contributing to the quality of life in a country. The main influences on life expectancy are:

» the incidence of disease, e.g. malaria (Figure 8.3)
» physical environmental conditions, e.g. very low rainfall
» human environmental conditions, e.g. pollution
» personal lifestyle, e.g. smoking.

Rates of life expectancy in HICs and LICs have converged significantly during the last 50 years, in spite of a widening wealth gap. Table 8.2 shows that the lowest life expectancy by world region is in Africa, at 63 years. However, in 2014, Africa's life expectancy was 59 years. It is encouraging that some aspects of development can improve quite quickly!

Infant mortality

The **infant mortality rate** is regarded as one of the most sensitive indicators of socioeconomic progress. It is an important measure of health inequality between and within countries. There are huge differences in infant mortality around the world, despite the wide availability of public health knowledge. The main causal factors of such disparities are:

» contrasts in material resources
» differences in the efficiency of social institutions and health systems.

Indicators of development

▲ **Figure 8.3** A public health poster in a park in Buenos Aires, Argentina – disease control is an important aspect of development

Fortunately, infant mortality rates have fallen sharply in many MICs and LICs over the last 30 years. Nevertheless, Table 8.2 shows that the infant mortality rate in Africa is 11 times that in Europe. Sixteen countries had an infant mortality rate of 50/1000 or more in 2023 – this means one in 20 infants dying before their first birthday. All 16 of those countries are in Sub-Saharan Africa, with the highest levels of infant mortality being in Sierra Leone (75/1000) and Lesotho (69/1000). Table 8.3 shows the extent to which infant mortality varies by region in Africa. Some countries have registered very large reductions in infant mortality in recent decades, while others have seen reductions at a lower rate.

▼ **Table 8.3** Infant mortality rate by region in Africa, 2023

Region	Number of countries above 50/1000	Average infant mortality rate
Northern Africa	0	23
Western Africa	8	54
Eastern Africa	3	39
Middle Africa	4	51
Southern Africa	1	27

Infant mortality generally compares well with many other indicators of development, which is a good indication of its value as a measure of development.

Daily calorie intake

Daily calorie intake is viewed as an important measure because people need energy, which is provided by calories, for their bodies to function properly. While the average figures by world region are useful in showing general contrasts around the world, it is important to remember that the countries with the lowest average calorie intake are those that usually suffer the greatest degree of food insecurity. During times of drought and famine, daily calorie intake may only be a fraction of the 'average' figure. A general daily guideline for weight maintenance is 2000 calories for women and 2500 for men. However, excessive calorie intake can result in obesity and associated metabolic diseases.

Doctors per thousand people

The number of doctors per unit of population is an indication of the financial resources available to a country for the training and recruitment of doctors. Access to well-qualified medical professionals impacts directly on the wellbeing of patients. Figure 8.4 is a scatter graph that shows the relationship between medical doctors per thousand people and GDP per capita for a wide range of countries. The graph shows that most, but not all, HICs have more than three doctors per thousand people, while LICs generally have less than one. In fact many LICs are in the situation that they are nowhere near one doctor per thousand people – look how many LICs are very close to the baseline (the horizontal axis) on the diagram. Overall, the scatter graph shows that there is a strong correlation (relationship) between doctors per thousand people and GDP per capita.

> **Interesting note**
>
> Corruption is a major obstacle to economic and social development around the world. The Corruption Perceptions Index has been published annually since 1995 by Transparency International. It ranks the countries of the world for perceived levels of public sector corruption.

8.1 MEASURING DEVELOPMENT

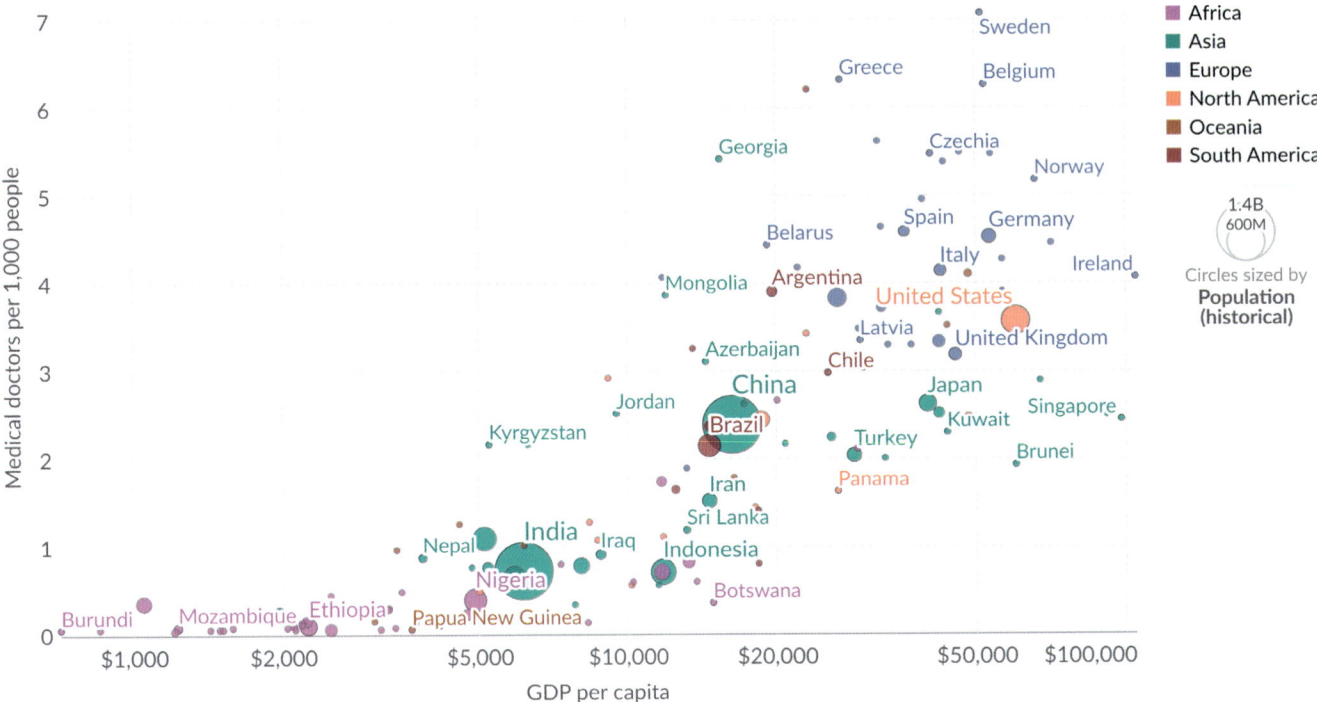

▲ **Figure 8.4** The relationship between doctors per 1000 people and GDP per capita

The relationship between different indicators of development

All indicators of human development have merits and limitations. The indicators discussed above have been widely used because their merits are generally considered to be greater than their limitations. Therefore, it is reasonable to assume that you might expect there to be a strong (but not perfect) relationship between any pair of these indicators, such as that illustrated by Figure 8.4.

> ### ▶ Activities
> 1 Define adult literacy.
> 2 Which world regions have the highest and lowest infant mortality rates?
> 3 List the four main influences on life expectancy.
> 4 Using the data from Table 8.2 (page 180), draw a scatter graph to show the relationship between life expectancy at birth and infant mortality rate. Insert a line of best fit and comment on the relationship between the two variables.

The Human Development Index: a broader measure of development

The **Human Development Index (HDI)** was devised by the United Nations in 1990. It is a composite index, combining a number of key development indicators. Its aim was to extend the concept of development beyond wealth alone. The United Nations Development Programme (UNDP), which is responsible for producing the HDI, has stated that 'The HDI is a summary measure of average achievement in key dimensions of human development – a long and healthy life, being knowledgeable and having a decent standard of living'. Canada (Figure 8.5) has a very high level of human development by these measures, but the UNDP acknowledges that the HDI includes only part of what human development entails. For example, it does not include measures of inequality, poverty, human security or empowerment.

The Human Development Index: a broader measure of development

▲ **Figure 8.5** The Waterfront, Vancouver: Canada has a very high level of human development

The HDI has changed slightly in composition over the years. The current index combines four indicators of development (Figure 8.6):

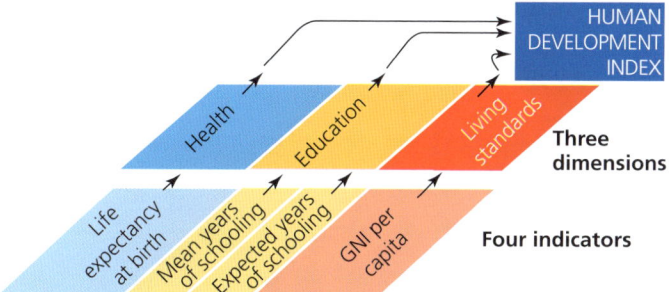

▲ **Figure 8.6** The components of the Human Development Index

- Life expectancy at birth
- Mean years of schooling for adults aged 25 years
- Expected years of schooling for children of school-entering age
- GNI per capita ($PPP).

The actual figures for each of these measures are converted into an index, which has a maximum value of 1.0 in each case. The index figures are then combined and averaged to give an overall HDI value. This also has a maximum value of 1.0 (Figure 8.7). The United Nations publishes an updated Human Development Report every year, which uses the HDI to rank all the countries of the world on their level of development. The HDI divides the countries of the world into four groups. The 2024 report, based on 2022 data, divided the countries of the world as follows:

- Very high HDI: The countries ranked 1–69
- High HDI: The countries ranked 70–115
- Medium HDI: The countries ranked 116–159
- Low HDI: The countries ranked 160–193.

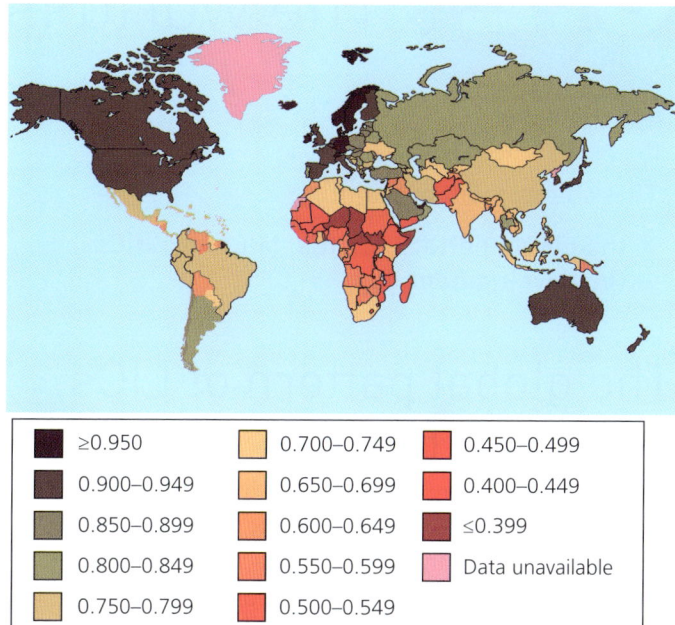

▲ **Figure 8.7** The Human Development Index scores in increments of 0.050, from the 2024 report, based on 2022 data

Table 8.4 shows examples of countries in each group in the 2024 report. Some countries have progressed from one group to another in a relatively short period of time.

▼ **Table 8.4** Examples of countries in each group in the 2024 report

Level of development	Example countries
Very high human development	Norway, Germany, Canada, UK, Japan, Australia
High human development	Cuba, Mexico, Brazil, China, South Africa, Indonesia
Medium human development	India, Morocco, Ghana, Pakistan, Philippines, Bangladesh
Low human development	Nigeria, Tanzania, Afghanistan, DR Congo, Yemen, Mozambique

Activities

1. Which indicators of development are combined to form the Human Development Index?
2. Look at Figure 8.7 and briefly describe the distribution of countries according to the Human Development Index.

8.2 The world is developing unevenly

This chapter will explain:
★ the global pattern of LICs, MICs and HICs
★ the development gap.

The global pattern of LICs, MICs and HICs

A reasonable division of the world in terms of economic development is shown in Figure 8.8, which identifies three broad global income groups – low-income countries (LICs), middle-income countries (MICs) and high-income countries (HICs). The MICs group of countries is often subdivided (as it is in Figure 8.8) into lower-middle income and upper-middle income. This is because of the very wide range of income and the number of countries involved in the MIC category.

As countries develop economically and socially they can move up the 'ladder of development', but under significant adverse circumstances they can also move down. Figure 8.9 shows the global pattern of countries by income group published by the World Bank in 2024, using 2022 data. The World Bank uses GNI per capita data in US dollars, converted from local currency using the World Bank Atlas method, which averages out exchange rate fluctuations. Examples of countries in each income group are given in Table 8.5.

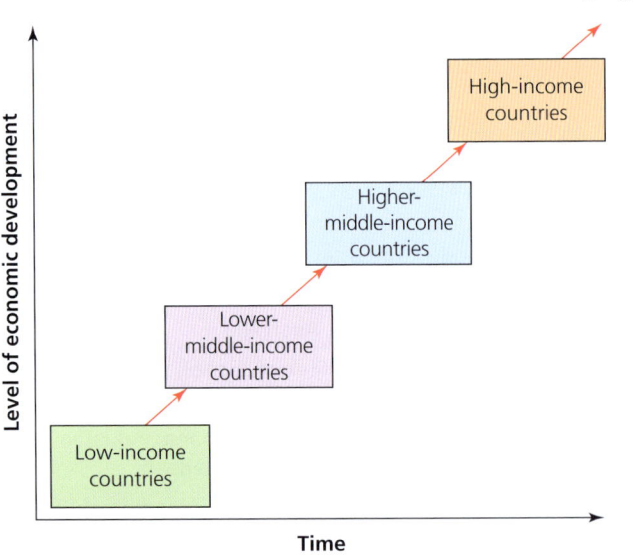

▲ Figure 8.8 The stages of development

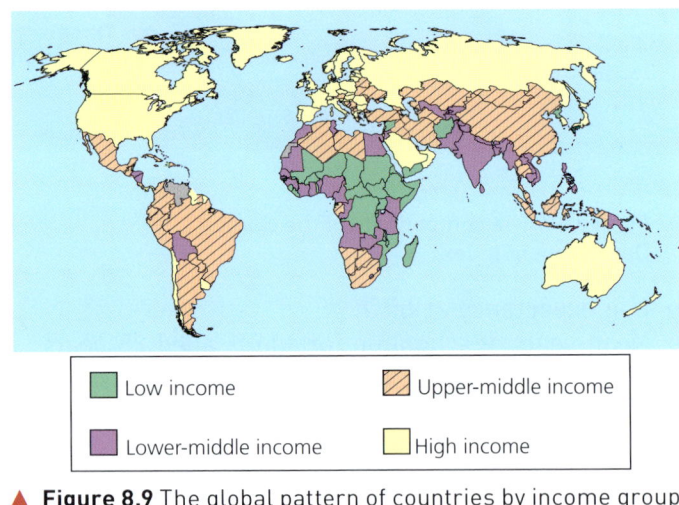

▲ Figure 8.9 The global pattern of countries by income group

▼ Table 8.5 Examples of countries in each income group in 2024

Income group	GNI per capita	Example countries
High income	$13,846 or more	USA, Canada, Germany, UK, Japan, Australia
Upper-middle income	Between $4466 and $13,845	Brazil, Argentina, Turkey, China, Indonesia, South Africa
Lower-middle income	Between $1136 and $4465	Bolivia, Algeria, Nigeria, India, Iran, Vietnam
Low income	$1135 or less	Ethiopia, Afghanistan, Sudan, DR Congo, Mozambique, Somalia

The global pattern of LICs, MICs and HICs

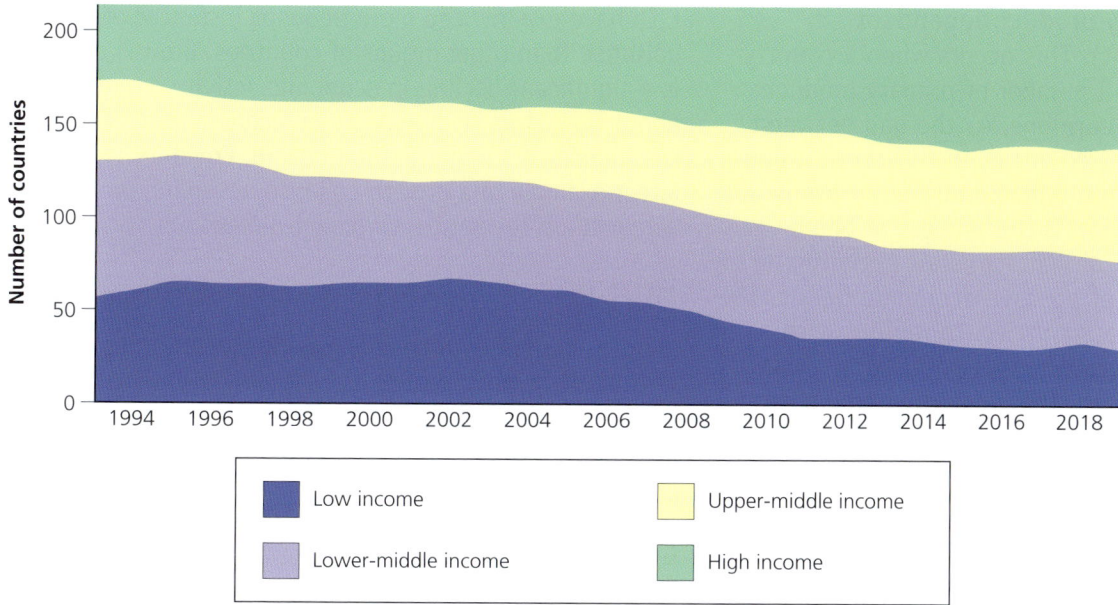

▲ **Figure 8.10** Changes in the number of countries by income group, 1993–2019

Figure 8.10 shows that between 1993 and 2019:

» The number of high-income countries increased considerably.
» The number of low-income countries declined sharply, particularly from the early 2000s.
» The number of middle-income countries changed to a much lesser extent, as countries transitioned both into and out of this group.

In the 1990s, more than six in 10 people in the world lived in low-income countries. Today it is less than one in 10. However, critics of the World Bank system of income classification say that with only a small proportion of the world's population classified as low income and as income thresholds have not changed since 1988 (in real terms), these income groups are somewhat arbitrary and dated.

Low-income countries

LICs are located predominantly in Africa, with most of the rest in Asia (Figure 8.9). With a total population of over 600 million, LICs accounted for just over 1.1 per cent of global GNI in 2023. In the same year, LICs contributed about 1 per cent to world trade. Table 8.6 shows World Bank data comparing LICs with MICs and HICs on a range of indicators.

▼ **Table 8.6** World Bank data comparing LICs, MICs and HICs

Indicator	Year	LICs	MICs	HICs
Population total (billion)	2023	0.723	5.9	1.4
Population growth (% p.a.)	2023	2.7	0.8	0.5
Poverty ratio (% pop.)	2022	44.5	N/A	0.3
Life expectancy at birth (years)	2022	63	71	80
Unemployment (%)	2023	5.3	5.1	4.4
CO_2 emissions (tonnes per capita)	2020	0.3	3.7	7.8
Access to electricity (% pop.)	2022	44.9	94.9	100.0
People using safely managed sanitation services (% pop.)	2022	24	55	91

8.2 THE WORLD IS DEVELOPING UNEVENLY

LICs are often **primary product dependent** (commodity dependent). This occurs when a country relies on one or a small number of primary products for most of its export earnings. As the gap between the richest and poorest countries in the world widens, LICs are being increasingly **marginalised** in the world economy. Marginalisation is the process of being pushed to the edge of global economic activity and of being largely left out of positive economic trends. The term can be applied to countries, to regions within countries, and to groups within a population – the poorest, the unemployed, the most vulnerable (Figure 8.11). There are great concerns that marginalisation has increased under the process of globalisation. The problems of LICs are often made worse by major geographical handicaps and natural and human-made disasters. Examples include the following:

- The catastrophic earthquake in Haiti in 2010 that killed an estimated 220,000 people and caused massive damage. Another severe earthquake hit Haiti in August 2021.
- Bangladesh experiences flooding every year, and when flooding is severe it can impact 75 per cent of the country. About three-quarters of Bangladesh is less than 10 metres above sea level.
- Small Island Developing States (SIDS) are the most disaster-prone countries in the world. They are regularly impacted by severe storms and other natural hazards. According to the United Nations Conference on Trade and Development (UNCTAD), this causes on average an annual damage of 2.1 per cent of GDP.

▲ **Figure 8.11** Poor transport infrastructure in remote regions is a major obstacle to development

The UN has called for increased investment in the **productive capabilities** of the LIC countries. This refers to the productive resources, entrepreneurial capabilities and production links that together determine a country's capability to produce goods and services, and to enable it to grow and develop.

LICs have experienced more frequent instances of **growth collapse** than other groups of countries. Growth collapse is a significant decline in economic activity brought about by factors such as a global economic shock or by the impact of a major natural hazard. Covid-19 is the latest example of a growth collapse, reversing progress achieved on several development dimensions.

Middle-income countries

Middle-income countries (emerging economies) are nations that have moved up the development ladder, having at different times in the past been considered LICs. MICs account for about 75 per cent of the world's total population and around 62 per cent of those in poverty. Representing about a third of global GDP, the most dynamic MICs are the fastest-growing economies in the world.

The first countries/territories to make the transition from low to middle income (in the 1960s) were South Korea, Singapore, Taiwan and Hong Kong (Figure 8.12). The media referred to them as the 'four Asian tigers'. A **'tiger economy'** is one that is growing rapidly. The reasons for the success of these countries/territories were:

- a good initial level of infrastructure
- a skilled but relatively low-cost workforce
- cultural traditions that revere education and achievement
- governments that welcomed foreign direct investment from transnational corporations
- distinct advantages in terms of geographical location
- the ready availability of bank loans for businesses, often extended on government orders and at attractive interest rates.

▲ **Figure 8.12** Hong Kong was one of the newly emerging economies that began to develop in the 1960s

The success of the four Asian tigers provided a model for others to follow, such as Malaysia, Brazil, China and India. South Korea and Singapore have developed so much that they are now considered to be HICs.

High-income countries

The high-income group of countries covers an interesting range. The core of this group is the traditional developed countries, long recognised by the United Nations – Northern America, Western Europe, Japan, Australia and New Zealand. It also includes the following countries/territories:

- Emerging economies/newly industrialised countries that have developed since the 1960s, so far and so fast that they have reached high-income status. Examples are Singapore, South Korea and Hong Kong.
- Small island nations within the European Union. Examples are Cyprus and Malta.
- Eastern European countries that have joined the European Union since the disintegration of the Soviet Union.
- Countries that have become rich through the exploitation of natural resources, particularly oil and gas. Examples are the UAE, Bahrain and Guyana.
- Tax havens such as the Cayman Islands and the British Virgin Islands.
- Small countries/territories where tourism has expanded rapidly and profitably, such as the Seychelles and Aruba.

> **Activities**
>
> 1. a Define primary product dependency.
>   b What are the disadvantages of a country being primary product dependent?
> 2. What do you understand by the terms:
>   a productive capabilities
>   b growth collapse?
> 3. Compare LICs, MICs and HICs for three of the indicators in Table 8.6 (page 185).
> 4. a What is a tiger economy?
>   b Identify the factors responsible for the development of the first generation of tiger economies in the 1960s.
> 5. In what ways do some of the HICs identified by the World Bank differ from the traditional developed countries, long recognised by the United Nations?

Explaining the development gap

There has been much debate about the causes of development and the extent of the **development gap**. The variations between countries are due to a range of factors. Physical geography can have a major impact on development.

- **Landlocked countries** (for example Niger and Nepal) have generally developed more slowly than coastal nations. Lacking direct access to sea ports increases the cost of importing and exporting goods as these need to be transported through one or more neighbouring countries.
- Small island countries (for example Kiribati and São Tomé and Príncipe) face considerable disadvantages in development. They often have limited raw materials and are highly dependent on tourism. Most are also highly vulnerable to the effects of climate change.
- Tropical countries have grown more slowly than those in temperate latitudes, reflecting the costs of poor health and unproductive farming in more challenging physical environments. Poor health is a result of high prevalence rates of a range of tropical diseases, along with poor nutrition due to low agricultural productivity.
- A generous allocation of natural resources has spurred economic growth in a number of countries, but there are examples where resource wealth has been wasted (for example Nigeria, a major oil producer).

Appropriate, realistic and well-applied economic policies can be an important factor in the development process:

- **Open economies** that encouraged foreign investment have developed faster than closed economies. Some MICs and LICs have established special economic zones to attract foreign direct investment. However, political manipulation by developed nations has sometimes caused developing nations to open their economies too fast or in other unhelpful/unsustainable ways.
- Fast-growing countries tend to have high rates of saving and low spending relative to GDP. A high rate of saving can provide money for investment in infrastructure for transport, energy, health, education, housing and water supply.

8.2 THE WORLD IS DEVELOPING UNEVENLY

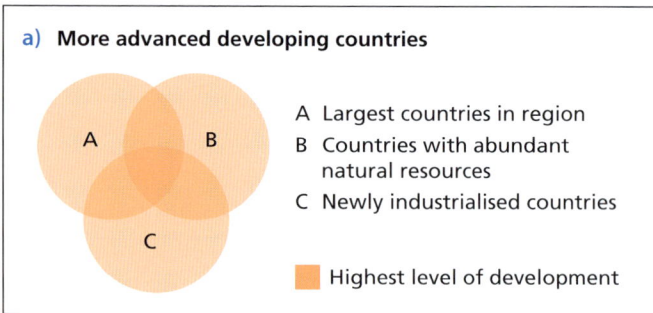
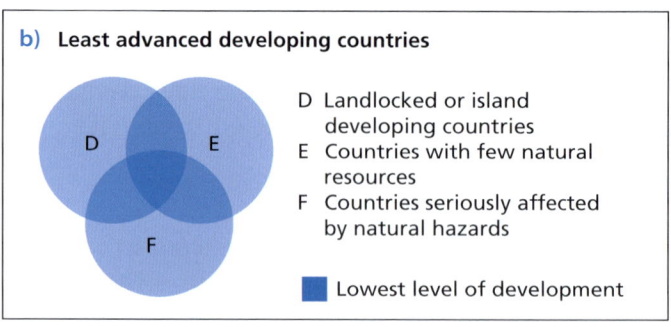

▲ **Figure 8.13** Venn diagrams showing fast and slow development in developing countries

» **Institutional quality** in terms of good government, law and order, and lack of corruption generally result in a high rate of growth. These positive attributes also encourage foreign direct investment.

Progress through demographic transition is a significant factor, with the highest rates of economic growth experienced by those nations where the birth rate has fallen the most, due to women being able to access education and be included in the workforce.

Figure 8.13 combines a range of factors to explain differences in development. For example, in diagram (a), Brazil would satisfy all three criteria. It is by far the largest country in South America, it has abundant natural resources, and it is an emerging economy. In contrast, countries such as Haiti and Niger would be affected by all three of the negative factors in diagram (b), based on a history of colonisation, which affected their land and borders, infrastructure development and debts.

Colonialism and neo-colonialism

Colonialism occurs when one country takes control of another and dominates it politically and economically. It involves people from the colonial power invading and settling in the newly controlled country. The age of 'modern colonialism' began around 1500, following the European discovery of both America and a sea route around Southern Africa. Portugal and Spain dominated the early period of colonialism, but were later followed by the Netherlands, Britain, France, Germany and Belgium.

Colonialism spread at a particularly fast pace in the eighteenth and nineteenth centuries. A period of special note was the 'Scramble for Africa' (or the Partition of Africa) between 1880 and the beginning of the First World War (1914). This involved the colonisation and division of most of Africa by the seven European powers named above. By the early twentieth century, a large majority of the rest of the world had been colonised by the European powers at some point in time. These actions by colonial powers brought about massive negative changes in many countries. Resistance to the colonisers was overwhelmingly met with violence and intimidation.

The European countries exploited their colonies for economic gain (both in terms of raw materials and labour), causing enormous suffering and future damage as a result of their actions. The imposition of unequal trading relationships has had a long-lasting effect. However, slavery was undoubtedly the most brutal aspect of colonialism.

In the modern world, the term '**neo-colonialism**' is used to describe how HICs can still dominate LICs in an economic and political sense.

Dependency theory was developed to help explain the impact of these processes. In the late 1960s, the economist A.G. Frank produced a theory that was critical of **capitalism**. He argued that:

» Poverty in the developing world was not an original condition, but that it developed through the spread of world capitalism. Many countries had been prosperous before the arrival of European colonists.
» The process of absorption into the capitalist system sowed the seeds of underdevelopment.
» The development of European and other dominant countries was achieved by the extraction of resources from developing countries.
» Developing countries became even more dependent on the West through substituting the production of export goods where once local food crops prevailed.
» The stronger the links to the developed world, the worse the level of development in LICs.

Frank used a simple model to explain how the '**economic core**' (the developed world) exploited the '**economic periphery**' (the developing world). His model (Figure 8.14) shows a chain of exploitation. It begins with small towns in the periphery exploiting resources from their surrounding rural areas. This process of exploitation works its way up the urban hierarchy in the periphery until, finally, the largest settlements are exploited by cities in the colonial power. The intensity of poverty increases with the number of stages down the chain of exploitation. As with most theories that try to explain complex situations, dependency theory has its advocates and critics.

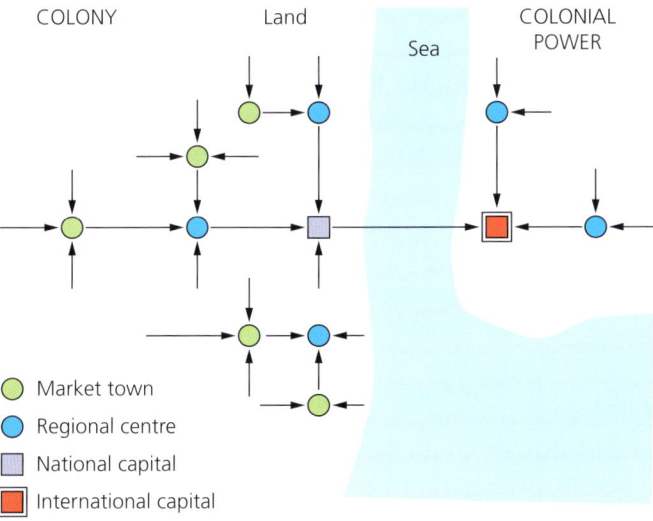

▲ **Figure 8.14** Dependency theory

The roles of trade and investment in development

Trade and investment play a key role in economic development. Investment in a country is the key to it increasing its trade. Some developing countries have increased their trade substantially in recent decades. Examples are China, India, Brazil and Mexico. Such countries have attracted the bulk of foreign direct investment. The emergence of newly industrialised countries (emerging economies) has been the biggest success of globalisation.

However, about 2 billion people live in countries where trade has fallen in relation to national income. In a sense, these countries have become less globalised rather than more! The share of world trade for many of the world's low-income countries has fallen in recent decades. Non-governmental organisations (NGOs) argue strongly that trade is the key to real development, being worth many times more than international aid. These NGOs continue to criticise the ways in which world trade operates to the detriment of developing countries. Sometimes, when a developing country increases the volume of its exports, its revenues have declined due to a fall in prices on the world market.

Political and economic policies

Open economies such as the UK and Singapore that encourage foreign investment have developed faster than countries that have much greater controls on foreign investment. Large-scale investment can create significant economic **multiplier effects**. Closed economies engage in only limited trade and other contacts with the outside world. Their development tends to suffer as a consequence. North Korea is the most extreme example of a closed economy. However, the strong market **protectionism** practised by some Western countries undermines them being perceived as open economies.

Until the early 1990s, India was a relatively closed economy. There were very high tariffs on imports, along with other restrictions. Reducing barriers to trade was an important part of the economic reforms the country made at the time. As India became more integrated into the global economy, the volume of both its exports and its imports rose sharply.

Social investment

Countries that have prioritised investment in education and health have generally developed at a faster rate than nations that have invested less in these sectors. High standards of education and health are vital for sustainable development. When Sierra Leone gained independence in 1963, it reduced investment in its health system. Since then the country's economy has struggled.

In contrast, Botswana has been described by the UNDP as 'one of Africa's veritable economic and human development success stories'. Instead of squandering the wealth generated by its diamond and tourist industries, it invested heavily into health and education. Primary school attendance up to age 13 is almost 90 per cent and 85 per cent of the population now live within 3 miles of a health facility. A healthy and well-educated population attracts investment and therefore encourages development. However, even affluent countries can struggle to

8.2 THE WORLD IS DEVELOPING UNEVENLY

provide enough investment for healthcare, as the Covid-19 pandemic illustrated.

As countries develop, fertility declines. Greater availability and knowledge of contraception and family planning is key to lower fertility. In addition, female education can create greater social awareness and opportunities for employment.

Technology and development

There are big differences between countries in their capacity to create and use technology for economic growth and development. Technology has significant potential to help deliver the Sustainable Development Goals (see page 192). However, it can also be at the root of exclusion and inequality. The development of human skills is fundamental to technological capacity. It is important to harness the benefits of advanced technologies for all. **Technology transfer** from developed nations to developing countries can play an important role in reducing the technology gap and aiding the development process.

Digital technology has the potential to tackle poverty by providing access to basic services such as e-health, online education and business advice. It can also be used by governments to better connect to their populations through e-government tools. The number of internet users worldwide increased from 2.08 billion in 2012 to 4.95 billion in 2022. The '**digital divide**' is a term used to refer to inequalities in access to ICT. Such inequality can be seen between:

- developed and developing countries
- urban and rural areas
- ethnic and socioeconomic groups in the same country
- different age groups
- males and females.

The **internet penetration rate** is the percentage of a total population that use the internet. It is a measure often used to illustrate the technology gap between world regions (Table 8.7). For individual countries in 2022, South Korea, the UK and Switzerland all reported internet penetration rates of 98 per cent. However, it is not just internet connection in itself, but also the speed of connection that has become increasingly important.

▼ **Table 8.7** The internet penetration rate by world region, 2022

Region	Estimated internet penetration rate, 2022 (%)
World	66.2
Africa	43.1
Asia	64.1
Oceania	70.1
Middle East	76.4
Latin America/Caribbean	80.4
Europe	88.4
North America	93.4

Culture and development

In recent decades, **culture** has come to be viewed as an important factor in development. Culture can be defined as the total of the inherited ideas, beliefs, values and knowledge that constitute the shared basis of social action. It is the way of life of a particular society or group of people. For example, materialism and consumer culture are clearly more important in some countries than in others. Societies that are largely secular may appear to have different priorities to those that are not.

A 2018 British Council report stated, 'When people or communities are given the opportunity to engage with, learn from and promote their own cultural heritage, it can contribute to social and economic development.' A number of UN reports have highlighted the important contribution of indigenous cultures to sustainable development. Respect for cultural diversity helps to promote inclusive societies. The cultural and creative sector of an economy can make an important contribution to GNI and employment. It can increase the attractiveness of a location as a destination to live, visit and invest in.

> ### Activities
> 1. What are the particular disadvantages for development in landlocked countries and small island countries?
> 2. a Define neo-colonialism.
>   b How can neo-colonialism hinder development in LICs?
> 3. Briefly discuss the importance of trade and investment to economic development.
> 4. Describe social investment.
> 5. Draw a diagram to illustrate dependency theory.
> 6. How important is access to modern technology for the development process?

8.3 Achieving sustainable development

This chapter will explain:

★ what sustainable development is
★ the strategies to try to achieve sustainable development
★ the techniques used by selected sectors that promote sustainability
★ the strategies to reduce uneven development
★ detailed specific example: Indonesia: an emerging economy.

What is sustainable development?

Sustainable development is development that meets the needs of the present without compromising the ability of future generations to meet their own needs. The concept of sustainability was established at the first UN Conference on the Environment in 1972, but it did not really take shape until 1987 when the report 'Our Common Future' identified in detail the objectives of sustainable development.

The concept of sustainability can be applied in various ways:

» On the full range of scales, from the individual to the world as a whole. Governments are increasingly reminding individuals and households about their carbon footprint and how this can be reduced. At the largest scale, sustainability focuses on the total carrying capacity of the planet.
» To different geographical environments, such as rainforests, temperate grasslands and urban areas. Satellite photography has enabled a major advance in our ability to see what is happening over large land areas. It has allowed short-term changes to be recognised quickly.
» To individual economic activities such as tourism, agriculture and forestry. Each sector has its own impact on the environment, which can be modified by careful management.

Sustainability need not require a reduction in quality of life, but it does require a change in attitudes and values toward a less consumptive lifestyle. These changes must embrace global interdependence, environmental stewardship, social responsibility and economic viability. Environmental sustainability in a country or region is difficult to achieve without economic and social sustainability because of the strong interconnectedness between these three vital aspects of life (Figure 8.15). Economic sustainability involves maintaining income and employment. Social sustainability means maintaining social capital, including that devoted to health, education, housing and the rule of law.

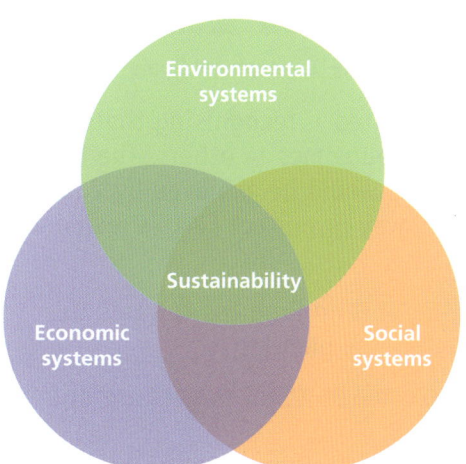

▲ **Figure 8.15** Social, economic and environmental sustainability

Sustainability is the foundation for the 2030 Agenda for Sustainable Development and its **Sustainable Development Goals**. The UN's 17 Sustainable Development Goals (SDGs) with their 169 targets were adopted by UN Member States in 2015 (Figure 8.16). The overall objectives were to:

a end poverty
b protect the planet
c ensure all people enjoy peace and prosperity.

8.3 ACHIEVING SUSTAINABLE DEVELOPMENT

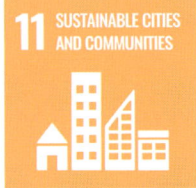

▲ **Figure 8.16** The 17 Sustainable Development Goals

The 17 SDGs are 'integrated' – recognising that action in one area will affect outcomes in others. The hope was that all these aims would be achieved by 2030. This would go a long way to protect Earth's resilience for future generations.

However, the 2024 Report on Progress towards the SDGs painted a pessimistic picture:

» Only 17 per cent of the 169 SDG targets are on track to be achieved.
» Nearly half are showing only minimal or moderate progress.
» Progress on over a third of the targets has stalled or even regressed.

The 2024 Report on Progress recognised the slow but steady progress in the early years (post-2015) of SDG implementation, but highlighted the mix of global problems since 2019 that have seriously hampered further progress, namely:

» the Covid-19 pandemic
» a growing number of major conflicts
» geopolitical and trade tensions
» the ever-worsening effects of climate change.

All these major issues were additional to the long-lasting problems of:

» significant shortcomings in global economic and financial systems
» historical injustices.

The 2024 Report on Progress views strong global solidarity as vital to achieving meaningful sustainability. It raised serious concerns about the deterioration in global cooperation.

The 2024 Environmental Performance Index

The **Environmental Performance Index (EPI)** uses 58 performance indicators across 11 categories to score and rank countries around the globe on:

» climate change performance
» environmental health
» ecosystem vitality.

These indicators provide a measure at the national scale of how close countries are to reaching environmental policy targets. The 2024 EPI Report concludes that countries' wealth is a strong predictor of overall EPI scores, with good governance being another extremely important factor. In the 2024 EPI overall ranking, European countries occupy the top 20 positions.

▼ Table 8.8 World map showing rankings in the 2024 Environmental Performance Index

Rank	Country	EPI	Rank	Country	EPI
1	Estonia	75.7	171	Afghanistan	31.0
2	Luxembourg	75.1	172	Iraq	30.3
3	Germany	74.5	173	Madagascar	30.1
4	Finland	73.8	174	Eritrea	29.0
5	United Kingdom	72.6	175	Bangladesh	28.1
6	Sweden	70.3	176	India	27.6
7	Norway	69.9	177	Myanmar	27.1
8	Austria	68.9	178	Laos	26.3
9	Switzerland	67.8	179	Pakistan	25.5
10	Denmark	67.7	180	Vietnam	24.6

> **Activities**
>
> 1 Define sustainable development.
> 2 What does Figure 8.15 (page 191) show?
> 3 Look at Figure 8.16 (page 192). Research two of the Sustainable Development Goals and then write a brief analysis of your findings.
> 4 How was the 2024 Environmental Performance Index calculated?
> 5 Look at Table 8.8. Describe the main characteristics of the countries with the highest Environmental Performance Index (Ranks 1–10) compared to those that have the lowest Environmental Performance Index (Ranks 171–180).

The strategies to try to achieve sustainable development

Environmental

Environmental sustainability centres on the preservation and protection of the natural environment over time. It involves appropriate practices and policies to meet present-day needs without compromising the availability of resources in the future. The key objectives of environmental sustainability are as follows:

» To reduce GHG emissions in all sectors of the economy.
» To transition to renewable sources of energy.
» To adopt sustainable practices in agriculture.
» To increase awareness of the issues involved in environmental sustainability.
» To promote the **circular economy** (Figure 8.17), which aims to keep raw materials in a closed loop. In this way, the maximum use is made of raw materials, resulting in a reduced need for new raw materials. Waste is reduced and the lifecycle of products is extended. The idea is that the waste of today becomes the raw materials of tomorrow.

All of these objectives combine to conserve and sustainably manage natural resources including water, soil, forests, wildlife and natural habitats.

Social

Social sustainability centres on the wellbeing of people and the communities they live in. Systems and policies should be directed towards tackling:

» poverty and inequality
» discrimination, prejudice and social exclusion
» lack of access to resources
» insecurity at all scales, from local to global
» poor governance.

The overall objective of social sustainability is to ensure equitable access to opportunities and resources for all people.

8.3 ACHIEVING SUSTAINABLE DEVELOPMENT

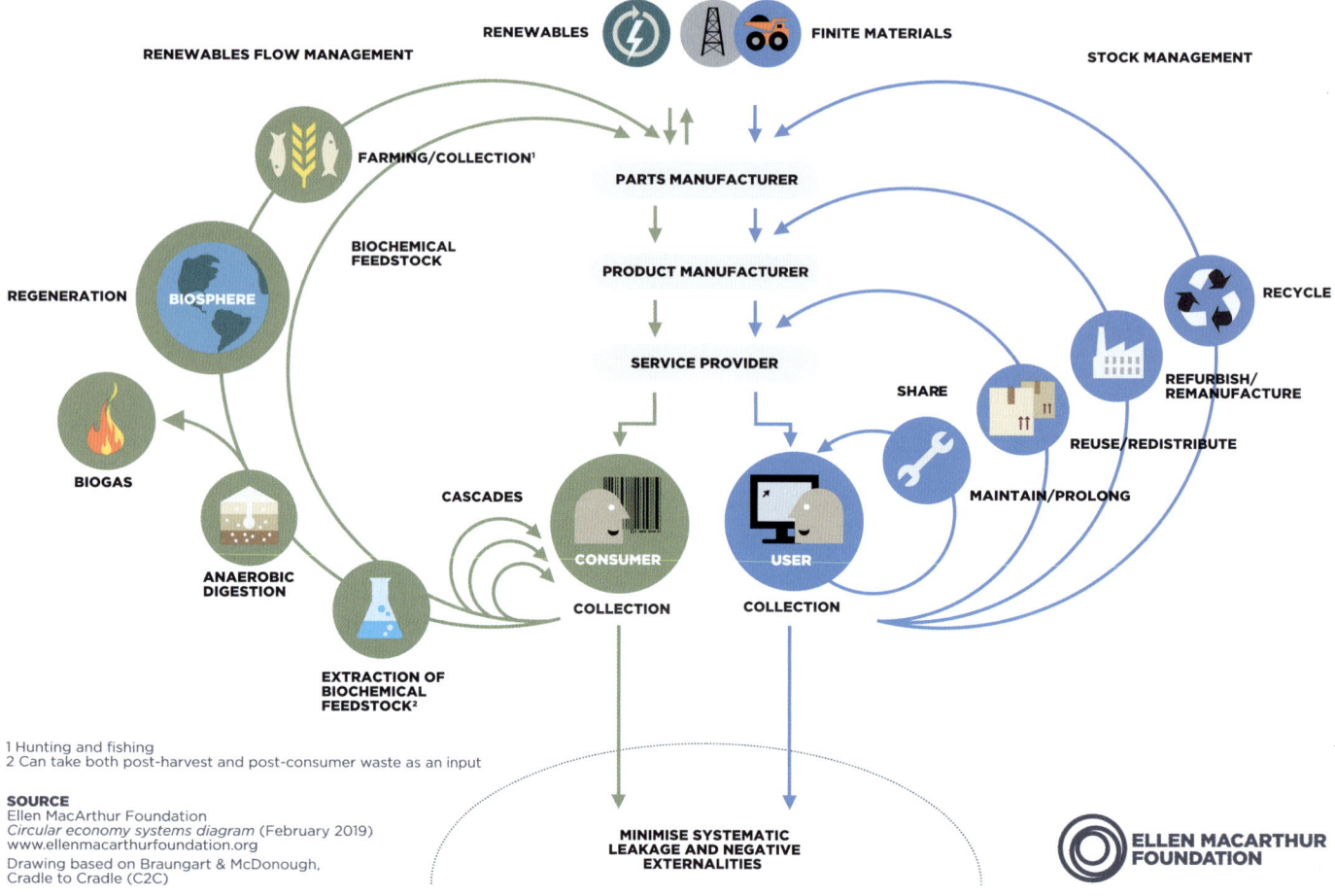

▲ Figure 8.17 The circular economy

Economic

Economic sustainability involves conducting economic activities in a way that preserves and promotes long-term economic wellbeing. Economies become more sustainable by:

» transitioning away from fossil fuels and generating as much energy as possible from renewable sources
» adopting policies and regulations that maximise energy efficiency
» promoting economic models based on the circular economy that reduce waste and limit resource exploitation
» encouraging international cooperation and partnerships between public administration and private enterprises
» ensuring corporate responsibility.

> ### Activities
> 1 a Define environmental sustainability.
>   b What are the main objectives of environmental sustainability?
> 2 Explain the concept of the circular economy.
> 3 Discuss the difference between social sustainability and economic sustainability.

The techniques used by selected sectors that promote sustainability

Agriculture

The range of management strategies that can be employed to reduce soil degradation in agriculture, including mechanical methods and cropping techniques, can be found in Chapter 10.3 on page 264.

The techniques used by selected sectors that promote sustainability

Rock and mineral extraction

Quarrying and mining represent a major human impact on the landscape. However, a range of sustainable practices can reduce the environmental impact of this essential industry. For example:

- Before extraction begins: If there is a reasonable choice of locations, selecting sites strategically close to major transport hubs is an important step in reducing the environmental impact.
- During the active extraction time period: Advanced quarrying methods such as water jetting and laser cutting are more accurate than traditional techniques. They minimise material wastage, produce lower levels of dust and do not involve the use of hazardous chemicals or explosives.
- After the quarry has closed: Phased restoration has become more mainstream in recent decades. This involves returning parts of the quarried area to a more natural state as quickly as possible after operations have ceased on a part of the site. Such a strategy allows ecosystems to begin their recovery process sooner.

Urban areas

Sustainable practices in combating urban pollution have become more commonplace globally in recent decades. The health and other costs associated with doing the bare minimum have become all too obvious in most cities. A range of techniques can be used to reduce air pollution from vehicles:

- In the UK, the Electric Car Scheme is a government tax incentive to make electric cars more financially accessible for the average person. EV charging stations are appearing in many sites across the UK.
- In Taipei (Taiwan), smart traffic lights form part of an intelligent adaptive traffic signal system. The objective is smooth traffic flow to reduce the delays and engine idling that contribute to pollution.
- In Copenhagen (Denmark), the Green Wave technology initiative coordinates traffic lights for cyclists, encouraging the use of cycling as an alternative to driving.
- In London, the **Ultra Low Emission Zone (ULEZ)** has significantly reduced the number of older, high-polluting vehicles entering the city (Figure 8.18). ULEZ was introduced in 2019 and initially confined to Central London. It was extended to broadly include inner London in 2021, and then

▲ **Figure 8.18** London's Ultra Low Emission Zone, introduced in 2019

further extended to cover all of Greater London in 2023. The scheme is claimed to be the largest clean air zone in the world.

Sustainable techniques introduced to improve the quality of life in other aspects of urban living include the following:

- Urban gardens: Many cities have tried to increase their areas of green space. Plants absorb pollutants and release oxygen, helping to improve air quality. Greater tree and ground vegetation coverage can ease cooling needs and reduce the use of air conditioning, which is very energy intensive. Higher vegetation cover can help mitigate the urban heat island effect.
- Domestic combustion: Smoke Control Areas (SCAs) are in place across the UK, with similar schemes in other countries. This legislation places restrictions on fuels used and smoke released by households. Gas boilers, which are a significant source of nitrogen oxides (NOx) emissions, are coming under close scrutiny, with encouragement to change to alternatives such as electric boilers and heat pumps.
- Construction: The construction industry is undertaking sustainable changes in the way that it operates in construction sites and in the design and operation of the buildings themselves. Sustainability measures on construction sites include:
  - implementing sustainable design, engineering and construction practices to reduce emissions throughout a project's lifecycle

8.3 ACHIEVING SUSTAINABLE DEVELOPMENT

» using logistics processes that optimise deliveries to reduce mileage, emissions and carbon footprint
» operating equipment in an energy-efficient manner.

In terms of construction's end product, finished buildings, the UN set ambitious targets for sustainability at the beginning of this decade. For example, the target reduction rate for energy intensity per m² in buildings is 30 per cent by 2030.

» Waste management: At the household scale, people have been encouraged to reduce, reuse and recycle (Figures 8.19 and 8.20). Minimising and repurposing help to preserve the environment. For those with gardens, composting can reduce land pollution.

At the municipal scale, **sanitary landfills** are being increasingly used in and around urban areas for land disposal of solid waste. These facilities are designed to control leachate and methane, thus minimising the risk of land pollution from solid waste disposal. Sanitary landfill sites are prepared with impermeable bottom liners to collect leachate and prevent contamination of groundwater.

▲ **Figure 8.19** The various categories of recycling

▲ **Figure 8.20** Green waste recycling in the UK

Industry

In developed countries in particular, industrial emissions have reduced substantially since the 1990s due to a combination of stricter emission standards and fuel changes to greener energy. Specific changes include the increasing use of electric boilers, electric arc furnaces and induction furnaces, and large-scale industrial heat pumps.

Industry has spent increasing amounts on research and development to reduce its environmental impact – the so called greening of industry. In general, after a certain stage of economic development, the level of pollution will decline (Figure 8.21). This is because countries have become more aware of their environmental problems, and they have also created the wealth to invest in improving the environment.

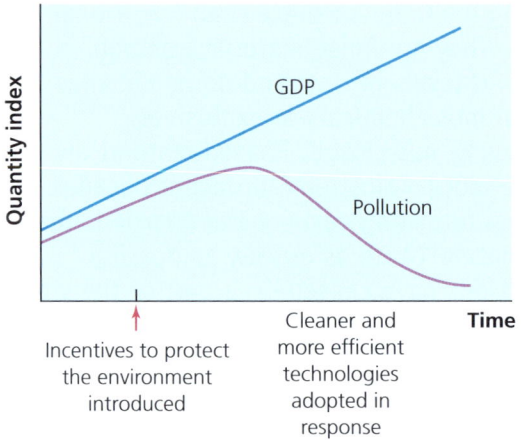

▲ **Figure 8.21** The relationship between GDP and pollution

Tourism

A consideration of sustainable development relating to tourism can be found in Topic 9 on page 241.

> **Activities**
> 1 Explain the benefits of three sustainable techniques used in agriculture.
> 2 Comment on the examples given on page 195 to improve sustainability in rock and mineral extraction.
> 3 What are sanitary landfills and why are they important?

The strategies to reduce uneven development

The development gap is such a major global issue that it is not surprising there are different views held by governments, organisations and individuals. For example:

» Some countries contribute a much higher proportion of their GDP to international development than others.
» The political character of a government can influence its attitude to international development.

- Development economists can differ in their views about the best ways to encourage development.
- The strength of **global civil society** in a country can be a big factor in shaping development policy.
- The respective benefits of prioritising trade or aid have been debated for a long time.

Trade

The **United Nations Conference on Trade and Development (UNCTAD)** was established in 1964 to promote the interests of developing countries in world trade. UNCTAD recently stated, 'Globalisation, including a phenomenal expansion of trade, has helped lift millions out of poverty. But not nearly enough people have benefited. And tremendous challenges remain.' There is a strong relationship between trade and development. UNCTAD has stated that even small changes in agricultural employment opportunities or farm prices can have major socioeconomic effects in developing countries.

UNCTAD's Trade and Development Report 2021 highlighted the development potential of:

- the 'vertical shift' in production from primary to secondary, and on to higher-end tertiary GDP and employment (Figure 8.22)
- the 'horizontal shift' of resources from lower to higher productivity. Here all sectors of the economy benefit from becoming more capital intensive and thus less labour intensive.

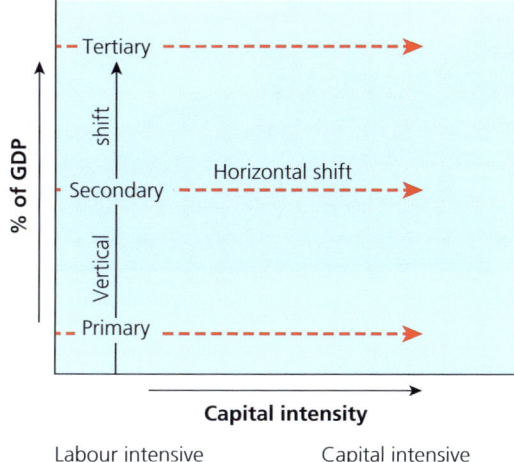

▲ **Figure 8.22** Sector and productivity change

The Report stated, 'Together, these processes have, in almost all successful development experiences, facilitated a more diversified structure of economic activity, raised productivity and led to an improvement across a broad set of social indicators, including poverty reduction.' More diversified economies are less vulnerable to external shocks.

Reforming the World Trade Organization

The World Trade Organization (WTO) deals with the rules of world trade. Its main aim is to ensure that world trade flows as freely as possible. The WTO was set up in 1995 with far greater powers than its predecessor (the General Agreement on Tariffs and Trade – GATT) to settle trade disputes. Its headquarters is in Geneva, Switzerland. The WTO has over 150 member countries, accounting for more than 97 per cent of world trade.

Today, average tariffs are only about a tenth of what they were in 1947 when the GATT was formed, and world trade has been increasing at a faster rate than GDP. However, in some areas protectionism is still an issue, particularly in the sectors of clothing, textiles and agriculture. Although most countries accept the principle of reducing trade barriers, they may worry that important domestic industries may not be able to compete with lower-cost goods being imported from abroad. As a result, trade agreements can often be very difficult to broker, and can take a long time to achieve.

Critics of the WTO complain it is dominated by the largest economies in the world, with little regard for the trade difficulties faced by many less affluent countries. Organisations such as the Global Policy Forum argue that world trade is very unequal, noting that:

- agricultural subsidies and trade barriers in the USA, EU and other HICs prevent LICs from gaining proper access to the most important markets in the world
- at the same time, LICs have been expected to open up their markets to competition from more advanced nations
- trade is heavily dominated by transnational corporations (TNCs), which often benefit the most from new trade rules
- giving developing countries greater access to markets in developed countries would be a major boost to the incomes of LICs.

Many non-governmental organisations (NGOs) have criticised the ways in which the global trading system operates. Table 8.9 summarises these complaints.

8.3 ACHIEVING SUSTAINABLE DEVELOPMENT

▼ **Table 8.9** Criticisms of the WTO

Dumping	Dumping is the export of products (for instance clothing) at a price less than the normal value of that product. This unfair trade practice can have a big negative impact on developing countries through these imported goods being cheaper than domestically produced goods.
Market access	The conditions attached to IMF–World Bank programmes have often forced developing countries to open their markets, regardless of the impact. Developing countries have often found it difficult to access markets in developed countries such as the EU and the USA.
Patent rights	The high level of international protection on intellectual property rights on essential products such as computer software and medicines has created 'unnecessary' costs for developing countries.
Corporate practices	The objective of many TNCs is to produce as cheaply as possible. This often results in the exploitation of people and environments in developing countries.
Uneven influence	Developing countries have long argued that their concerns about the ways in which the global trading system operates are largely overlooked.

Fairtrade

Fairtrade is a system designed to help small-scale producers in developing countries achieve sustainable and fair trade relationships. It recognises the small proportion of the final price of a product that goes to producers. For example, the great majority of the money generated by the tea industry goes to the post-raw-material stages (processing, distributing and retailing), usually benefiting companies in wealthy countries rather than the producers in LICs.

Many supermarkets and other large stores in developed countries now stock some 'fairly traded' products (Figure 8.23). Many are agricultural products such as bananas, orange juice, nuts, coffee and tea. However, the market in non-food goods such as textiles and handicrafts has increased in recent years. The Fairtrade system operates as follows:

» Small-scale producers group together to form a cooperative.
» These cooperatives deal directly with companies such as large supermarkets in more affluent countries. Dealing directly cuts out intermediaries, such as wholesalers.
» These companies pay more than the world market price for the products traded.
» The higher price achieved by the cooperatives provides both a better standard of living and some money to invest in the farms of producers.

Advocates of the Fairtrade system say that it is a model of how world trade should be organised to tackle global poverty. They also stress the high environmental standards set by the Fairtrade system. This system of trade began in the 1960s with Dutch consumers supporting Nicaraguan farmers. Today more than 1.9 million farmers and workers are in Fairtrade-certified producer organisations – there are 1880 of these organisations in 71 countries. Currently, Fairtrade is less than 1 per cent of total world trade. NGOs would like to see this system increased in scale.

As you might expect, the Fairtrade system has not been without its problems and critics. For example, some retailers in developed countries have withdrawn from the system in favour of organising their own way of helping producers in LICs.

▲ **Figure 8.23** An example of a popular Fairtrade product

International aid

International aid is assistance given to countries in need, in the form of grants or cheap loans. Most developing countries have been keen to accept foreign aid. This is because many:

» lack the hard currency to pay for vital imports
» have insufficient savings to invest in industry and infrastructure
» lack important technical skills.

The origins of international aid date back to the Marshall Plan of the late 1940s. This was when the USA set out to reconstruct the war-torn economies of Western Europe and Japan as a means of containing the international spread of communism. By the mid-1950s the battle for influence between West and East in the developing world began to have a marked effect on the geography of aid. Thus, it is not surprising that reservations are often expressed about the directions of international aid and the reasons behind it.

The basic distinction in international aid is between official government aid and voluntary aid (Figure 8.24):

» Official government aid is where the amount of aid given and who it is given to is decided by the government of the country/countries giving the aid.
» Voluntary aid is run by NGOs or charities, which raise money in various ways from individuals and organisations. However, an increasing amount of government money goes to NGOs because of their special expertise in running aid programmes efficiently.

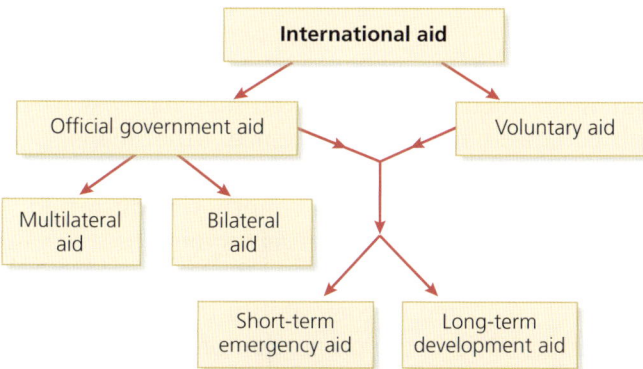

▲ **Figure 8.24** The different types of international aid

Official government aid can be divided into:

» bilateral aid, which is given directly from one country to another. A significant proportion of bilateral aid is 'tied', despite continual efforts to end this process
» multilateral aid, which is provided by many countries and organised by an international body such as the UN.

Aid supplied to lower-income countries is of two types:

» Short-term emergency aid, often called 'relief aid', is provided to help countries cope with unexpected disasters such as earthquakes, volcanic eruptions and tropical cyclones.
» Long-term development aid is directed towards continuous improvement in the quality of life in an LIC.

There is no doubt that many countries have benefited from international aid. All the countries that have developed from low income to middle income have received international aid. However, their development has been due to other reasons too. It is difficult to be precise about the contribution of international aid to the development of each country. The main concerns about aid are that:

» a significant proportion is **tied aid**
» too often aid fails to reach those that need it most
» the use of aid on large capital-intensive projects may worsen local poverty
» aid may delay the introduction of reforms, for example the substitution of food aid for land reform
» aid can create a culture of dependency.

Figure 8.25 relates to Sub-Saharan Africa, showing the clash between perceived goals and the obstacles to development in the post-colonial era. The author of the diagram, Dr Jong-Dae Park, argues that aid, although well-intentioned, created a culture of dependency in many countries in Africa. Many argue that this state of dependency was imposed on low-income nations as an expression of neo-colonialism. As a result, the hoped-for pace of development was rarely achieved. Dr Park notes:

» Humanitarian aid/disaster relief tends to be successful where donors and recipients have very easy and clear-cut relationships.
» Development aid tends to have problems where relationships between donor and recipient are far more complicated.

He adds that cultural differences and misunderstandings can cause considerable difficulties in achieving goals, particularly with large-scale projects.

Historically, India has been the biggest recipient of foreign aid. This has been because of the country's size and the scale of its development problems. However, such aid has declined rapidly in recent years as the country has developed. India itself now sends aid to other countries such as Bhutan, Nepal, the Maldives, Sri Lanka and Afghanistan. India is not the only emerging country to change from being a recipient of aid to an aid donor. Some advocates of international aid say this shows how successful aid can be in the development process. Nonetheless, the challenges for development will vary between countries and this route may not work for all nations.

8.3 ACHIEVING SUSTAINABLE DEVELOPMENT

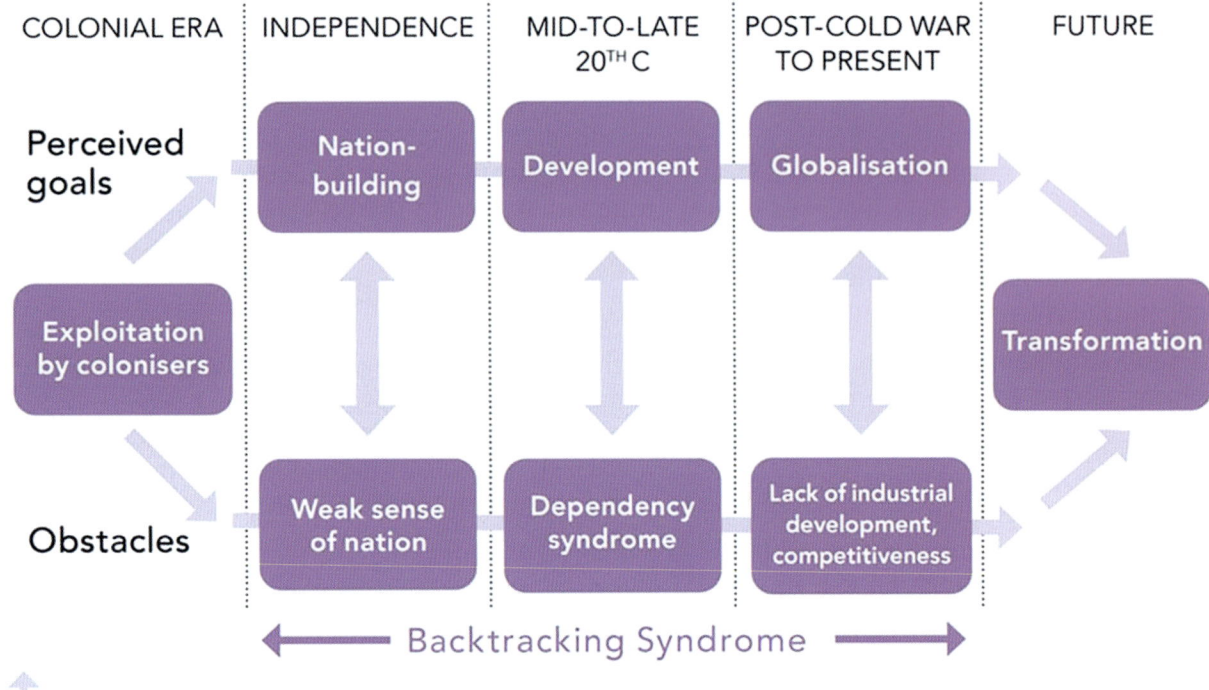

▲ Figure 8.25 The post-colonial period in Sub-Saharan Africa

Many development economists argue that there are two issues more important to development than aid:

- Changing the terms of trade so that low-income countries get a fairer share of the benefits of world trade.
- Writing off the debts of LICs, which were exploitative and impossible to pay back because they were at such a high level.

NGOs have often been much better than government agencies at directing aid towards sustainable development. The selective nature of such aid has targeted the poorest communities, using **intermediate technology** and involving local people in decision-making.

Debt relief

Many development organisations have singled out **debt** as the main problem for the world's LICs. The term 'debt' generally refers to foreign debt. This is money owed to creditors outside the country. This debt has two components:

- Debt outstanding, which is the total amount owed.
- Debt service, which is the interest to be paid each year.

Some of these debts were imposed as 'payback' for what Western countries perceived they had lost when their colonies gained independence. Other loans were set at levels of interest that were effectively unserviceable. Debt affects a country's credit rating and thus its overall economic vulnerability. The **debt service ratio** of many LICs is at a very high level compared to their ability to pay. The debt service ratio is the proportion of a country's export earnings that it needs to use to meet its debt repayments.

Restructuring the debt of LICs began in a limited way in the 1950s. The Heavily Indebted Poor Countries (HIPC) Initiative was established in 1996 by the IMF and World Bank. In 2006, the Multilateral Debt Relief Initiative (MDRI) was launched to provide extra support to HIPCs to reach the Millennium Development Goals. **Debt relief** frees up resources for social spending.

Debt relief is part of a much larger process that includes international aid, which is designed to address the development needs of LICs. Unfortunately, indebtedness grew across many regions from the start of the Covid-19 pandemic. Debt burdens are too high and export revenues too low in many parts of the developing world.

Microcredit and social business

The development of the Grameen Bank in Bangladesh illustrates the power of **microcredit** in the battle against poverty. The idea of microcredit came from the economist Muhammad Yunus, who opened the Grameen Bank in 1983. The Grameen Foundation advocates microfinance and innovative technology to fight global poverty and bring opportunities to those in extreme poverty. The bank provides tiny loans and financial services to these people to start their own businesses. Women are the beneficiaries of most of these loans. A typical loan might be used to buy a cow in order to sell milk to fellow villagers, or to purchase a piece of machinery that can be hired out to other people in the community. The concept has spread well beyond Bangladesh to reach millions of families in over 25 countries (Figure 8.26).

In his book, *Creating a World Without Poverty*, Yunus highlighted '**social business**' as the next phase in the battle against poverty. He presents a vision of a new business model that combines the operation of the free market with the quest for a more humane world.

However, microcredit has faced increasing criticism in recent years because:

» interest rates are often very high
» recipients often use loans to finance consumption, meaning they do not generate any new income that they can use to repay their loan.

▲ **Figure 8.26** Kathmandu, Nepal – financial cooperatives are very important for small businesses to develop

> ### Activities
> 1 Explain Figure 8.22 (page 197).
> 2 State one reason why most LICs have been keen to accept foreign aid.
> 3 What are the two components of debt?
> 4 List three characteristics of the Fairtrade system.
> 5 What are the advantages and disadvantages of microcredit?

Detailed specific example

Indonesia: an emerging economy

Indonesia's current level of economic development

According to the World Bank's income classification, Indonesia is an upper-middle-income country. Indonesia's significant economic indicators include:

- an annual GDP of over US$1 trillion
- an increasing, economically active population, with a national median age of 29
- a large and growing middle-class consumer base
- expanding commodity exports
- policies aiming to attract foreign direct investment
- a new capital city under construction in Kalimantan.

Geographical location

Indonesia is the world's largest **archipelagic country** (Figure 8.27), lying between the Indian Ocean and the Pacific Ocean and consisting of over 17,000 islands (Figure 8.28). It lies across the equator, spanning a distance of one-eighth of the Earth's circumference. Indonesia is strategically located along major shipping lanes connecting East Asia, South Asia and Oceania. The country is characterised by active volcanoes and is bounded to the south and west by a series of deep-ocean trenches.

8.3 ACHIEVING SUSTAINABLE DEVELOPMENT

▲ **Figure 8.27** Indonesia lies between the Indian and Pacific Oceans and consists of over 17,000 islands

▲ **Figure 8.28** A small, uninhabited island in the Ceram Sea, Indonesia

The capital city, Jakarta, is located near the northwestern coast of Java. About 70 per cent of the country's population live in coastal areas, relying on the surrounding seas for both income and nutrition. Indonesia contains:

- the third-largest tropical rainforest in the world
- the world's largest tropical peatlands
- the world's largest mangrove forests.

All of these important physical environments store huge amounts of carbon that help mitigate climate change impacts.

Population and economic status

With a population of almost 279 million (Table 8.10), Indonesia is the fourth most populous country in the world after India, China and the USA. Indonesia has the youngest population in the Southeast Asia region, with 24 per cent of its population under 15 years of age. However, the country's rate of population growth has fallen considerably in recent decades. With a birth rate of 17/1000 in 2023, Indonesia is now in stage 3 of demographic transition. Fifty-six per cent of Indonesia's population live on the island of Java, especially in the capital city, Jakarta. Java is the economic core of the country, accounting for 58 per cent of national GDP. Indonesia has the largest Muslim population in the world. It is a democratic country that has endured periods of autocracy in the past.

▼ **Table 8.10** Indonesia fact file for 2023

Population	278.9 million
Birth rate	17/1000
Death rate	6/1000
Infant mortality rate	15/1000
Life expectancy at birth	75 years
Population living in urban areas	59 per cent
Urban population in informal settlements	19 per cent

▼ Table 8.10 (continued)

Population per km² of arable land	1060
GNI per capita ($PPP)	14,250
Human Development Index category	High, 112th in the world

In economic terms, Indonesia is the tenth-largest economy in the world by purchasing power parity ($PPP). By 2028, it is projected to be the world's sixth-largest economy by $PPP (Figure 8.29). Indonesia is the largest economy in Southeast Asia and is generally viewed as an emerging economy. The country was invited to join the BRICS (Brazil, Russia, India, China and South Africa) group of countries in 2023, but declined the offer. However, it has not ruled out joining in the future. Indonesia is a member of G20, the group of the world's twenty largest economies. It assumed the rotating G20 Presidency in 2022.

The poverty rate in Indonesia has fallen by more than half since 1999, to under 10 per cent in 2023.

Raising Indonesia's level of development

The government of Indonesia has set a goal for the country to achieve 'high-income' status by 2045. This is consistent with the government's strategy in recent decades to raise the country's levels of development, quality of life and standard of living. The country has been following a 20-year development plan since 2005 (2005–25), which is segmented into five-year Medium-Term National Development Plans (RPJMN), each with different development priorities.

The 2023 Human Development Index ranked Indonesia as a country of medium human development. The quality of health in Indonesia has risen with the government focusing on improving the general health of people in poverty through the use of Universal Health Coverage (UHC) Policies. Examples of other significant improvements in the quality of life have been in the areas of economic stability, food security and the gender gap.

Economic transformation

In the 1980s and 1990s, Indonesia experienced rapid industrialisation through growth in manufacturing exports. This went hand-in-hand with a high rate of urbanisation. Millions of workers moved from agricultural to non-agricultural jobs, particularly in the expanding manufacturing sector. This 'first wave' of **structural transformation** substantially increased average incomes, resulting in a large reduction in the number of people below the poverty line.

However, the nature of structural transformation changed as industrialisation stalled after the 1997–98 Asian financial crisis. Indonesia's economy shifted into a natural resource-based growth model. Although workers continued to move away from agricultural employment in this 'second wave' of structural transformation, they mostly found employment in low-paid service sectors, rather than in better-paid manufacturing jobs. While low-paying jobs in the service sector are generally more productive than agricultural activities, the difference is much less than between agricultural and manufacturing employment.

The perceived lack of middle-income jobs in Indonesia today is mainly due to the more limited productivity gains in the second wave of structural transformation. In this latter period, most of the improvement in labour productivity occurred due to efficiency gains within economic sectors (agriculture, manufacturing, services) rather than the movement between sectors.

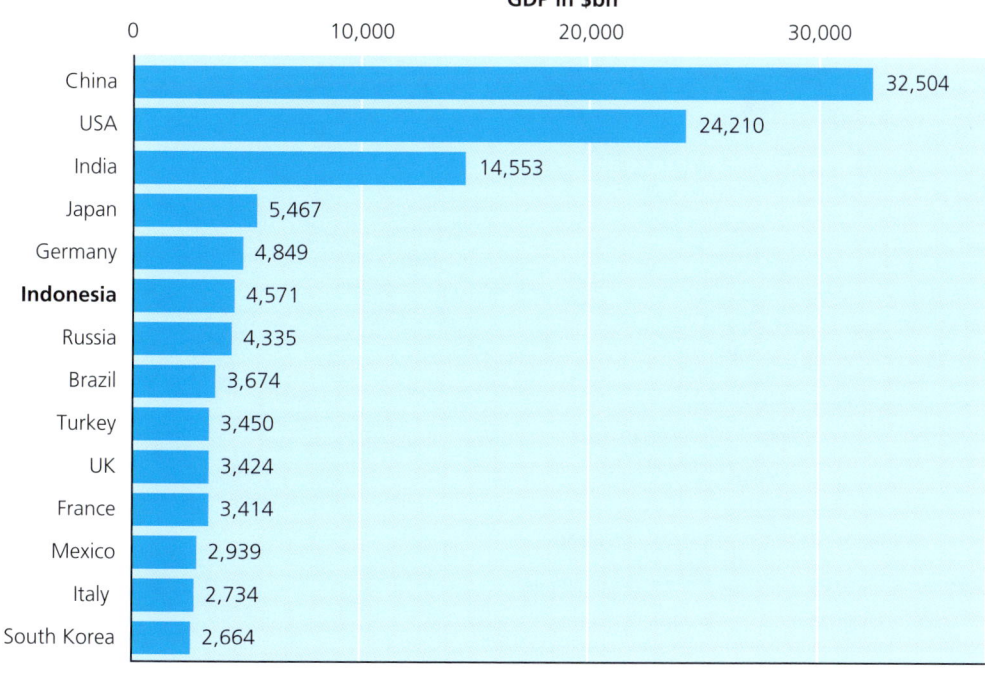

▲ Figure 8.29 The projected largest economies in the world in 2028 (GDP in $bn)

8.3 ACHIEVING SUSTAINABLE DEVELOPMENT

Between 2014 and 2023, Indonesia's economy maintained a growth rate of about 5 per cent (except during the Covid-19 pandemic), with GDP per capita rising to about US$5000 in 2023 (Figure 8.30). However, despite the significant decline in the incidence of poverty, data from the World Bank suggests that 53 per cent of Indonesia's population in 2018 either lived in poverty or were 'economically vulnerable'. In comparison, neighbouring Malaysia had an economically vulnerable population of only 3.7 per cent in 2018. While income inequality remained relatively stable during the first wave of economic transformation, it increased considerably during the second wave.

Economic partners

Indonesia's main trading partners are mainland China, the USA, Japan, Singapore, South Korea and Hong Kong. The country has longstanding trade and investment relationships with Japan, Singapore and South Korea.

In 2022, Indonesia's exports were valued at US$320 billion, up from US$190 billion in 2017. The main export destinations were China (21.2 per cent), the USA (9.9 per cent), Japan (8.3 per cent) and India (7.9 per cent). The main products exported were coal briquettes (15.9 per cent), palm oil (9 per cent), ferroalloys (4.3 per cent), petroleum gas (3.7 per cent), copper ore (2.8 per cent) and lignite (2.6 per cent).

In the same year, imports were valued at US$250 billion. The leading sources of imports were China (28.5 per cent), Singapore (9 per cent), Japan (6 per cent), Malaysia (5 per cent) and Thailand (4.2 per cent). The main products imported were refined petroleum (9.3 per cent), crude petroleum (4 per cent), motor vehicles, parts and accessories (1.7 per cent), and broadcasting equipment (1.6 per cent).

Indonesia had a current account surplus in 2021 for the first time in a decade, and this increased in 2022. However, maintaining a surplus could become a challenge if commodity prices decline.

Focus on digitisation and sustainable industries

Indonesia's digital economy grew from US$41 billion in 2019 to US$77 billion in 2022, with strategies to increase electronic transactions and digital banking activity. A key element of this strategy was the One Data Initiative, a drive to digitise government processes to create more efficient, secure and interconnected data flows between departments. The objectives of this strategy were to:

- increase economic growth through improvements in efficiency
- give potential foreign investors more confidence to begin or expand their operations in Indonesia.

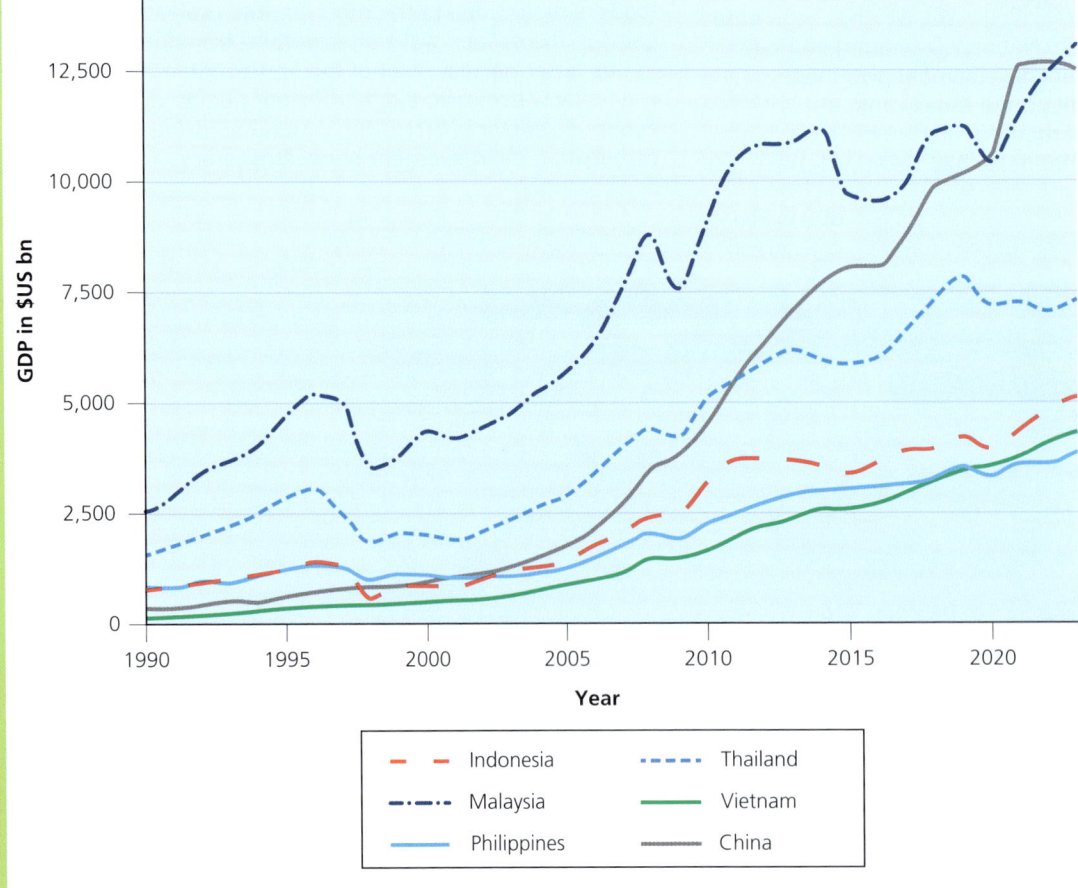

▲ **Figure 8.30** GDP per capita for Indonesia and five other Asian countries, 1990–2023

Digitisation is seen by the government as an important factor in the increase in the number of **unicorns** (start-ups), which is valued at least US$1 billion in the country's fintech, e-commerce and logistics sectors. Fintech is a portmanteau (combination) of the words 'financial' and 'technology'. It describes the use of technology to deliver financial services and products to consumers.

Indonesia has prioritised electric vehicle (EV) battery production as a key industrial sector. The country has the advantage of the world's largest nickel reserves and the third-largest cobalt reserves. Both of these raw materials are essential in the production of batteries for electric vehicles. Indonesia is shifting away from exporting these raw materials, to producing EV batteries domestically, aiming to diversify its economy and attract further foreign direct investment. This policy also forms part of its net-zero strategy. This is an element of a wider focus to attract value-adding industries to Indonesia's large reserves of natural resources. This trend broadens the country's manufacturing export base and adds resilience to its current account balance.

In 2020, Indonesia banned the export of nickel ore, which served to encourage companies such as China's Tsingshan, South Korea's LG and Brazil's Vale to set up more factories in the country in order to access its raw materials. This involves not just more nickel being refined in Indonesia, but also foreign companies building more of their supply chains in the country. Indonesia defied a ruling by the World Trade Organization that the ban was unfair. The country has also introduced export controls on bauxite, the ore used to produce aluminium, and plans to introduce export controls on copper.

In 2023, foreign investment grew 13.7 per cent to reach a record high of US$47.3 billion, after increasing by 46.6 per cent in 2022. About a third of FDI, predominantly from China, has gone into the metals and mining industries since 2021.

Developing infrastructure

The current administration has spent more on infrastructure than any previous government. High levels of investment have gone into roads, railways, dams and ports across the country. The money allocated to infrastructure rose from US$9.9 billion in 2014 to US$27 billion in 2024. Between 2014 and 2024, Indonesia built 16 new airports, 18 new sea ports, 36 new dams and 2100 km of toll roads. Lower logistics and distribution costs are an important incentive for foreign direct investment. However, critics argue that state-owned companies have built up substantial debts with this increase in expenditure on infrastructure.

Geopolitics

Indonesia has been a long-term member of the international Non-Aligned Movement (NAM). Member countries try to maintain good relations with both the 'Western' economies led by the USA, and what has become known as the 'Global South' under the 'leadership' of China. Indonesia's non-aligned stance is at least part of the reason that it turned down the invitation to join the BRICS in 2023.

Capital city: from Jakarta to Nusantara

A new capital city called Nusantara (which means *archipelago* in Indonesian) is in the early stages of construction in the heart of the rainforest in Kalimantan (Borneo). This is a US$45 billion project, announced in 2019, to move the government 1236 km away from the overcrowded and subsiding megacity of Jakarta. Jakarta's metropolitan area is home to 30 million people. Construction of the new city began in July 2022 in an area of forests and oil palm plantations, 30 km inland from the Makassar Strait. The first stage of development involves constructing government facilities and other buildings for the expected initial population of 500,000. Construction is planned to be completed by 2045.

> ### Activities
>
> 1 What is Indonesia's current level of economic development?
> 2 Suggest why Indonesia's relatively young population structure is viewed as an economic advantage.
> 3 Describe Indonesia's two waves of structural transformation.
> 4 Why is Indonesia investing heavily in digitisation and sustainable industries?
> 5 Describe and explain the importance for Indonesia of investing in infrastructure.

8.3 ACHIEVING SUSTAINABLE DEVELOPMENT

Practice questions

1 Figure 8.31 shows the world's least-developed countries.
   a Identify the least developed country in the Americas. [1]
   b Describe the distribution of the least developed countries. [3]
   c Suggest why islands such as Kiribati and Tuvalu are among the least developed countries. [3]

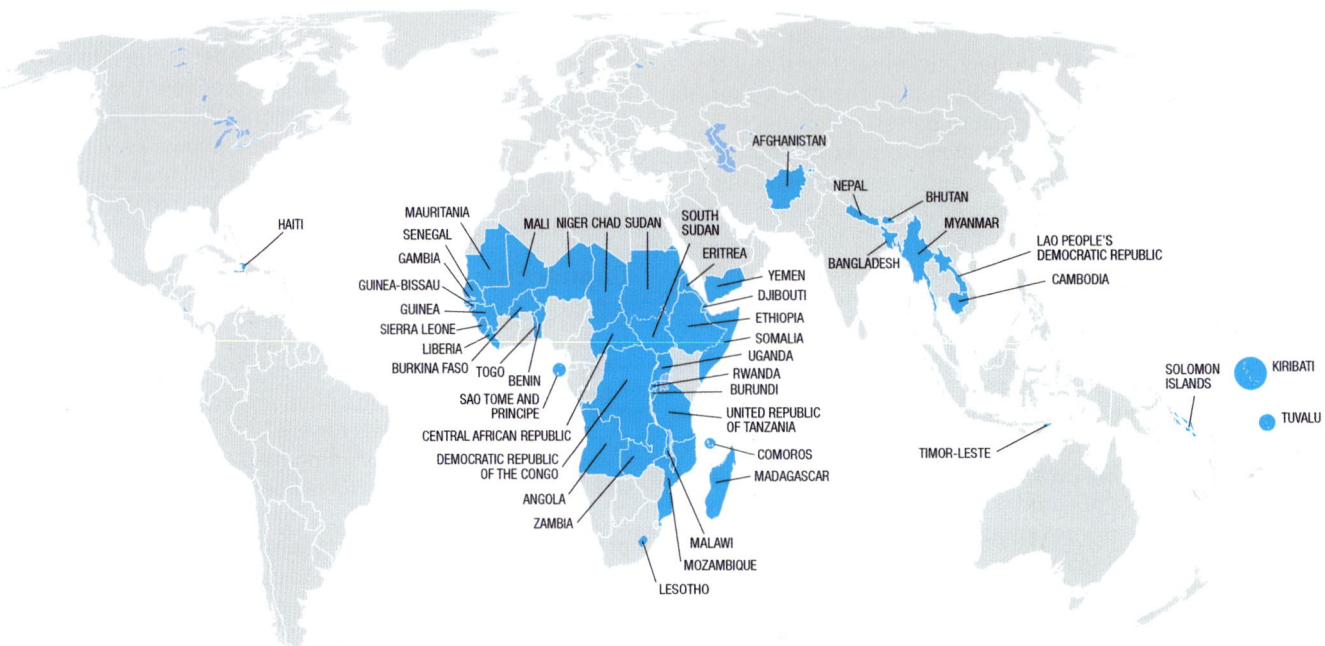

▲ Figure 8.31 The world's least-developed countries

2 Table 8.11 shows life expectancy (years) for selected countries.
   a Calculate the difference in life expectancy between the country with the highest life expectancy and the country with the lowest life expectancy. [2]
   b Compare the characteristics of countries with high life expectancy with those with low life expectancy. [4]

3 a Identify two economic indicators of development. [2]
   b Suggest how improving levels of literacy may raise levels of development. [3]
4 Using named examples, evaluate ways in which uneven development may be reduced. [7]

▼ Table 8.11 Life expectancy (years) for selected countries

Highest life expectancy (2020–25)	Lowest life expectancy (2020–25)
Monaco 89.4	Central African Republic 54.4
Hong Kong 85.3	Chad 55.2
Japan 85.0	Lesotho 55.6
Macau 84.7	Nigeria 55.8
Switzerland 84.3	Sierra Leone 55.9

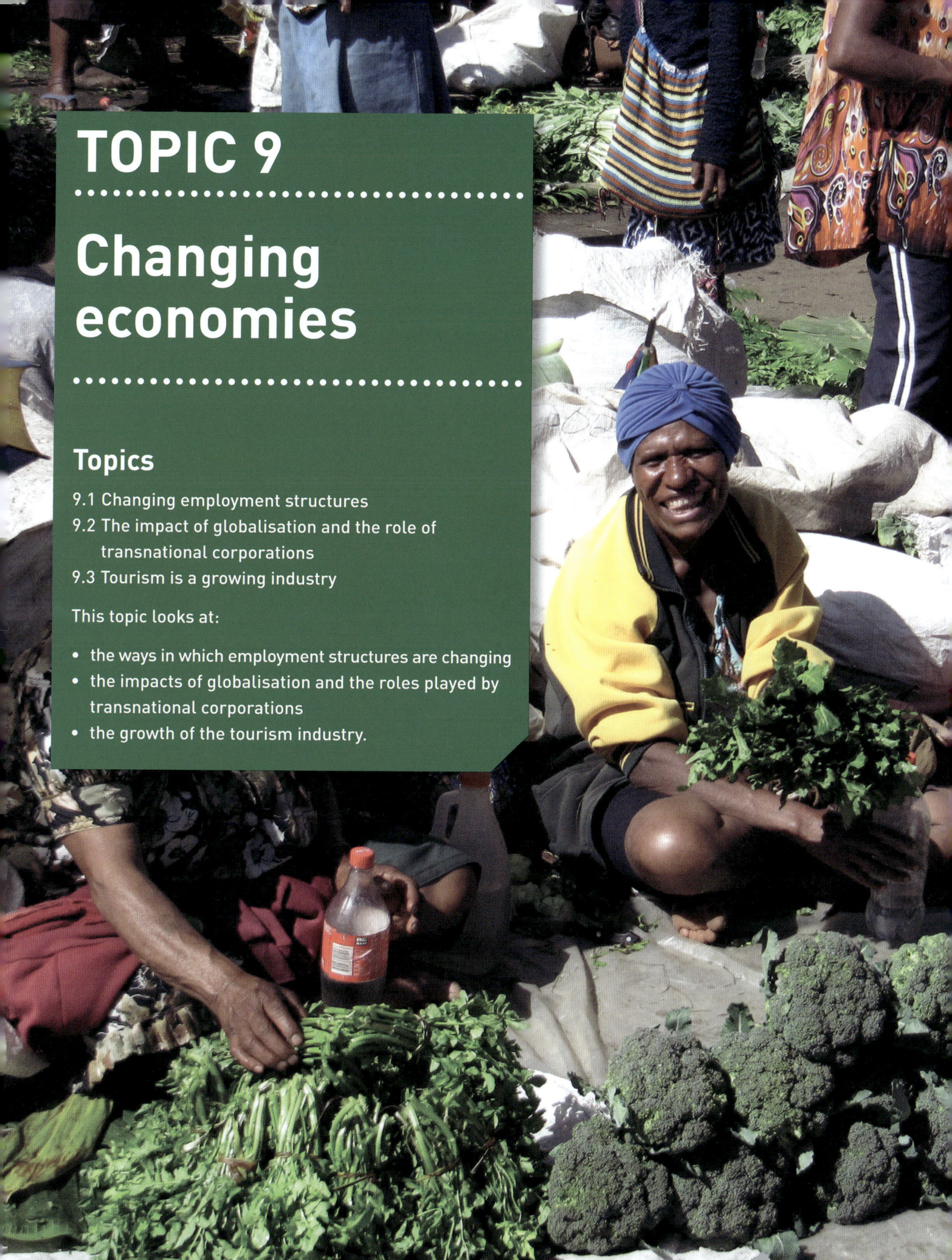

TOPIC 9

Changing economies

Topics

9.1 Changing employment structures
9.2 The impact of globalisation and the role of transnational corporations
9.3 Tourism is a growing industry

This topic looks at:

- the ways in which employment structures are changing
- the impacts of globalisation and the roles played by transnational corporations
- the growth of the tourism industry.

9.1 Changing employment structures

This chapter will explain:

★ how production is classified into different economic sectors
★ how employment structure varies and changes over time in LICs, MICs and HICs
★ the factors that affect the location and distribution of industries.

Classifying production into different economic sectors

In all modern economies people are employed across hundreds – and in some cases thousands – of different jobs. This wide range of jobs comprises four broad economic sectors:

» Primary
» Secondary
» Tertiary
» Quaternary.

The **primary sector** exploits raw materials. Farming (Figure 9.1), fishing, forestry and mining make up most jobs in this sector. Some primary products are sold directly to the consumer, but most go to secondary industries for processing. For example, fresh fish and fresh farm products are available to customers in supermarkets and other food retailers. However, large volumes of fish and farm products are sent to food processing factories to be frozen, tinned, packaged and so on. The drilling for crude oil is another example of a primary activity. However, this crude oil is of no use until it has been processed in an oil refinery into products such as petrol for motor vehicles and aviation fuel.

The **secondary sector** manufactures primary materials into finished products. Activities in this sector include the production of processed food (Figure 9.2), motor vehicles, clothing, furniture and building construction. Secondary products are classed as either consumer goods (produced for sale to the public) or capital goods (produced for sale to other industries).

The **tertiary sector** provides services to people and to businesses. Retail employees, delivery drivers, architects, nurses and teachers are examples of occupations in this sector. Transportation is a major tertiary industry that moves both people and freight over a wide range of distances. Figure 9.3 shows a freight train carrying sawn timber to be used directly in the construction industry or to be processed into furniture or other products. A modern economy could not function without a reasonably efficient transport sector.

▲ **Figure 9.1** The primary sector: a dairy farm, North Island, New Zealand

▲ **Figure 9.2** The secondary sector: a grain processing factory, Chicago, USA

Classifying production into different economic sectors

▲ Figure 9.3 The tertiary sector: a freight train on the Trans-Siberian Railway, carrying sawn timber from eastern Russia to Moscow

Banking is another important industry within the tertiary sector. Figure 9.4 shows a branch of the National Bank of Kenya in Nairobi's central business district, where many banks, both Kenyan and international, have branches.

▲ Figure 9.4 The tertiary sector: the National Bank of Kenya in Nairobi, Kenya

The **quaternary sector** uses high technology to provide information and expertise to all sectors of an economy. It is sometimes referred to as the 'knowledge-based part of the economy'. Jobs in this sector include aerospace engineers, research scientists (Figure 9.5), biotechnology workers and computer programmers. A well-functioning quaternary sector requires a highly educated workforce. The quaternary sector has only been recognised as a separate economic sector since the late 1960s. Before then jobs that are now classed as quaternary were placed in either the secondary or tertiary sectors, depending on whether or not they produced a tangible product. However, even today much of the available information on employment does not consider the quaternary sector.

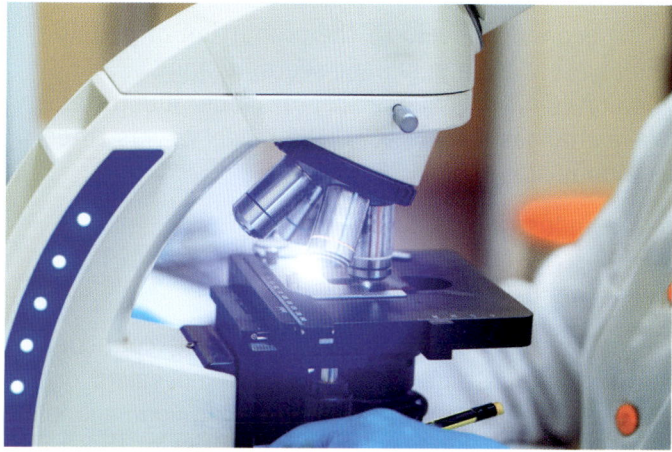

▲ Figure 9.5 The quaternary sector: a research scientist

A **product chain** can be used to show the relationship between the four sectors of employment. It is the full sequence of activities needed to turn raw materials into a finished product. The food industry (Figure 9.6) provides a good example. Some large companies are involved in all four stages of the food product chain.

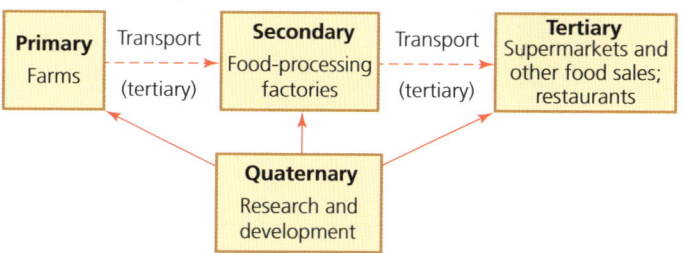

▲ Figure 9.6 The food industry's product chain

> ### Activities
> 1 Define the terms:
>   a Primary sector
>   b Secondary sector
>   c Tertiary sector
>   d Quaternary sector.
> 2 Give two examples of jobs in each of the four economic sectors.
> 3 Describe the food industry product chain shown in Figure 9.6.
> 4 What job do you want to do when you complete your education? In which sector of employment is this job?

9.1 CHANGING EMPLOYMENT STRUCTURES

LICs, MICs and HICs – how employment structure varies and changes over time

As an economy develops, the proportion of people employed in each economic sector changes. The Clark-Fisher economic sector model illustrates these changes (Figure 9.7). The model covers three economic stages:

» Pre-industrial (LICs)
» Industrial (MICs)
» Post-industrial (HICs).

In the process of economic development, countries move from one stage to another. Such movement occurs at different times and speeds, depending upon economic conditions in individual countries. Critics of the model question its linear view of development, which was based on the experience of Western economies. They argue that a country's level of development cannot be solely based on its major industries.

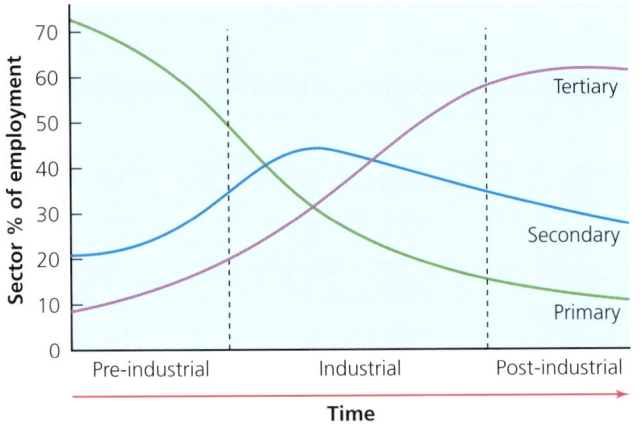

▲ **Figure 9.7** The Clark-Fisher economic sector model

LICs – pre-industrial

This is when the economy of a country is based mainly on primary activities. In this stage, most people work in agriculture and other primary industries, which are very **labour intensive**. These are largely rural economies, dominated by **subsistence agriculture**. The secondary and tertiary sectors are very limited in scope, and are also labour intensive. Many countries in this stage have faced difficulties in developing their secondary and tertiary sectors, due to historical and ongoing challenges such as poor access to markets of HICs and limited representation within important international organisations.

Many countries are still at this stage, such as are Papua New Guinea (Figure 9.8), Nepal and Afghanistan.

▲ **Figure 9.8** Papua New Guinea: a pre-industrial economy – here small farmers are selling produce in a local market in the Western Highlands

MICs – industrial

This stage begins when an individual country starts its '**Industrial Revolution**'. The UK was the first country in the world to undergo such a transformation, which began in the late eighteenth century. This was due to a combination of technological development in the UK and low-cost imports of important commodities from locations such as India and the Caribbean, over which the UK held colonial control. Some economists such as A.G. Frank have argued that colonisation held back social and economic development in many countries that were colonies.

Other leading nations such as Germany and the USA followed the industrialisation of the UK within a few decades. The developed countries in the world today began to exit the industrial economic stage from the 1960s. However, the 1960s was also the time when an increasing number of developing countries became industrialised, and were considered '**newly industrialised countries**' (emerging economies).

» Four countries stood out as the first generation of newly industrialised countries – South Korea, Taiwan, Hong Kong and Singapore (Figure 9.9).
» Such was the rapid rate of economic growth in these countries, that they were referred to as the 'Asian tigers'. A tiger economy is one that is experiencing very rapid growth.
» The emergence of newly industrialised countries is often associated with the changing nature of their relationships with Western economies in the post-colonial era.

- Within a few decades, a second generation of newly industrialised countries emerged, including Malaysia, Indonesia and Thailand.
- Within an even shorter time interval (10 to 20 years), the economies of a third group of nations began to expand rapidly, led by China and India.
- In emerging economies, production and employment in manufacturing has increased rapidly in the last 30 to 60 years.

▲ **Figure 9.9** The central business district, Singapore

HICs – post-industrial

The HICs of today do not include just the traditional developed regions of the world, but also the first generation of newly industrialised countries. This is because their industries have moved up the 'technology ladder', and their tertiary and quaternary sectors have expanded rapidly.

Countries such as the USA, the UK, Japan (Figure 9.10) and Germany are **post-industrial societies**, with most people employed in the tertiary sector. The UK exemplifies the way in which most HICs have experienced all three stages:

▲ **Figure 9.10** Japan's high-speed Bullet Train, in Tokyo Central Railway Station

- In 1800, about 75 per cent of employment in the UK was in the primary sector, with about 15 per cent in the secondary sector and 10 per cent in the tertiary sector.
- The Industrial Revolution resulted in massive changes in the UK economy. By 1900, the respective figures in the UK's economic sectors were 30 per cent primary, 55 per cent secondary and 15 per cent tertiary.
- As the twentieth century progressed, technological advance resulted in increasing **mechanisation** in primary industries, reducing the demand for workers in this sector. By 2022, only 1.2 per cent of UK workers were employed in the primary sector.
- While the share of workers employed in the secondary sector varied somewhat in the early part of the twentieth century, the period of most significant and continuous decline did not begin until the mid-1960s. Unsurprisingly, this was also the beginning of a rapid increase in the percentage employed in the tertiary sector.
- The decline in manufacturing employment was due to a range of factors, the most important of which were **deindustrialisation**, globalisation, automation and government policies. In terms of automation, most manufacturing plants became more **capital intensive**.
- The tertiary sector has also changed as computer networks and other technical advances such as artificial intelligence have reduced the number of people required in some occupations. But, for some service industries such as healthcare and tourism, employment has increased in many countries.
- Employment in the quaternary sector has become more important, which can be seen as a good measure of how advanced an economy is.

Table 9.1 compares the employment structure of an HIC, an MIC and an LIC. The contrasts are considerable, particularly between the contributions of the primary and tertiary sectors, and are a reasonable fit with the sector model presented in Figure 9.7 (page 210).

A comparison of Tables 9.1 and 9.2 shows that there is a very clear link between employment structure and socioeconomic development.

9.1 CHANGING EMPLOYMENT STRUCTURES

▼ Table 9.1 The employment structure of an HIC, an MIC and an LIC (data for 2022)

Country	% primary	% secondary	% tertiary
France	2.6	19.3	78.2
Brazil	8.7	20.5	70.8
Bangladesh	36.9	21.9	41.4

▼ Table 9.2 Development indicators for France, Brazil and Bangladesh

Country	GNI per capita (PPP)	Infant mortality rate	Life expectancy at birth
France	57,090	3.2	82
Brazil	17,260	11	73
Bangladesh	7690	25	74

A graphical method often used to compare the employment structure of countries is the **triangular graph** (Figure 9.11).

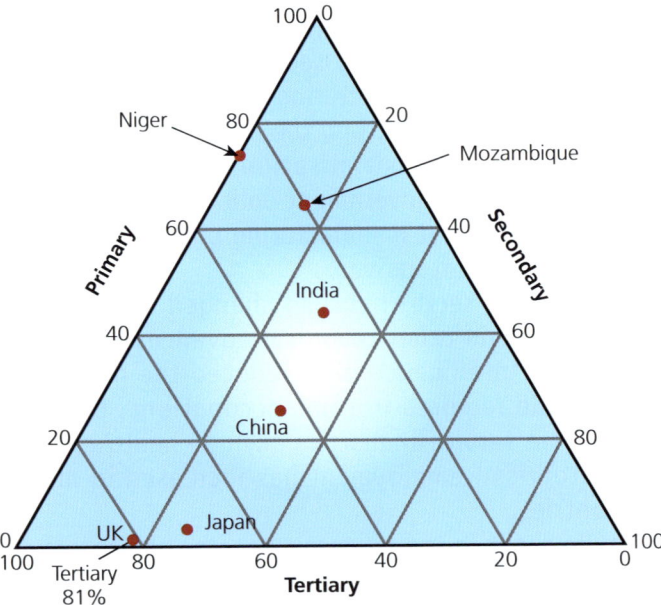

▲ Figure 9.11 Triangular graph

» One side (axis) of the triangle is used to show the data for each of the primary, secondary and tertiary sectors.
» Each axis is scaled from 0 to 100 per cent.
» The indicators on the graph show how the data for the UK can be read.

Figure 9.11 shows data for two HICs, two MICs and two LICs.

> **Activities**
> 1 Look at Figure 9.7 (page 210). Describe and explain how the employment structure of developed countries has changed over time.
> 2 Define the terms:
>   a Post-industrial society
>   b Deindustrialisation.
> 3 Look at Table 9.1. Plot the employment data for France, Brazil and Bangladesh on a copy of the triangular graph shown in Figure 9.11.

The factors affecting the location and distribution of industries

Every day, decisions are made about where to locate industrial premises, ranging from small workshops to huge industrial complexes. In general, the larger the company, the greater the number of real alternative locations available. For each possible location, a wide range of factors can have an impact on total costs and thus influence the decision-making process. The factors affecting industrial location differ from industry to industry and their relative importance is subject to change over time. These factors can be broadly subdivided into physical and human (Figure 9.12). They relate both to individual factories and to industrial zones.

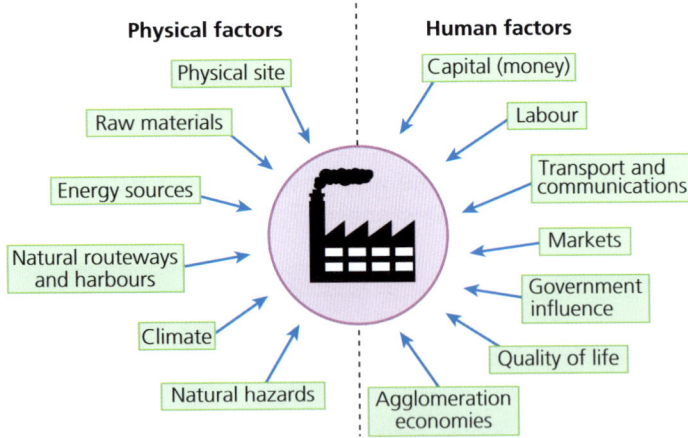

▲ Figure 9.12 The physical and human factors influencing industrial location

The factors affecting the location and distribution of industries

The physical factors influencing industrial location

Physical site

The availability and cost of the land area required for the site of a factory or other business is a fundamental factor in location. Large factories need flat, well-drained land on solid bedrock. An adjacent water supply may be essential if very large quantities of water are required for industrial production. The space requirements can vary enormously for different industries. Technological advance has made modern industry much more space-efficient than in the past. However, modern industry is more horizontally structured (on one floor) as opposed to, for example, the textile mills of the nineteenth century, which were on four or five floors. For modern developments such as **industrial estates**, retail parks and leisure centres, a large site and a highly accessible location are important.

In the UK during the Industrial Revolution, entrepreneurs faced relatively few legal restrictions on location compared with today. However, over time more and more areas have been placed off-limits to industry, mainly to conserve the environment. Areas such as National Parks, Country Parks and Areas of Special Scientific Interest now occupy a significant portion of the UK. In urban areas, land-use zoning restricts location and **green belts** often prohibit location at the edge of urban areas.

Raw materials

Industries requiring heavy and bulky raw materials tend to locate as close as possible to the raw material deposits. This is mainly to minimise transport costs. The processes involved in turning raw materials into a manufactured product usually result in **weight loss**. This means that the weight/bulk of the finished product leaving the factory is much less than the weight/bulk of the raw materials needed to produce it. Therefore, the transport costs of bringing the raw materials to the factory will be greater than the cost of moving the finished product to market (Figure 9.13). The clearest examples of this influence are where only one raw material is used. Figure 9.14 shows a very large, modern pulp and paper mill in British Columbia, Canada. It is located within an enormous forested area that supplies the mill with wood. In the UK, sugar beet refineries are centrally located in crop-growing areas because there is a 90 per cent weight loss in manufacture.

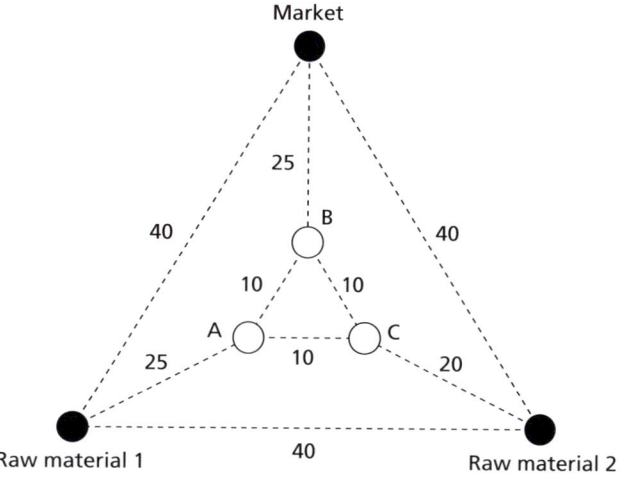

▲ **Figure 9.13** An example of weight loss in manufacturing

▲ **Figure 9.14** A pulp and paper mill, British Columbia, Canada, located within the forests that provide its raw material

9.1 CHANGING EMPLOYMENT STRUCTURES

Large and well-equipped sea ports are good locations for industries that use significant quantities of imported raw materials. Examples include flour milling, food processing, chemicals and oil refining. Such locations are **break-of-bulk points**, where cargo is unloaded from bulk carriers and transferred to smaller units of transport for onward transportation. If raw materials are processed at the break-of-bulk point, it can result in significant savings in transport costs.

> ### ▶ Activity
>
> Look at Figure 9.13 (page 213), which illustrates the concept of weight loss. The triangle shows the location of:
>
> - the two raw materials used to manufacture a product
> - the single market in which the product is sold
> - the road network and the distance in kilometres between road junctions
> - the three possible locations for the factory, A, B or C.
>
> Two tonnes of raw material 1 and one tonne of raw material 2 are required to manufacture one tonne of the finished product. Transport costs are £1 per tonne per km. Construct a table to calculate which of the three potential factory locations, A, B or C, is the lowest transport-cost location.

Energy sources

The Industrial Revolution in the UK and many other countries was based on the use of coal as a fuel. The coal was often much more costly to transport than the raw materials required for processing. It is therefore not surprising that outside of London, most of the UK's industrial towns and cities developed on coalfields or ports close to coalfields. However, during the twentieth century the construction of national electricity grids and gas pipeline systems made energy a virtually ubiquitous resource in HICs. However, even in HICs there are some industries that are constrained in terms of location because of a very high energy requirement. For example, the lure of low-cost hydroelectric power has resulted in a large concentration of electro-metallurgical and electrochemical industries in southern Norway.

Natural routeways and harbours

Many modern roads and railways follow natural routeways. A good example is the gaps eroded by rivers in chalk escarpments. Settlements with accompanying economic activity are typically located at both the scarp slope (the steep slope of an escarpment) and the dip slope (the more gentle slope of an escarpment) entry points of these natural routeways.

A natural harbour is a landform where part of a body of water is protected from storm conditions and is deep enough to provide an anchorage. Sydney Harbour (Figure 9.15), Australia is generally recognised to be the largest and deepest natural harbour in the world.

▲ **Figure 9.15** Sydney, Australia

Climate

Some industries, such as the aerospace and film industries, benefit directly from a sunny climate. For example, a climate that has a large number of cloud-free days provides the aerospace industry with more opportunities to carry out test flights. It is no coincidence that the Hollywood film industry is in southern California. Indirect benefits include lower heating bills and a more favourable quality of life. Locations with extreme climates and those regions experiencing climatic hazards usually struggle to attract business investment unless there is a very compelling reason to locate in such an environment.

Natural hazards

Regions that experience devastating natural hazards are likely to lose out on investment to other regions where severe natural hazards do not occur. Apart from loss of life, infrastructure, workplaces and vital public services can be destroyed and transport links

The factors affecting the location and distribution of industries

severed. When such disasters occur, companies may decide to rebuild in an environment with less risk. Natural hazards and the ability to deal with the consequences of natural hazards can limit economic development in LICs.

> ### Activities
> 1 Give two examples of locations where companies would be unlikely to be granted planning permission to build a factory or retail park.
> 2 What are the advantages for some types of manufacturing industry in locating at or close to a large international port?
> 3 Give two examples of industries where climate might be an important location factor.
> 4 Why do countries subject to serious natural hazards often find it hard to attract foreign direct investment?

▲ **Figure 9.16** A large capital input was required to build this container port in Indonesia

The human factors influencing industrial location

Capital (money)

Some areas are more likely to attract business investment than others. Countries and regions within countries with a reputation for business success are much more likely to attract investment in the secondary and tertiary sectors compared to locations perceived to be less successful. For example, Indonesia is an emerging economy that has attracted a high level of **foreign direct investment** (Figure 9.16). This is due to its large population of over 275 million, which is a huge internal market, along with an attractive business environment encouraged by government, abundant natural resources and a skilled population.

It is often more difficult to attract investment in LICs than in HICs and MICs because of the perceived risk. Political unrest and armed conflict deter investment.

Labour

The quality and cost of labour are important location factors (Figure 9.17). Although all industries have become more capital intensive over time, labour still accounts for about 20 per cent of total costs in manufacturing industries.

▲ **Figure 9.17** Tapestry making is labour-intensive production in Hanoi, Vietnam

Variations in labour costs can be identified at different scales. Table 9.3 provides data for eight of the 27 countries within the European Union to show how much hourly labour costs varied in 2022. In general, labour costs within the EU are higher in Northern and Western Europe than they are in Eastern and Southern Europe. By far the greatest disparity is at the global scale. One of the major reasons transnational corporations have been attracted to emerging and developing economies since the 1960s has been the relatively low labour costs in these economies. The reputation, turnover, mobility and quantity of labour are other relevant factors.

9.1 CHANGING EMPLOYMENT STRUCTURES

▼ **Table 9.3** Variations in labour costs within the European Union, 2022

Country	Hourly labour costs (US$)
Luxembourg	53.39
France	42.96
Germany	41.60
Italy	30.96
Portugal	16.95
Poland	13.17
Hungary	11.26
Bulgaria	8.60

Transport and communications

Although once a major locational factor, the share of industry's total costs accounted for by transportation has fallen steadily over time. However, the quality of transport links and their subsequent reliability remains an important issue in locational decision-making. Transport costs are particularly important for heavy and bulky goods. **Accessibility** to airports, ports, motorways and so on may be crucial for some industries, such as those producing perishable goods.

Markets

Where a company sells its products may be a big influence on where a factory is located. Not all production processes result in weight loss. Industries that add substantial amounts of water to the other raw materials required, such as soft drinks and brewing, are examples of **weight-gain** industries. These need to be located close to their markets.

However, there are other reasons for market location. Industries where fashion and taste are variable need to be able to react quickly to changes demanded by customers. One of the reasons that the major vehicle manufacturing companies are spread around the world is to respond to customers' demands, which can vary substantially in different parts of the world.

Government influence

Government policies and decisions can have a large direct and indirect impact on the location of industry. In centrally planned communist countries, such as China and Cuba, the influence of government on industry is very strong. In other countries, for instance in Northern America or the European Union, the significance of government intervention depends on:

» the degree of public ownership of an economy – for example, in some countries industries such as water, energy and railways are run by public corporations, while in other countries they are run by independent companies responsible to their shareholders
» the strength of regional policy in terms of restrictions and incentives.

Governments influence industrial location for economic, social and political reasons. Government at national and regional levels may use grants to attract major international companies. There is a high level of competition both between and within countries to attract inward investment. Significant improvements in infrastructure may be important in attracting companies to locate in a country or region. Examples of major government spending policies include building new high-speed railways and airport expansion. Such decisions can affect levels of employment in different sectors of an economy.

National governments may also establish special economic zones, such as **freeports**, in which imported goods can be held or processed free of customs duties before re-export. Figure 9.18 illustrates the multiplier effect that a successful freeport can have on a regional economy. There are many examples of freeports around the world, including Shanghai (China), Barcelona (Spain), Cork (Ireland), Teesside (UK), Manaus (Brazil) and Bangkok (Thailand).

Quality of life

Highly skilled workers, who because of their qualifications have a reasonable choice about where they live and work, will favour areas where the quality of life is high. Influencing factors include:

» proximity to attractive physical environments for recreation
» highly rated schools
» low crime rates
» good quality housing
» a high level of accessibility – roads, railways, airports.

This means that organisations and industries wishing to attract highly skilled workers are likely to take these factors into account when considering where to locate.

Agglomeration economies

Agglomeration is when similar or related industries cluster together in the same location so that ideas and activities can be combined and exchanged. This brings benefits in terms of cost savings, such as reduced transport costs. Figure 9.18 shows that in locations 1, 3 and 7, the three factories are too far away from each other to benefit from any significant cost reductions. In locations 2, 4 and 6, different pairs of factories can benefit from close proximity. However, the area of maximum mutual benefit is location 5, where all three enterprises are in very close proximity.

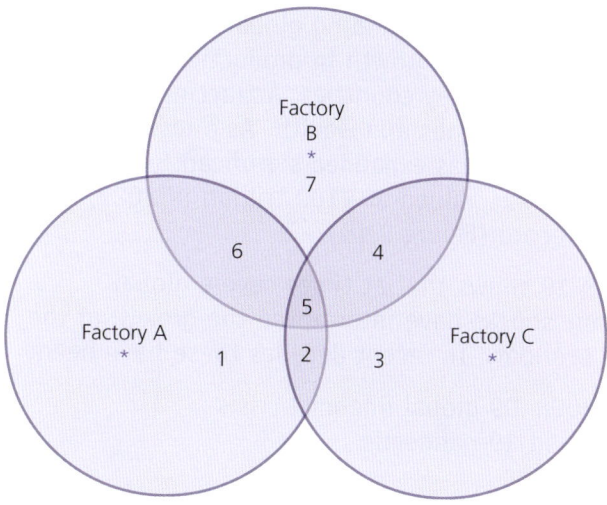

▲ **Figure 9.18** Agglomeration economies

Agglomeration economies, also known as external economies of scale, can be subdivided into the following:

- Urbanisation economies: The cost savings resulting from urban location, due to factors such as the range of services available and the investment in infrastructure already in place.
- Localisation economies: When a firm locates close to its suppliers (backward linkages) or to firms to which it supplies (forward linkages).

Increasingly, manufacturing and large retail companies have located together on industrial estates or **business/retail parks**. These are usually located close to good quality transport infrastructure, often in outer suburban or urban fringe locations.

Technology

Advances in technology have affected all sectors of employment over the last century. For example, large farms and manufacturing industries can now be run by a smaller workforce due to advanced mechanisation and automation. The development of the internet has allowed large companies to outsource production and manage complex operations all over the world. Many jobs in the service industries that were once located in HICs have been outsourced to emerging economies, such as call centres and medical transcription services. Technological advances can also reduce the amount of raw materials required or change the balance between the different raw materials used. Recycling is an important way of reducing the amount of new raw materials used in manufacturing. Such changes can affect location decision-making.

Containerisation

Containerisation has been central to most other transport developments. This system, which has been operating worldwide since the latter part of the last century, involves the use of standardised shipping containers for transport by road, rail and sea. Containerisation has been key to the development of modern intermodal transport terminals. Compared to the traditional way of moving goods from one form of transport to another, the use of containers has substantially reduced costs and the time required for transfers, along with reducing the risk of damage and theft.

A new generation of **deep-sea ports** have been built in recent decades to accommodate the increasing scale of **ultra-large container ships**. Such vessels have a capacity of at least 14,000 TEU (twenty-foot equivalent units). Container ships have increased in size because of the economies of scale that can be achieved. The HMM *Algeciras* class ships (23,964 TEU) are currently the largest container ships in the world. They are 400 m long and 61 m wide.

> ### ▶ Activities
>
> 1 Which aspects of labour are most important when selecting an area in which to locate?
> 2 State three factors relating to quality of life that might attract highly skilled labour to a location.
> 3 Give an example of the way in which technological advance might impact on industrial location.

9.2 The impact of globalisation and the role of transnational corporations

This chapter will explain:

★ globalisation and its key features
★ the impacts of globalisation on trade, transport, culture, communications and technology
★ the role, location and costs and benefits of transnational corporations
★ detailed specific example: globalisation and TNCs in Mongolia.

Globalisation and its key features

'**Globalisation**' is a term that describes the rapidly increasing and more complex global links that have developed since the 1960s. These links are economic, cultural and political. The global economy today is more extensive and complicated than it has ever been. The key features of globalisation are summarised in Figure 9.22 (page 221). A hierarchy of **global cities** has emerged to act as the command centres of the global economy. New York, London and Tokyo are at the highest level in this hierarchy. These are major financial and decision-making centres. The number of global cities has increased as the process of globalisation has advanced.

Competition within and between the different levels of the global hierarchy is intensifying because of the enormous wealth that such business and financial centres can create. Expanding global connectivity has encouraged the growth in production of goods and services in many countries. Attracting more business creates jobs and wealth. As a result, the global economy has expanded significantly in recent decades. It was valued at $100 trillion in 2022, an all-time record (Figure 9.19).

Figure 9.19 shows that at times severe global economic shocks have interrupted the growth of the global economy. In recent decades these have been:

» the 2008–09 global financial crisis
» the Covid-19 pandemic.

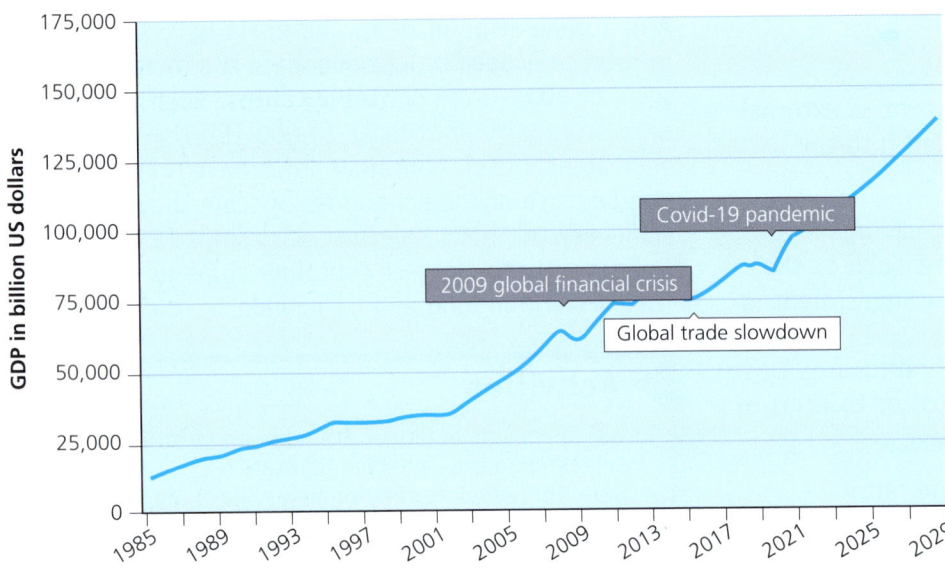

▲ **Figure 9.19** Global GDP at current price, 1985–2022, and estimates to 2029

Globalisation and its key features

Spatial distribution

Figure 9.20 illustrates the spatial distribution of the global economy in 2021. In this year the USA and China together accounted for more than 42 per cent of global GDP. The USA has been the world's largest economy since 1871. However, by 2030, China's GDP is projected to surpass that of the USA. The economic rise of China and other large emerging economies

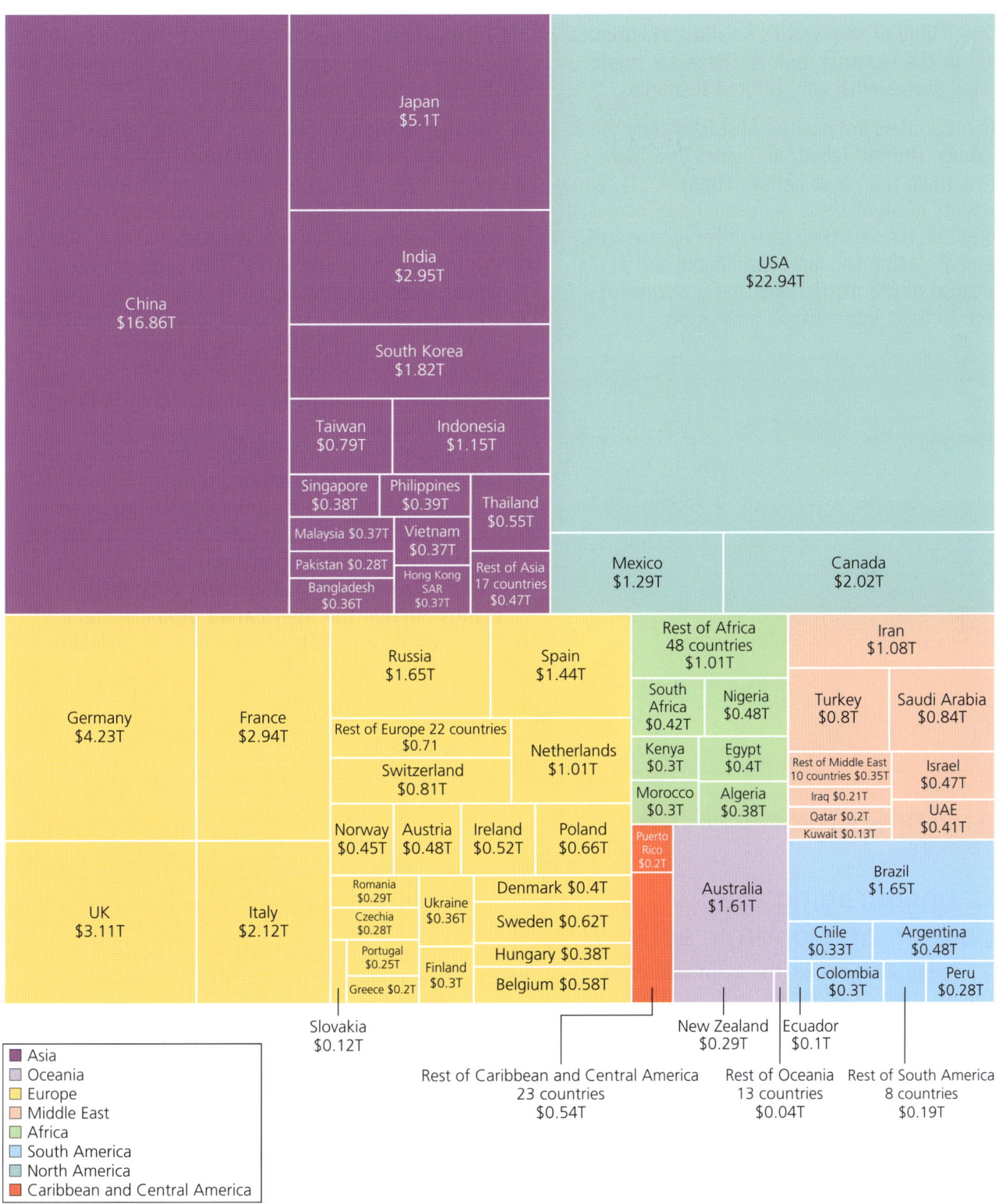

▲ **Figure 9.20** The spatial distribution of the global economy, 2021

9.2 THE IMPACT OF GLOBALISATION AND THE ROLE OF TRANSNATIONAL CORPORATIONS

has been rapid. Asia is now the largest continental economy in the world, accounting for 39 per cent of world GDP in 2021, up from 15.1 per cent in 1970. The rapid growth of emerging economies has brought about a changing world economic order. Some countries have benefited much more than others in this process. Many of the world's smallest economies are located in the Oceania region. These are small Pacific island states with very limited resources.

Most political borders are not the obstacles they once were, so goods, capital, labour and ideas flow more freely across them than ever before (Figure 9.21). While the movements of all of these elements have increased significantly, the barriers they face differ in strength. The 'movement' facing the most formidable barriers is labour, because of the restrictions that governments put in place to limit international migration.

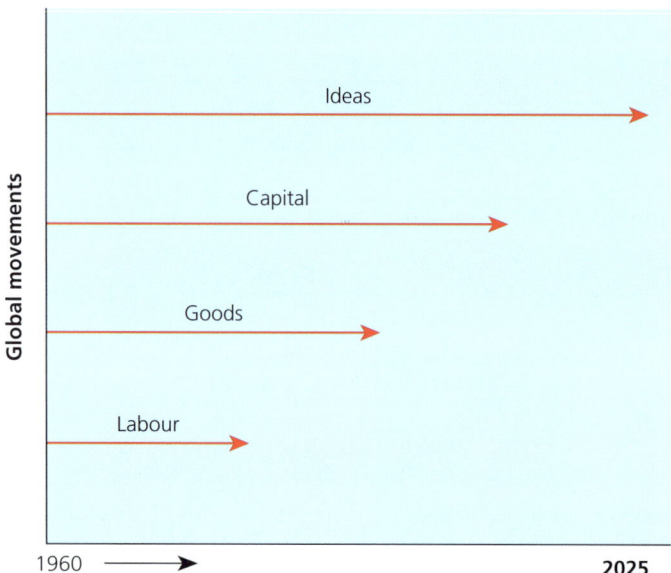

▲ **Figure 9.21** The ability to overcome political borders in the globalisation era

Factors encouraging the globalisation of economic activity

Figure 9.22 (page 221) shows the main influences on the globalisation of economic activity. Until the post-1950 period, industrial production was mainly organised within individual countries. This has changed rapidly in the last 70 years or so, with the emergence of a **new international division of labour (NIDL)**, which has developed hand in hand with the global increase in foreign direct investment. The NIDL divides production into different skills and tasks that are often spread across a number of countries. The following are other significant factors responsible for economic globalisation:

» The increasing complexity of international trade flows as the NIDL has developed.
» Major advances in trade liberalisation under the World Trade Organization. The barriers to world trade (tariffs, quotas and regulations) are much lower today than in the past. This means there is more incentive to trade.
» The emergence of free-market governments in the USA and UK around 1970. The economic policies, such as privatisation, that were developed by these governments influenced policy-making in many other countries.
» The emergence of an increasing number of newly industrialised countries (emerging economies).
» The integration of the former Soviet Union and its Eastern European satellites into the capitalist system after the fall of communism in the late 1980s. Now no significant group of countries stands outside the free-market global system.
» The opening up of other economies, particularly those of China and India, as these countries wanted to benefit from the process of globalisation.
» The deregulation of world financial markets, allowing a much greater level of international competition in financial services.

However, there are concerns that the era of globalisation has at the very least come to a period of stagnation, and some writers assert that a level of deglobalisation is already occurring. This is because of the growing geopolitical tensions between the Western economies and China and Russia.

The impacts of globalisation

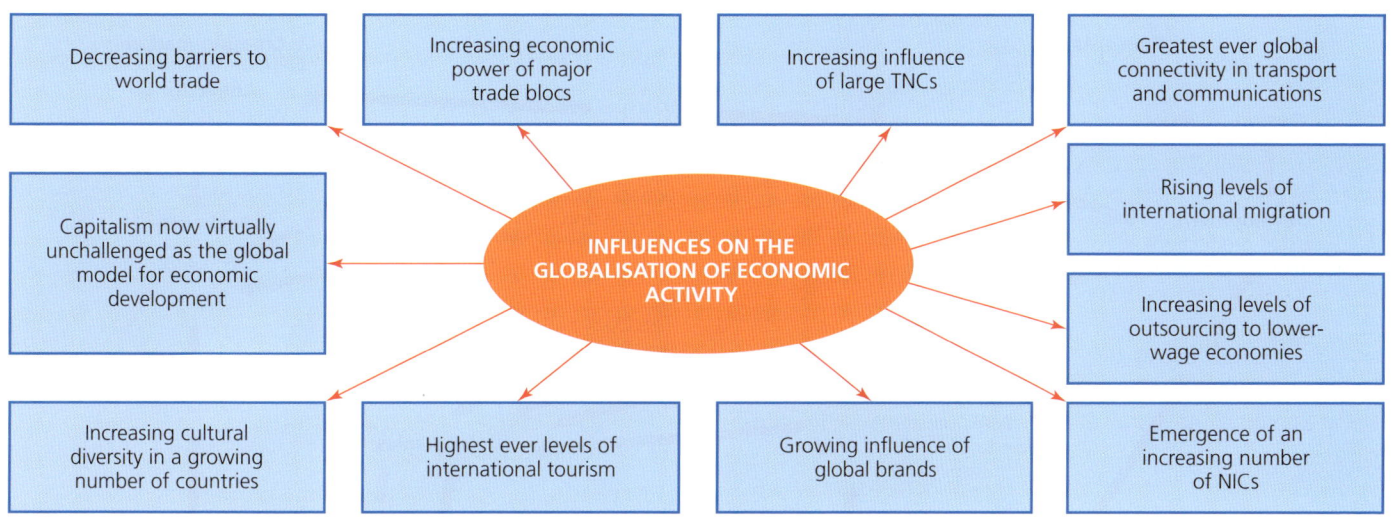

▲ Figure 9.22 Factors encouraging the globalisation of economic activity

Labour and labour migration

The large pools of relatively low-cost labour in emerging and developing countries have been a major attraction for transnational companies to manufacture in such countries. The increasing skills of these pools of labour often set off a cycle of investment, attracting production higher up the value chain. As the cost of labour has risen in the fastest-growing emerging and developing countries, companies have often looked to relocate to lower-wage economies, for example from China to Vietnam in recent years.

Fast-growing economies often experience shortages of workers. This situation can be at least partially solved by labour migration. Potential employees in lower-income countries are attracted to the jobs available in higher-income countries. The international mobility of highly skilled workers has increased substantially since the 1990s. TNCs often stress the importance of being able to attract highly skilled workers from around the world.

> ### Activities
> 1 Define globalisation.
> 2 What is a global city?
> 3 Explain the new international division of labour (NIDL).
> 4 What is foreign direct investment?
> 5 Discuss three factors (other than those covered in the previous activities) encouraging the globalisation of economic activity.

The impacts of globalisation
Trade

The development of **commodity chains** (production chains) has been an important aspect of globalisation. Each stage of manufacturing is usually completed in the location where production costs are lowest. Value is added at each stage of the process. An example is the production of footwear by companies such as Nike and Adidas. Figure 9.23 illustrates Nike's international linkages.

The barriers to **international trade** (tariffs, quotas, regulations) are much lower today than in the past. The World Trade Organization (WTO) and its predecessor the General Agreement on Tariffs and Trade (GATT) have chaired many rounds of international trade talks to reduce the barriers to trade. Between 1945 and 2021, there was a 40-fold increase in international trade. In 2023, global trade was valued at $31 trillion, down from an all-time high of $32 trillion in 2022.

The development of **trade blocs** has been an important part of this process. A trade bloc is a group of countries that share trade agreements between each other. Since the 1950s, there have been many examples of groups of countries joining together to stimulate trade between themselves and to obtain other benefits from economic cooperation. The following forms of increasing economic integration between countries can be recognised: free trade areas, customs unions, common markets and economic unions.

9.2 THE IMPACT OF GLOBALISATION AND THE ROLE OF TRANSNATIONAL CORPORATIONS

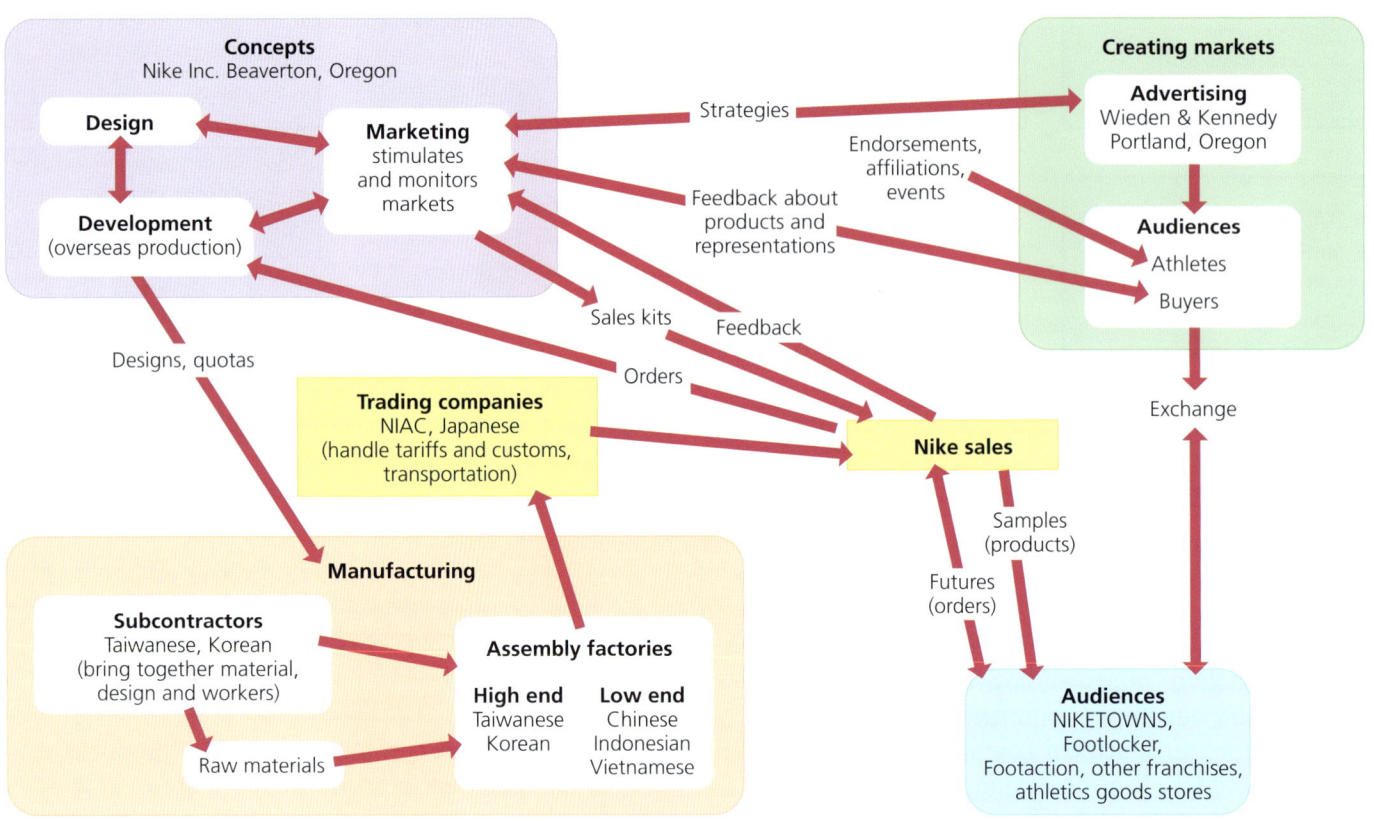

▲ Figure 9.23 Nike's commodity linkages

Longstanding examples of trade blocs include the European Union (EU) and the North American Free Trade Agreement (NAFTA). The most recent example is the African Continental Free Trade Area, which was formed in 2018 and is comprised of 54 nations. Apart from trade blocs, there are a number of looser trade groupings aiming to foster the mutual interests of member countries.

There is a strong relationship between trade and development. In general, countries that have a high level of trade are richer than those with lower levels of trade. Countries that can produce goods and services that are in demand elsewhere in the world will benefit from strong inflows of foreign currency and from the employment their industries provide. Foreign currency allows a country to purchase goods and services from abroad that it either does not produce itself or does not produce in large enough quantities.

An important economic grouping is the Group of Seven (G7), which holds meetings on a regular basis to discuss all aspects of the global economy. The members of G7 are the USA, Canada, Japan, the UK, Germany, France and Italy. A very significant and more recent change was the establishment of the Group of 20 (G20) in the wake of the 1997 Asian financial crisis, to bring together advanced and emerging economies to stabilise the global financial market. Every continent is represented in the G20 group. However, smaller countries still feel they have little say in policy formation. African countries have noted that the continent has only one country (South Africa) representing it.

Advances in transportation

Major advances in transportation have reduced the geographical barriers separating countries and peoples. Transport systems are the means by which materials, products and people are transferred from place to place. **Transport hubs** are crucial for global connectivity and economic growth. The expansion of international trade has created global supply chains that rely on efficient transportation and logistics systems. The world's largest transport hubs are massive global operations (Table 9.4).

The impacts of globalisation

▼ **Table 9.4** The largest transport hubs in the world: air, sea and ground

Location	Characteristics
Air: Hong Kong International Airport	• High-tech facilities and technology, such as automated storage and retrieval systems, temperature-controlled storage and advanced cargo-tracking systems • The location of over 80 cargo carriers, creating a network to over 220 destinations worldwide. In 2023, the airport processed more than 4.3 million tonnes of cargo • It is a key centre for shipments between Asia and the rest of the world
Sea: Port of Shanghai, China	• In 2021, the port processed over 49 million TEU (twenty-foot equivalent units) of cargo in containers, which was approximately one-third of China's total container throughput • It has links to more than 2000 ports in over 160 countries • It benefits from the most advanced technology for handling, tracking and distribution
Ground: FedEx World Hub, Memphis, Tennessee, USA	• FedEx is the world's largest cargo airline in terms of both fleet size and freight tonnes flown • Located at Memphis International Airport (with an area of 3.5 km^2), it is the main centre for the FedEx Express delivery network, which connects to over 220 countries and territories globally • It processes millions of packages and documents daily, with sophisticated equipment, such as automated guided vehicles, advanced scanning and tracking systems, and real-time data analytics. This ensures the smooth running of operations throughout the site

There have been significant developments in logistics, with the aim of making the shipment of freight from one form of transport to another as seamless as possible. Digital and smart transportation systems have advanced efficiency and safety, and have been instrumental in the rapid development of e-commerce and online marketplaces. E-commerce has resulted in the development of new delivery models, such as same-day delivery.

Air freight has increased in importance in recent decades, particularly for high-value and time-sensitive goods such as electronics, pharmaceuticals and perishable goods. However, most goods are still carried by sea. In many countries, deep-sea ports have been built to accommodate ultra-large container ships, which allow for the most extensive economies of scale.

Autonomous vehicles may well become significant in moving freight as well as people in the not too distant future. An autonomous vehicle is one that can drive itself from a starting point to a predetermined destination in an 'autopilot' mode using various in-vehicle technologies and sensors. Significant developments in this area could transform transportation and further integrate global markets.

However, while the pace of transport infrastructure development has been stunning in many HICs and emerging economies, many LICs still face the challenge of severe infrastructure limitations. Upgrading infrastructure to a high standard is an expensive process, and the global gap in infrastructure capability is widening rather than narrowing for many LICs. This will further hinder their connectivity and integration into the global economy.

Culture

The mixing of cultures is a major dimension of globalisation. **Cultural diffusion** is the process of the spreading of cultural traits from one place to another. Cultures can move in both real space and cyberspace. This has occurred through:

» migration, which circulates ideas, values and beliefs around the world and plays a significant role in cultural diffusion (Figure 9.24)
» the rapid spread of new ideas and fashions through the mass media, trade and travel
» the growth of global brands, such as Coca-Cola and McDonald's, which serve as common reference points
» the internet, which has allowed individual and mass communication on a scale never available before
» the transport revolution, which has facilitated the mass movement of people and products around the world.

9.2 THE IMPACT OF GLOBALISATION AND THE ROLE OF TRANSNATIONAL CORPORATIONS

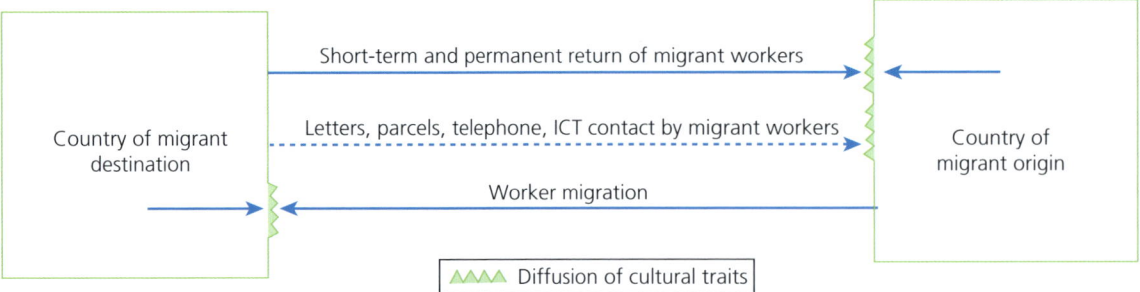

▲ **Figure 9.24** Worker migration and cultural diffusion

The spread of a global **consumer culture** has been important to the success of many major companies. The mass media have been used very effectively to encourage consumers to 'want' more than they 'need'. The power of brands and their global marketing strategies cannot be underestimated. This is particularly so for food, beverages and fashion.

Globalisation has undoubtedly had a significant impact on the increasing uniformity of landscapes. Figure 9.25 illustrates the main ways in which common urban characteristics have diffused around the world. Many developments will of course encompass local traits as well, but the strong global elements will be clear to see. Industrial, agricultural, touristic and transport landscapes have also become more uniform as similar processes worldwide have influenced their development. Some people find this disappointing as more and more places lose their 'uniqueness'.

Communications

Communications systems are the ways in which information is transmitted from place to place in the form of ideas, instructions and images. As time has progressed, the diffusion of new ideas has speeded up so that a technical breakthrough in one part of the world can have an impact on other parts of the world much more quickly than ever before.

Major advances in telecommunications systems have significantly improved global connectivity and have influenced all aspects of economic activity. The internet has been essential to the development and speed of globalisation. It is the fastest-growing mode of communication ever. It has been argued that the massive advances in information and communications technology over the last 40 years or so have been the main driver of globalisation.

The number of internet users around the world increased from 361 million in 2000 to over 5 billion in 2024. According to Statistica.com, as of April 2024 there were 5.44 billion internet users worldwide – 67.1 per cent of the global population. Of this total, 5.07 billion people were social media users.

The internet has allowed companies to manage complex operations all over the world, and to talk directly to their customers. Companies can react more quickly than ever before to changing consumer demand.

Technology

Advances in technology have affected all aspects of global economic activity. Robotics technology is gathering pace. Examples include driverless cars and drones. The impact on business organisations and many sectors of employment will be considerable.

▲ **Figure 9.25** Aspects of global urban uniformity

According to the International Federation of Robotics (IFR), the top five global robotics trends in 2024 are as follows:

- Artificial Intelligence (AI) and machine learning: Robot manufacturers are developing technology that allows users to program robots using natural language instead of code.
- Robots expanding to new applications: Advances in sensors, vision technologies and smart grippers allow robots to respond in real time to changes in their environment to work safely alongside human workers.
- Mobile manipulators: Automating material-handling tasks in industries such as automotive, aerospace and logistics.
- Digital twins: The twin exists purely as a computer model, which can save costs by being stress-tested and modified with no safety implications.
- Humanoid robots: The human-like design allows these robots to work flexibly in environments that were created for humans.

China aims to mass-produce humanoid robots by 2025. The Chinese Ministry of Industry and Information Technology (MIIT) predicts humanoids will become a **disruptive technology**, like computers and smartphones did before them. This could transform the production of goods, as well as the way that human societies live.

> ### Activities
> 1. Write a brief summary of the linkages illustrated by Figure 9.23 (page 222).
> 2. List the barriers to international trade.
> 3. Explain the processes illustrated by Figure 9.24 (page 224).
> 4. Find out more about one of the robotic trends listed above. Give a brief class presentation of your research.

Transnational corporations

Transnational corporations (TNCs) and nation states (countries) are the two main elements of the global economy. The governments of countries individually and collectively (through global institutions) set the rules for the global economy, but the bulk of investment is through TNCs. Thus, TNCs have a substantial influence on the global economy in general, and on the countries in which they choose to locate in particular. Table 9.5 shows the ten largest global corporations by revenue. The list is dominated by the USA and China, which each have four of the top ten global corporations.

▼ Table 9.5 The ten largest global corporations by revenue

Rank	Company	Country	Revenues (US$ million)	Profits (US$ million)
1	Walmart	USA	572,754	13,673
2	Amazon	USA	469,822	33,364
3	State Grid	China	460,617	7138
4	China National Petroleum	China	411,693	9638
5	Sinopec Group	China	401,314	8316
6	Saudi Aramco	Saudi Arabia	400,399	105,369
7	Apple	USA	365,817	94,680
8	Volkswagen	Germany	295,820	18,187
9	China State Construction Engineering	China	293,712	4444
10	CVC Health	USA	292,111	7910

> ### Interesting note
> Walmart, the US retail giant, has been the world's largest company by revenue since 2014. It operates a chain of hypermarkets, discount department stores and grocery stores in the USA and 23 other countries, employing 2.1 million people around the world.

The role of TNCs

TNCs play a major role in world trade in terms of what and where they buy and sell. A considerable proportion of world trade is intra-firm, taking place within TNCs. The organisation of the major car manufacturing companies (Figure 9.26) exemplifies **intra-firm trade**, with engines, gearboxes and other key components produced in a number of countries and exported for assembly elsewhere. Apart from their direct ownership of productive activities, many TNCs are involved in a web of collaborative relationships with other companies across the globe.

9.2 THE IMPACT OF GLOBALISATION AND THE ROLE OF TRANSNATIONAL CORPORATIONS

▲ **Figure 9.26** Japanese investment in the UK: Suzuki in Crawley

The increase in the importance of **global value chains (GVCs)** in recent decades has become a major feature of economic globalisation. This is the trend to sub-divide the value chain to produce goods in stages in a number of locations, adding value at each stage. The overall objective is to produce the final product at the lowest total cost.

TNCs are involved in all economic sectors (primary, secondary, tertiary and quaternary). The 100 largest TNCs represent a significant proportion of total global production. Manufacturing industry at first, and more recently services, have relocated in significant numbers from developed countries to selected emerging and developing countries as TNCs have taken advantage of lower labour costs and other ways to reduce costs. This process has resulted in the emergence of an increasing number of newly industrialised countries (emerging countries) since the 1960s. Major examples are China, India and Brazil. The share of emerging economies in global foreign direct investment (FDI) inflows has risen significantly in recent decades.

TNCs are the main source of FDI. TNCs invest to make profits and are the driving force behind economic globalisation. Low- and middle-income countries often rely heavily on TNCs (which can be from both developed countries and emerging economies) to provide investment to develop large businesses and infrastructure. Countries generally welcome investment from TNCs, and there is often competition between countries for such investment. This is probably because many developing countries may not be able to attract other sources of investment, and feel that the advantages of hosting TNCs outweigh the disadvantages.

Such FDI has increased rapidly over the last 60 years. Many major global companies have operations in a large number of countries. For example, India's Tata Group has operations in more than 100 countries and receives more than 60 per cent of its revenue from outside India.

TNCs from emerging economies

Thirty years ago the vast majority of the world's TNCs had their headquarters in Northern America, Western Europe and Japan. However, since then, emerging economies have been accounting for an increasing slice of the global economy. Much of this economic growth has been achieved through the expansion of their own most important companies, first domestically (as national corporations) and more recently on an international basis (as TNCs).

The global organisation of TNCs

TNCs vary widely in overall size and international scope. Variations include:

- the number of countries
- the number of subsidiaries
- the share of production accounted for by foreign activities
- the division of research activities and routine tasks by country
- the balance of advantages and disadvantages to the countries in which they operate.

Large TNCs often exhibit three organisational levels:

- Headquarters: The headquarters of a TNC will typically be in the city where the company was established.
- Research and development: This is likely to be located in the city where the company was established, or in other areas within the same country.
- Branch plants: These are often the first to be located in other countries.

However, some of the largest and most successful TNCs have divided their business operations into world regions, each with its own research and development facilities and a high level of independent decision-making. Figure 9.27 shows the locational changes that tend to occur as TNCs develop.

Transnational corporations

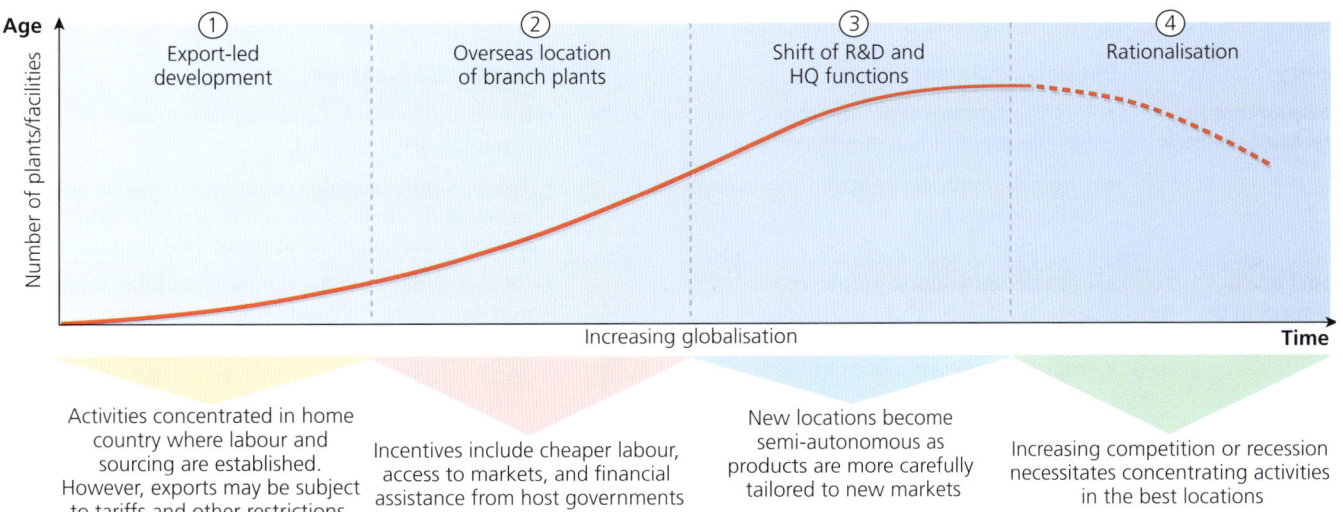

▲ Figure 9.27 The development of TNCs: locational changes

The benefits and costs to countries hosting TNCs

There are many advantages for large companies in working at the global scale:

» Sourcing raw materials and components on a global basis reduces costs.
» TNCs can seek out the lowest-cost locations for labour.
» Selling goods and services to a global market allows TNCs to achieve large economies of scale.
» Global marketing helps to establish brands with huge appeal all around the world.

Figure 9.28 summarises the possible positive and negative effects of TNCs locating in developing countries. The balance of issues will vary considerably in each individual case, and much depends on the ethical standards of individual companies. TNCs can of course also link developed nations (for example German companies operating in the USA) and emerging and developing countries (Chinese companies in South Africa).

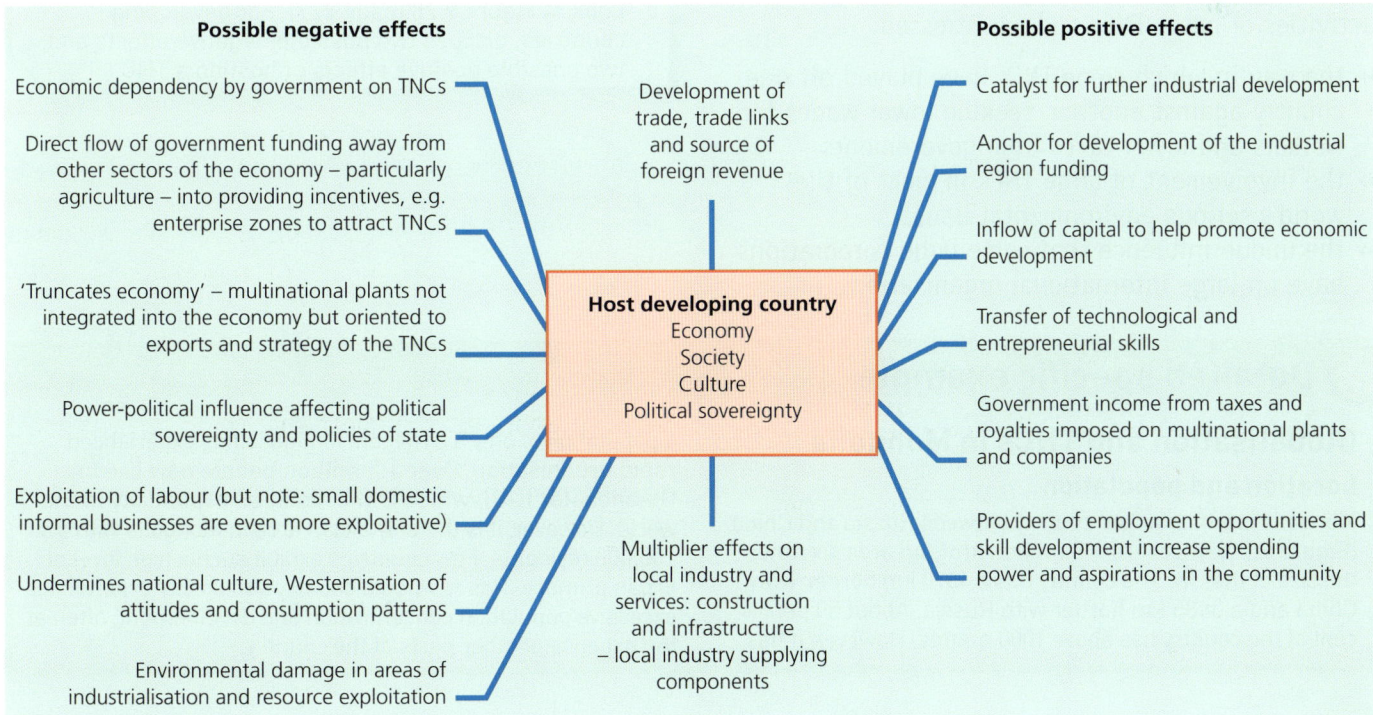

▲ Figure 9.28 The impact of transnational corporations on host developing countries

9.2 THE IMPACT OF GLOBALISATION AND THE ROLE OF TRANSNATIONAL CORPORATIONS

▼ Table 9.6 The potential advantages and disadvantages of TNCs

Country	Possible advantages	Possible disadvantages
Headquarters in developed country	• Positive employment impact and stimulus to the development of high-level skills • Direct and indirect contribution to local and national tax base	• Indirect loss of jobs and negative impact as products are imported • Trade unions complain of an uneven playing field because of the big contrast in working conditions between developing and developed countries
Impact of outsourcing in developing country	• Creates substantial employment • Pays higher wages than local companies • Improves the skills base of the local population • The success of a global brand may attract other TNCs, setting off the process of cumulative causation • Exports are a positive contribution • Sets new standards for home-grown companies • Contribution to local tax base helps pay for improvements to infrastructure	• Concerns over the exploitation of cheap labour and poor working conditions • Company image and advertising may help to undermine national culture • Concerns about the political influence of large TNCs • The knowledge that investment could be transferred quickly to lower-cost locations

TNCs exert power in host countries because of their size in terms of the number of people they directly and indirectly employ, the resources they use, and the capital flows they direct. The way TNCs operate has come under much criticism for some time now. This relates not just to the impact on developing economies, but also to the impact on the countries of origin of TNCs. Table 9.6 summarises the case of the US sportswear firm Nike.

The anti-globalisation movement has questioned the activities of many TNCs and has stressed:

» the way in which some TNCs have played off one country against another, seeking lower wages for workers and lower taxes from governments
» the involvement of large TNCs in most of the world's serious environmental issues
» the undue influence that some large corporations have on large international organisations.

> ### Activities
> 1 Define a transnational corporation.
> 2 Why have TNCs been so important in the process of globalisation?
> 3 What are global value chains and how are they related to TNCs?
> 4 Study Figure 9.27 (page 227). What are the locational changes often characteristic of the development of TNCs?
> 5 Look at Figure 9.28 (page 227). For developing countries, discuss two possible negative effects and two possible positive effects of hosting a TNC.

Detailed specific example

Globalisation and TNCs in Mongolia

Location and population

Mongolia is a landlocked country between Russia and China (Figure 9.29). It covers 1.5 million km², an area six times the size of the UK. The country has a 4670 km border with China and a 3485 km border with Russia. About 80 per cent of the country lies above 1000 metres. However, it has a population of only 3.3 million. Mongolia has experienced rapid urbanisation. Over 1.5 million people now live in Ulaanbaatar (UB), which is the 'coldest' capital city in the world. However, it is the only urban area in Mongolia with over 100,000 residents. Few countries exhibit such a high level of urban primacy. This is when a country's main city experiences excessive population concentration and development, often at the expense of other parts of the country.

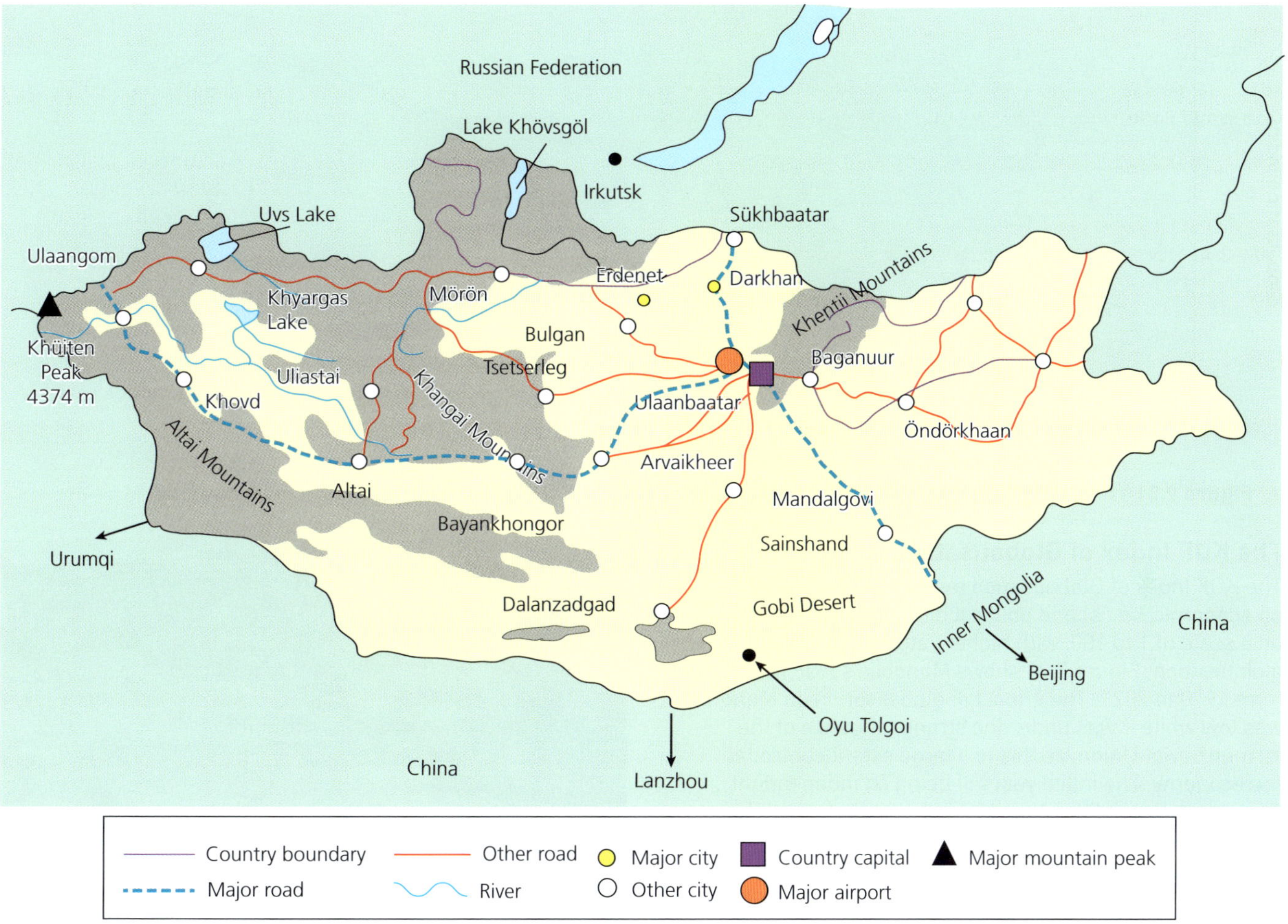

Transnational corporations

Figure 9.29 A map of Mongolia

The rest of the population is spread over vast, often inaccessible areas. Here, movement associated with herding is the norm. Mongolia has the lowest national population density in the world, at 2.13 persons per square kilometre. Funding infrastructure in such a large and sparsely populated country is a significant development problem.

The impacts of globalisation on Mongolia

Like most countries, Mongolia has been impacted by globalisation. As a satellite state of the former Soviet Union, Mongolia was a communist country until after the collapse of the Soviet Union in 1989. Mongolia peacefully transitioned to an independent democracy in 1990, and in 1992 adopted a new constitution that established a free-market economy. Its membership of a wider range of international institutions followed, including:

- the World Trade Organization in 1997
- UN peacekeeping missions from 2002
- the International Organization for Migration in 2008.

Mongolia also established stronger relations with a number of major financial institutions, including the World Bank and the Asian Development Bank (ADB). The latter is Mongolia's largest multilateral development partner. In terms of bilateral aid (aid from a single donor country), Japan has been its major supporter (Figure 9.30). In 2019, ADB committed to a $38 million loan to develop ecotourism in two national parks.

Figure 9.30 A Japanese bilateral aid project in Mongolia

9.2 THE IMPACT OF GLOBALISATION AND THE ROLE OF TRANSNATIONAL CORPORATIONS

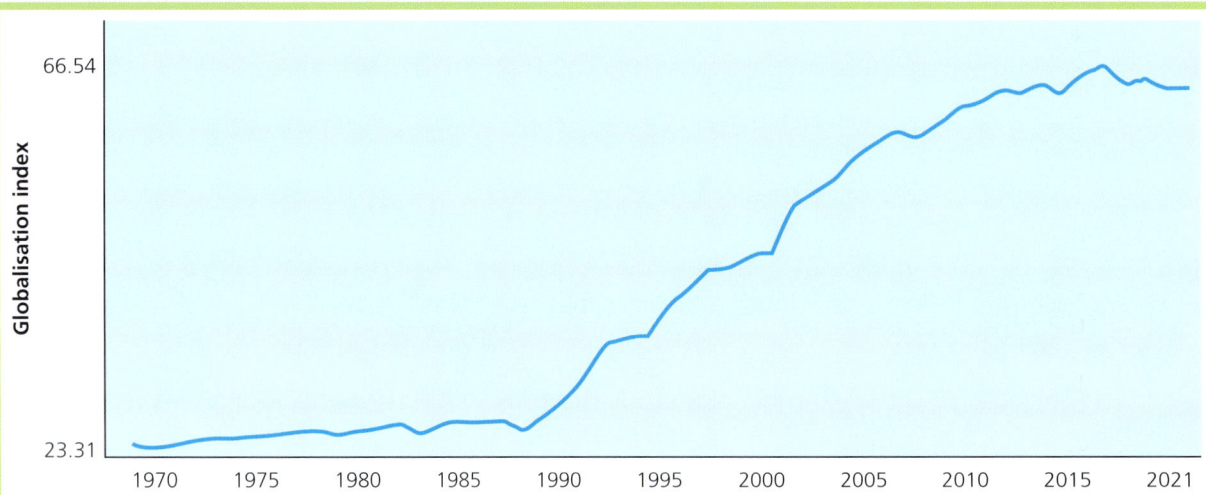

▲ **Figure 9.31** The changes in Mongolia's KOF Globalisation Index, 1970–2021

The KOF Index of Globalisation

The KOF Index of Globalisation measures globalisation on economic, social and political dimensions. The index is on a scale of 1 to 100, with higher values denoting greater globalisation. Figure 9.31 shows Mongolia's KOF Index from 1970 to 2021. The impact of globalisation on Mongolia was low while it was under the strong influence of the former Soviet Union, as this to a large extent controlled the economy. The initial years of being an independent democracy were difficult as Mongolia had relied on economic assistance from Moscow for a long time. However, as Mongolia gradually opened up to the wider world economically, socially and politically, the country's increasing global connections encouraged incoming foreign direct investment. Figure 9.31 shows the very significant change in Mongolia's globalisation index from the 1990s onwards.

Mongolia transitioned from an LIC to a lower-middle-income country in 2020. Its increase in GNI and other indicators of human development were mainly due to:

- FDI, which was almost entirely directed into the mining sector
- government investment funded by mineral royalties and external borrowing.

Economic characteristics

Mongolia relies heavily on the export of minerals. The valuable foreign exchange that this brings in allows the country to pay for the range of imports it requires. The country's extractive industries are based on large deposits of copper (Figure 9.32), gold, coal, tin, molybdenum, fluorspar, uranium and tungsten. Mongolia also has substantial known reserves of rare earth elements, but these have yet to be developed on any significant scale. The global production of rare earth elements is heavily dominated by China. Mongolia's known reserves add significantly to the country's strategic importance. The expansion of mining has led the transformation of the economy from its traditional

▲ **Figure 9.32** A copper mine near Erdenet, northern Mongolia – copper is a major mineral resource in high demand around the world

dependence on herding and agriculture. The mining sector accounts for about 21 per cent of GDP.

The impact of Rio Tinto and the Oyu Tolgoi mine

Rio Tinto is a British–Australian TNC. It is recognised as the world's second-largest metals and mining company, after BHP. It operates in 35 countries and its revenue in 2023 amounted to US$54 billion. In a strategic partnership with the government of Mongolia, Rio Tinto is currently developing the Oyu Tolgoi copper and gold mine in the southern Gobi region of the country (Figure 9.33). The mine is about 80 km north of Mongolia's border with China. It is the largest mining complex in Mongolia. Rio Tinto has 66 per cent ownership, and the government of Mongolia has the remaining 34 per cent. This form of joint ownership has become much more common in recent years in Mongolia in an attempt to ensure that the country obtains as fair a share as possible of the benefits of raw material exploitation in the country. Rio Tinto runs the mine of behalf of the partnership.

Oyu Tolgoi is one of the largest copper and gold deposits in the world. The mine is a combined open pit and underground mining project. The site was discovered in 2001, with construction beginning in 2010. Open pit mining began at Oyu Tolgoi in 2011. The copper concentrator, the largest industrial complex ever built in Mongolia, began processing mined ore into copper concentrate in 2013. The first batch of copper concentrate was shipped to customers in July 2013. In 2023, underground production began. This is where the greatest value of the project is located. At peak production, Oyu Tolgoi is expected to produce 500,000 tonnes of copper concentrate a year. When complete, the underground operation will have 200 km of tunnels, with the deepest shaft reaching 11.3 km. The project is using the leading underground mining technology known as 'block caving' to maximise the safety of employees. The project has been designed to be one of the most water-efficient copper mines in the world. The current infrastructure will allow the mine to operate for decades to come.

▲ **Figure 9.33** The Gobi Desert in south Mongolia

Socioeconomic benefits to Mongolia

As a partner in a joint venture, Mongolia is assured of a significant share of future profits, which are likely to be substantial. The global demand for copper is expected to remain very strong for many years to come as it is a material essential for decarbonisation and electrification. Oyu Tolgoi is expected to account for as much as 33 per cent of Mongolia's GDP once full commercial ore production starts.

In terms of value to the local and regional communities, a number of safeguards were negotiated between Rio Tinto and the national and regional government in Mongolia, including:

- Oyu Tolgoi's Cultural Heritage Management System, to ensure the management and protection of tangible and intangible cultural heritage
- a policy to create a supply chain in Mongolia with a particular focus on the southern Gobi region.

This includes a 'Made in Mongolia' procurement strategy to source products manufactured locally, such as miners' footwear. Oyu Tolgoi's spend in the southern Gobi region increased from $0.5 million in 2010 to $888 million in 2021.

Public opinion in Mongolia

Negotiations over the Oyu Tolgoi project were lengthy and subject to breakdown. A general feeling had developed in Mongolia over decades that:

- the country's natural resources were being exploited with insufficient benefits to the Mongolian population
- the environmental impact of mining operations was too high.

With an economy so reliant on mining, Mongolia had the difficult task of searching for the right balance between environmental conservation and natural resources development. It seems that while there was strong support for the economic opportunities associated with globalisation, there were significant concerns about the distribution of the benefits. Concerns about the fair distribution of economic gains were strongest in rural areas, where the 'trickle down' of wealth had been weaker and less obvious. There were concerns also about the erosion of traditional values and the spread of the worst aspects of Western culture.

Challenges ahead

- Mongolia continues to experience its three main geographical challenges – that it is landlocked, remote and vast (although to a lesser extent than in the past). These issues are compounded by infrastructure and logistics bottlenecks.
- Mongolia has attempted to maintain good relations with its two giant neighbours, Russia and China, as well as having a good dialogue with the USA and other countries of strategic importance. The China–Mongolia–Russia Economic Corridor has become an important element of China's Belt and Road Initiative and a major component of cooperation between China and Russia.
- With such extremes of climate and environment, Mongolia is very sensitive to climate change. Mongolian scientists say that the average temperature has increased by over 2°C since 1940.

▶ Activities

1. Describe the geographical location of Mongolia.
2. Study Figure 9.31 (page 230). What were the reasons for the increase in Mongolia's KOF Globalisation Index from the early 1990s?
3. Write a bulletpoint summary of the ownership and development of the Oyu Tolgoi mine.
4. How important is revenue from the mine expected to be when it is fully operational?

9.3 Tourism is a growing industry

This chapter will explain:

★ the growth of international tourism
★ the benefits and problems caused by tourism
★ the strategies and techniques used to sustainably manage tourism
★ detailed specific example: tourism in Jamaica.

The growth of international tourism

Over the last 50 years, **tourism** has developed into a major global industry that is still expanding rapidly. Tourism is defined as travel away from the home environment, whether for leisure, recreation and holidays, to visit friends and relatives, or for business and professional reasons.

Globally, travel and tourism's direct contribution to GDP was approximately US$7.7 trillion in 2022, accounting for 7.6 per cent of global GDP. **International tourist arrivals** were forecast to be about 1.5 billion in 2024 (Figure 9.34), which will be slightly above the pre-pandemic record in 2019. It was only in 2012 that international tourist arrivals exceeded 1 billion for the first time. In 1950, there were only 25 million international tourists.

Figure 9.34 also shows how differently the economic shocks of the SARS epidemic of 2003, the global financial crisis of 2008–09 and the Covid-19 pandemic affected international tourism. 2020 has been described as the worst year in tourism history. **International tourist receipts** in 2020 were only 37 per cent of those of the previous year, followed by 40 per cent in 2021 and 74 per cent in 2022. In 2023 the figure exceeded the pre-pandemic record of 2019.

Table 9.7 shows international tourist arrivals by world region and the percentage share of each world region for 2005 and 2023. Europe remains the world region with the greatest number of both tourist arrivals and tourist receipts. People from HICs still dominate global tourism, but many emerging economies such as China and India have shown very fast growth rates in recent years. When people can afford to travel, they invariably do. **Tourist-generating countries** have a big impact on the flow of money around the world.

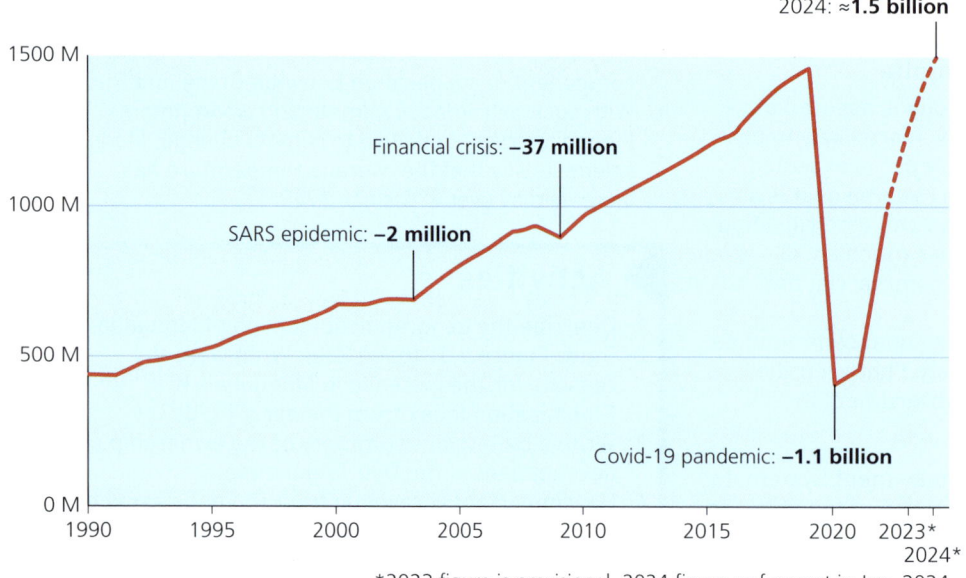

▲ Figure 9.34 The change in international tourist arrivals, 1990–2024

The growth of international tourism

▼ **Table 9.7** International tourist arrivals by world region – percentage share, 2005 and 2023

World region	2005 (%)	2005 (millions)	2023 (%)	2023 (millions)
Europe	56.0	452.7	54.5	709.4
Americas	16.4	133.3	15.4	200.2
Asia and the Pacific	19.0	154.1	18.1	237.2
Middle East	4.2	33.7	6.8	87.0
Africa	4.3	34.8	5.1	66.3

International tourism receipts worldwide amounted to around $1.48 trillion in 2023. Table 9.8 shows the countries with the highest international tourist receipts. Data is provided for both 2019 and 2022, and figures for the latter year show the tourism industry in some countries had recovered better from the pandemic than others. Seven of the 13 countries are in Europe, although the USA in 2022 had close to twice the international tourist receipts of Spain, the second-ranking country.

In terms of outbound tourism expenditure, the USA and China were virtually neck and neck in 2022, but in 2019 China was way ahead. In 2022, 36 per cent of outbound tourist departures from the USA were to Mexico. Canada and the UK followed in second and third place at 12 per cent and 4 per cent respectively.

▼ **Table 9.8** The countries with the highest international tourist receipts, 2019 and 2022 (figures are in $billion)

Country	2019	2022
USA	199	175.9
Spain	79.7	92
UK	58.4	73.9
France	63.5	68.6
Italy	49.5	55.9
UAE	38.4	51.9
Turkey	34.3	49.5
Australia	45.5	46.6
Canada	29.8	39.2
Japan	46.1	38.6
Germany	41.8	37.4
Saudi Arabia	16.4	36
Macau	40.1	32.6

Reasons for the growth of global tourism

Tourism is an increasingly important contributor to economic growth and employment in a significant number of countries. A range of factors have been responsible for the growth of global tourism. Table 9.9 subdivides these factors into economic, social and political, and also includes factors that can reduce levels of tourism, at least in the short term. Some of these factors have been active for a longer time period than others.

The medical profession was largely responsible for the growth in people taking holidays away from home. During the seventeenth century, doctors increasingly began to recommend the benefits of mineral waters, and by the end of the eighteenth century there were hundreds of spas in existence in the UK. Bath (Figure 9.35) and Tunbridge Wells were among the most famous. The second stage in the development of holiday locations was the emergence of the seaside resort. Sea bathing is usually said to have begun at Scarborough, UK, in about 1730.

▲ **Figure 9.35** The historical mineral waters in the spa town of Bath, UK

The annual holiday for the masses, away from work, was a product of the Industrial Revolution, which brought big social and economic changes. However, until the latter part of the nineteenth century only the very rich could afford to take a holiday away from home.

The first **package tours** were arranged by Thomas Cook in 1841. These tours took travellers from Leicester to Loughborough in the UK – a distance of 19 km. At the time, it was the newly laid railway

9.3 TOURISM IS A GROWING INDUSTRY

▼ Table 9.9 Factors affecting global tourism

Economic	• Steadily rising real incomes – tourism grows on average 1.3 times faster than GDP
	• The decreasing real costs (with inflation taken into account) of holidays
	• The widening array of destinations within the middle-income range
	• The heavy marketing of shorter foreign holidays, aimed at those who have the time and disposable income to take an additional break
	• The expansion of budget airlines
	• 'Air miles' and other retail reward schemes aimed at travel and tourism
	• 'Globalisation' has increased business travel considerably
	• Periods of economic recession can considerably reduce levels of tourism
Social	• An increase in the average number of days of paid leave
	• An increasing desire to experience different cultures and landscapes
	• Raised expectations of international travel with increasing media coverage of holidays, travel and nature
	• High levels of international migration over the last decade or so mean that more people have relatives and friends living abroad
	• More people are avoiding certain destinations for ethical reasons
Political	• Many governments have invested heavily to encourage tourism
	• Government backing for major international events such as the Olympic Games and the World Cup
	• The perceived greater likelihood of terrorist attacks in certain destinations
	• Government restrictions on inbound/outbound tourism
	• Calls by non-governmental organisations to boycott countries such as Myanmar

network that provided the transport infrastructure for Cook to expand his tour operations. Of equal importance was the emergence of a significant middle class, who had the time and money to spare for extended recreation.

By far the greatest developments have occurred since the end of the Second World War, arising from the substantial growth in leisure time, affluence and mobility enjoyed in HICs. However, it took the jet plane to herald the era of international mass tourism. In 1970, when Pan Am flew the first Boeing 747 from New York to London, scheduled planes carried 307 million passengers. By 2019, according to the International Air Transport Association (IATA), the number had reached 9.4 billion.

Travel motivators are the reasons that people travel. All the major tourism organisations recognise three major categories (Figure 9.36).

Many LICs and MICs have become more open to foreign direct investment (FDI) in tourism than they were about three decades ago. In general, there are now fewer restrictions on foreign investment in tourism in LICs and MICs than for many other economic activities.

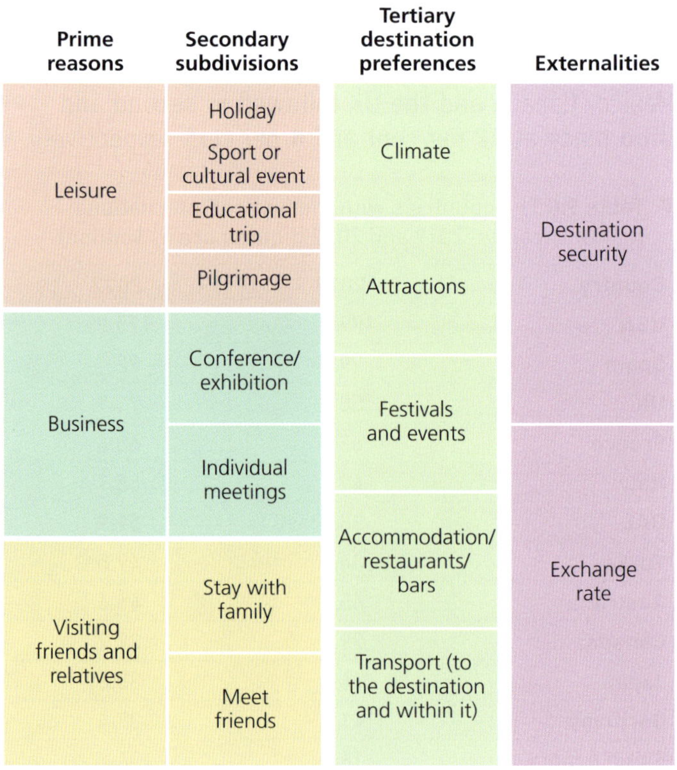

Prime reasons	Secondary subdivisions	Tertiary destination preferences	Externalities
Leisure	Holiday	Climate	Destination security
	Sport or cultural event		
	Educational trip	Attractions	
	Pilgrimage		
Business	Conference/ exhibition	Festivals and events	
	Individual meetings		
Visiting friends and relatives	Stay with family	Accommodation/ restaurants/ bars	Exchange rate
	Meet friends	Transport (to the destination and within it)	

▲ Figure 9.36 Key travel motivators

The growth of international tourism

In fact, many governments in LICs and MICs have very actively promoted a range of:

» 'soft' measures, such as tourism internet sites and support for trade fairs
» 'hard' measures, such as providing incentives for foreign investors.

Tourism is one of the top five export categories for over 80 per cent of countries and is the main source of foreign exchange for at least a third of countries.

The Butler Model

Butler's model of the evolution of tourist areas (Figure 9.37) attempts to illustrate how tourism develops and changes over time. In the first stage, the location is explored independently by a small number of visitors. If visitor impressions are good and local people perceive that real benefits are to be gained, then the number of visitors will increase as the local community becomes actively involved in the promotion of tourism. In the development stage, holiday companies from HICs take control of organisation and management, with package holidays becoming the norm. Eventually, growth ceases as the location loses some of its former attraction. At this stage, local people will have become all too aware of the problems created by tourism. Finally decline sets in, but because of the perceived economic importance of the industry, efforts will be made to re-package the location. If these are successful, they may either stabilise the situation or result in renewed growth ('rejuvenation').

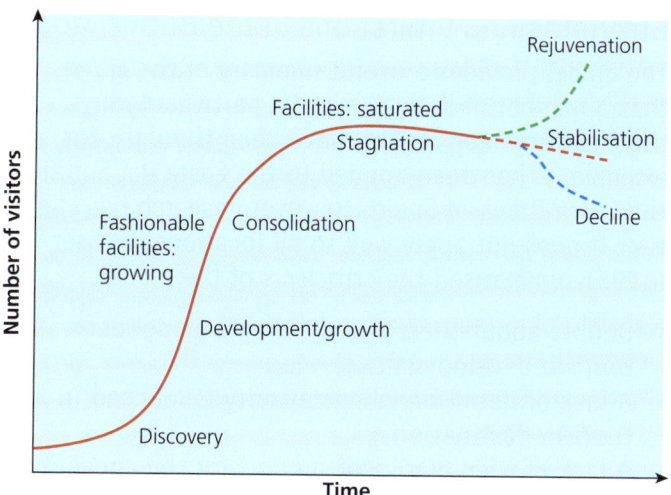

▲ **Figure 9.37** Butler's model of the evolution of tourism in a region

The Butler Model has been applied to many resorts on the Mediterranean coast of Spain, as well as in other locations. The coastal town of Sitges is located on the Costa Dorado about 30 km southwest of Barcelona. The surrounding mountains provide a sheltered environment, which has around 300 days of sunshine a year. Table 9.10 correlates the evolution of tourism in Sitges with the Butler Model.

▼ **Table 9.10** The evolution of tourism in Sitges in terms of the Butler Model

Stage of the Butler Model	Characteristics
1 Discovery	Towards the end of the nineteenth century Sitges attracted artists and wealthy people, many escaping the increasing pollution in Barcelona. The arrival of the railway in 1881 increased the town's popularity for the population of Barcelona. In 1916, the first purpose-built hotel, the Hotel Subur, opened.
2 Development/growth	In the 1960s, the advent of jet aircraft and package holidays attracted international tourists in noticeable numbers for the first time. This resulted in the construction of large hotels on the seafront and smaller hotels within the town, along with an increasing number of restaurants, bars and shops catering for tourists.
3 Consolidation	With an increasing number of tourist arrivals, Sitges became a significant tourist resort. It became an even more popular place to live and to buy a second home in. Tourism infrastructure expanded to meet the increase in demand.
4 Stagnation	As more tourist resorts expanded along Spain's coastlines and elsewhere, the increasing competition for custom impacted many smaller resorts such as Sitges. Some holiday companies removed Sitges from their brochures in favour of larger (and newer) resorts elsewhere.
5 Rejuvenation or decline	From the late 1990s, Sitges attempted to rebrand and develop more sustainable strategies. It marketed itself as an all-year destination and invested in a range of environmental improvements. Its relative proximity to Barcelona Airport and the low-cost airlines that use it has made Sitges a popular destination for short breaks and for those who mainly want to visit Barcelona, but want to stay in an attractive coastal location nearby. The Garraf Tunnels built in the 1990s made it possible to drive to Barcelona Airport in 25 minutes and the city centre in 35 minutes.

9.3 TOURISM IS A GROWING INDUSTRY

Strengths and limitations of the Butler Model

The model provides a useful summary of the stages that a number of holiday resorts, particularly in the Mediterranean region, have been through. For example, it has been applied to the Costa del Sol and the Costa Brava in Spain. However, research has shown that it does not apply well to all locations. Prosser (1995) summarised the criticisms of the model:

- Doubts about there being a single model of tourism development.
- A lack of detail on when capacity is reached in a tourism destination.
- A lack of wide empirical support for the concept.
- The limited practical use of the model.

Also, it does not include the possible role of local and national governments in the destination country or the impact of, say, a low-cost airline choosing to add a destination to its network.

> **Activities**
> 1. Define tourism.
> 2. Study Figure 9.34 (page 232). Describe the change in the number of international tourist arrivals between 1990 and 2024.
> 3. What were the factors responsible for the early development of tourism?
> 4. Why is an understanding of travel motivators important to travel and tourism companies?

The benefits and problems caused by tourism

Economic impact

It is easy to underestimate the economic impact of tourism. What is commonly thought of as the tourism industry is only the tip of the iceberg. Figure 9.38 shows both the direct and indirect economic impacts of tourism. Tourism undoubtedly brings valuable foreign currency to LICs and MICs, as well as a range of other obvious benefits, but critics argue that its value is often overrated. This is due to the following:

- **Economic leakages** (Figure 9.39) from LICs and MICs to HICs run at a rate of between 60 and 75 per cent. With cheap package holidays, by far the greater part of the money paid by tourists stays in the country where the holiday was purchased.
- Tourism is labour-intensive. It provides a range of jobs, especially for women and young people. However, most local jobs created are menial, low paid and seasonal. Overseas labour may be brought in to fill middle and senior management positions.
- Money borrowed to invest in the necessary infrastructure for tourism increases the national debt of countries.
- At some destinations tourists spend most of their money in their hotels or resorts, with minimum benefit to the wider community. This is particularly the case with all-inclusive holidays.
- Tourism might not be the best use for local resources. These could create a larger multiplier effect in the future if used by a different economic sector. A multiplier effect is when jobs and money multiply as a result of tourism development.
- Locations can become overly dependent on tourism, and therefore be badly affected if there is a downturn in the industry.
- International trade agreements, such as the General Agreement on Trade in Services (GATS), are a major impetus to globalisation and allow the global hotel giants to set up in most countries. Even if governments favour local investors, there is little they can do.

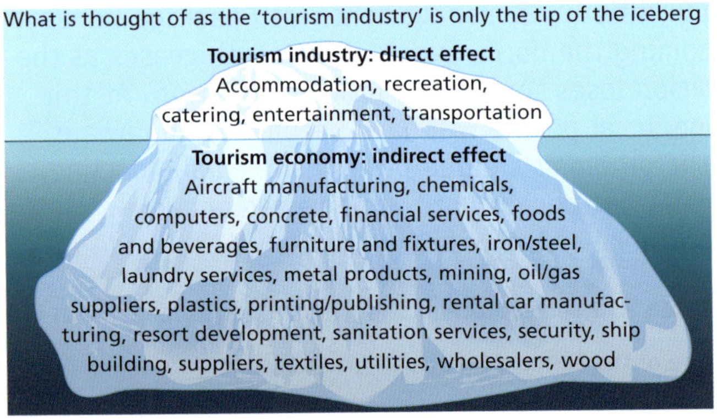

What is thought of as the 'tourism industry' is only the tip of the iceberg

Tourism industry: direct effect
Accommodation, recreation, catering, entertainment, transportation

Tourism economy: indirect effect
Aircraft manufacturing, chemicals, computers, concrete, financial services, foods and beverages, furniture and fixtures, iron/steel, laundry services, metal products, mining, oil/gas suppliers, plastics, printing/publishing, rental car manufacturing, resort development, sanitation services, security, ship building, suppliers, textiles, utilities, wholesalers, wood

▲ **Figure 9.38** The direct and indirect economic impacts of the tourism industry

DEVELOPING WORLD TOURIST DESTINATION
Total money spent on tourism to this destination

- Transport costs paid to airlines and other carriers
- Payments to foreign owners of hotels and other facilities
- The cost of goods and services imported for the tourist industry
- Remittances sent home by foreign workers
- Foreign debt relating to tourism
- Payments to foreign companies to build tourist infrastructure

LEAKAGES

▲ **Figure 9.39** Economic leakages from LICs and MICs to HICs

However, supporters of the development potential of tourism make the following arguments:

- Tourism benefits other sectors of the economy, providing jobs and income through the supply chain; this is a multiplier effect.
- It is an important factor in the balance of payments of many nations.
- It provides governments with considerable tax revenues.
- By providing employment in rural areas, it can help to reduce rural-to-urban migration.
- A major tourism development can act as a growth pole, stimulating the economy of the larger region.
- It can create openings for small businesses, in which start-up costs and barriers to entry are generally low.
- It can support many jobs in the informal sector, where money goes directly to local people (Figure 9.40).

▲ Figure 9.40 A beach artist in Agadir, Morocco – an example of informal-sector employment

Seasonality

Destinations with high fluctuations in **seasonality** frequently face challenges in both peak seasons and in 'shoulder' and low seasons. In the former, issues include overcrowding, high prices and inadequate infrastructure. In the latter, unemployment, underemployment and a lack of services are generally the main issues. Seasonality is a widespread phenomenon, with significant social and economic impacts. Definitions of the shoulder season can vary, but typically it means the period of time between a location's peak season and the low or off-season. For most locations this would be all or part of spring and autumn.

Social and cultural impact

The traditional cultures of many communities in LICs and MICs have suffered because of the development of tourism. Table 9.11 summarises some of the possible social and cultural advantages and disadvantages of tourism. The disadvantages listed in Table 9.11 are mainly caused by foreign (predominantly Western) tourist demands and behaviour. The tourist industry and the various scales of government in host countries have become increasingly aware of these problems and are now using a range of management techniques in an attempt to mitigate such effects. Education is the most important element, so that visitors are made aware of the most sensitive aspects of the **host culture**.

▼ Table 9.11 The social and cultural advantages and disadvantages of tourism

Advantages	Disadvantages
An increase in the range of social facilities for local people	The loss of locally owned land
A greater understanding between people of different cultures	The abandonment of traditional values
Wider understanding and appreciation of the historical legacy of destination countries	The displacement of people in the construction of tourist infrastructure
The development of foreign language skills in destination communities	The weakening of some traditional community structures
May encourage migration to major tourism-generating countries	Human rights abuses
Major international events such as the Olympic Games can promote a country's national and global image, however some major global events have led to big financial losses	The increasing availability of alcohol and drugs
	Crime, sometimes involving children
	Visitor congestion at key locations
	Denying local people access to beaches and other sites of interest
	The loss of housing for local people as more visitors buy second homes

9.3 TOURISM IS A GROWING INDUSTRY

Attitudes to tourism can change over time. The industry is usually seen as very beneficial initially, but can eventually become the source of considerable irritation. In countries at all income levels, the loss of housing for local people as more visitors buy second homes in popular tourist areas has become a major issue (Figure 9.41). In recent years there have been mass protests in a number of major tourism destinations about the adverse impacts of over-tourism, including in the Balearic Islands (Spain) in summer 2024.

▼ Table 9.12 Doxey's index of irritation caused by tourism

1 Euphoria	• Enthusiasm for tourist development • Mutual feeling of satisfaction • Opportunities for local participation • Flows of money and interesting contacts
2 Apathy	• Industry expands • Tourists taken for granted • More interest in profit-making • Personal contact becomes more formal
3 Irritation	• Industry nearing saturation point • Expansion of facilities required • Encroachment into local way of life
4 Antagonism	• Irritations become more overt • Environment has changed irreversibly • The resource base has changed and the type of tourist has also changed
5 Final level	• The destination will continue to thrive if it is large enough to cope with mass tourism • The tourist is seen as the harbinger of all that is bad • Mutual politeness gives way to antagonism

▲ Figure 9.41 Entrance to a national park in Andalucia, Spain – the graffiti refers to the number of foreigners buying up houses in the nearby village of Frigiliana

Tourism can also introduce a clash of cultures, with parents in particular often fearful of the impact that 'outside' cultures may have on their children. Table 9.12 shows George Doxey's 'index of irritation', which he published in 1975.

At its very worst, the impact of tourism amounts to gross abuse of human rights. For example, the actions of the military regime in Myanmar (Figure 9.42) – forcing people from their homes to make way for tourism developments, and using forced labour to construct tourist facilities – have brought condemnation from all over the world.

▲ Figure 9.42 Tourists visiting a Buddhist shrine in Myanmar in 2018

Changing community structure

Communities that were once socially and economically very close may be weakened considerably due to a major outside influence such as tourism. The traditional hierarchy of authority within the community can be altered as those whose incomes are enhanced by employment in tourism gain higher status in the community. The age and sex structure may change as young people in particular move away to be closer to work in tourist enclaves. Changing values and attitudes can bring conflict to previously settled communities. The close ties of the extended family often diminish as the economy of the area changes and material wealth becomes more important.

Environmental impact

The tourism industry has a huge appetite for basic resources, which often impinge heavily on the needs of local people. A long-term protest against tourism in Goa, India, highlighted how one five-star hotel consumed as much water as five local villages, with the average hotel resident using 28 times more electricity per day than a local person. In such situations, tourist numbers may exceed the **carrying capacity** of a destination by placing too much of a burden on local resources. The concept of carrying capacity has sometimes been taken beyond just the ability of the physical environment to accommodate tourists/visitors without resultant deterioration and degradation. One classification has identified four elements of the concept:

- Physical: The overall impact on the physical environment, for example footpath erosion (Figure 9.43).
- Ecological: The number of tourists who can be accommodated without significant impact on the flora and fauna.
- Economic: The number of tourists a destination can take without significant adverse economic implications.
- Perceptual: The attitudes of the local people in terms of how they view increasing tourist numbers.

Tourism has reached such a large scale and is so extensive in many parts of the world that its impact

▲ **Figure 9.43** Tackling severe informal footpath erosion on Mount Vesuvius, Italy

on the natural environment must be carefully managed. Tourism now accounts for nearly 60 per cent of air travel. By 2030, a 25 per cent increase in CO_2 emissions from tourism is expected compared to 2016. The industry can have a major impact on the depletion of local natural resources, often in places where resources are already scarce. The environmental organisation The World Counts states that an average golf course in a tropical country uses as much water as 60,000 rural villagers. The tourism industry generally overuses water resources as well as generating a high volume of wastewater.

Other examples of the negative environmental impact of tourism include the following:

- The destruction of natural ecosystems. For example, coral reefs are damaged by tourist boats and water sports. In both Belize and Costa Rica, coral reefs have been destroyed to create better conditions for water sports.
- Disturbance of wildlife and flora, particularly in areas where 'nature tourism' is a major activity. Such activities can significantly increase pressure on endangered species.
- Congestion and overcrowding at 'honeypot' locations. Intense trampling can cause soil erosion over a considerable area, and waste disposal and littering also affect these environments.

9.3 TOURISM IS A GROWING INDUSTRY

- The destruction of sand dunes in coastal areas by trampling, intensifying coastal erosion and destroying wildlife habitats (Figure 9.44). Many coastal wetlands have been drained to provide more land for hotels and other tourism infrastructure. Such actions can cause critical damage to local ecosystems.
- Air and noise pollution, not just from aircraft but also from tourist road traffic (coaches, buses, cars, motorcycles), and from recreational vehicles such as snowmobiles and jet skis. All can cause annoyance, stress and even hearing loss for both people and wildlife. For example, noise from snowmobiles has been found to cause animals to alter their natural activity patterns.

management of protected areas such as national parks and national forests.

> **Activities**
> 1. Compare the direct and indirect economic impact of the tourism industry.
> 2. With reference to Figure 9.39 (page 236), explain the concept of economic leakages.
> 3. Discuss three economic benefits of tourism.
> 4. Explain three negative social/cultural aspects of tourism.
> 5. Comment on Doxey's index of irritation (Table 9.12, page 238).

▲ **Figure 9.44** Sand dune restoration works, County Kerry, Ireland

Education about the environment in the area visited can have a big impact on tourist behaviour. Scuba divers in the Ras Mohamed National Park in the Red Sea, who were made to attend a lecture on the ecology of local reefs, were found to be eight times less likely to bump into coral (the cause of two-thirds of all damage to the reef), let alone deliberately to pick a piece.

Positive environmental impacts

The environmental impact of tourism is not always negative. Landscaping and sensitive improvements to the built environment have significantly improved the overall quality of some areas. On a larger scale, tourist revenues can fund the designation and

Strategies and techniques to sustainably manage tourism

The type of tourism that does not destroy what it sets out to explore has come to be known as '**sustainable tourism**'. The term comes from the 1987 UN Report on the Environment, which recommended the kind of development that meets the present needs without compromising the prospects of future generations. Following the 1992 Earth Summit in Rio de Janeiro, the World Travel and Tourism Council (WTTC) and the Earth Council drew up an environmental checklist for tourist development, which included waste minimisation, reuse and recycling, energy efficiency and water management. The WTTC has since established a more detailed programme called 'Green Globe', which is designed to act as an environmental blueprint for its members.

Sustainable tourism is tourism organised in such a way that its level can be sustained in the future without creating irreparable environmental, social and economic damage to the receiving area. It emphasises the important issues of equity and local control, which are difficult to achieve for a number of reasons:

- Governments are reluctant to limit the number of tourist arrivals because of the often desperate need for foreign currency.
- Local businesses cannot compete with foreign multinationals on price and marketing.
- It is difficult to force developers to consult local people.

Environmental groups are keen to make travellers aware of their **destination footprint**. This is the environmental impact caused by an individual tourist on holiday in a particular destination. The actions they urge people to adopt include:

- Flying less and staying longer: Replacing business trips with video conferences, and using trains and buses for regional travel if possible. Regional flights emit more carbon per kilometre travelled than long-haul flights. According to the environmental group Transport and Environment, 1 per cent of the global population is responsible for 50 per cent of emissions from flying.
- Carbon-offsetting their flights: These are voluntary schemes where people pay to make up for the emissions their flights produce. The money collected often funds the planting of new forests that act as carbon sinks.
- Considering **slow travel**.

Tourists might consider the impact of their activities both for individual holidays but also in the longer term. For example, they may decide that every second holiday will be in their own country (not using air transport). They could also stay in locally run guesthouses and small hotels as opposed to hotels run by international chains. This enables more money to remain in local communities.

As the level of global tourism increases rapidly, it is becoming more and more important for the industry to be responsibly planned, managed and monitored. Tourism operates in a world of finite resources and its impact is becoming of increasing concern to a growing number of people. A travel report published in 2021 concluded, from the available evidence, that less than 20 per cent of people around the world have ever flown and only 5–10 per cent of the world's population flies in any one year. However, this is undoubtedly going to increase over time.

Virtually every aspect of the industry now recognises that tourism must become more sustainable. Examples of best practice include:

- Costa Rica's emphasis on ecotourism and providing education and training for local people in tourism-related activities – this highlights the potential for tourism to improve broader socioeconomic development
- New Zealand's collaborative approach to tourism planning involves communities in decision-making – this tries to align tourism development with community aspirations and needs
- Bhutan's high-value, low-impact model imposes a daily fee on tourists to support healthcare, education and infrastructure development
- Amsterdam has strict regulations on short-term tourist rentals to protect housing for residents.

Ecotourism is at the leading edge of this movement. Ecotourism is a specialised form of tourism in which people experience relatively untouched natural environments such as coral reefs, tropical forests and remote mountain areas, and ensure that their presence does no further damage to these environments.

Protected areas

Over the course of the last 130 years or so, more and more of the world's most spectacular and ecologically sensitive areas have been designated for protection at various levels. The world's first National Park was established at Yellowstone in the USA in 1872. Now there are well over 1000 worldwide (Figure 9.45). Many countries have National Forests, Country Parks, Areas of Outstanding Natural Beauty, World Heritage Sites and other designated areas that merit special status and protection. Wilderness areas with the greatest restrictions on access have the highest form of protection.

In many countries and regions there are often differences of opinion when the issue of special protection is raised. For example, jobs in mining, forestry and tourism in some areas may depend on developing presently unspoilt areas. So it is not surprising that values and attitudes can differ considerably when big decisions about the future of environmentally sensitive areas are being made. Often, a clear distinction has to be made between the objective of preservation and that of conservation. **Preservation** is maintaining a location exactly as it is and not allowing development. Conservation is allowing for developments that do not damage the character of a location.

9.3 TOURISM IS A GROWING INDUSTRY

▲ **Figure 9.45** Tarangire National Park, Tanzania – an important protected area

Quotas

Quotas seem to be one of the best remedies on offer. The UK Centre for Future Studies has suggested a lottery-based entrance system, an idea endorsed by Tourism Concern. This would mean the number of visitors would not be allowed to exceed a sustainable level. This is an idea we are likely to hear much more about in the future. In 2022, the island of Corsica in France adopted daily quotas to protect the environments of its most popular destinations. The introduction of quotas is much more of an option in more affluent tourism destinations than it is in less affluent destinations as these are heavily reliant on tourism income.

Governance

Leo Hickman, in his book *The Final Call*, claims: 'The net result of a widespread lack of government recognition is that tourism is currently one of the most unregulated industries in the world, largely controlled by a relatively small number of Western corporations such as hotel groups and tour operators. Are they really the best guardians of this evidently important but supremely fragile global industry?' Hickman argues that most countries only have a junior minister responsible for tourism rather than a secretary of state for tourism, which is what the size of the industry in most countries would justify.

There have been many examples in recent years where first public opinion and later local authority action has placed restrictions on tourism. In cities like Barcelona, Amsterdam and Venice, anti-tourist sentiment has built up in response to unyielding tourist growth. Too often, the tourism supply chain stimulates demand with little consideration of the capacity of destinations and the effects on the wellbeing of local communities.

Tourist hubs

The concept of tourist hubs or clusters is a model that has been applied in a number of locations. The idea is to concentrate tourism and its impact in one particular area so that the majority of the region or country feels few of the negative impacts of the industry. Benidorm in Spain and Cancún in Mexico are examples where this model was adopted, but both locations show how difficult it is to confine tourism within preconceived boundaries as the number of visitors increases and people want to travel beyond the tourist enclaves.

> ### Activities
> 1 Define sustainable tourism.
> 2 What do you understand by the term 'destination footprint'?
> 3 Which environments in the region in which you live are protected, and why?
> 4 What do you think of the idea of quotas for visitor numbers at certain locations?
> 5 What do you understand by the concept of 'slow travel'?

Detailed specific example

Tourism in Jamaica

Reasons for the growth of tourism in Jamaica

Situated in the Caribbean Sea, Jamaica has an attractive warm tropical climate (Figure 9.46). It has good beach weather all year round, with daytime temperatures consistently reaching an average of 30°C, while night-time temperatures rarely dip below 20°C. It is the warmest of the Caribbean islands. The peak season is from mid-December to mid-April. This is when Jamaica has its clearest skies and driest weather.

There are numerous highly rated beaches with soft white sands fringing clear waters. Very popular examples are the Seven Mile Beach in Negril and Doctor's Cave Beach in Montego Bay. Jamaica's National and Marine Parks are particular attractions (Figure 9.47).

Jamaica boasts a wide range of flora and fauna, along with a diverse array of attractive landscapes. Cultural attractions include the fascinating history of the island, its music (the birthplace of reggae) and nightlife, and its food.

The Jamaica Tourist Board (JTB) is responsible for marketing the country abroad. Apart from highlighting the island's physical attractions, the JTB also promotes the positive aspects of Jamaican culture – for example, the Bob Marley Museum in Kingston has become a popular attraction. Such attractions are an important part of Jamaica's objective of reducing seasonality. The physical attractions of Jamaica almost sell themselves, so the government is putting much effort into trying to boost the island's human attractions.

Tourism has become an increasingly vital part of Jamaica's economy in recent decades. It has brought considerable opportunities to its population, although it has not been without its problems. Jamaica has been

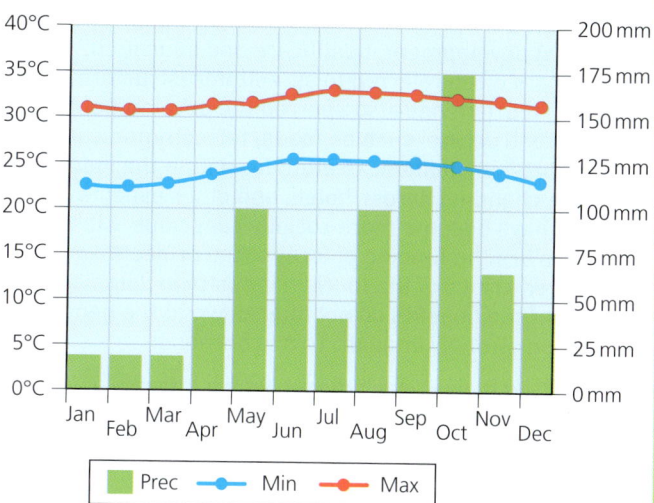

▲ Figure 9.46 Climate graph for Kingston, Jamaica

determined to learn from the mistakes of other countries and ensure that its population will gain real benefits from the growth of tourism. Jamaica's tourism policies have increasingly emphasised the importance of sustainable development.

The benefits of tourism

In 2022, Jamaica attracted 3.3 million visitors, with foreign exchange earnings of about US$3.7 billion. Tourism is the country's largest generator of foreign exchange, and a major contributor to GDP and employment. The tourism industry employs 175,000 people directly and generates indirect employment for another 354,000 people. Most visitors arrive from the USA and the UK.

▲ Figure 9.47 Jamaica's National and Marine Parks

9.3 TOURISM IS A GROWING INDUSTRY

The revenue from tourism plays a significant part in helping central and local government fund economic and social policies. Special industry taxes have gone directly into social development, healthcare and education, all of which are often referred to as 'soft infrastructure'. However, tourism has also spurred the development of 'hard infrastructure' such as roads, telecommunications and airports. In addition, as attitudes within the industry itself are changing, larger hotels and other aspects of the industry have become more socially conscious. Classic examples are the funding of local social projects.

However, critics of the tourism industry in Jamaica argue that the benefits do not filter down to the general population strongly enough. This is because:

- the all-inclusive resort model, which attracts a significant proportion of tourists to the country, has a limited positive impact on local economies
- the island's location and economic structure mean that many goods have to be imported
- a large proportion of the hotels are owned by multinational chains.

The problems caused by tourism

The high or 'winter' season runs from mid-December to mid-April, and this is when hotel prices are highest. The rainy season extends from May to November. It has been estimated that 25 per cent of hotel workers are laid off during the off-season. This has an adverse impact on the standard of living of households that are reliant on the tourist industry. It also of course means that expensive tourism infrastructure is underused for part of the year.

Although seasonality is seen as the major problem associated with tourism in Jamaica, other negative aspects include:

- the environmental impact of tourism, which includes traffic congestion and pollution at popular locations, and the destruction of the natural environment to make way for tourism infrastructure
- the heavy use of resources, particularly water, by hotels
- the underuse of facilities in the off-season
- the challenges of overfishing, waste management and increased coastal development have impacted on the health of ecosystems such as coral reefs and mangrove forests.

Sustainably managing tourism

Jamaica has established over 200 protected areas on the island. These include the National Park, Marine Parks, Forest Reserves, Special Fishery Conservation Areas and National Heritage Sites. The Jamaica Protected Areas Trust is a public–private initiative that seeks to protect and enhance the country's natural resources and biodiversity.

Figure 9.47 shows the location of Jamaica's National and Marine Parks. The Blue and John Crow Mountains National Park is a UNESCO World Heritage Site. It covers almost 42,000 hectares and includes Jamaica's highest point, Blue Mountain Peak, at 2256 metres. The Jamaican government sees the designation of the parks as a positive environmental impact of tourism. Entry fees to the parks pay for conservation. The desire of tourists to visit these areas and the need to conserve the environment to attract future tourism drives the designation and management process.

The Marine Parks (Figure 9.48) are attempting to conserve the coral reef environments off the coast of the island. They are at risk due to damage from overfishing, industrial pollution and mass tourism. The Jamaica Conservation and Development Trust is responsible for the management of the National Park, while the National Environmental Planning Agency has overseen the government's sustainable development strategy since 2001.

Ecotourism is a developing sector of the industry, with the greatest attractions for this type of holiday in the Blue Mountains. For example, raft trips on the Rio Grande are increasing in popularity. Tourists are taken downstream in very small groups. The rafts, which rely solely on manpower, leave singly with a significant time gap between them to minimise any disturbance to the peace of the forest. Camp sites are located in areas supervised by the Jamaican Forestry Department. A permit is required to use one of the camp sites.

Considerable efforts are being made to promote **community tourism** so that more money filters down to the local population and small communities. The Sustainable Communities Foundation through Tourism (SCF) programme has been particularly active in central and southwest

▲ **Figure 9.48** A beach fringed with palm trees in Montego Bay Marine Park

Jamaica. Community tourism is seen as an important aspect of **pro-poor tourism**. This is tourism that results in increased net benefits for people with low incomes.

As a Caribbean SIDS (small island developing state), Jamaica aims to sustainably manage the ocean area under its jurisdiction, which covers 272,000 km². For island states such as Jamaica, coastal and marine tourism is an important aspect of economic growth and job creation.

Jamaica faces a difficult balancing act: it needs both to grow its economy, but also to protect its environment. In terms of tourism, the two objectives are intertwined. Jamaica has a high level of legislation regarding environmental protection, but critics point to a poor level of enforcement of environmental policies.

> **Activities**
> 1 Explain the importance of tourism to the economy of Jamaica.
> 2 Describe the location and importance of Jamaica's National and Marine Parks.
> 3 Describe an example of ecotourism in Jamaica.
> 4 Define:
>   a Community tourism
>   b Pro-poor tourism.
> 5 Discuss the main problems associated with tourism.

Practice questions

1 Table 9.13 shows changes in South Korea's economy between 1970 and 2022.

▼ **Table 9.13** Changes in South Korea's economy, 1970–2022

	Primary industries	Secondary industries	Tertiary industries
1970	50.4 per cent	14.3 per cent	35.3 per cent
2022	5.4 per cent	24.5 per cent	70.1 per cent

   a Define the terms:
      i Secondary industries [1]
      ii Tertiary industries. [1]
   b Describe the changes in South Korea's employment structure between 1970 and 2022. [3]
   c Suggest reasons for the change in employment structure in South Korea between 1970 and 2022. [3]
2 a Define the term 'globalisation'. [1]
   b Explain how TNCs influence globalisation. [3]
3 Assess the impact of globalisation at a local scale. [5]
4 Using named examples, evaluate the impacts of TNCs on host countries. [7]

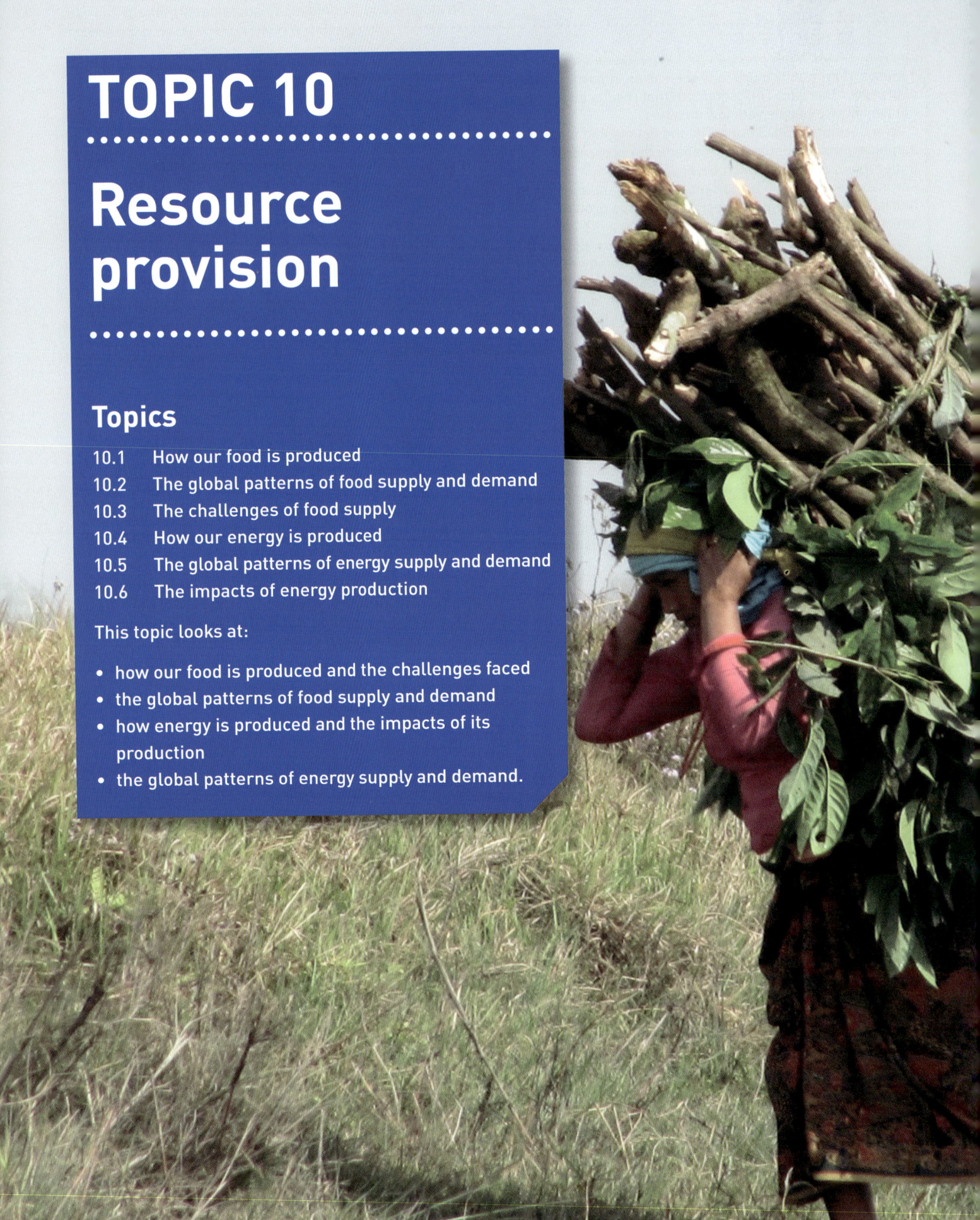

TOPIC 10

Resource provision

Topics

10.1 How our food is produced
10.2 The global patterns of food supply and demand
10.3 The challenges of food supply
10.4 How our energy is produced
10.5 The global patterns of energy supply and demand
10.6 The impacts of energy production

This topic looks at:

- how our food is produced and the challenges faced
- the global patterns of food supply and demand
- how energy is produced and the impacts of its production
- the global patterns of energy supply and demand.

10.1 How our food is produced

This chapter will explain:
★ the different types of farming
★ the systems used in farming.

Farming types

Arable, pastoral and mixed farming

Arable farms (Figure 10.1) cultivate crops and are not involved with livestock. An arable farm may concentrate on one crop (monoculture) such as wheat, or may grow a range of different crops. The crop grown on an arable farm may change over time. For example, if the market price of potatoes increases, more farmers will be attracted to grow this crop.

▲ **Figure 10.1** Arable farming in the Nile valley, with the pyramids in the background

Pastoral farming (Figure 10.2) involves keeping livestock such as dairy cattle, sheep, goats and pigs.

Mixed farming involves cultivating crops and keeping livestock together on a farm. Usually on a mixed farm at least part of the crop production will be used to feed the livestock.

▲ **Figure 10.2** Pastoral farming – goats feeding from a bowl (because the ground is frozen) in Central Asia

Subsistence and commercial farming

Subsistence farming is the most basic form of agriculture, in which the produce is consumed entirely or mainly by the family who work the land or tend the livestock. Figure 10.3 shows a smallholding in northern India with less than half a hectare of land. The family have two cows and grow crops for themselves and to feed their cows. The manure from the cows is dried and used as a fuel for cooking (and for heating in the cold months). If a small surplus is produced by subsistence farming, it may be sold or traded. Examples of subsistence farming are shifting cultivation and nomadic pastoralism. Subsistence farming is generally small scale and labour intensive with little or no technological input.

In contrast, the objective of **commercial farming** is to sell everything that the farm produces. The aim is to maximise yields in order to achieve the highest profits possible. Commercial farming can vary from small scale (Figure 10.4) to very large scale. The very largest farms are often owned by TNCs. Figure 10.5 shows how farming types can change with levels of development.

247

10.1 HOW OUR FOOD IS PRODUCED

▲ **Figure 10.3** A smallholding in northern India

▲ **Figure 10.4** A small rice farm in Sri Lanka

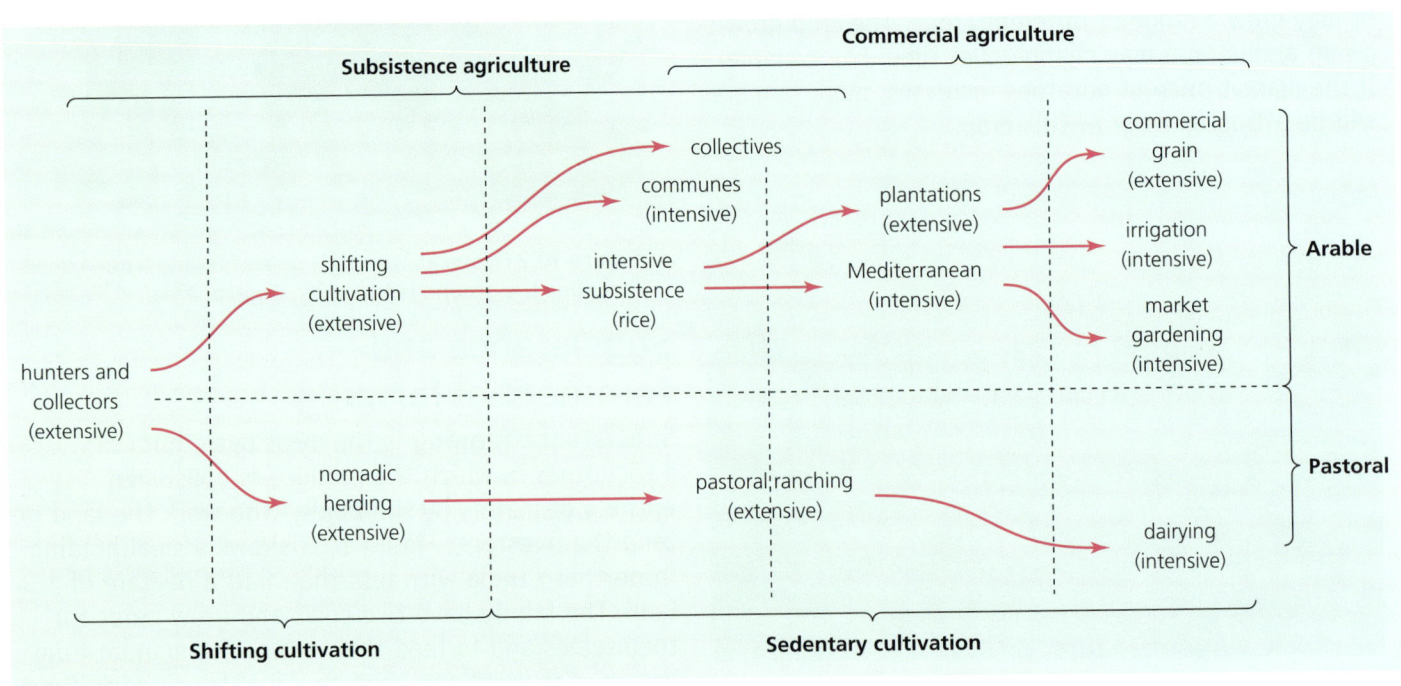

▲ **Figure 10.5** Farming types and levels of development

Extensive and intensive farming

Extensive farming is where a relatively small amount of agricultural produce is obtained per hectare of land, meaning such farms tend to cover large areas of land. Inputs per unit of land are low. Extensive farming can be arable or pastoral in nature. Examples of extensive farming are sheep farming in Australia and wheat cultivation on the Canadian Prairies (Figure 10.6).

In contrast, **intensive farming** is characterised by high inputs per unit of land to achieve high yields per hectare. Examples of intensive farming include market gardening, viticulture (wine production), dairy farming and horticulture. Intensive farms tend to be relatively small in terms of land area. Countries well-known for intensive farming include the Netherlands and Denmark.

▲ **Figure 10.6** Extensive farming in the Canadian Prairies (the province of Manitoba), where most farms are over 1000 hectares in size

Organic farming

Organic farming does not use manufactured chemicals, so production does not involve the use of chemical fertilisers, pesticides, insecticides or herbicides. Instead, animal and green manures are used, along with mineral fertilisers such as fish and bone meal. Organic farming therefore requires a higher input of labour than mainstream farming. Weeding is a major task in this type of farming. Organic farming is less likely to result in soil erosion and is less harmful to the environment in general. For example, there will be no nitrate runoff into streams and so much less harm to wildlife.

Organic farming tends not to produce the 'perfect' potato, tomato or carrot. However, because of the increasing popularity of organic produce, it commands a substantially higher price than mainstream farm produce.

Aeroponics, hydroponics and aquaponics

Advanced farming systems such as **aeroponics** and **hydroponics** allow crops to be grown in places where traditional farming cannot be practised. Neither of these relatively new systems uses soil for crop production.

» In aeroponics, plants are never placed into water and crops grow suspended in the air. Nutrients are supplied from a mist that is sprayed on to their roots. The entire process is enclosed, so the nutrient mist is able to remain around the plants for longer. This helps plants grow more quickly than on a traditional outdoor farm.

» In hydroponics, plants may be suspended in nutrient water all the time or fed by an intermittent flow of water. Because the plants' roots are submerged, they do not receive as much oxygen as in aeroponics systems, leading to a generally smaller plant and crop yield.

Aquaponics is a system that combines fish farming (aquaculture) with hydroponics. Fish waste is naturally converted into nutrients for plants, while the plants filter the water for the fish.

Aeroponics and hydroponics are frequently practised in vertical farming systems (Figure 10.7), which can be established in relatively small areas of land. Locations are often within urban areas. The initial investment for these types of farming can be substantial, but the potential is growing for vertical farming to become a profitable venture practised on a large scale.

▲ **Figure 10.7** Vertical farming

10.1 HOW OUR FOOD IS PRODUCED

Farming systems

Individual farms and general types of farming can be seen to operate as a **system**. A farm requires a range of **inputs**, such as labour and energy, to enable the **processes** that take place on the farm, such as ploughing and harvesting, to be carried out. The aim is to produce the best possible **outputs**, such as milk, eggs, meat and crops. A profit will be made only if the income from selling the outputs is greater than expenditure on the inputs and processes. Figure 10.8 shows an input–process–output (IPO) diagram for a wheat farm. Different types of agricultural system can be found within individual countries and around the world.

Agricultural systems are dynamic human systems that change as farmers attempt to react to a range of physical and human factors.

> ### Activities
> 1. a Explain the difference between arable farming and pastoral farming.
>   b What is mixed farming?
> 2. Discuss the differences between:
>   a commercial farming and subsistence farming
>   b intensive farming and extensive farming.
> 3. Describe the characteristics of organic farming.
> 4. Describe the inputs, processes and outputs for a wheat farm (Figure 10.8).

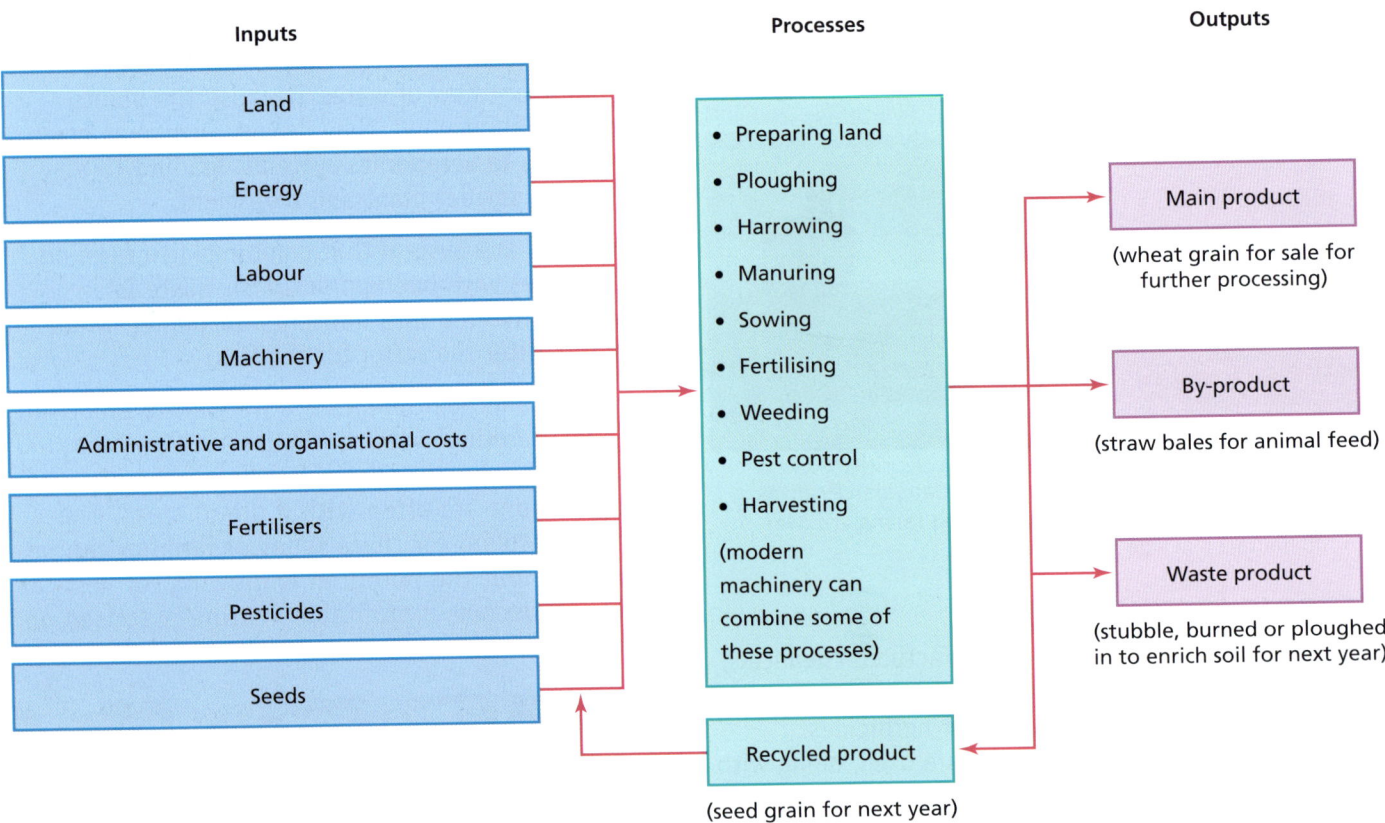

▲ Figure 10.8 The operation of a wheat farm

10.2 The global patterns of food supply and demand

This chapter will explain:

★ global variations in calorie intake and the reasons for this
★ the changing global production and consumption of food
★ the strategies used to increase food supply
★ the reasons for the globalisation of food supplies
★ the impacts of the globalisation of food supplies.

Global variations in calorie intake and the reasons for this

Per capita calorie supply is the main international indicator used to show variations in global access to food supply. It has limitations – a wholesome diet requires more than just energy and supply figures do not include consumption-level waste (food that is wasted at household, restaurant and retail levels). However, it is the best indicator available. Figure 10.9 shows the global variation in the supply of calories for 2021, measured in kilocalories per person per day. The data comes from the Food and Agriculture Organization of the United Nations. It extends back to 1961 and is updated annually.

Table 10.1 shows the changes by world region in per capita daily kilocalorie supply for 1980, 2000 and 2021.

» Since 1960 there has been a consistent increase in global per capita calorie supply, but this has varied across the world's regions.
» There has been a very significant increase in calorie supply in Africa and Asia in recent decades, resulting in the convergence of global trends in calorie supply.

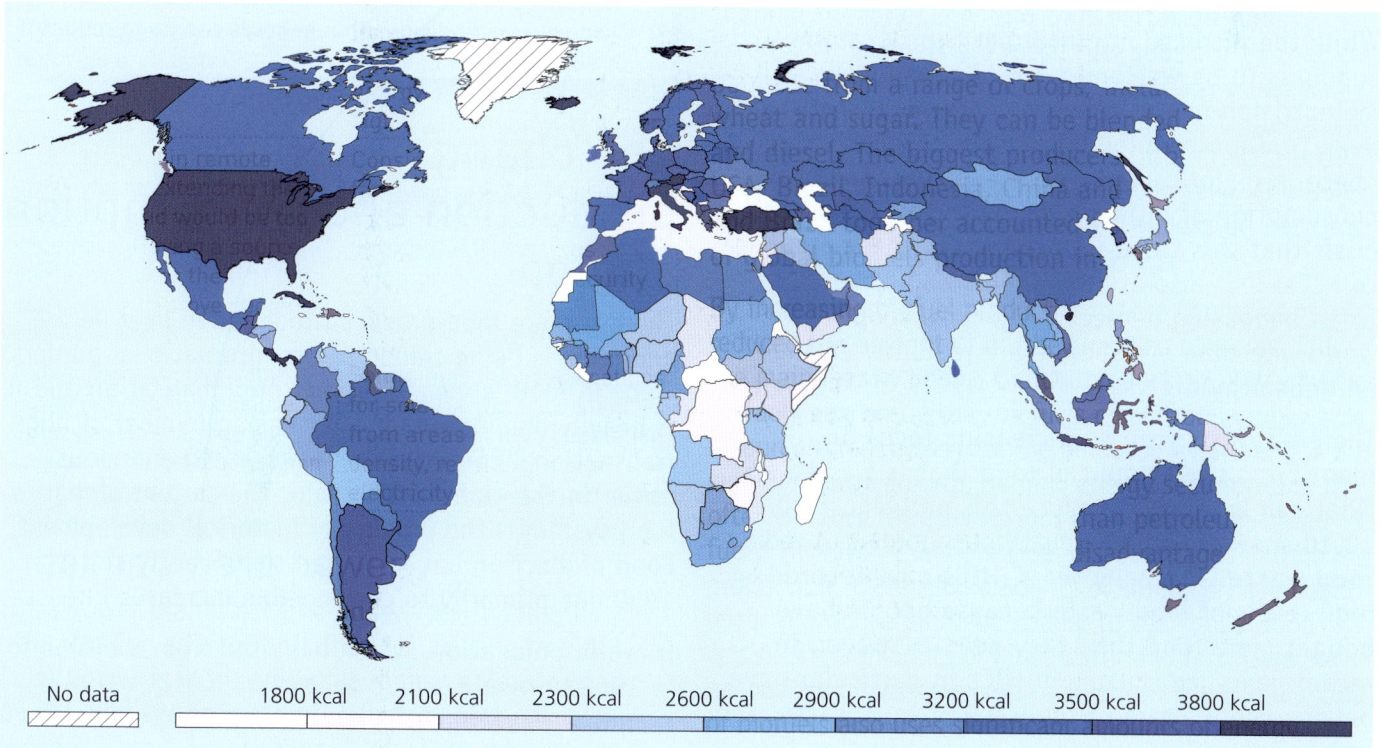

▲ Figure 10.9 The daily supply of calories per person, 2021

10.2 THE GLOBAL PATTERNS OF FOOD SUPPLY AND DEMAND

Variations in per capita global calorie supply have traditionally illustrated unequal access to food around the world. However, the strength of this relationship has weakened over time with the convergence of global trends in calorie supply. The relationship is still strong for lower-income and middle-income countries. Table 10.1 shows clearly that Africa and Asia are still significantly below the world average, but to a much lesser extent than previously. The variations between the world regions that are above the world average have much less to do with calorie supply in terms of the lack of calories, but relate more to overeating.

▼ **Table 10.1** Per capita kilocalorie supply from all foods per day, 1980, 2000 and 2021

Region	1980	2000	2021
World	2459	2669	2959
Europe	3348	3234	3456
North America	3102	3448	3578
Oceania	3013	3014	3087
South America	2658	2780	3108
Africa	2234	2405	2573
Asia	2186	2579	2927

Obesity and food waste

While the increase in global per capita calorie supply is to be welcomed for reducing levels of malnutrition in less affluent countries in particular, over 1 billion people globally are now classed as obese. A major cause of obesity is the increasing consumption of ultra-processed foods. The health crisis that this trend is creating is not confined to HICs; it is also affecting MICs and some LICs to an increasing degree. Many more people worldwide than ever before are regularly eating an unhealthy **diet**.

There has also been an increasing focus on identifying the amount of food that is wasted and what can be done to reduce this problem. Figure 10.10 shows how individual households can reduce **food waste**. For many years, HICs have recorded high levels of food waste because people have bought more food than they need. However, in recent years, consumers in HICs in particular have become more aware of the issue of food waste, and as a result consumer behaviour is beginning to change. For example, there have been campaigns to get supermarkets to stop offers of 'buy one, get one free' and instead to reduce the price of individual items.

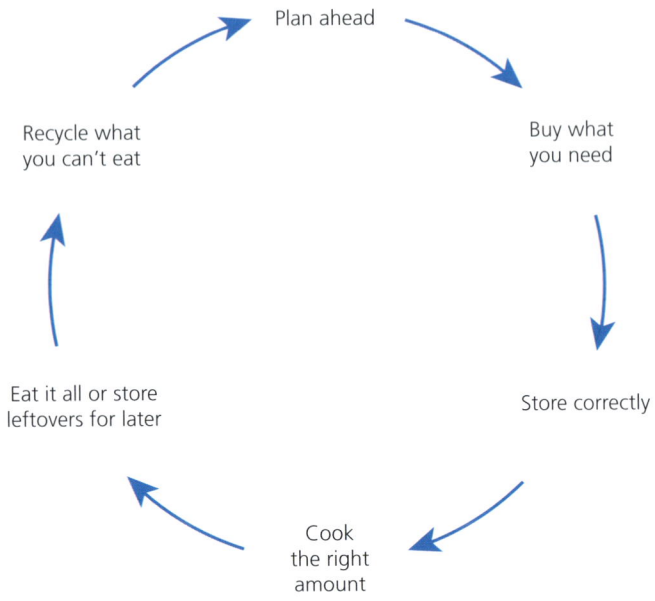

▲ **Figure 10.10** Methods of reducing food waste

> ### Activities
> 1 Study Table 10.1. Describe the main changes illustrated by the table.
> 2 Discuss what individual households can do to reduce food waste.

The changing global production and consumption of food

There is more food produced today than ever before, and more is being traded across international borders. Trade in food is a major aspect of globalisation and it is heavily dominated by a decreasing number of huge transnational corporations, which wield enormous power in the global food chain. The current global food system is the product of historical development. Food production has increased significantly since 1950 due primarily to considerable increases in:

- world population, which has risen from 3 billion in 1960 to over 8 billion today
- global wealth, particularly in emerging economies with a rapidly increasing 'middle class'

» urbanisation of the global population – with urbanisation and associated rising incomes, households often eat greater and more diverse quantities of food.

These factors have driven a substantial increase in the demand for food, along with a change in dietary preferences towards more resource-intensive foods, particularly meat. The growth of Asia's economies, especially China and India, has created new consumer markets as a growing number of people who can be described as middle class demand better quality and more varied food. The demand for certain specialist types of food, for example organic and halal foods, have become clearly recognisable trends.

A small number of very large companies dominate the global food chain. For example, two companies control 40 per cent of the global commercial seed market, a much higher concentration compared to the year 2000. Other major aspects of the global food industry have become similarly concentrated. This is a trend aided by the increasing use of 'big data' and artificial intelligence. Many people view this high level of control by a relatively small number of companies as worrying and anti-democratic.

Agro-industrialisation, or industrial agriculture, is the form of modern farming that refers to the industrialised production of livestock, poultry, fish and crops. This type of large-scale, capital-intensive farming originally developed in HICs, and has spread to MICs and some LICs since the beginning of the Green Revolution in the late 1960s (page 254). The characteristics of agro-industrialisation include:

» very large farms
» concentration on one or a small number of farm products
» a high level of mechanisation
» low labour input per unit of production
» heavy use of fertilisers, pesticides and herbicides
» sophisticated ICT management systems
» highly qualified managers
» often owned by large agribusiness companies
» often vertically integrated with food processing and retailing.

Not all farms and regions where agro-industrialisation is important will display all of these characteristics. For example, intensive market gardens may be relatively small, although the capital inputs are extremely high. Industrial agriculture is heavily dependent on oil for every stage of its operation. The most obvious examples of the use of oil are fuelling farm machinery, transporting produce and producing fertilisers and other farm inputs.

Agro-industrialisation is a consequence of the globalisation of agriculture, the profit ambitions of large agribusiness companies and the drive for cheaper food production. Vertical integration has become an increasingly important process, with growing linkages between the different stages of the food industry (Figure 10.11).

Regions where agro-industrialisation is clearly evident on a large scale include:

» the corn and wheat belts in the USA
» the Paris basin in France
» the Pampas in Argentina
» the Mato Grosso in Brazil.

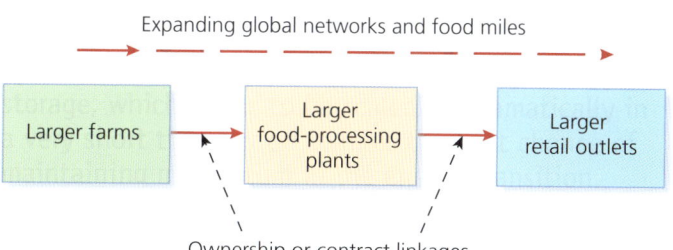

▲ Figure 10.11 Agro-industrialisation: increasing vertical integration

Food choices

High levels of meat consumption have been the norm in HICs for some time. However, as average incomes have increased in emerging economies in particular, the global demand for meat has risen rapidly. This is placing much greater demands on global food production, resulting in soil degradation and other environmental problems in various parts of the world. Figure 10.12 compares the efficiency of energy conversion in arable and pastoral farming, while Figure 10.13 looks at the environmental impact of the increasing demand for meat. The average efficiency of livestock at converting plant feed to meat is less than 3 per cent. Agricultural practice is not necessarily at fault, but maybe our choice of food is.

However, it should, of course, be remembered that livestock can produce more than just meat. Milk production is an important aspect of farming in many countries, and animals may also provide other valuable products such as wool and hides. In LICs, livestock may also provide an important source of labour.

10.2 THE GLOBAL PATTERNS OF FOOD SUPPLY AND DEMAND

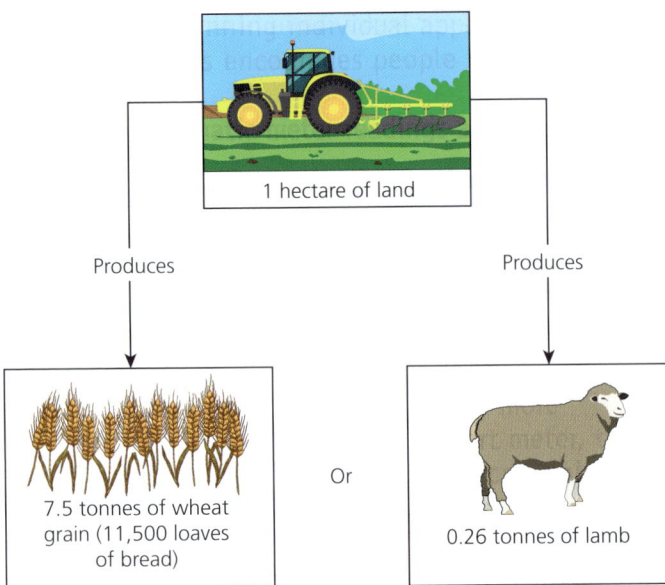

▲ Figure 10.12 A comparison of the efficiency of energy conversion in arable and pastoral farming

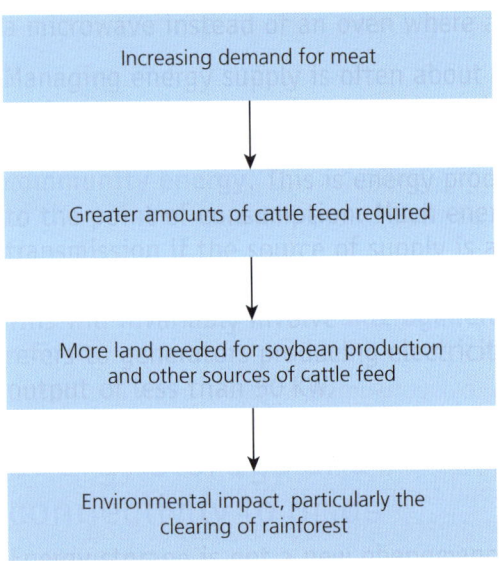

▲ Figure 10.13 The environmental impact of the increasing demand for meat

Activities

1. State the three main reasons for the increase in global food production and consumption.
2. a Define agro-industrialisation.
   b Give three characteristics of agro-industrialisation.
   c State three examples of regions where large-scale agro-industrialisation occurs.
3. Explain Figure 10.12.

The strategies used to increase food supply

The Green Revolution

The package of agricultural improvements generally known as the **Green Revolution** was seen as the answer to the food problem in many parts of the developing world in the post-1960 period. India was one of the first countries to benefit, when its High Yielding Variety Programme (HYVP) started in 1966–67. In terms of production this was a turning point for Indian agriculture, which had virtually reached stagnation. The programme introduced new hybrid varieties of five cereals: wheat, rice, maize, sorghum and millet. All were drought-resistant (with the exception of rice), were very responsive to the application of fertilisers, and had a shorter growing season than the traditional varieties they replaced (Figure 10.14). Although the benefits of the Green Revolution are clear, serious criticisms have also been made. The two sides of the story can be summarised as follows.

▲ Figure 10.14 Green Revolution crops being harvested in Brazil

The advantages of the Green Revolution

» Yields are two to four times greater than for traditional varieties.
» The shorter growing season has allowed the introduction of an extra crop in some areas.
» Farming incomes have increased, allowing the purchase of machinery, better seeds, fertilisers and pesticides.
» The diet of rural communities is now more varied.
» Local infrastructure has been upgraded to accommodate a stronger market approach.

- Employment has been created in industries supplying farms with inputs.
- Higher returns have justified a significant increase in irrigation.

The disadvantages of the Green Revolution

- High inputs of fertiliser and pesticide are required to optimise production. This is costly in both economic and environmental terms. Rural indebtedness has risen sharply in some areas.
- HYVs require more weed control and are often more susceptible to pests and diseases.
- Middle- and higher-income farmers have often benefited much more than the majority on low incomes, thus widening the income gap in rural communities. Increased rural-to-urban migration has often been the result.
- Mechanisation has increased rural unemployment.
- Some HYVs have an inferior taste.
- The problem of salinisation has increased along with the expansion of irrigated areas.

In recent years a much greater concern has arisen about Green Revolution agriculture. The problem is that the high-yielding varieties introduced during the Green Revolution are usually low in minerals and vitamins. Because the new crops have displaced the local fruits, vegetables and legumes that traditionally supplied important vitamins and minerals, the diet of many people in the developing world is now extremely low in zinc, iron, vitamin A and other micronutrients.

The Green Revolution has been a major factor in enabling global food supply to keep pace with population growth, but with growing concerns about a new food crisis, new technological advances may well be required to improve the global food security situation.

The New Green Revolution

Agricultural economists have talked about the concept of a **New Green Revolution** in recent decades, which draws upon the best of the agricultural technologies currently available. It emphasises improved farm management in order to minimise environmental damage from external inputs and aims to benefit low-income farmers and marginal areas bypassed by the original Green Revolution. Examples of this approach include:

- breeding crop varieties that can cope with adverse conditions, such as salt-tolerant rice and more drought-resistant sorghums and millets
- seeds produced to improve the nutritional value of crops as well as to provide more calories
- soil nutrient cycling through crop rotation and biomass recycling
- reliance on genetic resistance to pests and disease to replace chemical and mechanical pest control.

This new process aims to blend farmers' traditional techniques with new knowledge in crop and animal husbandry. For example, integrated pest management (IPM) uses natural predators such as spiders and wasps to fight pests, reducing both costs to farmers and environmental damage.

> **Activities**
> 1 Define the Green Revolution.
> 2 Discuss three advantages and three disadvantages of Green Revolution farming.
> 3 In what ways is the New Green Revolution introducing new farming practices?

The reasons for the globalisation of food supplies

Reducing the barriers to trade in agricultural products

While trade in manufactured goods was gradually liberalised (made easier by reducing the barriers to trade – tariffs, quotas, regulations) after the General Agreement on Tariffs and Trade in 1947, it was not until the World Trade Organization Agreement on Agriculture in 1995 that agriculture as a sector was explicitly included in trade liberalisation at the multilateral level. As Figure 10.15 shows, trade in food and agricultural products increased significantly after 2000.

- Brazil, China and other emerging economies increased their share of global trade in food and agricultural products.
- The share of global exports originating in LICs and MICs increased from about 30 per cent in 1995 to 40 per cent in 2011, and since then has remained fairly constant.
- HICs account for about 60 per cent of food and agricultural exports.

10.2 THE GLOBAL PATTERNS OF FOOD SUPPLY AND DEMAND

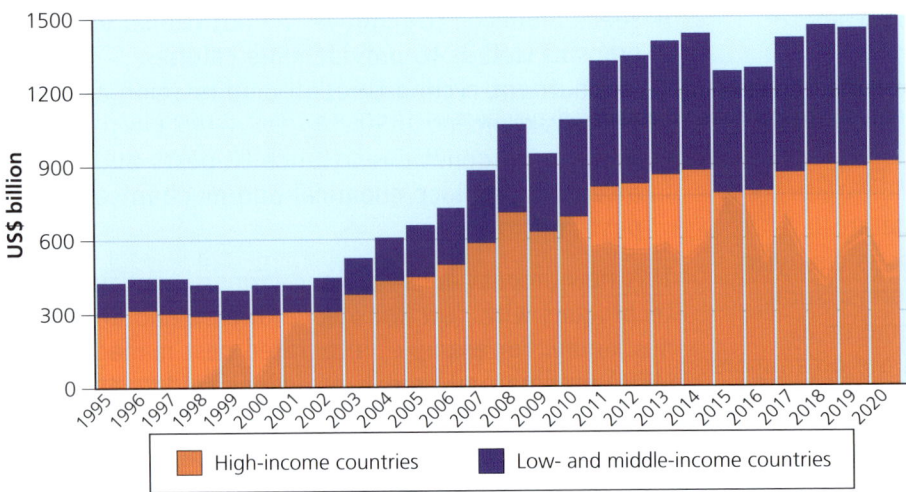

▲ Figure 10.15 Global food and agricultural trade, 1995–2020

The increase in food and agriculture trade since 2000 was also due to increased connectivity between countries, with more countries expanding their participation in global food and agricultural trade. Today countries are connected to more trade partners than ever before in terms of food and agriculture.

The increased efficiency of seaborne transport

Seaborne shipping is by far the most important mode of transport in the international trade in agricultural products. An increasing part of this trade is carried in containers. The majority of world containers are moved through liner shipping services, which are regular transport services provided by global shipping companies. These services provide a dense network that connects ports and countries around the world. The improved logistics of seaborne shipping and of **intermodal transport hubs** has been beneficial to the international trade in food and agricultural products.

The rising demand for international food products

Tourism and global marketing have introduced people in an increasing number of countries to a much wider range of global foods than previous generations had. This is reflected in the eclectic range of international foods stocked in large supermarkets in many countries. Major restaurant and takeaway companies such as KFC and McDonald's have expanded their global networks in terms of both ownership and franchising. Certain countries and cultures are internationally known for their food – the evidence can be seen in restaurants with national identities. Italian, French, Mexican, Chinese, Indian and Thai restaurants are associated with a particular country and are also frequently found in other countries. As a class, you may be able to think of the particular foods that are associated with such restaurants.

> **Interesting notes**
>
> ★ The largest food-exporting countries are the USA, the Netherlands and China.
> ★ The largest net exporters, which export more than they import, are Brazil, Argentina and Spain.
> ★ The largest net importing countries are China, Japan and the UK.

> ▶ **Activity**
>
> Study Figure 10.15. Describe and explain the changes illustrated by the graph.

The impacts of the globalisation of food supplies

Globalisation has had profound effects on the world food system. As major transnational companies have increased their ownership and control of food production, distribution and consumption, more food products have crossed international borders. A wider range of foods has become available to consumers around the world. In many countries this has resulted in the year-round availability of seasonal and non-native foods.

The globalisation of the world food system pushed forward major changes in food production practices, with a big shift towards monoculture and larger-scale farming. While these changes have improved productivity and lowered costs, there has been an increasing negative impact on environments around the world. The impact on local agricultural systems has been overwhelming in many countries and regions. Here, the historic cultivation of local foods for the resident population has frequently given way to production for export markets. Such a massive change can leave communities vulnerable to fluctuations in global food prices and dependent on imported foods.

The power and influence of global food brands (Figure 10.16) has led the way in the expansion of Western dietary patterns. This has impacted both on cultural food heritage and on health in terms of obesity and other forms of ill-health associated with a significant consumption of ultra-processed food.

The carbon emissions associated with larger cross-border movements of food have increasingly contributed to climate change. The term 'food miles' was first used in the 1990s to draw attention to this expanding environmental impact. **Food miles** can be defined as the distance that food travels from the farm where it is produced to the plate of the final consumer. It is an indication of the environmental impact of food consumption. However, it is important to remember that many other processes also contribute to the carbon footprint of food, including:

» farming methods
» processing and packaging
» refrigeration.

The intensification of agricultural practices brought about by the Green Revolution in MICs and LICs, along with a range of other advances that characterise modern agriculture, has resulted in serious negative consequences, particularly:

» ecological degradation
» unsustainable resource consumption
» dependency on non-renewable resources such as fossil fuels.

> ### Activities
> 1 How has globalisation impacted traditional agricultural production in many parts of the world?
> 2 a Define food miles.
>   b Why are food miles an important environmental concept?

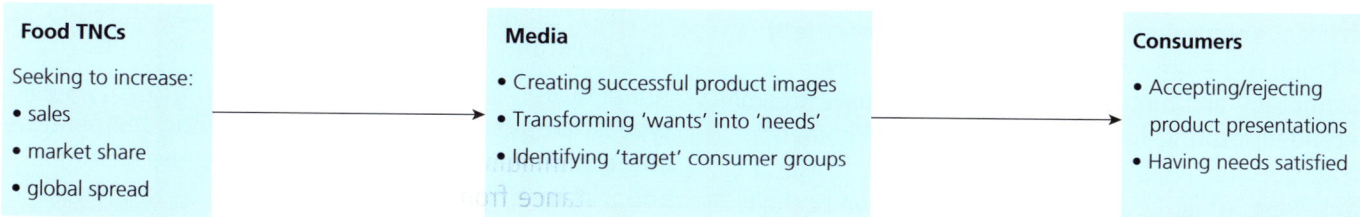

▲ Figure 10.16 Food TNCs, the media and consumers

10.3 The challenges of food supply

This chapter will explain:

★ the physical (natural) and human factors that negatively affect food supply
★ strategies to increase food supply
★ the problems caused by food insecurity
★ the sustainable strategies and techniques used to manage desertification and soil erosion
★ detailed specific example: food insecurity in Nigeria.

The physical (natural) and human factors negatively affecting food supply

A wide range of factors combine to influence agricultural land use and practices on farms. These can be placed under the general headings of physical (or natural) factors and human factors, including economic, social/cultural and political factors.

Physical factors

Physical factors set broad limits on:

» whether agriculture is feasible in an environment
» if agriculture is possible, how productive it might be.

In addition to these factors, the general agricultural productivity of an environment might be drastically altered for a certain time period by extreme weather or tectonic events.

Northern America, for example, has many different physical environments. This allows a wide variety of crops to be grown and livestock kept. New technology and high levels of investment have steadily extended farming into more difficult environments. Irrigation has enabled farming to flourish in the dry southwest, while new varieties of wheat have pushed production northwards in Canada. However, the physical environment remains a big influence on farming. There are certain things that technology and investment can do little to alter. So relief, climate and soils set broad limits as to what can be produced. This leaves the farmer with some choices, even in difficult environments. The farmer's decisions are then influenced by human (economic, social/cultural and political) factors.

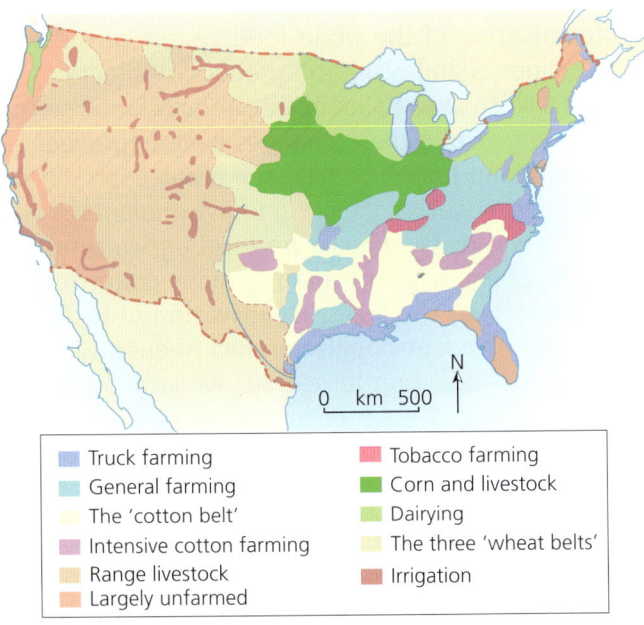

▲ Figure 10.17 The agricultural regions of the USA

Figure 10.17 shows the **agricultural regions** of the USA. Look at relief and climate maps of the USA online or in an atlas to see how agricultural regions vary according to different physical conditions.

Temperature is a critical factor in crop growth as each type of crop requires a minimum growing temperature and a minimum growing season. Latitude, altitude and distance from the sea are the major influences on temperature. Precipitation is equally important. This is not just the annual total, but the way it is distributed throughout the year. Long, steady periods of rainwater to infiltrate into the soil are best, making water available for crop growth. In contrast, short heavy downpours can result in surface runoff, leaving less water available for crop growth and causing soil erosion.

The physical (natural) and human factors negatively affecting food supply

Soil type and soil fertility have a huge impact on **agricultural productivity**. Areas that have never been cleared for farming are often ignored because soil fertility is poor or is perceived to be poor. In some regions wind can have a serious impact on farming, for example causing bush fires in some US states such as California. Locally, the land's aspect (the direction it faces) and the angle of slope may also be important factors in deciding how to use the land.

Cotton, for example, needs a frost-free period of at least 200 days. Rainfall should be over 625 mm a year, with not more than 250 mm in the autumn harvest season. Cotton production is now highly mechanised. Irrigation has allowed cotton to flourish in the drier western states of California, New Mexico and Texas. In contrast, the cotton-growing area has fallen considerably in the southern states. A crop pest called the boll weevil, which caused great destruction to cotton crops in the past, has been a big factor in the diversification of agriculture in the southern states.

In contrast, corn is grown further north than cotton. Corn needs a growing season of at least 130 days. For the crop to ripen properly, summer temperatures of 21°C are needed, with warm nights. Precipitation should be over 500 mm, with at least 200 mm in the three summer months.

In Canada, farming is severely restricted by climate. Less than 8 per cent of the total area of the country is farmed. Seventy per cent of Canada lies north of the **thermal limit for crop growth**. Other high-latitude countries such as Russia, Sweden and Finland also face considerable climatic restrictions on agriculture.

Water is vital for agriculture. **Irrigation** is an important factor in farming in many parts of the world. Table 10.2 compares the main types of irrigation. This is an example of the **ladder of agricultural technology**, with surface irrigation being the most traditional method and subsurface (drip) irrigation the most advanced technique. Figure 10.18 shows an irrigation canal in northern Spain. Irrigation canals of this size can transport large volumes of water, which can then be filtered down to smaller volumes so that a large area of land can be irrigated.

▼ **Table 10.2** The types of irrigation

Type of irrigation	Efficiency (%)
Surface: Used in over 80 per cent of irrigated fields worldwide:	
a Furrow: A traditional method; cheap to install; labour-intensive; high water losses; susceptible to erosion and salinisation	20–60
b Basin: Cheap to install and run; needs a lot of water; susceptible to salinisation and waterlogging	50–75
Aerial (using sprinklers): Used in 10–15 per cent of irrigation worldwide; costly to install and run; low-pressure sprinklers preferable	60–80
Sub-surface ('drip'): Used in 1 per cent of irrigation worldwide; high capital costs; sophisticated monitoring; very efficient	75–95

▲ **Figure 10.18** An irrigation canal in northern Spain

Human factors

The human factors influencing agriculture can be divided into economic, social/cultural and political.

Economic

Economic factors include transport, markets, capital and technology. The cost of growing different crops or keeping different livestock varies. The market prices for agricultural products will also vary and can change from year to year. The necessary investment in buildings and machinery can mean that some changes in farming activities are very expensive. These would be more difficult to achieve than other, cheaper changes. Thus it is not always easy for farmers to react quickly to changes in consumer demand.

10.3 THE CHALLENGES OF FOOD SUPPLY

In most countries there has been a trend towards fewer but larger farms. Large farms allow economies of scale to operate, which reduce the unit costs of production. As more large farms are created, small farms find it increasingly difficult to compete and make a profit. Selling to a larger neighbouring farm may be the only economic solution. The EU is an example of a region where average farm size varies significantly. Those countries with a large average farm size generally have more efficient agricultural sectors than countries with a small average farm size.

Agricultural technology is the application of techniques to control the growth and harvesting of animal and vegetable products. The development and application of agricultural technology requires investment and thus it is an economic factor. The status of a country's agricultural technology is vital for its food security and other aspects of its quality of life. The transfer of agricultural technology from more advanced to less advanced countries is therefore an important form of aid.

Social/cultural factors

What a particular farm and neighbouring farms have produced in the past can be a significant influence on current farming practices. There is a tendency for farmers to stay with what they know best and often a sense of transgenerational responsibility to maintain a family farming tradition. Tradition matters more in some farming regions than others.

Land tenure means the ways in which land is or can be owned. In the past inheritance laws have had a huge impact on the average size of farms. In some countries it has been the custom on the death of a farmer to divide the land equally between all his sons, but rarely between daughters. Also, dowry customs may include the giving of land with a daughter on marriage. The reduction in the size of farms by these processes often reduces them to operating at only a subsistence level.

In most societies women have very unequal access to, and control over, rural land and associated resources. It is now generally accepted that those societies with well-recognised property rights are also the ones that thrive best economically and socially.

Political factors

The influence of government on farming has steadily increased in many countries. For example, in the USA the main parts of government farm policy over the past half-century have been as follows:

- » Price support loans: Loans that help farmers until they sell their produce.
- » Production controls: These limit how much surplus a farmer can produce of crops.
- » Income supplements: These are cash payments to farmers for major crops in years when market prices fail to reach certain levels.

Thus the decisions made by individual farmers are heavily influenced by government policies such as those listed above. An agricultural policy can cover more than one country, as evidenced by the EU's Common Agricultural Policy.

Adverse influences on food supply and distribution

Table 10.3 summarises some of the current adverse influences on food supply and distribution.

▼ Table 10.3 Adverse influences on global food production and distribution

Nature of adverse influence	Effect of adverse influence
Economic	• Demand for cereal grains has outstripped supply in recent years
	• Rising energy prices and agricultural production and transport costs have pushed up costs all along the farm-to-market chain
	• Serious underinvestment in agricultural production and technology in LICs has resulted in poor productivity and underdeveloped rural infrastructure
	• The production of food for local markets (Figure 10.19) has declined in many LICs as more food has been produced for export
Ecological	• Significant periods of poor weather and a number of severe weather events have had a major impact on harvests in key food-exporting countries
	• There have been increasing problems of soil degradation in both HICs and LICs
	• Declining biodiversity may impact food production in the future

▼ Table 10.3 (Continued)

Nature of adverse influence	Effect of adverse influence
Sociopolitical	• The global agricultural production and trading system, built on import tariffs and subsidies, creates great distortions, favouring production in HICs and disadvantaging producers in LICs • There is an inadequate international system of monitoring and deploying food relief • There are disagreements over the use of transboundary resources such as river systems and aquifers • Major international conflicts can significantly decrease the supply of important food products to the global market

▲ Figure 10.19 Selling local produce in a Vietnamese food market

Climate loss in food production

Climate loss is pre-harvest food loss. These are the losses farmers experience from:

» not planting a crop, or at least not the desired crop
» changing crops due to weather
» extreme weather reducing or wiping out a crop before harvest
» increased predation, pests and diseases that are due to climate change.

The UN Food and Agriculture Organization report 'The Impact of Disasters on Agriculture and Food Security 2023' stated that over the previous 30 years, an average of US$123 billion per year was lost due to disasters in agricultural crop and livestock production:

» Losses were highest in LICs and lower MICs, totalling as much as 15 per cent of their agricultural GDP.
» Small island states were more affected than average.

Activities

1 List the main physical factors that can influence farming.
2 Summarise the information presented in Table 10.3.
3 Why has the size of farms steadily increased in many agricultural regions?
4 Briefly state the importance of advances in agricultural technology.
5 Give an example of how a social/cultural factor can have an impact on farming.
6 How can political factors influence farming?

The problems caused by food insecurity

The definition of **food security** is when all people at all times have access to sufficient, safe, nutritious food to maintain a healthy and active life. Food security has four components:

» Availability: This is the existence of sufficient food within a community. It can become an issue when resources decline, for example water for irrigation and the degradation of soil.
» Access: Access to food can be affected by factors such as rising prices, household proximity to food retailers and poor infrastructure.
» Utilisation: Food needs to be of sufficient quality to provide the required levels of nutrition and households need the necessary knowledge and tools to properly select, prepare and store foods.
» Stability: Access, availability and utilisation of food needs to remain relatively stable over time.

The total amount of food produced around the world today is enough to provide everyone with a healthy diet. The problem is that while some countries produce a food surplus or have enough money to

10.3 THE CHALLENGES OF FOOD SUPPLY

buy it elsewhere, other countries are in food deficit and lack the financial resources to buy enough food abroad. About one in nine of the world's population remains chronically undernourished.

The current food crisis presents three fundamental threats. These threats are:

» pushing more people into poverty
» eroding development gains that have been achieved in many countries in recent decades
» presenting a strategic threat by endangering political stability in some countries.

The impact of food insecurity has been felt most intensely in developing countries, where adequate food stocks to cover emergencies affecting food supply usually do not exist. However, developed countries have not been without their problems and are not immune to the physical problems that can cause food shortages. For example, in recent years both the USA and Australia have suffered severe drought conditions. However, developed countries invariably have the human resources to cope with such problems, so actual food shortages do not generally occur. However, in some HICs, an increasing number of people have had to rely on **food banks** to maintain a reasonable level of food intake.

Short-term and long-term effects

The effects of food shortages are both short-term and longer-term. **Malnutrition** can affect a considerable number of people, particularly children, within a relatively short period when food supplies are significantly reduced. With malnutrition, people are less resistant to disease and more likely to fall ill. Diseases connected to malnutrition include beriberi (vitamin B1 deficiency), rickets (vitamin D deficiency) and kwashiorkor (protein deficiency). People who are continually starved of nutrients never fulfil their physical or intellectual potential. Malnutrition reduces people's capacity to work, so land may not be properly tended or other forms of income successfully pursued. This is threatening to lock parts of the developing world into an endless cycle of ill-health, low productivity and underdevelopment.

The 2023 Global Hunger Index (Figure 10.20) noted:

» little progress had been made in reducing hunger at a global scale since 2015
» hunger remained serious or alarming in 43 countries
» South Asia and Sub-Saharan Africa are the world regions with the highest hunger levels
» the fight against hunger has been impeded by overlapping crises – Covid-19, multiple violent conflicts and climate disasters around the world.

Strategies and techniques used to increase food supply

The United Nations Environment Programme has argued that increasing food energy efficiency provides a critical path for significant growth in food supply without compromising environmental sustainability.

Options with short-term effects:

1 Price regulation on commodities and larger cereal stocks to decrease the risk of highly volatile prices.
2 To reduce/remove subsidies on biofuels to discourage farmers from using cropland for energy production rather than food production.

Options with mid-term effects:

1 To reduce the use of cereals and food fish in animal feed.
2 To support farmers in developing diversified eco-agricultural systems that provide critical ecosystem services (for example water supply and regulation) as well as adequate food to meet local and consumer needs.
3 To increase trade and improve market access by improving infrastructure and reducing trade barriers.

Options with long-term effects:

1 To limit global warming, including promoting climate-friendly agricultural production systems and land-use policies at a scale to help mitigate climate change.
2 To raise awareness of the pressures of increasing population growth and consumption patterns on sustainable ecosystem functioning.

See page 254 for reference to the Green Revolution and the New Green Revolution.

The problems caused by food insecurity

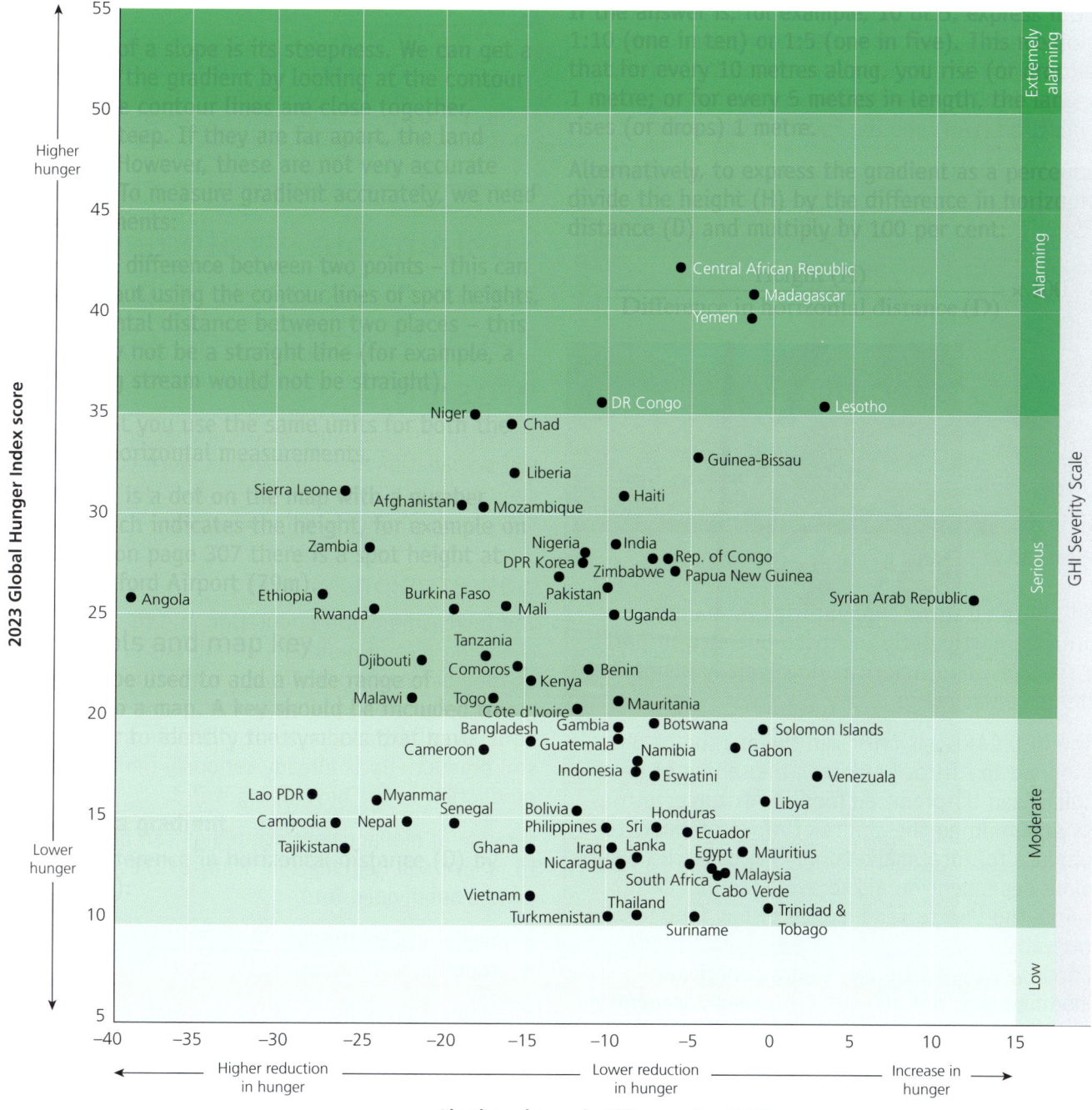

▲ **Figure 10.20** The 2023 Global Hunger Index scores and progress since 2000

The role of food aid in improving food security

In the short term, and in some instances the medium term, food aid is absolutely vital to cope with food shortages (Figure 10.21). When disaster strikes, there is no alternative to this strategy. According to the NGO ActionAid there are three types of food aid:

» Relief food aid, which is delivered directly to people in times of crisis.
» Programme food aid, which is provided directly to the government of a country for sale in local markets (this usually comes with conditions from the donor country).
» Project food aid, which is targeted at specific groups of people as part of longer-term development work.

10.3 THE CHALLENGES OF FOOD SUPPLY

The USA and the EU together provide about two-thirds of global food aid deliveries. At the international level the main organisations are the UN World Food Programme (WFP), the UN Food and Agriculture Organization (FAO) and the Food Aid Convention.

▲ **Figure 10.21** Food aid being delivered in Haiti

Food aid is vital for communities in many countries, particularly in Africa but also in parts of Asia and Latin America. However, it is not without controversy:

- The NGO CARE has criticised the method of US food aid to Africa. CARE sees the selling of heavily subsidised US-produced food in African countries as undermining the ability of African farmers to produce for local markets, making countries even more dependent on aid to avoid famine. CARE wants the USA to send money to buy food locally instead.
- Friends of the Earth says that a genetically modified rice, not allowed for human consumption and originating in the USA, has been found in food aid supplied to West Africa.
- Food aid is very expensive, not least because of the high transport costs involved.

There have been recent concerns that food aid may be required for even more people in the future. In recent years, the term 'global food crisis' has been used more and more by the media. Steep increases in the price of food have caused big problems in a number of countries. Major protests about the price of food have taken place in countries including Haiti, Indonesia, the Philippines and Egypt. The World Bank has warned that progress on development could be destroyed by rapidly rising food costs.

> ### ▶ Activities
> 1 List the four components of food security.
> 2 What is malnutrition and how can it impact people and countries?
> 3 Describe the different types of food aid.
> 4 Produce a brief analysis of Figure 10.20 (page 263).

Sustainable strategies and techniques used to manage desertification and soil erosion

The environmental and socioeconomic consequences of soil degradation are considerable. Such consequences can occur with little warning as damage to soil is often not perceived until it is far advanced.

Soil erosion

The increasing world population and the rapidly changing diet of hundreds of millions of people as they become more affluent is placing more and more pressure on land resources. Some soil and agricultural experts say that a decline in long-term soil productivity is already seriously limiting food production in the developing world.

A range of management strategies can be employed to reduce soil degradation. Many of these strategies can be subdivided into:

- mechanical methods
- cropping techniques.

Mechanical methods focus on preventing or slowing down the movement of rainwater downslope and reducing the impact of wind on the soil. Afforestation (planting trees on areas that have not previously been forested) or reforestation have been widely used over sizeable areas or as part of a wider package of measures. The selection of the areas to be forested is of crucial importance. Steeper slopes are the natural starting point but other factors must be taken into account, such as the predominant wind direction.

Contour ploughing is a tried and trusted technique that prevents or diminishes the downslope movement of water and soil. Such ploughing ensures that ridges and furrows are at right angles to the slope. Where slopes are too steep for contour ploughing, **terracing**

(Figure 10.22) may be practised. Here the steep slope is converted into a series of flat steps with raised outer edges (bunds). The monsoon regions of Southeast Asia exhibit widespread terracing.

▲ **Figure 10.22** Terracing in a valley in the Andes Mountains, Peru

The planting of trees in **shelterbelts** and the use of hedgerows can do much to dissipate the impact of strong winds, reducing the wind's ability to disturb topsoil and erode particles.

Various cropping techniques can be employed to reduce soil degradation:

- Converting land from arable to pastoral uses. The planting of grass helps to bind soil particles together, reducing the action of wind and rain.
- Including grasses in crop rotations.
- Leaving unploughed grass strips between ploughed fields.
- Keeping a crop cover on the soil for as long as possible, thus minimising the 'bare soil' period.
- Increasing the organic content of the soil by applying animal manure, compost or sewage sludge. This enables soil to hold more water, preventing aerial erosion and stabilising soil structure.

- Selecting and using farm machinery carefully, in particular avoiding where possible the use of heavy machinery on wet soils to prevent damage to the soil structure and using low ground pressure set-ups on machinery when possible.
- Leaving the stubble and root structure in place after harvesting.
- Using reduced or shallow cultivation to maintain or increase near-surface organic matter.
- Shepherding livestock and moving forage areas to avoid overgrazing.
- Using the most modern and efficient means of irrigation (Figure 10.23).

▲ **Figure 10.23** A sprinkler system irrigating crops in northern Spain

Many degraded environments require substantial investment to bring in realistic solutions. Such finance is beyond the means of many LICs. However, there may be a choice between low-cost and high-cost schemes, as Table 10.4 illustrates.

▼ **Table 10.4** Technical and institutional costs in resource management by and for those in poverty

	High institutional costs	Low institutional costs
Relatively high technical costs	• Large-scale irrigation • Arid or semi-arid land reforestation or pasture improvement • Sodic or saline land reclamation • Mangrove reforestation • Integrated river-basin management • Shaping or changing many transboundary resources, e.g. international rivers, air quality	• Small-scale hill irrigation • Food crop systems on difficult soils • Localised water-harvesting structures • Centralised provision of energy services • Solar energy for individual households • Pipe sewer systems

10.3 THE CHALLENGES OF FOOD SUPPLY

▼ Table 10.4 (Continued)

	High institutional costs	Low institutional costs
	• Resettlement schemes • Water-pollution reduction programmes • Rural road maintenance • Ocean fisheries management	• Emissions-reduction devices • Improved public transport
Relatively low technical costs	• Aquifer management • Protection of critical areas • Coastal fisheries management • Coral-reef management • Pasture management • Land-reform programmes • Integrated pest management • Wild game management	• Treadle pump irrigation • Humid tropics reforestation • Small water-harvesting systems • Joint forest management regimes • Improved cooking stoves and cooking energy for low-income families • Sloping agricultural land technology (SALT) • Small-scale quarrying • Household-based sanitation systems

Managing salt-affected soils

Periodic soil testing and treatment, combined with proper management procedures, can improve the conditions in salt-affected soils (Figure 10.24) that contribute to poor plant growth. There are three ways to manage saline soils:

» Salts can be moved below the **root zone** by applying more water than the plant needs. This is known as the leaching requirement method.
» Where soil moisture conditions dictate, the leaching requirement method can be combined with artificial drainage.
» Salts can be shifted away from the root zone to locations in the soil, other than below the root zone, where they are not harmful. This is called managed accumulation. Sometimes, salt-tolerant crops may need to be selected in addition to managing soils.

▲ **Figure 10.24** Infertile saline soil in the south of France

The rise of no-till farming

The traditional practice of turning the soil before planting a new crop is a leading cause of soil degradation. An alternative is no-till farming, which minimises soil disruption. Here, farmers leave crop residue on the fields after harvest, where it acts as a mulch to protect the soil and provide nutrients. To sow seeds, the farmers use seeders that penetrate through the residue to the undisturbed soil below. This important sustainable approach to farming is spreading, but so far it has been mainly confined to major farming nations because of the high equipment costs involved in changing from traditional practices.

The advantages of no-till farming are that it reduces soil erosion, conserves water, improves soil health, cuts fuel and labour costs, reduces sediment and fertiliser pollution in lakes and streams, and sequesters carbon. The drawbacks, apart from costly new machinery, are greater reliance on herbicides, and that more nitrogen fertiliser may initially be required. There may be greater prevalence of weeds and other pests, and yield may be reduced.

Desertification

Desertification is the gradual transformation of habitable land into desert. It is arguably the most serious environmental consequence of soil degradation. Desertification is usually caused by climate change and/or by destructive use of the

Sustainable strategies and techniques used to manage desertification and soil erosion

land. The natural causes of desertification include temporary drought periods of high magnitude and long-term climate change towards aridity. The main human causes are:

- overgrazing
- overcultivation
- deforestation.

Desertification occurs when already fragile land in arid and semi-arid areas is over-exploited. It is a considerable problem in many parts of the world, for example on the margins of the Sahara Desert in Northern Africa and the Kalahari Desert in Southern Africa. In semi-arid areas such as the edge of the Kalahari Desert, a combination of low and variable precipitation, nutrient-deficient soils and heavy dependence on subsistence farming make soil degradation a significant threat. At present, 25 per cent of the global land territory and nearly 16 per cent of the world's population are threatened by desertification. There are many ways of combating this, depending on the perceived causes (Table 10.5).

▼ Table 10.5 The strategies for preventing desertification and their disadvantages

Cause of desertification	Strategies for prevention	Problems and drawbacks
Overgrazing	• Improved stock quality: Vaccination programmes and the introduction of better breeds enable yields of meat, wool and milk to be increased without increasing the herd size • Better management: Reducing herd sizes and grazing over wider areas both reduce soil damage	• Vaccination programmes improve survival rates, leading to bigger herds • Population pressure often prevents these measures
Overcultivation	• The use of fertilisers: These can double yields of grain crops, reducing the need to open up new land for farming • New or improved crops: Many new crops or new varieties of traditional crops with high-yielding and drought-resistant qualities could be introduced • Improved farming methods: The use of crop rotation, irrigation and grain storage can all increase, which would reduce pressure on land	• The cost to farmers can be high • Artificial fertilisers may damage the soil • Some crops need expensive fertiliser • There is a risk of crop failure • Some methods require expensive technology and special skills
Deforestation	• Agroforestry: This combines agriculture with forestry, allowing the farmer to continue cropping while using trees for fodder, fuel and building timber. Trees protect, shade and fertilise the soil • Social forestry: Village-based tree-planting schemes involve all members of a community • Alternative fuels: Oil, gas and kerosene can be substituted for wood as sources of fuel	• There is a long growth time before the benefits of trees are realised • Expensive irrigation and maintenance may be needed • Expensive, special equipment may be needed

▶ Activities

1. What is soil erosion and to what extent is it a major global problem?
2. Describe three techniques commonly used to combat soil erosion.
3. Define desertification.
4. With reference to Table 10.5, explain one strategy for preventing desertification for each of the three causes of desertification.

10.3 THE CHALLENGES OF FOOD SUPPLY

 Detailed specific example

Food insecurity in Nigeria

With a population over 220 million, Nigeria is the most populous country in Africa. The country achieved lower-middle-income status in 2014. Agriculture is an important economic sector in Nigeria, contributing about 23 per cent to the country's GDP, and accounting for just over 50 per cent of all employment. However, about 37 per cent of the population live below the poverty line. In 2023, the Nigerian government declared a state of emergency on food security. In early 2024, the World Food Programme projected that 26.5 million people across Nigeria would face acute hunger in the June–August lean season.

Factors affecting food supply and food insecurity

Nigeria's food supply problems are due to a combination of physical and human factors. The country is affected by periodic floods and droughts that have a major impact on agricultural output. Such situations increase the vulnerability of populations, particularly in rural areas. Widespread flooding in the 2022 rainy season damaged more than 675,000 hectares of farmland, impacting heavily on national food supply. More extreme weather patterns influencing food security are expected in the future.

Land degradation and desertification have reduced production in a number of regions. Fifteen out of the 36 states of Nigeria are affected by desertification. Other factors impacting Nigeria's food crisis include the following:

- A relatively low level of irrigation, which limits agricultural production.
- Poor access to markets.
- The longstanding conflict in Nigeria's northeast region has displaced over 2 million people and created a particularly high level of food insecurity in this region. Agricultural production is frequently interrupted by attacks from insurgents, with important infrastructure often destroyed. More recent conflict has also affected the northwest and north-central regions.
- Clashes between herders and farmers over the use of land deter investment in the areas most affected.
- A high level of inflation is driving rising food prices. Food consumption makes up about 50 per cent of average expenditure for Nigerian households.

Figure 10.25 illustrates a recent estimate of the main causes of the food crisis in Nigeria.

The Nigerian government sees the country's current food systems as being unsatisfactory and in urgent need of improvement. Agricultural yields in the country are some of the lowest in the world.

- Post-harvest losses are high, with more than 10 per cent of production lost post-harvest for several crops.
- Food storage facilities are often lacking in capacity and quality, resulting in a high level of food waste.
- Food processing centres often lack automation.
- Poor road infrastructure and connectivity causes delays in the transportation and distribution of food, resulting in poor food safety.
- Agricultural practices have contributed to a high level of soil degradation and erosion.

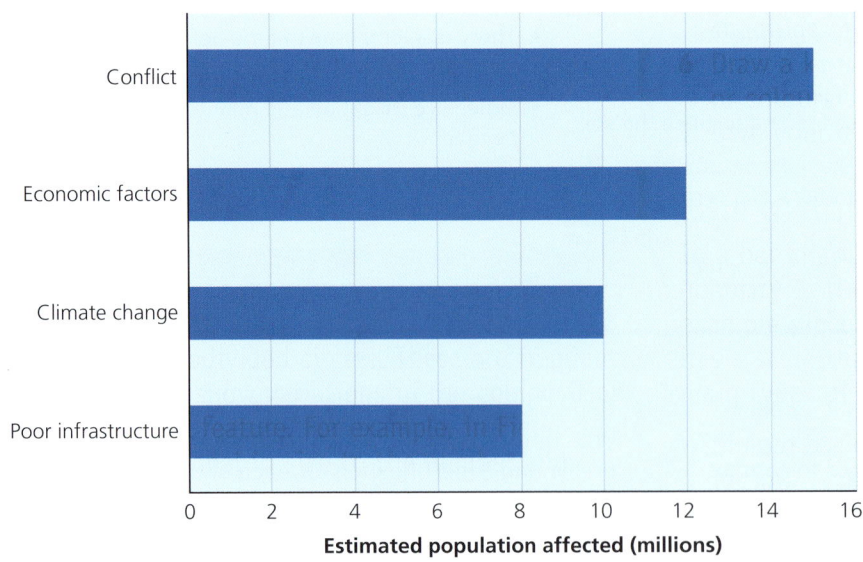

▲ Figure 10.25 The primary causes of the food crisis in Nigeria

Problems caused by food insecurity

People who experience food insecurity are more likely to suffer malnutrition, which compromises the immune system and can result in chronic conditions such as diabetes and heart disease. Malnutrition stunts growth and impairs cognitive development. The long-term consequences can extend into adult life, continuing a cycle of poverty and ill-health. Malnutrition is at a high level in Nigeria, especially in the more remote conflict-affected areas in the northeast. In a number of regions, nutrition stabilisation centres have been essential to keeping children alive. Having a significant number of people in this situation places intense pressure on medical services and also affects labour productivity in agriculture and other economic sectors. Children are the most vulnerable to food insecurity.

Strategies to increase food supply

Food supply challenges exist in all parts of the country, but they vary in intensity across states and within states. Therefore there is a need for location-specific interventions. Recent strategies have built upon a variety of existing programmes and include:

- modernising agricultural infrastructure, including storage facilities and processing plants
- increasing investment in research and development
- expanding mechanisation and modern farming techniques
- improving access to credit and finance for farmers, including microfinance schemes
- strengthening agricultural education and training
- promoting sustainable agricultural practices
- diversifying agricultural crops to produce a wider range of food products
- investment in agro-processing, as the value added will move up the value chain. This will increase the value of Nigeria's agricultural exports.

Sustainable strategies need to result in higher incomes and reduced poverty among smallholder farmers. Scaling up more environmentally friendly practices, such as increasing the number of cropping cycles in a year and raising yields, will be essential to increasing food production in Nigeria. A number of national and international organisations have stressed the need for land tenure reforms in Nigeria. Farmers frequently face problems in the acquisition of suitable land for farming activities. In addition, increasing urbanisation and other built development has resulted in the sale of agricultural land for construction.

Nigeria has established a number of important agricultural policies over the years, but many have not delivered the desired objectives due to a combination of inadequate funding, poor coordination and arguably a lack of political will.

The World Food Programme

The World Food Programme (WFP) is playing an important role in combating malnutrition. It places a strong emphasis on providing specialised nutritious food to children under five at risk of malnutrition. In collaboration with the Nigerian government and other agencies, it has set up livelihoods programmes and income-generating activities such as helping people acquire skills and employment in areas such as food processing, aquaculture and vegetable gardening. WFP also provides:

- technical assistance to the national home-grown school feeding programme
- air transport to the humanitarian community working in hard-to-reach isolated areas
- internet connectivity to key agricultural advisors via its Emergency Telecommunications Service.

> ### Activity
> Discuss the factors affecting food supply and food insecurity in Nigeria.

10.4 How our energy is produced

This chapter will explain:

★ non-renewable and renewable energy sources
★ transforming the global energy mix to achieve a better balance between renewable and non-renewable sources of energy.

Non-renewable and renewable energy sources

Non-renewable energy sources are the fossil fuels (coal, oil, natural gas) and nuclear fuel. Eventually, these non-renewable resources could become completely exhausted. The burning of fossil fuels creates pollution and is the major source of greenhouse gas emissions. Climate change due to these emissions is the biggest environmental problem facing the planet.

Renewable energy resources are mainly forces of nature; these are sustainable and usually cause little or no pollution. Renewable energy includes hydroelectricity, biomass, wind, solar, geothermal, tidal energy and wave power. Hydroelectricity is the one renewable source of energy that is sometimes described as a traditional source of energy because water power has been used to generate electricity for over 100 years.

It is estimated that globally around 750 million people do not have access to electricity to light their homes and provide other services that most other

▲ Figure 10.26 Fuel station on the Amazon River, Brazil

people take for granted. In developing countries about 2.5 billion people rely on **fuelwood** (Figure 10.27) as their main source of energy. Wood and charcoal are collectively called fuelwood, and this accounts for just over half of global wood production. Fuelwood can be classed as either a renewable or a non-renewable source of energy, depending on whether or not the system of consumption allows the vegetation to renew itself to its previous density.

▲ **Figure 10.27** Woman carrying fuelwood from a local forest, Nepal

Transforming the global energy mix to achieve a better balance between renewable and non-renewable sources of energy

Fuelwood provides much of the energy needs for many countries in Africa. It is also the most important use of wood in Asia. Wood is likely to remain the main source of fuel for those in poverty around the world for the foreseeable future. The transition from fuelwood and animal dung to 'higher level' sources of energy, known as the **energy ladder**, occurs as part of the process of economic development. It requires a high level of investment, for example constructing an electricity transmission corridor (pylons and cables) to connect remote areas of a country that did not previously have electricity.

In LICs **energy poverty** has a big impact on people's lives and is a major obstacle to development. Energy poverty differs from **fuel poverty**. Fuel poverty is when people, usually in affluent countries, struggle to pay rising energy bills in harsh winters.

At present, non-renewable resources still dominate global energy supply. The challenge is to transform the global **energy mix** to achieve a better balance between renewable and non-renewable sources of energy. This is often referred to as the '**clean air transition**'.

Countries are eager to harness renewable energy resources in order to:

» reduce their reliance on domestic fossil fuel resources
» lower their reliance on costly fossil fuel imports
» improve their energy security
» cut greenhouse gas emissions.

The main drawback in the past to the new alternative energy sources was that they usually produced higher-cost electricity than traditional sources. However, the cost gap with wind and solar power closed rapidly in the first two decades of this century, and now these sources of energy are generally less costly than traditional energy, and without the climate-changing environmental costs.

> ### ▶ Activities
> 1 Explain the difference between renewable and non-renewable energy sources.
> 2 What do you understand by the terms 'energy mix' and 'clean air transition'?
> 3 Briefly discuss two reasons why countries are keen to develop renewable sources of energy.

10.5 The global patterns of energy supply and demand

This chapter will explain:

★ the increasing global production and consumption of energy
★ trends in the consumption of energy
★ energy surplus and deficit and the importance of energy security
★ the reasons for variations in the types of energy produced and used.

The increasing global production and consumption of energy

The 2024 Statistical Review of World Energy noted that 2023 was a record year for:

» the consumption of fossil fuels – global consumption of crude oil broke through the 100 million barrels per day level for the first time ever
» global GHG emissions from energy – global energy-related GHG emissions exceeded 40 gigatonnes for the first time ever
» the generation of electricity from renewable sources, driven by increasingly competitive wind and solar energy – the renewables share of total primary energy reached 14.6 per cent, while fossil fuel consumption as a proportion of primary energy continued to fall to 81.5 per cent.

The demand for and production of energy has changed over time due to the following factors.

Population growth

World population was 3 billion in 1960. It doubled to 6 billion in 1999, a phenomenal increase in just under 40 years. It then reached 8 billion in late 2022. Such a large population increase has created huge extra demand for energy in itself, without considering any other factors. Add to this:

» the factors such as increasing wealth that are pushing up the demand for energy
» the considerable variations in the rate of population growth around the world, placing different national and regional pressures on energy production and demand.

Increased wealth

There is a strong correlation between per capita national income and the consumption of energy. However, this relationship is not perfect as other factors, climate in particular, affect the demand for energy. Generally, as average incomes increase, living standards improve. This involves the increasing use of energy and the use of a greater variety of energy sources. For example, when they can afford to do so people are likely to buy refrigerators, washing machines, dishwashers and cars. Demand will increase in all sectors of the economy as average incomes continue to increase.

Technological advances

Technological advances involve both technical improvements in the energy sector itself and technological developments in the rest of the economy that affect energy production and demand. For example, in the energy sector:

» nuclear electricity has only been available since 1954
» oil and gas can now be extracted from much deeper waters than in the past
» renewable energy technology is advancing at an incredible speed. The costs of onshore wind and solar power fell by around 40 per cent and 55 per cent respectively between 2016 and 2020.

In terms of the wider economy, new technology can change the nature of energy demand in a big way. At one time, most of the world's trains were powered by coal. Today, more trains around the world are either electric or diesel. In recent years, the production of electrically powered cars has increased rapidly, replacing petrol and diesel models. As the world moves to a low-carbon economy, new technologies and digitisation are creating huge changes in the way energy is created and used.

Changes in price

The relative prices of different types of energy can influence demand. Electricity production in the UK began switching from coal to gas over 25 years ago. Initially, the main reason was that power stations were cheaper to run on natural gas. At the time, the environmental gain was a secondary factor.

Environmental factors and public opinion

Public opinion can influence decisions made by governments. People today are much better informed about the environmental impact of energy resources and other economic activities than they were in the past. There is now more comprehensive legislation regarding the development of major energy projects, including detailed **environmental impact assessments**. However, the requirements and their enforcement vary from country to country. Canada is an example of a country that has comprehensive environmental impact assessments.

The impact of Covid-19

During the Covid-19 pandemic, the imposition of 'lockdowns' around the world had an astonishing impact on energy demand. The BP Statistical Review of World Energy 2021 described it as 'one of the most tumultuous years for global energy in modern history'. The drop in energy consumption was mainly due to the sudden downturn in economic activity, which decimated transport-related demand. This led to falling demand for oil, which accounted for almost three-quarters of the decline. Overall, global energy demand fell by 4.5 per cent in 2020, the largest decline since the end of the Second World War. Carbon emissions from energy use fell by 6.3 per cent, to their lowest level since 2011. However, by 2023 the demand for energy had not only returned to pre-pandemic levels, but was beginning to exceed it.

> **Activity**
>
> Describe and explain two factors that have affected the demand for and production of energy.

Trends in the consumption of energy

Global energy consumption (Figure 10.28) increased by about 50 per cent between 2000 and 2023. Fossil fuels still dominate the global energy mix but their contribution is falling. Their relative contribution to primary energy consumption in 2023 was: oil 31.7 per cent, coal 26.5 per cent and natural gas 23.3 per cent. Together, the three fossil fuels accounted for 81.5 per cent of global primary energy consumption, down from 84.7 per cent in 2018. In contrast, hydroelectricity accounted for 6.4 per cent, nuclear energy 4.0 per cent and renewable energy 8.2 per cent. The latter was up from 4.1 per cent in 2018. Even in just a five-year time period, significant changes can occur.

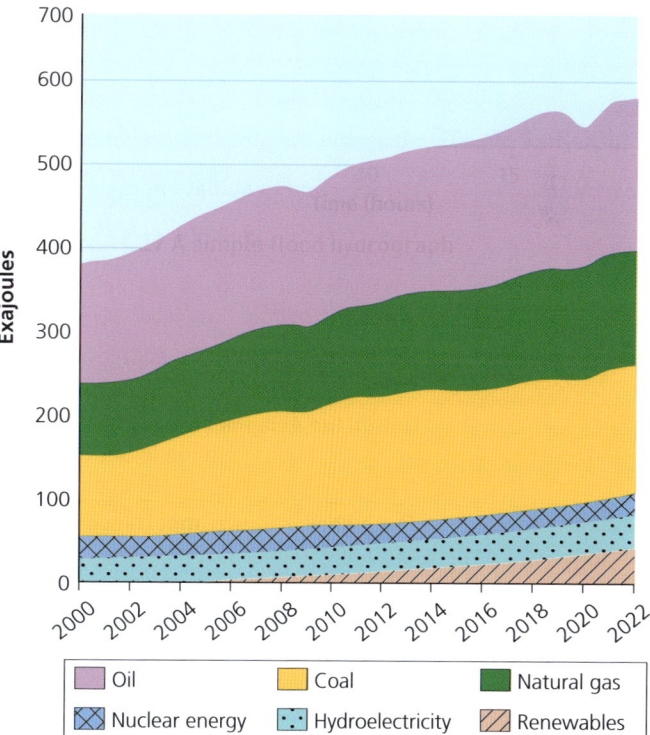

▲ **Figure 10.28** Global consumption of primary energy by source, 2000–22

10.5 THE GLOBAL PATTERNS OF ENERGY SUPPLY AND DEMAND

Figure 10.29 provides more detail about the changing contributions of the different sources of global primary energy. The declining share of oil is very clear, as is the rapidly rising contribution of renewable energy. Despite renewed interest in further development of nuclear energy, the declining global contribution of this source of energy is clear.

Figure 10.30 shows the regional pattern of energy consumption for 2022. Consumption by type of fuel varies widely by world region. Table 10.6 highlights some of the major contrasts between world regions. In 2023, the leading countries in total primary energy consumption were China, the USA, India, Russia and Japan. In terms of global share, the relevant figures were: China 27.6 per cent, the USA 15.2 per cent, India 6.3 per cent, Russia 5.1 per cent and Japan 2.8 per cent.

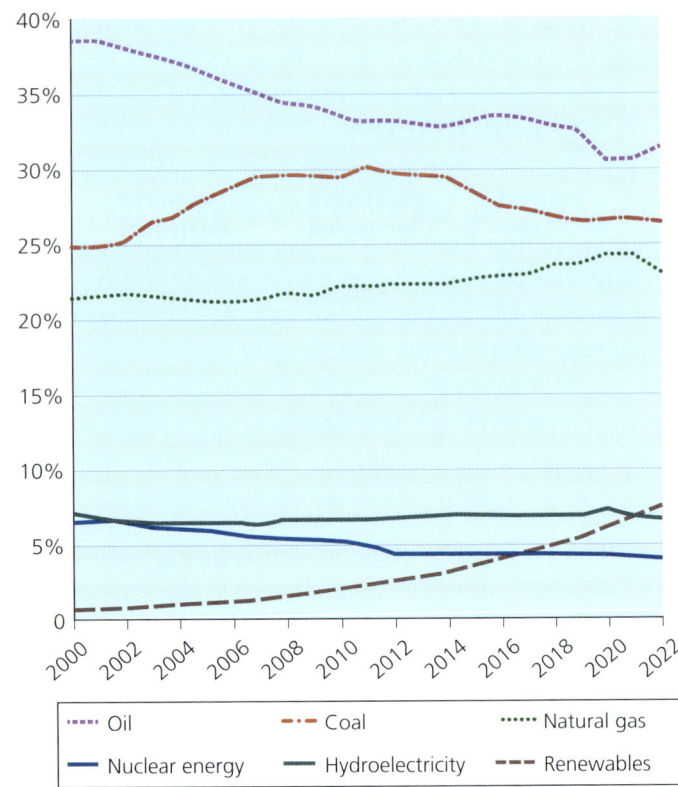

▲ **Figure 10.29** The changing share of global primary energy by source, 2000–22

▼ **Table 10.6** Major contrasts in the global energy mix

Energy source	Contribution to global energy mix in 2023
Oil	Only in the CIS (a grouping of Russia and eight other former Soviet Union countries) and Asia Pacific regions is the contribution of oil to total energy consumption less than 30 per cent. It is the main source of energy in four of the seven regions shown in Figure 10.30. Oil makes its largest contribution to the energy mix in the Middle East (45.2 per cent)
Natural gas	Natural gas is the main source of energy in the Middle East (51.4 per cent) and in the CIS (52.7 per cent). Its lowest share of the energy mix is in Asia Pacific (11.3 per cent)
Coal	Only in the Asia Pacific region is coal the main source of energy. In contrast, it accounts for only 3.7 per cent of consumption in South and Central America, and 0.9 per cent in the Middle East. China was responsible for 56 per cent of global coal consumption in 2023
Nuclear energy	The highest relative importance of nuclear energy is in Europe (8.5 per cent), Northern America (7.0 per cent) and the CIS (5.1 per cent)
Hydroelectricity	The relative importance of hydroelectricity is greatest in South and Central America, where it accounts for 22.4 per cent of the energy mix. Elsewhere its contribution varies from 7.7 per cent in Europe to less than 1 per cent in the Middle East
Renewable energy	Consumption of renewable energy other than hydroelectric power is rising rapidly, but from a low base. Renewable energy makes the largest contribution to energy consumption in Europe (15.2 per cent) and in South and Central America (13.0 per cent)

Trends in the consumption of energy

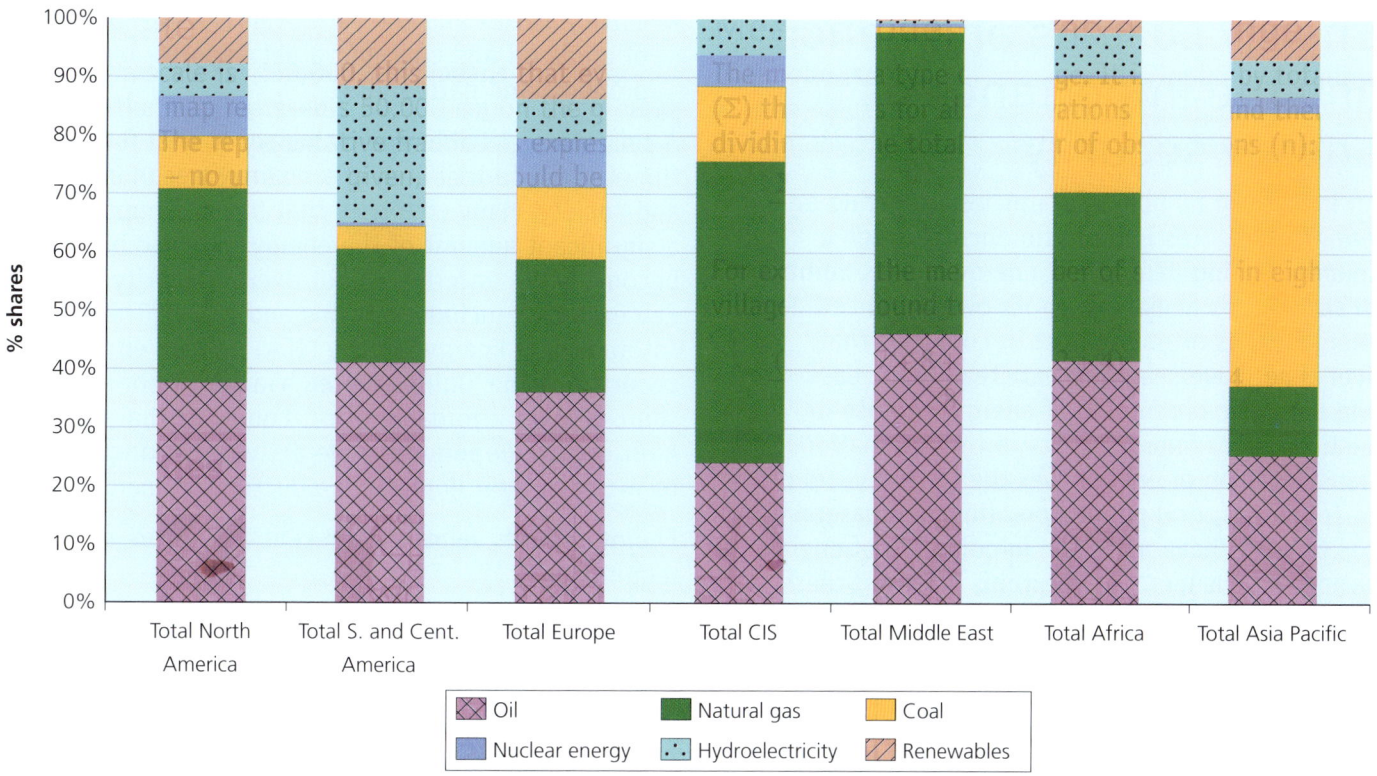

▲ Figure 10.30 Regional consumption of fuel by type, 2022

Table 10.7 shows per capita energy consumption by world region in 2013 and 2023. In 2023, average consumption in Northern America was 16 times that of Africa. However, on an individual country basis there are a number of countries that have much higher consumption figures. Table 10.8 shows all countries that have a per capita consumption over 300 gigajoules.

▼ Table 10.7 Primary energy consumption per capita, 2013 and 2023

World region	Gigajoules (2013)	Gigajoules (2023)
World	74.1	77.0
North America	244.0	230.0
South and Central America	59.5	58.3
Europe	130.2	115.2
CIS	155.8	163.7
Middle East	136.8	142.9
Africa	15.2	14.3
Asia Pacific	54.9	67.3

▼ Table 10.8 Countries with the largest energy consumption per capita – over 300 gigajoules per capita

Country	Per capita consumption (2023)
Qatar	816.7
Singapore	577.0
UAE	539.4
Trinidad and Tobago	384.3
Kuwait	365.9
Norway	363.7
Canada	359.7
Oman	333.4
Saudi Arabia	313.9

> ### Activities
>
> 1 Describe the changes in world energy consumption shown in Figure 10.29 (page 274).
> 2 To what extent do the types of energy consumption vary by world region?
> 3 Summarise the differences and trends in Table 10.7.
> 4 Suggest reasons for the very high levels of energy consumption in the countries listed in Table 10.8.

10.5 THE GLOBAL PATTERNS OF ENERGY SUPPLY AND DEMAND

Energy surplus and deficit and the importance of energy security

An increasing number of countries are facing an '**energy gap**'. This is the difference between a country's rising demand for energy and its ability to satisfy this demand from its own resources. The energy gap is often greatest in countries that, over time, have seen demand steadily increase and domestic production of energy stabilise or decline. Major improvements in the international transportation of energy over the last 50 years or so (pipelines, grid connectors, liquefied natural gas tankers) have made the importation of energy easier and cheaper. To an extent, however, these improvements in energy infrastructure may have deterred some countries from investing in the development of their own potential sources of energy. In 2023, the total international trade in oil, gas and coal was 53 per cent higher than it was in 2000.

- The Middle East is crucial to the functioning of the global energy market, although it has lost some of its dominance in recent decades. In 2022, five of the top ten oil producers and three of the top twenty natural gas producers were located in the region.
- Since the 1980s, Europe has consistently been a net importer of energy. The biggest energy deficit for Europe is oil. In 2023, Europe's oil production met only 23 per cent of its demand.
- The Asia Pacific region has been a net importer of energy since the 1980s. In 2023, its largest deficit was in oil as production met only 19 per cent of consumption.
- Northern America's energy system has changed significantly in recent decades, with rapid growth in the production of unconventional oil and gas beginning in the early 2000s. As a result, this region changed from being a net importer of energy to a net exporter. In 2023, oil production in Northern America was 16 per cent higher than its domestic consumption.
- Latin America and the Caribbean is a net exporter of crude oil and coal, but a net importer of oil products and natural gas.

A secure and reliable energy supply is vital for the efficient functioning of a country's economy and society. Many countries that have to import a significant amount of the energy they use have become very concerned about energy security. Key examples include Germany and Japan. The International Energy Agency (IEA) defines **energy security** as the 'uninterrupted availability of energy sources at an affordable price'. Countries that have a high **import dependency** are actively looking at ways to increase their domestic production of energy.

Energy security relates to present and future energy requirements:

- Short-term energy security depends on the ability of an energy system to react quickly to sudden changes in the supply–demand balance.
- Long-term energy security requires careful planning and significant investment in line with economic development and energy needs.

Energy supply interruption can be caused by a number of factors, including:

- natural hazards, such as tropical storms and earthquakes
- civil internal disturbance and international war
- geopolitics, such as Russia reducing gas supply to European countries in 2022
- rapid resource depletion, for example oil and gas fields running out
- affordability
- environmental concerns.

The energy supply interruption itself is the first stage in a three-stage sequence (Figure 10.31). The next stage is the impact of the interruption on an energy system. The final stage is the consequences resulting from such an impact. Consequences can vary from manageable to catastrophic.

The reasons for variations in the types of energy produced and used

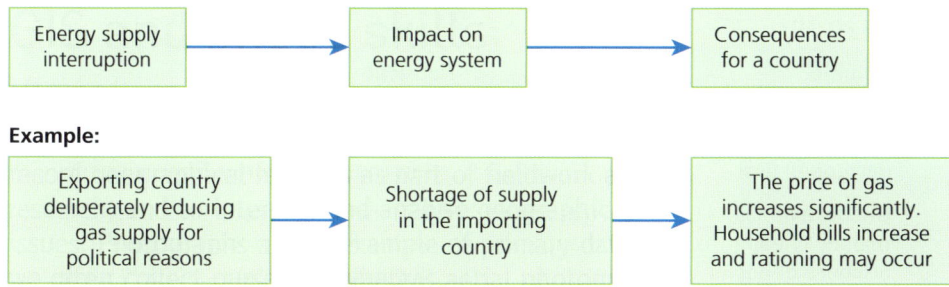

▲ **Figure 10.31** The three stages of energy supply interruption

Countries with surplus energy can generally be considered to be energy secure. Countries including Canada and Russia earn valuable foreign currency by exporting energy to countries with limited domestic energy supplies. Where countries become over-reliant on energy from other countries, there is the risk that energy supply might be used as a political weapon. Africa is the world's most energy-insecure continent. Lack of investment has resulted in insufficient power plant capacity, leading to low energy consumption and low access to power. Households connected to an energy grid often experience interruptions to supply.

Energy sources are affected by several factors including global climate change, population growth and economic development. In particular, extreme weather events such as drought, floods and rising temperatures are increasing demand for energy but also affecting supplies. Energy resources need to be safeguarded in order to meet the increased demands of population growth and economic development.

Energy supplies also depend on water. This is not just for hydroelectric schemes and biofuels but also for the production of fossil fuels and operating power plants. In addition, energy systems have become more vulnerable to cybercrime. Indeed, the energy sector is one of the UK's top targets for cybercrime – nearly one-quarter of all cyberattacks in the UK were in the energy sector. Without doubt, energy supply has become a prominent factor in **geopolitics**.

As there is little excess capacity to ease pressure on energy resources, energy insecurity is increasing. The USA is seriously concerned about the political leverage associated with importing oil: in 1977 it began the construction of a **strategic petroleum reserve**. The oil is stored in huge underground tanks in southern Louisiana and Texas, and could easily be linked up to pipelines and shipping routes if needed. Since then, many other countries have built up oil reserves as protection against future energy crises. The USA's strategic petroleum reserve was tapped after both Hurricane Katrina (2005) and Hurricane Gustav (2008) severely disrupted oil production in the US states bordering the Gulf of Mexico. Some countries also store reserves of coal and gas.

In recent decades concerns have developed about the security of key **energy pathways**, which might be blocked or destroyed as a result of political conflict. Most Middle East oil exports go by tanker through the Strait of Hormuz, a narrow body of water (55 km wide at its narrowest) between the Persian Gulf and the Gulf of Oman. Some countries have suggested that they might block the shipping route to prevent the oil getting through.

The reasons for variations in the types of energy produced and used

The reasons for variations in energy production can be subdivided into physical, economic and political factors. Table 10.9 shows examples for each of these groupings. The combination of factors operating in each country can vary considerably. Some countries are rich in domestic energy resources, while others are energy resource poor. The latter countries rely heavily on imports of energy. However, energy resources by themselves do not constitute production. Capital and technology are also required to exploit resources.

10.5 THE GLOBAL PATTERNS OF ENERGY SUPPLY AND DEMAND

▼ Table 10.9 Factors affecting the production of energy

Physical	Economic	Political
• Deposits of fossil fuels are found in only a limited number of locations • Large-scale hydroelectric power (HEP) development requires high precipitation, major steep-sided valleys and impermeable rock • Large power stations require flat land and geologically stable foundations • Solar power needs a large number of days each year with strong sunlight • Wind power needs high average wind speeds throughout the year • Tidal power stations require a very large tidal range • The availability of biomass varies widely according to climatic conditions	• The most accessible, and lowest cost, deposits of fossil fuels are invariably developed first • Onshore deposits of oil and gas are usually cheaper to develop than offshore deposits • Potential HEP sites close to major transport routes and existing electricity transmission corridors are more economical to build than those in very inaccessible locations • In LICs, FDI is often essential for the development of energy resources • When energy prices rise significantly, companies increase spending on exploration and development	• Countries wanting to develop nuclear electricity require permission from the International Atomic Energy Agency • International agreements such as the Kyoto Protocol can have a considerable influence on the energy decisions of individual countries • Potential HEP schemes on 'international rivers' may require the agreement of other countries that share the river • Governments may insist on energy companies producing a certain proportion of their energy from renewable sources • Legislation regarding emissions from power stations favours the use of, for example, low-sulphur coal, as opposed to coal with a high sulphur content

A country's **energy policy** can have a significant impact on demand if it focuses on efficiency and sustainability as opposed to concentrating on building more power stations and refining facilities. High levels of pollution can be a strong stimulus to develop a cleaner energy policy.

Developing countries face many energy challenges, including:

» poor security of supply
» insufficient generating capacity
» underdeveloped or non-existent grid infrastructure
» a lack of adequate monitoring and control equipment and skilled human resources.

The use of fuelwood as a source of energy is largely confined to low- and lower-middle-income countries. In terms of other sources of energy, countries at this stage of development will largely rely on any domestic resources of fossil fuels if they have been able to gain access to the often high levels of investment required to develop such resources. Limited access to investment also goes a long way to explain the relatively low level of development of renewable energy sources in most low- and lower-middle-income countries. If domestic energy production is limited, significant levels of imports may be required. The types of fuel imported will be dictated by existing energy infrastructure and by differences in price.

Upper-middle-income countries have generally expanded their energy infrastructure in line with economic development and their increasing demand for energy. Typically, countries in this income bracket obtain energy from a wider range of non-renewable sources and will also have a growing renewable energy sector.

High-income countries have had the longest timespan in which to develop their energy infrastructures and the greatest availability of capital to adapt to changing challenges and opportunities. Such countries have generally invested the most in renewable energy and in commercial nuclear power. High levels of affluence allow for significant imports of energy if necessary and the ability to construct energy connections with neighbouring countries.

▶ Activities

1. Explain the terms:
   a Energy gap
   b Energy security.
2. Write a summary of Figure 10.31 (page 277).
3. Why do many countries hold reserves of energy resources?
4. What are energy pathways and why is their security so important?
5. With reference to Table 10.9, examine the physical factors that can affect the production of energy in a country.

10.6 The impacts of energy production

This chapter will explain:
★ the advantages and disadvantages of non-renewable sources of energy
★ the advantages and disadvantages of renewable sources of energy
★ the strategies and techniques used to increase energy supplies
★ detailed specific example: energy resource management in Sweden.

The advantages and disadvantages of non-renewable sources of energy

Coal

Coal is the dirtiest of the fossil fuels and is the major contributor to air pollution and climate change (Figure 10.32). This is the main reason why coal consumption is being phased out in so many countries. To meet the Paris Agreement on climate change (2015), coal burning for power must decline rapidly. The objective is for coal power to cease in developed countries by 2030, and in developing countries by 2050. At the same time, coal has come under increasing competitiveness from renewable energy.

▲ **Figure 10.32** Polluting brick kilns, fired by burning coal, along the bank of the Mekong River, Vietnam

The UK was the first country to develop coal-fired power. Coal consumption in the UK has fallen dramatically over the last 30 years, from 108 million tonnes in 1990 to 6.1 million tonnes in 2022. The UK's last three deep underground coal mines closed in 2015. In June 2021, the UK government confirmed its intention to phase out coal-fired electricity generation entirely by October 2024.

In 2023, global coal production reached its highest ever level. The Asia Pacific region accounted for nearly 80 per cent of global output, with production concentrated in four countries – China, Indonesia, India and Australia. China is the world's largest coal producer (56 per cent share), consumer and importer. While China has not committed to phasing out coal, it is gradually working to lower its reliance on the fuel to improve air quality and limit the impact of climate change. Closing down surplus and inefficient coal and heavy industry plants is a priority under China's 13th Five-Year Plan.

However, significant investment in new coal projects remains. China, Japan, South Korea and India together account for an estimated US$22 billion of the US$24 billion of planned investment by G20 countries in new coal projects.

Oil

Even though oil is under pressure from climate change policies and the falling cost of renewable energy, it still contributed more than 31 per cent to global primary energy consumption in 2023. Global oil production reached a record level of over 96 million barrels per day in 2023 (with consumption at over 100 million barrels a day) (Figure 10.33). Some analysts believe this might mark peak global oil consumption. However, other analysts think that

10.6 THE IMPACTS OF ENERGY PRODUCTION

peak oil will occur later in the present decade. What appears certain is that oil will remain important to global energy supply in the short and medium term, although its relative contribution will fall.

Road fuel makes up half of global oil demand. This means that the shift from petrol/diesel to electric vehicles (EVs) will have a considerable impact on the demand for oil. However, although this shift is occurring at an increasing pace in HICs, it will take place more slowly in emerging and developing countries. The rate of change will depend to a large extent on the cost of EVs. There can be little doubt that mass manufacturing will eventually make the price of EVs much more attractive. The future availability of charging stations is an important linked issue, as is the longevity of EV batteries in terms of how far an EV can be driven before the battery needs to be recharged (as well as the lifetime of the battery itself).

▲ **Figure 10.33** An oil refinery and liquefied petroleum gas tanker in Japan

Other technological advances will also limit the demand for oil. These include:

- Sustainable aviation fuel (SAF), which is aviation biofuel from sustainable sources. SAF is chemically very similar to traditional jet fuel, and its efficacy has already been proved with commercial aircraft.
- Advances in the recycling of plastics and other oil-based products. Only about 12 per cent of plastics waste is currently reused or recycled.

Natural gas

Natural gas has become an increasingly important fuel in the global energy mix in recent decades. It is the least polluting of the fossil fuels and is often viewed as a 'transition fuel' from coal and oil to renewable energy. Natural gas has displaced higher-emitting fossil fuels in a number of countries. In so doing it has improved local air quality and reduced GHG emissions. The relative contributions of coal, oil and natural gas to total global fossil fuel emissions of CO_2 in 2021 were:

- coal: 45 per cent
- oil: 35 per cent
- natural gas: 20 per cent.

Natural gas is exported across land and shallow sea borders by pipeline, and across oceans in liquefied natural gas (LNG) tankers. The role of LNG in the international trade in energy has increased significantly in recent decades.

Shale oil and gas

The exploitation of shale oil has been a relatively recent development. It has been concentrated mainly in the USA. Shale oil is extracted from oil reserves, sometimes described as 'tight oil' reserves. These are held in shales and other rock formations, from which the oil will not naturally flow freely. Shale oil has become more accessible due to advances in technology. The rapid increase in the scale of production in the USA has fundamentally changed global energy markets, because it has enabled the USA to quickly regain much of its self-sufficiency in energy. Gas can also be obtained from shale.

Although the basic technology was originally developed in the USA in the 1940s, the more recent 'shale revolution' was the result of technological advances in horizontal drilling and hydraulic fracturing (fracking). This is a controversial technique, with concerns relating to groundwater pollution and seismic problems. As a result, some countries with commercial shale deposits, such as the UK, have decided to abandon plans to commence fracking.

Nuclear power

Over the last 60 years or so, no other source of energy has created such heated discussion as nuclear power. The main concerns about nuclear power are related to:

- power plant accidents, which could release radiation into air, land and sea
- radioactive waste storage/disposal – most concern is over the small proportion of 'high-level waste'
- terrorist use of nuclear fuel for weapons
- the high construction and decommissioning costs
- the possible increase in certain types of cancer near nuclear plants.

Before the late 1970s, the rise of nuclear power seemed unstoppable. It was generally viewed as an efficient and clean source of energy. However, incidents like the serious Chernobyl disaster in Ukraine in 1986 brought any growth in the industry to a virtual halt. As nuclear technology has advanced and safety has improved, new nuclear plants have been constructed but at a limited pace. However, the Fukushima disaster in Japan in 2011 created new doubts about the safety of nuclear reactors.

The advantages of nuclear power are:

- Zero emissions of greenhouse gases and reduced reliance on imported fossil fuels.
- It is not as vulnerable to fuel price fluctuations as oil and gas.
- Uranium, the fuel for nuclear plants, is relatively plentiful.
- In recent years, nuclear plants have demonstrated a very high level of reliability in terms of energy efficiency.
- Nuclear technology has spin-offs in fields such as medicine and agriculture.

A number of countries are considering the construction of small modular reactors (SMRs). These mini-nuclear power plants generate around 450 megawatts each, about 15 per cent the amount produced by traditional nuclear plants. As they only require the space of about two football pitches, they can be constructed relatively quickly.

> **Activities**
>
> 1 Why is coal the most important fossil fuel to phase out around the world as quickly as possible?
> 2 Why are so many countries still reliant on oil as a source of energy?
> 3 State three advantages and three disadvantages of nuclear power.

The advantages and disadvantages of renewable sources of energy

Hydroelectric power

Of the five traditional major sources of energy, hydroelectricity is the only one that is renewable. Five countries accounted for almost 58 per cent of the global total in 2023. These were:

- China: 28.9 per cent
- Brazil: 10.1 per cent
- Canada: 8.6 per cent
- The USA: 5.6 per cent
- Russia: 4.7 per cent.

Most of the best hydroelectric power (HEP) locations are already in use (Figure 10.34), so the scope for more large-scale development is limited. In many countries, though, there is scope for small-scale HEP plants to supply local communities.

Although HEP is generally seen as a 'clean' form of energy, it is not without its problems:

- Large dams and power plants can have a huge negative visual impact on the environment.
- The dams and power plants may obstruct the river for aquatic life, for example migrating fish.
- There may be a deterioration in water quality.
- A large area of land may need to be flooded to form the reservoir behind a dam.
- Submerging large forests without prior clearance can release methane, a greenhouse gas.

▲ Figure 10.34 A hydroelectric power plant along the River Danube in Romania

10.6 THE IMPACTS OF ENERGY PRODUCTION

Newer alternative energy sources

The first wave of interest in new alternative energy sources resulted from the energy crisis of the early 1970s. However, the relatively low price of oil from the 1980s to the opening years of the present century dampened interest in these energy sources. Growing concern about climate change and energy price rises, however, kickstarted the alternative energy industry again. Table 10.10 shows the leading countries in 2023 for renewable energy electricity generation.

▼ **Table 10.10** Renewable electricity generation by source in terawatt hours ('other renewables' includes geothermal, biomass, etc., but not HEP)

Country	Wind	Solar	Other renewables	Total
China	885.9	584.2	198.1	1668.2
USA	429.5	240.5	67.3	737.3
Germany	142.1	61.2	49.5	252.8
India	82.1	113.4	37.3	232.8
Brazil	95.5	51.1	55.8	202.4
Japan	10.0	97.0	42.0	149.0
UK	82.0	13.8	34.0	129.8

Wind power

In the last 10 to 15 years, wind energy has reached the 'take-off' stage both as a source of energy and as a manufacturing industry. In 2023, global-installed capacity reached 1017 gigawatts – 93 per cent onshore and 7 per cent offshore (Figure 10.35).

The main advantages of wind energy are that it can:

- generate significant amounts of electricity without emitting any carbon or creating polluting waste
- be harnessed to a reasonable degree in most parts of the world
- be sited in both onshore and offshore locations
- coexist with other land uses such as farming
- be cost effective – the cost per unit of electricity has fallen significantly over the last decade.

Apart from establishing new wind energy sites, **repowering** can also play an important role. This involves replacing first-generation wind turbines with modern multi-megawatt turbines, which give much better performance.

▲ **Figure 10.35** Renewable energy: a wind farm on the coast of northern Norway

On the other hand, as wind turbines have been erected in more areas of more countries, the opposition to this form of renewable energy has increased. For example:

- Some people feel that huge turbines located nearby look unattractive, disrupt beautiful scenery and have a significant impact on property values.
- There are concerns about the hum of the turbines disturbing both people and wildlife.
- Turbines can kill birds. In response, wind companies argue that they don't place turbines near migratory routes.

Solar power

The global-installed capacity of solar electricity has increased rapidly in recent years. Solar electricity is currently produced in two ways:

- Photovoltaic (PV) systems: These are solar panels that convert sunlight directly into electricity (Figure 10.36).
- Concentrating solar power (CSP) systems: These use mirrors or lenses and tracking systems to focus a large area of sunlight into a small beam. This concentrated light is then used as a heat source for a conventional thermal power plant.

In recent years, photovoltaic systems have come to dominate installed capacity. They are easier and cheaper to establish at a wide range of scales. The leading countries for installed PV systems are China, the USA, Japan, Germany, India and Italy.

The advantages and disadvantages of renewable sources of energy

▲ **Figure 10.36** Solar electricity being generated by photovoltaic panels in northern Spain

Table 10.11 outlines the advantages and disadvantages of solar power.

▼ **Table 10.11** The advantages and disadvantages of solar power

Advantages	Disadvantages
It is a completely renewable resource	The initial cost of solar farms is very high
It generates no significant noise or direct pollution	Solar power cannot be harnessed during storms, on cloudy days or at night
It requires very limited maintenance	It is of more limited use in countries with low annual hours of sunshine
Technology is improving and reducing unit costs	Large areas of land are required to capture the sun's energy in order to generate significant amounts of power
It can be used in remote areas where extending the electricity grid would be too expensive. Having a source of electricity for the first time can greatly improve people's quality of life	Considerable areas of farmland may be lost to accommodate very large solar farms, thus compromising food security
Public perception of solar power is generally positive. Solar panels are seen as less landscape intrusive than wind turbines	Sometimes the best locations for solar farms are distant from areas of high population density, resulting in high electricity transmission costs

Tidal and wave power

Tidal power plants act like underwater windmills, transforming sea currents into electrical current. Tidal power is more predictable than solar or wind power, and the infrastructure is less obtrusive. However, start-up costs are high. Thus, the 240 MW Rance Tidal Power Station in northwestern France was the only utility-scale tidal power system in the world for 45 years, until the Sihwa Lake Tidal Power Station opened in South Korea in 2011. It uses sea wall defence barriers complete with 10 turbines to generate 254 MW. Some smaller-scale systems are in operation at other locations.

Although currently in its infancy, a study by the Electric Power Research Institute estimated that as much as 10 per cent of US electricity could eventually be supplied by tidal energy. This potential could be equalled in the UK and surpassed in Canada. So for some countries, potential energy production from this source could be very high.

Wave energy is where generators are placed on the ocean's surface and energy levels are determined by the strength of the waves. The first experimental wave farm opened in Portugal in 2008 at the Aguçadoura Wave Farm. However, the facility was shut down two months after opening due to technical problems. A number of research projects are in operation, including one off the shores of Oregon in the USA. The costs and benefits of wave energy are broadly similar to those of tidal power.

Biomass

Biomass can be used to produce renewable energy, thermal (heat) energy and transportation fuels (biofuels). Biofuels are fossil fuel substitutes that can be made from a range of crops, including oilseeds, wheat and sugar. They can be blended with petrol and diesel. The biggest producers of biofuels are the USA, Brazil, Indonesia, China and Germany. The USA and Brazil together accounted for almost 60 per cent of global biofuels production in 2023.

By increasing biofuel production, these countries have reduced the amount of oil they need to consume. This is the main reason behind biofuel production. Advocates of biofuels also argue that biofuels come from a renewable resource (crops); they can be produced wherever there is sufficient crop growth, helping energy security; and they often produce fewer emissions than petroleum-based fuels. However, there are clear disadvantages in biofuel production. Increasing amounts of cropland have been used to produce biofuels, adding to the 'global food crisis'. Large amounts of land, water and fertilisers are needed for large-scale crop production. The manufacture of biofuels also uses significant amounts of energy, creating greenhouse gas emissions.

10.6 THE IMPACTS OF ENERGY PRODUCTION

Geothermal electricity

Geothermal energy is the natural heat found in the Earth's crust in the form of steam, hot water and hot rock (Figure 10.37). Rainwater may percolate several kilometres down in permeable rocks, where it is heated due to the Earth's geothermal gradient. This is the rate at which temperature rises as depth below the surface increases. The average rise in temperature is about 30°C per km, but the gradient can reach 80°C near plate boundaries. This source of energy can be used directly for industry, agriculture, bathing and cleansing. For example, in Iceland hot springs supply water at 86°C to 95 per cent of the buildings in and around Reykjavik.

Table 10.12 summarises the advantages and disadvantages of this form of electricity production.

▼ **Table 10.12** The advantages and limitations of geothermal power

Advantages	Limitations
As a non-carbon source of power, it has an extremely low environmental impact	There are few locations worldwide where significant amounts of energy can be generated – therefore, it is of only limited use
A geothermal plant occupies a relatively small land area (compared to a wind farm or coal-fired plant)	Some of these locations are far from where the energy could be used, that is, the areas of highest energy demand
Unlike wind or solar power, it is not dependent on weather conditions	Total global generation remains very small
Maintenance costs are relatively low	The installation costs of the plant and piping are relatively high since the required depth of piping below the surface is considerable

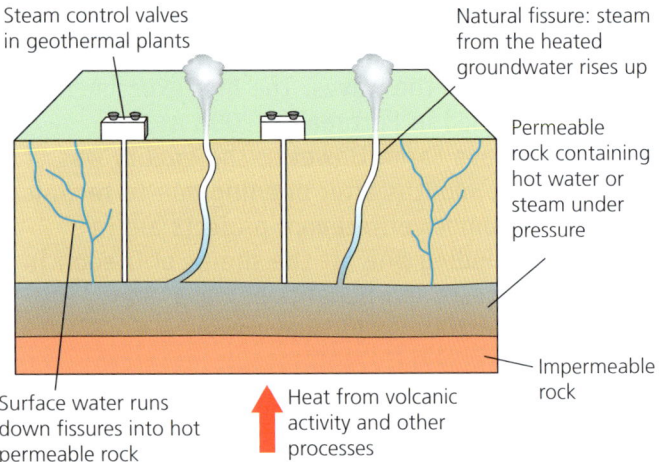

▲ **Figure 10.37** Geothermal power

The USA is the world leader in geothermal electricity. However, total production accounts for less than 1 per cent of the electricity that the USA uses. Other leading countries using geothermal electricity are Indonesia, the Philippines, Turkey, New Zealand (Figure 10.38), Mexico, Italy and Iceland.

▲ **Figure 10.38** A geothermal power plant, Wairakei, New Zealand

Hydrogen fuel cells

Investment in the commercial development of hydrogen fuel cells has risen rapidly in recent years. Hydrogen fuel cells can be used:

- to power buses and other vehicles
- to power generators on construction sites, replacing diesel generators
- for a range of other uses.

In terms of buses, it takes as little as five minutes each day to top up each hydrogen fuel cell at a fuelling station where hydrogen dispensers, about the size of a traditional petrol pump, are located.

> **Activities**
>
> 1 Suggest why the location of hydroelectric power is so spatially concentrated both between and within countries.
> 2 Give two advantages and two disadvantages for each of the following forms of renewable energy:
>   - Wind
>   - Solar
>   - Biofuels
>   - Geothermal
>   - Tidal.
> 3 For the country in which you live, find out which forms of renewable energy are used and how much they contribute to total energy production.

The strategies and techniques used to increase energy supplies

A range of strategies and techniques are being used by countries and energy companies to increase energy supply.

Improving the efficiency of existing fossil fuel power plants

The efficiency of older fossil fuel power plants, which mainly burn coal, becomes degraded over time. Lower efficiency reduces the power generated per unit of fuel input, and it also results in more CO_2 being emitted per unit of electricity generated. The options that are most often considered for increasing the efficiency of older fossil fuel power plants include:

- equipment refurbishment
- plant upgrades
- improved operation and maintenance schedules.

Fossil fuel power plants can reuse wasted heat by being adapted to combined-cycle systems. Reusing heat makes the most of the fossil fuels consumed. Some power plants have been adapted to burn biomass alongside fossil fuels. This process, which makes the consumed fossil fuels last longer, is known as co-firing.

Scale and efficiency of production

Large land (and sea) areas are required for wind and solar projects of a size that can contribute a significant increase in renewable electricity. The scale of wind and solar projects has increased considerably since the late twentieth century. Scale goes hand in hand with efficiency, as investment in technological improvements seeks to achieve more energy from each solar panel and wind turbine. In general, the larger the size of renewable energy projects, the greater the **economies of scale** that can be achieved. However, lagging well behind the significant scaling up of renewal energy production is battery energy storage, which needs to be scaled up dramatically in a very short time period to have the best chance of maintaining momentum in the energy transition.

Energy intensity is a measure of energy efficiency. It is defined as consumption per unit of output. It provides a useful indication of changes in energy efficiency over time. For example, between 2010 and 2019 domestic energy consumption per household in the UK declined by 21 per cent. Reasons for this include:

- changing to light-emitting diode (LED) and compact fluorescent light (CFL) bulbs
- upgrading to higher energy-rated boilers and domestic appliances
- more homes installing cavity wall and loft insulation, as well as double glazing, to reduce home heat loss
- the rollout of smart meters.

Smart power grids are being developed in many countries to improve the efficiency of power networks. This reduces power lost by inefficient operation. This is broadly similar to reducing leakages in water pipe networks. **Smart meters** are an integral part of smart grids. These are installed in individual homes and premises. A meter is only smart if it can communicate effectively with other equipment. Smart meters:

- enable households to monitor the energy they are using (and wasting) in real time by identifying

10.6 THE IMPACTS OF ENERGY PRODUCTION

the cost of running individual appliances and gadgets – this encourages people to adopt good energy habits
» provide automatic meter readings and billing.

Smart meters enable the implementation of Home Energy Management Systems (HEMS) and Building Energy Management Systems (BEMS), allowing visualisation of power usage in individual homes or entire buildings.

Energy conservation means reducing the consumption of energy by using energy more efficiently. Apart from installing a smart meter, there are other energy-saving measures that can make a difference to households. These include: loft and cavity wall insulation, energy-efficient windows, draught-proofing windows and doors, top-rated energy-saving appliances, energy-efficient light bulbs, washing at cooler temperatures, switching off 'standby', making the most of natural light, and using a microwave instead of an oven where appropriate.

Managing energy supply is often about balancing socioeconomic and environmental needs. Many countries are looking increasingly at the concept of **community energy**. This is energy produced close to the point of consumption. Much energy is lost in transmission if the source of supply is a long way away. Energy produced locally is much more efficient. This will invariably involve **microgeneration**, which refers to generators producing electricity with an output of less than 50 KW.

Energy storage and connectivity to grids

Energy storage is not a new phenomenon, as pumped hydro-storage has been used worldwide for decades. In contrast, modern **battery energy storage systems (BESS)** enable excess energy from renewables to be stored and then released when the power is needed most. BESS makes renewable energy more reliable and therefore more investable. The supply of wind and solar energy can fluctuate. BESS are able to 'smooth out' this flow to provide a continuous power supply. Lithium-ion batteries are currently the dominant storage technology for large-scale BESS.

Although the use of battery energy storage in power systems is increasing, it is doing so from a low base. Only 16 GW of new storage systems were deployed in 2022, compared to 192 GW of solar and 75 GW of wind energy installed in the same year. The World Economic Forum (WEF) has stated that annual additions of grid-scale battery energy storage need to increase to an average of 120 GW annually to meet agreed climate change targets. To date, most battery storage capacity has been in advanced economies. However, the World Bank's Energy Sector Management Assistance Program (ESMAP) is working to provide concessional financing for battery energy storage projects in developing countries.

Energy (electricity) grids are the framework for power distribution within a country or geographical area, and increasingly electricity grids link different countries. The US power grid is a vast interconnected network serving almost 400 million consumers across the continent. It is sometimes referred to as 'the world's largest machine'. This grid has been developed by linking five smaller interconnections.

The energy transition

The transition to renewable energy sources, coupled with economic growth, will result in an increasing demand for electricity. This is estimated to rise by 40 per cent between 2020 and 2030, and to double by 2050. Existing energy grids were designed for centralised, mainly fossil fuel generation. The best locations for wind and solar power do not always match the geographical pattern of existing grids. Also, grid infrastructure can struggle with the increased inflow of intermittent power sources (solar and wind), resulting in a growing need for complex balancing services. Many existing grids are showing signs of deterioration due to age.

Sustainable development is vital if the world is to limit the impact of climate change and to avoid the worst of so many other forms of environmental damage. 'Affordable and clean energy' is one of the UN's Sustainable Development Goals. In 2020 the UN made the following points about progress to achieving the objectives of Goal 7:

» There are encouraging signs that energy is becoming more sustainable and widely available. Access to electricity in LICs has begun to accelerate, energy efficiency continues to improve, and renewable energy is making impressive gains in the electricity sector.
» Financial flows to developing countries for renewable energy are increasing, but only 12 per cent goes to LICs.

The strategies and techniques used to increase energy supplies

» More effort is needed to improve access to clean and safe cooking fuels and technologies for 3 billion people, to expand the use of renewable energy beyond the electricity sector, and to increase electrification in many countries in Africa.

In December 2022, the International Monetary Fund, when discussing the enormity of the task ahead, stated that, 'the objective of this transition is not just to bring on new energy sources, but to entirely change the energy foundations of what today is a $100 trillion global economy.' Added to this is the very short timescale envisaged – little more than a quarter of a century. The International Energy Agency (IEA) sees the world moving from a fuel-intensive to a mineral-intensive energy system that will substantially increase pressure on the supply of critical minerals. The IEA estimates it takes 16 years from discovery of a mineral deposit to first production from a new mine, while some other estimates are longer.

Breaking the link between economic growth and greenhouse gas emissions will necessitate historic levels of capital expenditure in renewable energy, transport and industrial infrastructure. The predicted costs of the energy transition vary widely, with estimates ranging from US$100 trillion to US$300 trillion between 2020 and 2050. Economists say that the energy transition can be viewed as a series of shocks to the economy, both negative (such as an increase in energy costs) and positive (increased productivity created by investment in renewable technology). Economic growth could benefit significantly from the multiplier effect of this new investment.

Activities

1. How can the efficiency of existing fossil fuel power plants be improved?
2. Explain the meaning of 'energy intensity'.
3. Why are many countries investing heavily in the rollout of smart meters?
4. Why are battery energy storage systems so important in the energy transition?

Detailed specific example

Energy resource management in Sweden

Sweden is a large, elongated Nordic country stretching from a latitude of about 55 degrees north to 69 degrees north, which is beyond the Arctic Circle. Few countries consume more energy per capita than Sweden due to a combination of its northerly location and the high average income of its population.

Sweden's energy mix

Sweden does not have any significant reserves of fossil fuels. However, its carbon emissions are very low due to a very high level of renewable energy production in the country. Sweden has the second lowest CO_2 emissions per unit of GDP and per capita among IEA members. In 2022, Sweden's primary consumption energy mix was:

- oil: 21.9 per cent
- natural gas: 1.3 per cent
- coal: 3.1 per cent
- nuclear: 20.2 per cent
- hydropower: 28.5 per cent
- other renewable energy sources: 24.6 per cent.

Electricity use in Sweden peaked in 2001 and has declined somewhat since, following the pattern of a number of other developed countries. Energy use per capita declined from 229.8 gigajoules in 2010 to 215.7 gigajoules in 2022 (Figure 10.39). This was due largely to improvements in energy efficiency, with more energy-efficient insulation, domestic appliances and lighting. The residential and services sectors use the most electricity, followed by industry and then transport. Sweden has a long-term energy policy of a controlled transition to an entirely renewable electricity system. Its overall aims are ecological sustainability, competitiveness and security of supply.

10.6 THE IMPACTS OF ENERGY PRODUCTION

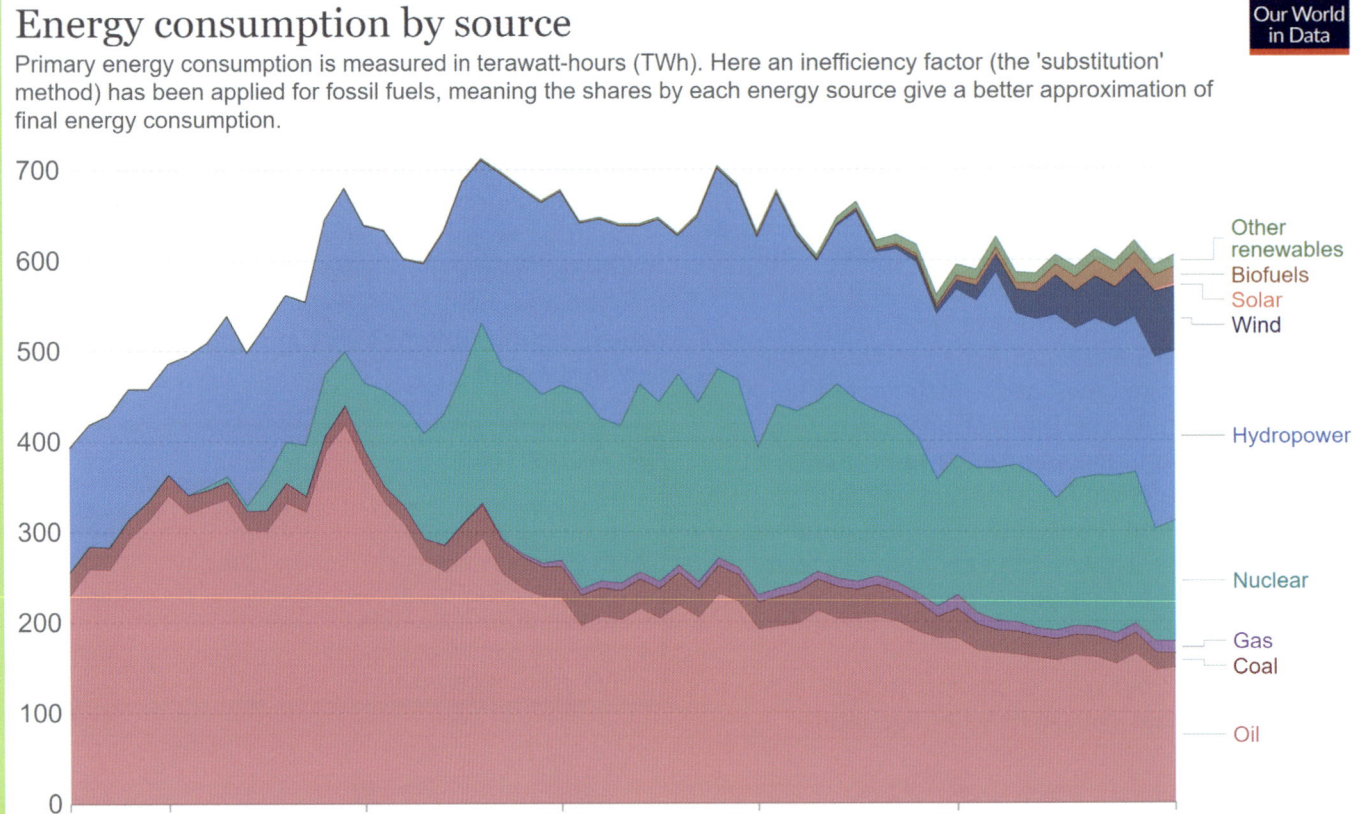

▲ Figure 10.39 Energy consumption by source in Sweden, 1965–2021

The impacts of the different types of energy used by Sweden

Hydropower and bioenergy are the top renewable resources. Bioenergy is electricity and gas generated from organic matter (biomass). Bioenergy contributes to heating in individual homes and for district heating, as well as in electricity production and for industry. Sweden is well-endowed with high volume, fast-flowing rivers, providing ideal conditions for the production of hydroelectric power.

Forests cover 63 per cent of Sweden's land area, giving the bioenergy sector a very strong resource base. There has been a rapidly rising trend of using biofuels in the transport sector. While the majority of Sweden's electricity is produced in the north of the country, most people live in the south. Good quality and highly efficient transmission corridors are important in making the most of the country's energy resources.

Nuclear energy and hydropower account for 80 per cent of electricity production. Sweden has three nuclear power plants, with a total of eight nuclear reactors in commercial operation. However, nuclear power is a controversial subject in Sweden. About 11 per cent of electricity production is from wind power, and this is set to increase sharply in the future. Combined heat and power (CHP) plants account for 9 per cent of electrical output. These plants are mainly powered by biofuels.

Solar power remains limited in capacity largely due to Sweden's northerly latitude, but it is growing with the aid of government funding. Between 2019 and 2020 the number of grid-connected PV systems increased by 50 per cent. By the end of 2020 total solar-installed capacity had reached 1090 MW. The number of heat pumps has increased sharply since the 1990s. These pumps use renewable energy by transferring heat, mainly from the ground, which reduces demand on electricity from the grid.

Strategies to manage Sweden's energy supplies

Carbon taxation has proved effective in energy change and efficiency. Sweden's carbon tax dates from 1991 and is levied on all fossil fuels in proportion to their carbon content. Pricing carbon is a way of applying the principle of 'the polluter pays'. This helps to reduce emissions in the most cost-effective way and encourages the use of new, clean technology.

Sweden operates an electricity certificate system to support and encourage renewable electricity. Electricity retailers must buy a proportion of 'green electricity' as part of their normal supply. Electricity producers receive

The strategies and techniques used to increase energy supplies

certification for the renewable energy they generate.

Around 500 local 'district heating' systems use excess heat to warm a large number of Swedish homes. These schemes date from the late 1940s when a power station's excess heat was first used to heat nearby buildings. Steam is forced along a network of pipes to reach people's homes.

Sweden has a high level of energy security. Apart from the factors already covered, it has electricity connections with a number of neighbouring countries, including Norway, Poland and Germany. Natural gas can be imported by pipeline from Denmark.

As a member of the EU, Sweden's energy policies stem from the broad objectives set by this multinational organisation. The long-term aims of Sweden's energy policy are:

- a transition to 100 per cent renewable electricity production by 2040
- a zero-carbon economy by 2045.

Sweden plans to achieve 50 per cent more efficient energy use by 2030. Compared to the 1980s, Sweden has 25 per cent more people and GDP has doubled, but it is now using less energy and electricity than it was then. Sweden also plans to be at the forefront of energy storage because of its large variations in seasonal demand for energy.

> **Activities**
> 1 Draw a graph to show Sweden's energy mix.
> 2 Why are hydropower and bioenergy Sweden's top two renewable energy resources?
> 3 Describe Sweden's carbon taxation system.
> 4 What are the long-term aims of Sweden's energy policy?

Practice questions

1 Figure 10.40 shows the daily supply of calories in relation to GNP.
   a i State the approximate calorie intake for the USA and for Niger. [2]
     ii Describe the relationship between calorie intake and GDP per capita as shown on the graph. [3]
   b i Explain one human factor and one physical factor that negatively affects food supply. [4]
     ii Explain two ways in which food production can be increased. [4]

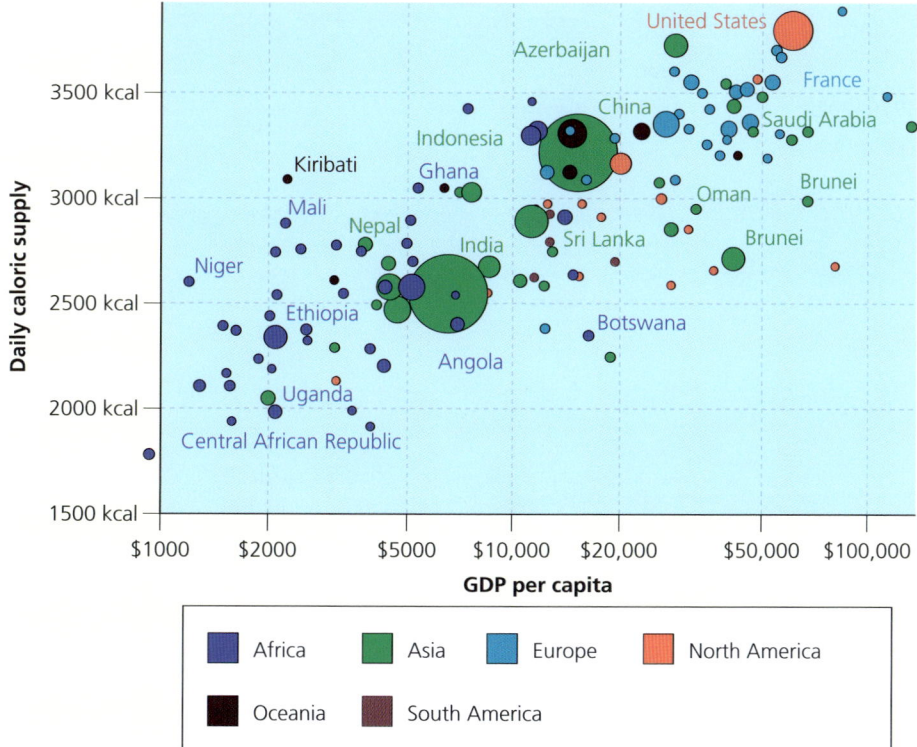

▲ Figure 10.40 The daily supply of calories in relation to GNP

10.6 THE IMPACTS OF ENERGY PRODUCTION

2 Figure 10.41 shows the world's different fuel sources.
   a Calculate the change in the size of the world's fuel sources between 1970 and 2030 (predicted). [3]
   b Identify the fuel source that is predicted to have increased most in absolute size, and the fuel source that is predicted to have increased most in relative size between 1970 and 2030. [2]
   c Assess the advantages and disadvantages of different types of energy sources. [7]

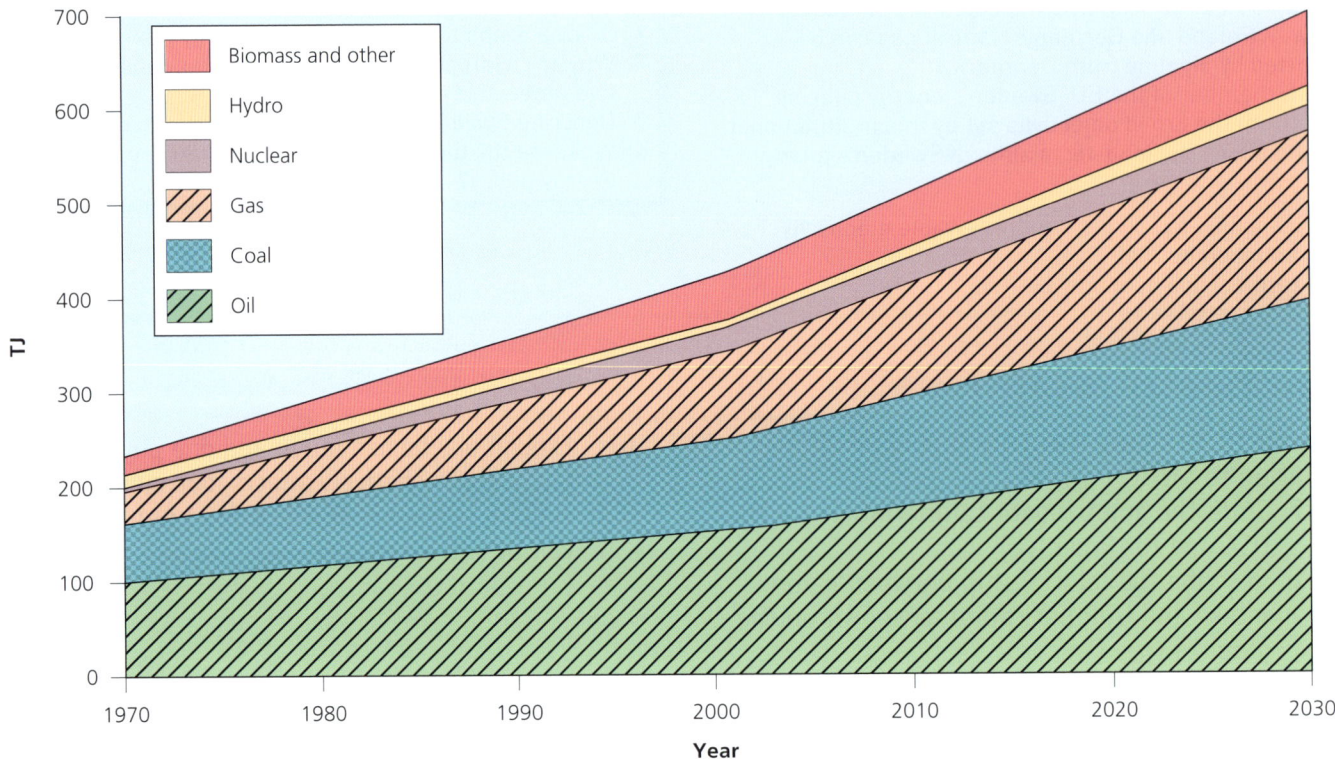

▲ **Figure 10.41** The world's fuel sources, 1970–2023

Geographical skills

This chapter will look at a range of skills that geographers should be able to use. These include:

★ mapwork (cartographic) skills, such as grid and square references, distance, direction and land use
★ graphical skills, such as scatter graphs, pie charts and triangular graphs
★ mathematical skills, such as means, ranges and inverse proportions
★ fieldwork skills, such as sampling, pilot studies and questionnaires.
★ GIS and image skills

Cartographic skills

Atlas maps

Atlas maps come in a variety of **scales**. For example:

» A double-page map of southern Britain may be shown at a scale of 1:1,200,000, where 1 cm represents 12 km.
» A full-page map of the UK may be at a scale of 1:4,000,000. Here, 1 cm is equivalent to 40 km.
» A map of Europe over two pages might be at a scale of 1:16,000,000, with 1 cm representing 160 km.
» A map of the entire world set over a double page could be at a scale of 1:77,500,000, so that 1 cm equals 775 km.

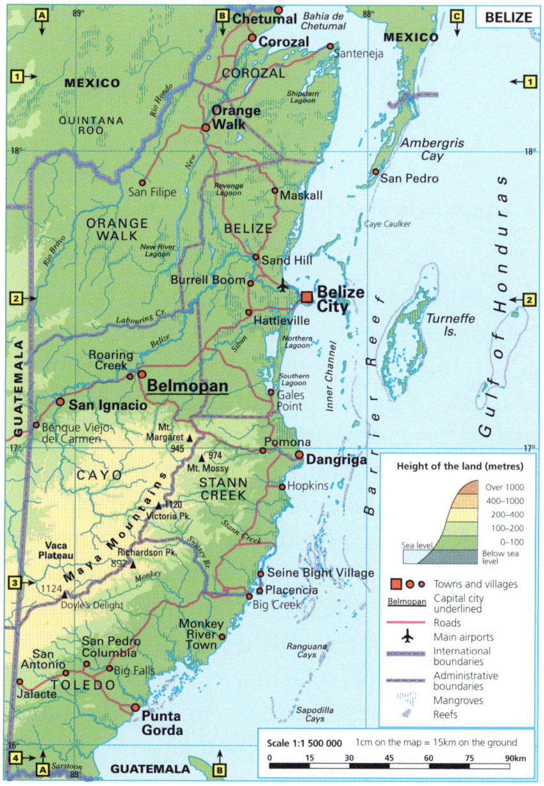

These examples were all taken from a standard school atlas. The scale of an atlas map is generally shown in the lower left-hand or right-hand corner of the page, but this can vary. It is customary for the scale to be given numerically and in a linear format. Figure 11.1 shows an atlas map of Belize.

An Ordnance Survey map extract of part of Wales in the UK, focusing on the city of St Davids, is shown in Figure 11.2.

▲ **Figure 11.2** A 1:50,000 map extract of St Davids, Wales
© Crown copyright and database rights 2025 Ordnance Survey (100036470). For Educational Use only

▲ **Figure 11.1** An atlas map of Belize

Latitude refers to the angular distance of a location north or south of the equator. In contrast, **longitude** refers to the angular location east or west of Greenwich, UK. On Figure 11.1 the latitudes 16°N, 17°N and 18°N are identified, with the longitudes 88°W and 89°W shown at the top of the map.

GEOGRAPHICAL SKILLS

Lines of latitude and longitude are divided into degrees, minutes and seconds. Each degree can be divided into 60 minutes, and each minute into 60 seconds. In most cases only degrees and minutes are used, for example Kingston, Jamaica has a latitude of 18° 0 minutes and London has a latitude of 51° 30 minutes.

> ### Activities
> 1 Look at Figure 11.1 (page 291).
>   a State the direction of Belize City from Belmopan.
>   b Approximately how far is Belmopan from Belize City?
>   c State the altitude of the Maya Mountains.
> 2 Look at Figure 11.2 (page 291). Estimate the size of the city of St Davids. (All grid squares on Ordnance Survey maps are 1 km by 1 km.)

Base maps

A **base map** at the simplest level may be merely an outline of a geographical area, showing just enough information to allow the reader to recognise the area concerned (Figure 11.3). Because of this, base maps are sometimes referred to as outline maps. In this course, you might have been given outline maps of the world, the UK and other countries and asked to add information to these maps. In investigative work, you might have used base maps that cover a relatively small geographical area, such as a stretch of coastline, a section of a river valley or the central business district of an urban area.

For example, a base map of a coastal area would show the coastline and a few other features to show the limits of the map in each direction. This allows you to be clear about the extent of the geographical area covered by the base map. Information can then be added on a range of other features, such as sand dunes, surface drainage, communications and settlements (Figure 11.4). Alternatively, a base map may come from a commercial source such as Multimap. Here, more information is presented, but there is still scope for annotation. Thus, you can either draw your own base map or search the internet for the most appropriate base map for the task in hand.

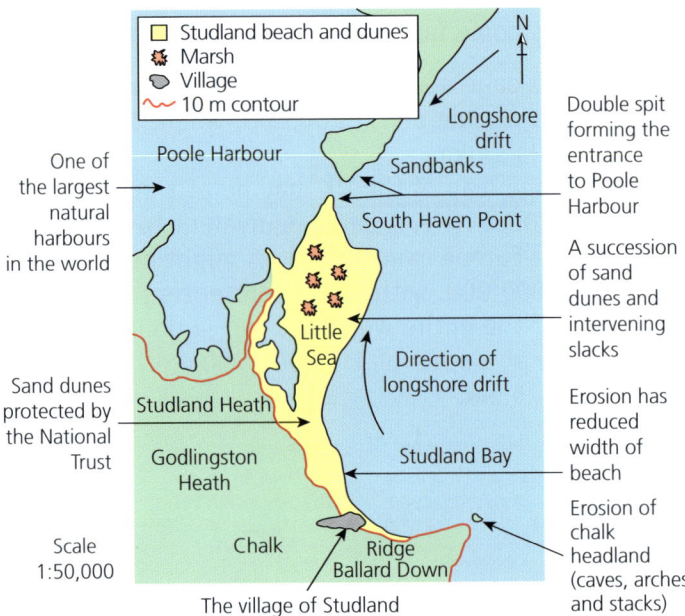

▲ Figure 11.4 A complete sketch map of Studland Beach and dunes

Photocopied base maps or maps downloaded from the internet need careful adaptation to give them your personal 'stamp'. It is likely that you will want to insert and label a number of features on a base map. **Annotation** takes this process a significant step further by the use of short sentences to add description and maybe some very brief explanation applied to a particular feature or features on a map, diagram or photograph. The objective is to produce an effective piece of geographical communication. The box below is an example of the increasing sequence of sophistication relating to sand dune succession. However, don't forget to include the basic requirements such as a title, scale and key.

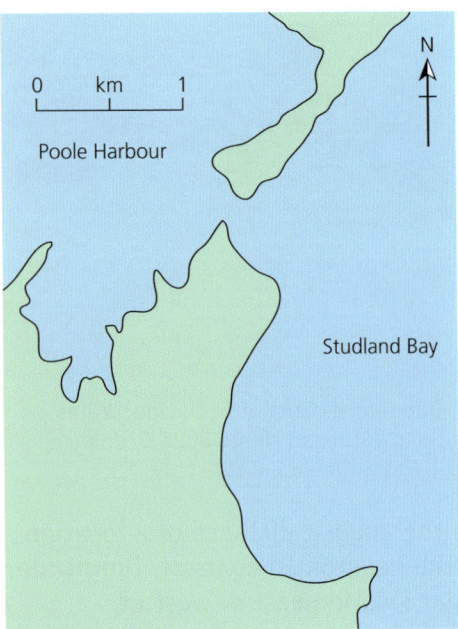

▲ Figure 11.3 A base map of Studland Beach and Bay

Cartographic skills

> **The sequence of annotation**
>
> **Label:** Sand dunes
>
> **Descriptive annotation:** A succession of sand dunes
>
> **Descriptive and explanatory annotation:** A succession of sand dunes with newer dunes formed by wind transport and deposition by the coastline.

In this example, the difference between the label and the descriptive annotation is only a matter of three words, but it adds important information. Likewise, the difference between the descriptive and explanatory annotations is the introduction of the two major processes of wind transport and deposition. These are important elements of explanation.

Every illustration you use should have a clear purpose. It should enhance the text and may also have the benefit of freeing up words if a word limit has been placed on the project or investigation you are carrying out. Refer clearly to illustrations in the text by giving them figure numbers. If it is the first illustration to be used, it would be 'Figure 1'. You should briefly comment on what the illustration shows. Base maps often provide the foundation for many or all of the other illustrations that are to follow. Base maps can be used with photographs. For example, a base map showing the course of a river might include small photographs of a waterfall and a meander.

> **▶ Activities**
>
> 1. a Draw a base map of your school. Show the different buildings, along with the different outdoor areas on the school site.
>   b Provide a label (name) for each building and outdoor area.
>   c For any four areas on the school site, add an annotation to provide extra information.
> 2. Look at the box above. Produce another sequence of annotation, beginning with the label 'Vegetation on the dunes'.

The features of maps

Scale

Most Ordnance Survey maps that we use are either at a 1:50,000 or a 1:25,000 scale. On a 1:50,000 map, 1 cm on the map relates to 50,000 cm on the ground. On a 1:25,000 map, every 1 cm on the map relates to 25,000 cm on the ground. In every kilometre there are 100,000 centimetres (1000 × 100 cm). This means that:

- on a 1:50,000 map, every 2 cm corresponds to a kilometre
- on a 1:25,000 map, every 4 cm corresponds to a kilometre.

A 1:25,000 map is more detailed than a 1:50,000 map and is therefore an excellent source for geographical enquiries. By contrast, a 1:50,000 map provides a more general overview of a larger area. You may also come across other scales, for example 1:10,000 and 1:2500.

Estimating area

To estimate area, count the number of squares. Each square on a 1:50,000 and 1:25,000 map is 1 km². To be more accurate, estimate the percentage of each square that makes up the feature you are measuring, for example the lake in Figure 11.29 on page 307.

Grid and square references

Measurement on maps is made easier by grid lines. These are the regular vertical and horizontal lines you can see on an Ordnance Survey map. The vertical lines are called eastings and the horizontal lines are called northings. They help pinpoint the exact location of features on a map.

Grid references are the **six-figure references** that locate precise positions on a map.

- The first three figures are the eastings and these tell us how far a position is across the map.
- The last three figures are the northings and these tell us how far up the map a position is.

An easy way to remember which way round the numbers go is by using the phrase 'along the corridor and up the stairs'. In Figure 11.5, the church at Rose Hill is located at 691045 and the church at Davis Town is found at 737043.

Sometimes a feature covers an area rather than a point, for example the villages and areas of woodland in Figure 11.5. A grid reference for these would be inappropriate, so we use **four-figure square references**.

- The first two numbers refer to the eastings.
- The last two numbers refer to the northings.

GEOGRAPHICAL SKILLS

The point where the two grid lines meet is at the bottom left-hand corner of the square. So in Figure 11.5, most of the village of Seafield is found in 7504. Some features may occur in two or more squares, for example Long Bay is found both in square 7006 and in 7106.

Direction

Directions can be expressed in two ways:

» Using compass points, for example southwest. Sixteen compass points are commonly used. Some of these are shown in Figure 11.6.

▲ Figure 11.5 A section of a 1:50,000 map of Jamaica

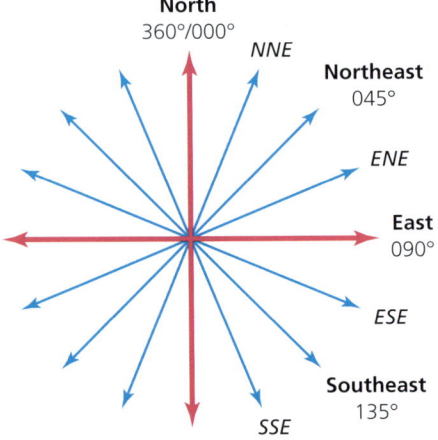

▲ Figure 11.6 Compass points

» Using compass bearings or angular directions, for example 45°. Compass bearings are more accurate than compass points but they can be quite confusing.
- Compass bearings show variations from magnetic north.
- This is slightly different from the grid north on an Ordnance Survey map – this is the way in which the northings go.
- True north is different again – this is the direction of the North Pole.

294

Gradient

The gradient of a slope is its steepness. We can get a rough idea of the gradient by looking at the contour pattern. If the contour lines are close together, the slope is steep. If they are far apart, the land is quite flat. However, these are not very accurate descriptions. To measure gradient accurately, we need two measurements:

» The vertical difference between two points – this can be worked out using the contour lines or spot heights.
» The horizontal distance between two places – this may or may not be a straight line (for example, a meandering stream would not be straight).

Make sure that you use the same units for both the vertical and horizontal measurements.

A spot height is a dot on the map with a number next to it which indicates the height, for example on Figure 11.29 on page 307 there is a spot height at 463153 by Oxford Airport (79m).

Map symbols and map key

Symbols can be used to add a wide range of information to a map. A key should be included to allow the user to identify the symbols that have been used.

Working out a gradient

Divide the difference in horizontal distance (D) by the height (H):

$$\frac{\text{Difference in horizontal distance (D)}}{\text{Height (H)}}$$

If the answer is, for example, 10 or 5, express it as 1:10 (one in ten) or 1:5 (one in five). This means that for every 10 metres along, you rise (or drop) 1 metre; or for every 5 metres in length, the land rises (or drops) 1 metre.

Alternatively, to express the gradient as a percentage, divide the height (H) by the difference in horizontal distance (D) and multiply by 100 per cent:

$$\frac{\text{Height (H)}}{\text{Difference in horizontal distance (D)}} \times 100\%$$

▲ **Figure 11.7** The cable car from Lago di Fedaia to Marmolada Glacier

Activities

Study the OS map shown in Figure 11.5 (page 294).

1. Calculate the distance from the school in Goodwill to the school in the middle of Dundee:
   a in a straight line
   b by road.
2. Estimate the length of the coastline shown on the map extract (to the nearest kilometre).
3. Estimate the length of the airstrip.
4. Calculate the width of:
   a the coral in Long Bay
   b the mangrove forest between Minto and Salt Marsh.
5. State the six-figure grid reference for:
   a the two schools at Dundee
   b Greenwood Great House.
6. Identify the feature found at grid reference 705023.
7. State the four-figure grid reference for:
   a Chatham
   b Davis Town.
8. Suggest reasons why there is an airstrip in grid square 6905.
9. State the direction of:
   a Long Bay from Davis Town
   b Goodwill from Rose Hill.
10. Make a copy of Figure 11.6 (page 294) and complete the missing compass points.

GEOGRAPHICAL SKILLS

▲ **Figure 11.8** A section of a 1:25,000 map of northern Montserrat in the Caribbean

Study Figure 11.8 (page 296) to answer questions 11–15.

11. State the height of:
    a Silver Hill (8658)
    b Baker Hill (8455).
    (Note that the contours on this map are drawn at 50-feet intervals; assume that 3 feet equals 1 m.)
12. State the direction in which Little Bay (8457) faces.
13. Estimate how steep the slope is between Silver Hill and the coastline at Thatch Valley (8659). Measure from the peak of Silver Hill to the nearest point of the coast in Little Redonda. Express your answer as a 'one in x slope'.
14. Describe the relief (height and gradient) of squares 8658 (Silver Hill), 8457 (Potato Hill) and 8655 (Judy Piece).
15. Following an eruption of the Soufrière volcano in 1997, much of the southern third of the island was evacuated. Plans were made to develop the northern part of Montserrat. Study the map in Figure 11.8.
    a Suggest the problems of trying to develop the northern part of the island.
    b Suggest, and justify, the best location to develop housing, services and economic activity.

Sketch maps and annotated photographs

You can label a photograph or diagram to make it very informative. It is important that you clearly label all the important features.

Many photographs that you will encounter in your studies are aerial views, which show industrial, residential, recreational and commercial land uses. In your projects, however, you are likely to use much simpler photos. If you study these carefully, you can find a number of interesting features.

Isoline diagrams

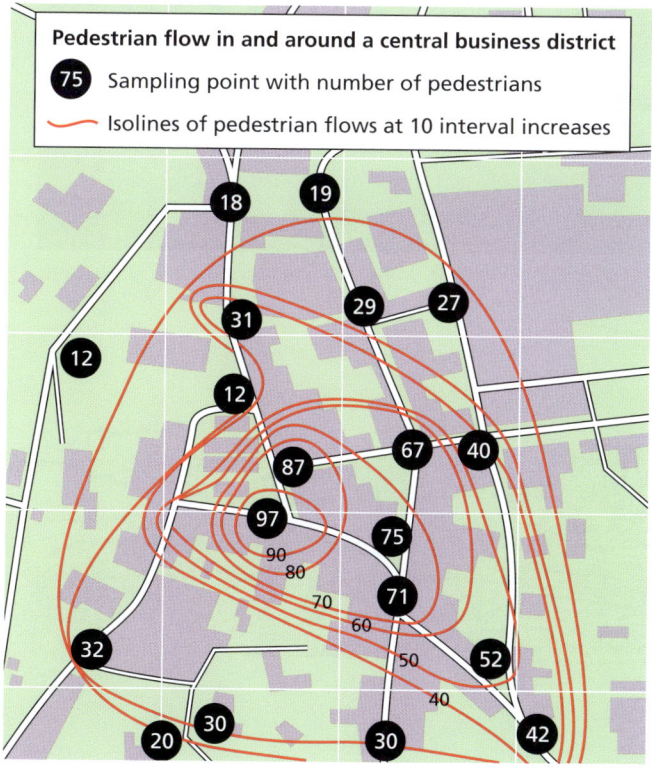

▲ **Figure 11.9** An example of an isoline map

Isoline diagrams join points of equal value on a map. They are similar to contours on an Ordnance Survey map. Isolines can only be drawn when the values under consideration change in a fairly gradual way over the area of the map. Data for quite a large number of locations are required in order to draw a good isoline map, as these maps are unsuitable for patchy data. Figure 11.9 is an isoline map showing pedestrian flow in and around a central business district.

Choropleth maps

Choropleth maps can use variations in colour or different densities of black and white shading. The following steps should be followed in the construction of a choropleth map:

» Look at the range of data and divide it into classes. There should be no fewer than four classes and no more than eight.
» Allocate a colour to each class. The convention is that shading gets darker as values increase.
» Now apply each colour to the applicable areas of the map.
» Provide a key, scale and north point.

The choropleth map is a popular technique, frequently used in atlases, textbooks and many other types of publication. It can convey a lot of information in a straightforward and visually appealing way.

The main disadvantage of the choropleth is that it can appear to show abrupt changes at boundary lines, when in reality the change is much more gradual. It also gives the impression of uniformity within individual areas on the map, when in reality a reasonable degree of variation may be present. Careful selection of class sizes can reduce this problem.

GEOGRAPHICAL SKILLS

Distribution maps
Figure 3.10 on page 76 shows an example of a distribution map on the distribution of tropical rainforests.

Route maps
See Figure 11.29 on page 307 for an example of a route map.

Sphere of influence maps
Figure 11.10 is a sphere of influence map showing the distances people will travel for different services.

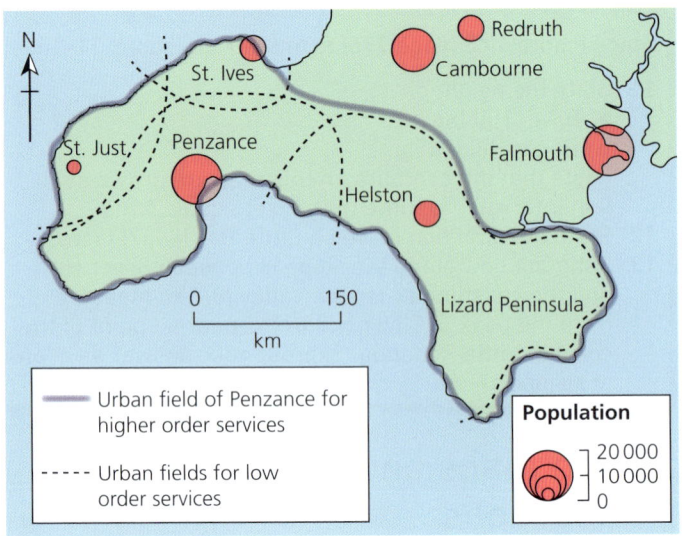

▲ **Figure 11.10** An example of a sphere of influence map

> ### Activities
> 1 Study Figure 11.11.

▲ **Figure 11.11** An aerial view of the Hyundai shipyard in Busan, South Korea

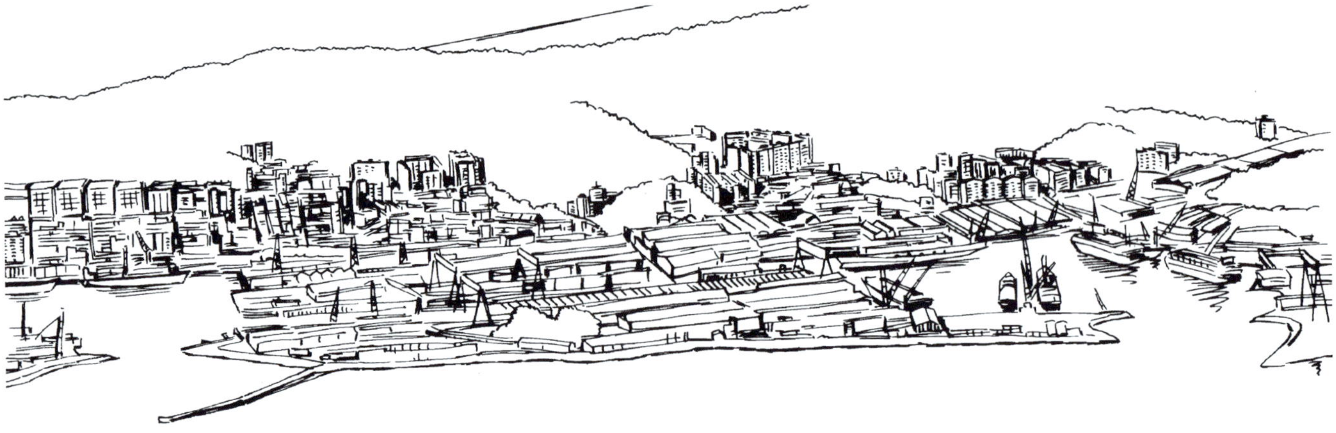

▲ **Figure 11.12** A sketch drawing of Figure 11.11

a Make a copy of Figure 11.12 and add the following labels in the most appropriate locations:
   • Harbour wall to reduce wave energy
   • Flat land for large-scale industrial development
   • Lack of development on steep ground
   • Deep estuary allows development of port industries
   • High-rise residential accommodation
   • Large docks for ships to be repaired or built
b Add two other labels based on your own observations.

2 Study Figure 11.13.

   Make a sketch diagram of the resort and add the following labels:
   • Purpose-built holiday resort
   • Easy access to the beach
   • Boat moorings
   • Fine, white sandy beach
   • Bay
   • Lagoon

▲ **Figure 11.13** Jolly Harbour resort in Antigua

3 Look at Figure 11.9 (page 297), which shows an isoline map. Suggest two more examples for which isoline maps could be drawn.

Graphical skills

You may come across the techniques demonstrated below in all parts of your Geography course. Guidance is given here on how to construct graphs for coursework and the Geographical investigations, as well as how to interpret them.

Pictograms

Pictograms (or picture graphs) use pictures or symbols to represent the data. The number of times a symbol occurs can represent the value or amount – in this way the pictogram acts very much like a bar chart (Figure 11.15, page 300).

Line graphs

A line graph shows points plotted on a graph, with the points connected to form a line. This type of graph is used to show continuing data. It shows the relationship between two variables. Many line graphs show changes over time, however time does not have to be one of the variables of a line graph. Examples of the use of line graphs include:

» temperature change during the course of a day
» pedestrian counts by time of day
» temperature change with altitude.

The axes of a line graph should begin at zero and the variable for each axis should be clearly labelled. Be careful with the choice of scale, as this will determine the visual impression given by the graph. On Figure 11.14 only one line has been drawn, but it is valid to show a number of lines so long as the course of each line is absolutely clear from start to finish.

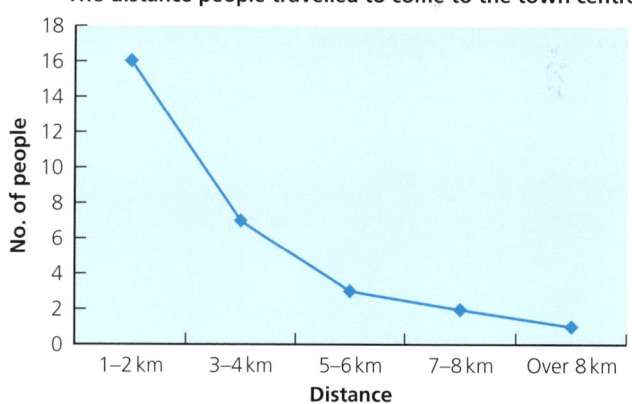

▲ **Figure 11.14** An example of a line graph

Line graphs can be simple (showing one feature) or they may be multiple (showing many features and trends, for example the demographic transition model, Figure 6.11 on page 132). Compound or stacked line graphs show the value of each category on top of the previous ones (for example the global consumption of primary energy by source, Figure 10.28 on page 273). This is done in this way to show the value of total energy consumption, which would not be clear from a multiple line graph.

GEOGRAPHICAL SKILLS

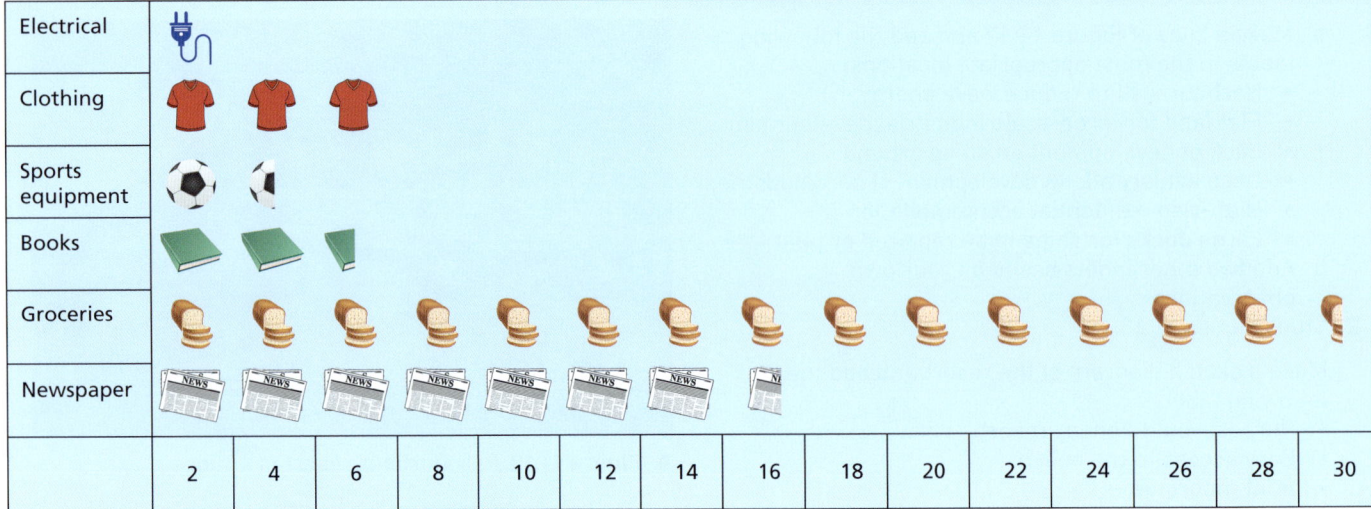

▲ Figure 11.15 What goods did you buy today?

Activities

1 Study Figure 11.15. How many people bought:
   a groceries
   b clothing
   on their visit?
2 Using the data in Table 11.1, draw a line graph to show world population growth, 1800–2024.

▼ Table 11.1 World population growth, 1800–2024

Year	Population (billions)
1800	1
1930	2
1960	3
1974	4
1987	5
1999	6
2011	7
2024	8

Pie charts

Pie charts are subdivided circles. These are frequently used on maps to show variations in the composition of a geographical feature. For example, in Figure 11.16 they are proportional in size to the number and proportion of people employed in primary, secondary and tertiary industries. The pie chart may also be drawn proportional in size to show an extra dimension, in this case the size of GRDP.

Plotting a pie chart

The following steps should be followed in the construction of a pie chart:

1 Convert the data into percentages.
2 Convert the percentages into degrees (by multiplying the percentage by 3.6 and rounding up or down to the nearest whole number).
3 Draw appropriately located circles on your base map.
4 Subdivide the circles into sectors using the figures obtained in step 2.
5 Differentiate the sectors by means of different shadings or colours.
6 Draw a key explaining the scheme of shadings and/ or colours.
7 Give your diagram a title.

Bar graphs

In a bar chart, the length of the bar represents the quantity of the component being measured, for example places or time intervals. The vertical axis has a scale that measures the quantity. There are four main types of bar chart:

» Simple bar chart: Each bar indicates a single factor.
» Multiple or group bar chart: Features are grouped together on one graph to help comparison, for example Figure 6.26 on page 148.

Graphical skills

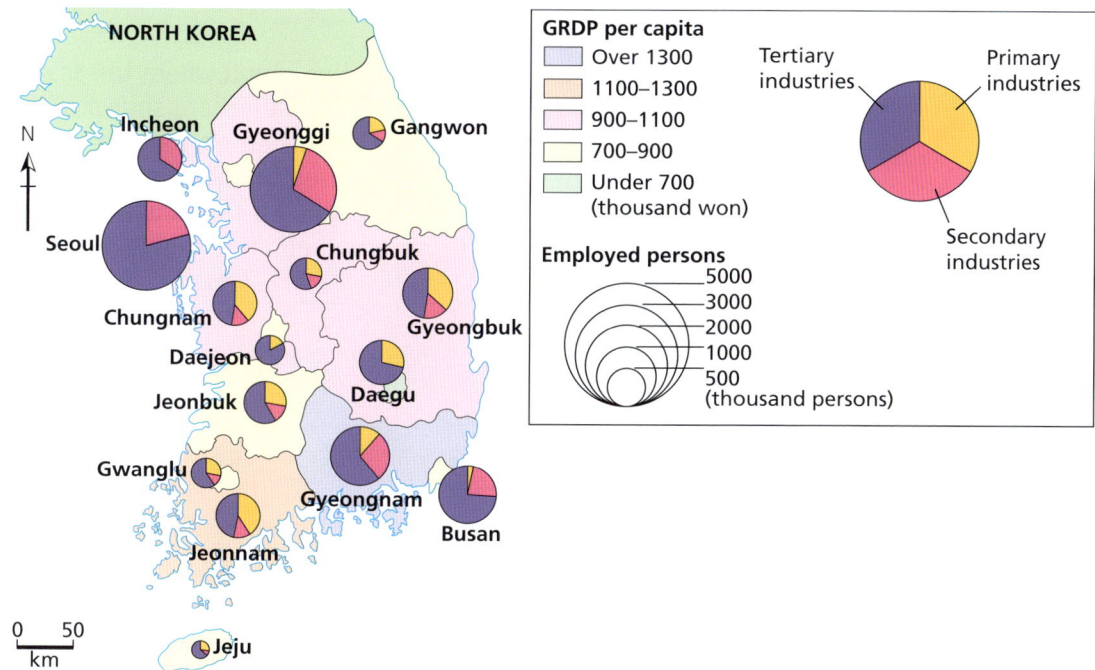

▲ Figure 11.16 An example of proportional pie charts

» **Compound bar chart:** Various elements or factors are grouped together on one bar. The largest or most stable element or factor (the one with least variation) is placed at the bottom of the bar to avoid confusion, for example Figure 10.28 on page 273.
» **Percentage compound bar chart:** This is a variation on the compound bar chart. It is used to compare features by showing the percentage contribution, for example Figure 10.30 on page 275. These graphs do not give a total in each category but compare relative changes in terms of percentages.

Median-line bar graphs

A median-line bar graph is useful when the objective is to show both positive and negative changes. The median line is set at zero, with the positive scale above the median line and the negative scale below it (Figure 11.17). This type of bar graph can create a very good visual impression. You can see instantly whether changes are positive or negative and exactly what the extent of the individual changes is.

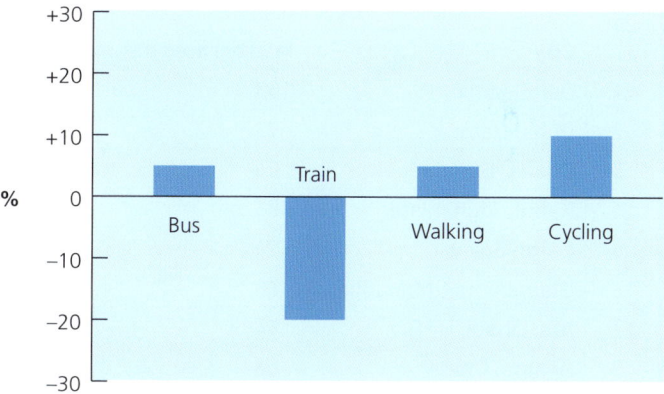

▲ Figure 11.17 An example of a median-line bar graph

Histograms

A histogram is a special type of bar graph (Figure 11.18, page 302) that shows the frequency distribution of data. The x-axis must be a continuous scale, with the values marked on it representing the lower and upper limits of the classes within which the data have been grouped. The y-axis shows the frequency with which the data fall into each of the classes. A vertical rectangle or bar represents each class. The bars must be continuous, without any gaps between them.

GEOGRAPHICAL SKILLS

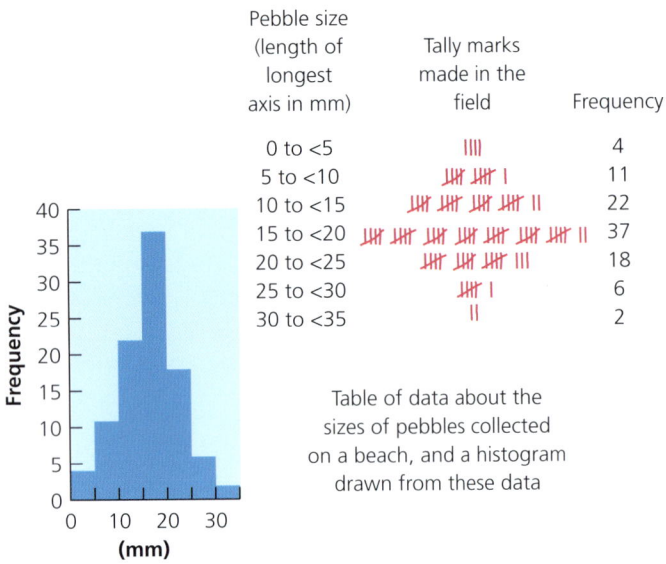

▲ Figure 11.18 An example of a histogram and the data used

Kite graphs

A kite graph is a form of line graph where the scale is split in two, so that half the values are shown on one side of the line and the other half on the other side. They are most commonly used to show vegetation distribution, for example along a sand dune or across a footpath and its surrounding area (Figure 11.19).

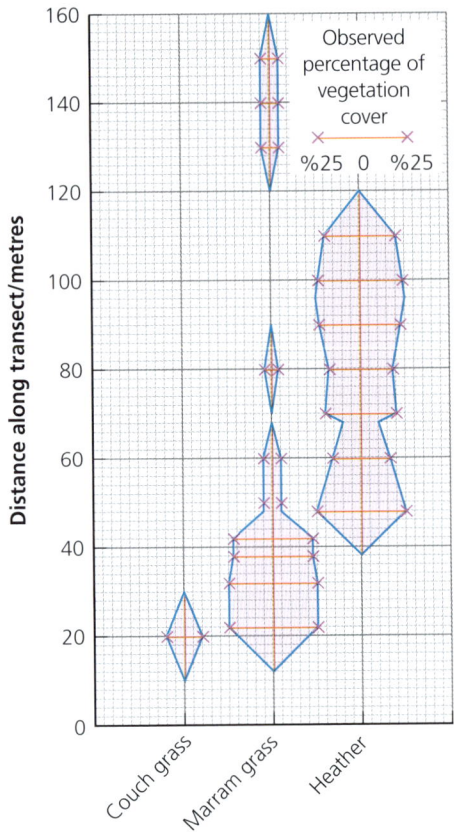

▲ Figure 11.19 An example of a kite diagram

Activities

1. Table 11.2 shows data for air pollution in ten cities. Construct a bar graph to show these data. Remember to add a scale and a title.

▼ Table 11.2 Air pollution in ten cities

	City	Micrograms per m³
1	Delhi, India	153
2	Doha, Qatar	93
3	Dhaka, Bangladesh	86
4	Kabul, Afghanistan	86
5	Cairo, Egypt	74
6	Ulaanbaatar, Mongolia	68
7	Abu Dhabi, UAE	64
8	Beijing, China	56
9	Kathmandu, Nepal	50
10	Accra, Ghana	49

2. Table 11.3 shows the contribution of the main economic sectors to Korea's GDP.
   a. Using the data in Table 11.3, draw a pie chart showing the contribution of each economic sector to Korean GDP.
   b. Comment on the chart you have drawn.

▼ Table 11.3 The contribution of the main economic sectors to Korea's GDP

Agriculture	Industry	Services
2.6%	39.2%	58.2%

Flow-line diagrams and maps

Flow-line diagrams and maps are used to illustrate movements or flows (Figure 11.20). One might be used to show the variation in volumes of traffic from different smaller settlements into a larger settlement. Straight lines are used, but the width of each individual flow line is proportional to the amount of traffic it represents. Thus, a line 10 mm wide may represent 500 vehicles an hour along a road. On the same scale, a line 2 mm wide would represent 100 vehicles an hour. Flow lines could also be used to show the number of buses coming into a town on a particular day.

Graphical skills

▲ Figure 11.20 A flow-line diagram showing the volume of water in the three main rivers that form the River Nile

Dispersion graphs

A dispersion graph is very useful for showing the range of a data set and its tendency to group or disperse, and for comparing two sets of data. It involves plotting the values of a single variable on a vertical axis. The horizontal axis shows the frequency. The resulting diagram shows the frequency distribution of a data set (Figure 11.21).

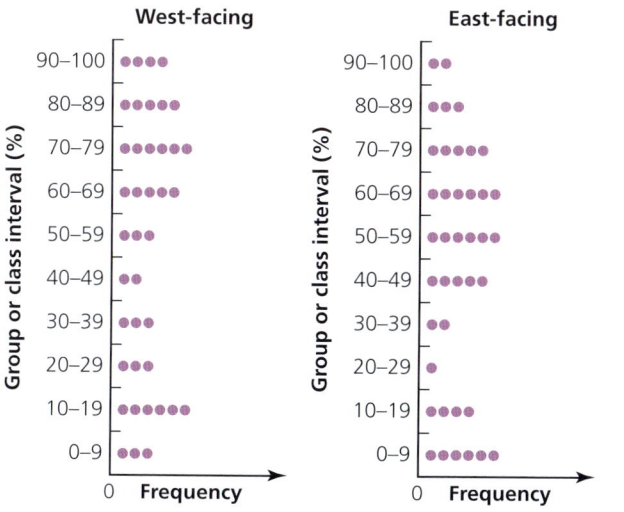

▲ Figure 11.21 An example of a dispersion diagram showing the percentage of lichen found on west- and east-facing gravestones

Ray diagrams

There are two main types of ray diagram: **wind roses** and **desire lines**. Ray diagrams are made up of straight lines (rays), which show a connection or movement between two places.

Figure 11.22 shows a wind rose diagram depicting the variations in wind direction for a certain time period. The direction of each ray relative to the centre is the direction from which the wind is blowing. Each ray is proportional in length to the number of days the wind blew from that direction.

Desire line diagrams show movement from one place to another. This type of diagram could be used to show where people live and the supermarket they use. If there are four supermarkets in an area then the rays would focus on four points rather than just one, as in a wind rose diagram. Desire line diagrams are therefore more complicated than wind rose diagrams.

Wind directions recorded for one year at a school weather station in Liverpool, UK

Direction of wind	N	NE	E	SE	S	SW	W	NW	Calm
Number of days per year	26	37	39	32	30	57	60	53	31

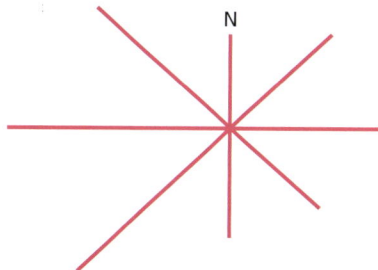

▲ Figure 11.22 An example of a wind rose diagram

Radial (circular) graphs

Radial (or circular) graphs (Figure 11.23, page 304) can be used to plot:

» a variable that is continuous over time, such as temperature data over the course of a year
» data relating to direction, using the points of the compass.

The two axes of a radial graph are the circumference of the circle and the radius. Values increase from the centre of the circle outwards.

GEOGRAPHICAL SKILLS

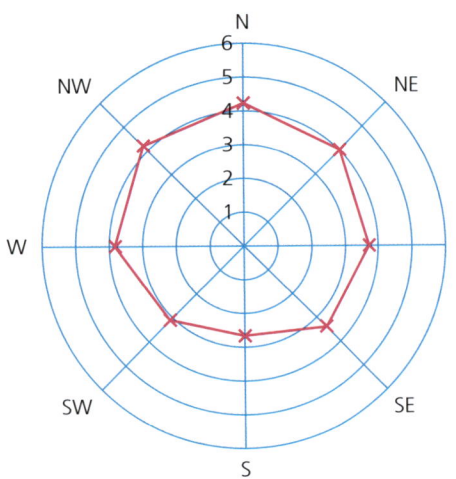

A radial graph to show the influence of aspect on gravestone weathering. The graph shows the mean of Rahn's index for each compass direction.

Rahn's index

Class	Description
1	Unweathered
2	Slightly weathered: faint rounding to corners of letters
3	Moderately weathered: gravestone rough, letters legible
4	Badly weathered: letters difficult to read
5	Very badly weathered: letters indistinguishable
6	Extremely weathered: no letters left, scaling

▲ Figure 11.23 An example of a radial graph

Scatter graphs

Scatter graphs show how two sets of data are related to each other, for example population size and number of services, or distance from the source of a river and average pebble size. To plot a scatter graph, decide which variable is independent (in these examples, this would be GDI) and which is dependent (total fertility rate). The independent variable is plotted on the horizontal or x-axis and the dependent variable on the vertical or y-axis. For each data point, project a line from the corresponding x- and y-axis, and where the two lines meet, mark the point with a dot or an X (Figure 11.24).

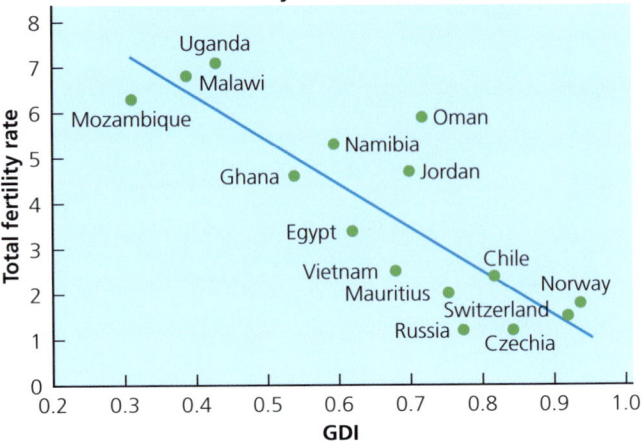

▲ Figure 11.24 An example of a scatter graph

When all the data are plotted, a **line of best fit** is drawn. This does not have to pass through the origin. It is useful to label some of the points, for example the highest and lowest anomalies (exceptions), especially if these are referred to in any later description.

Triangular graphs

Triangular graphs are used to show data that can be divided into three parts, for example soil (sand, silt and clay), employment (primary, secondary and tertiary), or population (young, adult and elderly) (Figure 11.25).

LDCs	Less developed countries
MDCs	More developed countries
UK	United Kingdom
Fr	France
Sw	Sweden
Jp	Japan
Bo	Bolivia

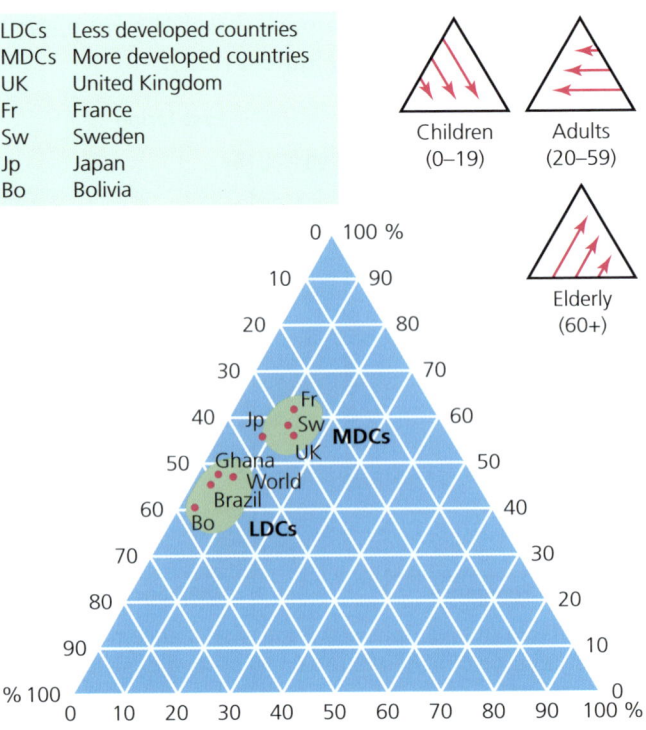

▲ Figure 11.25 An example of a triangular graph

They require that the data are in the form of a percentage and that the percentages total 100 per cent. The main advantages of triangular graphs are:

» they allow a large number of data to be shown on one graph (think how many pie charts or bar charts would be needed to show all the data in Figure 11.25)
» groupings are easily recognisable (in the case of soils, for example, groups of soil texture can be identified)
» dominant characteristics can be easily shown
» classifications can be drawn up.

Triangular graphs can be tricky and it is easy to get confused, especially if care is not taken, but they do provide a fast and reliable way of classifying large amounts of data that have three components.

Doughnut graphs

Another way to present data is by using a doughnut graph. An example is shown in Figure 11.26.

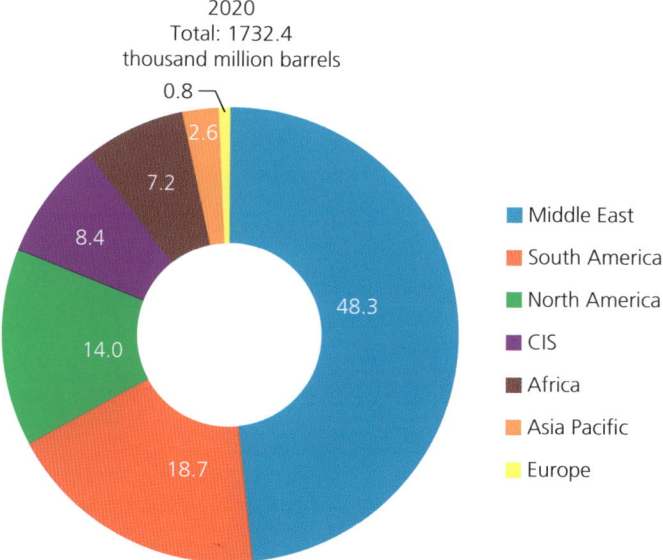

▲ **Figure 11.26** A doughnut graph showing the distribution of proven oil reserves by region

Proportional symbols

For examples of proportional symbols, see Figure 11.16 on page 301.

Venn diagrams

For examples of Venn diagrams, see Figure 1.37 on page 21 and Figure 8.15 on page 191.

Flood hydrographs

A flood hydrograph shows how the discharge of a river varies over a short time. See Figure 11.27 for an example.

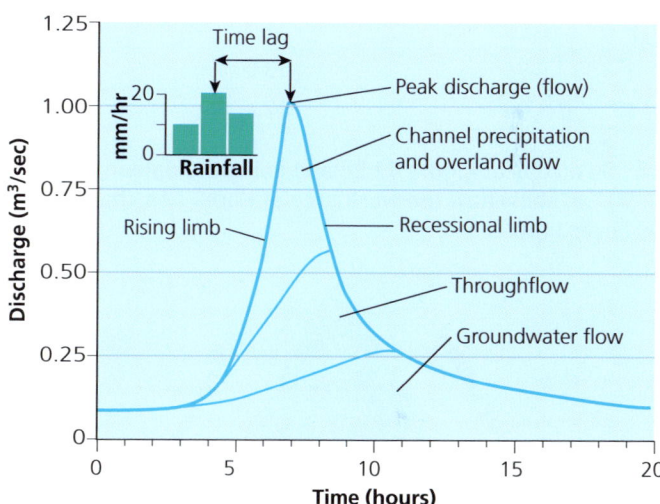

▲ **Figure 11.27** A simple flood hydrograph

GEOGRAPHICAL SKILLS

Activities

1 Construct a scatter graph using the data in Table 11.4.

▼ Table 11.4 Discharge and suspended load for ten sites

Site	Discharge (m³/sec)	Suspended load (g/m³)
1	0.45	10.8
2	0.42	9.7
3	0.51	11.2
4	0.55	11.3
5	0.68	12.5
6	0.75	12.8
7	0.89	13.0
8	0.76	12.7
9	0.96	13.0
10	1.26	17.4

2 On a copy of Figure 11.28 and using the data in Table 11.5, show how the workforce of Korea has changed over time.

▼ Table 11.5 The percentage of the Korean workforce employed in primary, secondary and tertiary industries, 1970–2015

	Primary industries	Secondary industries	Tertiary industries
1970	50.4	14.3	35.3
1980	34.0	22.5	43.5
1990	17.9	27.6	54.5
2000	10.9	20.2	68.9
2015	5.7	24.2	70.2

3 Find two examples of line graphs in a Geography textbook. What do the line graphs show in these examples?
4 What is the difference between a histogram and an 'ordinary' bar graph?
5 Discuss the merits and limitations of choropleth maps.
6 Draw a series of proportional circles using fieldwork data or data from a textbook.

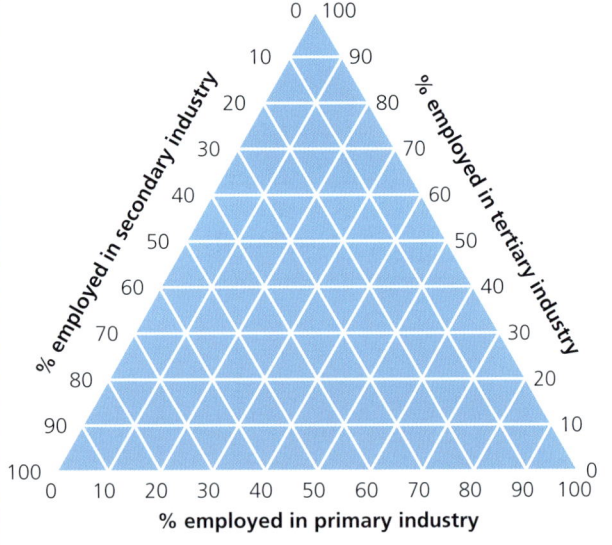

▲ Figure 11.28 An outline for a triangular graph

Mathematical skills

There are a number of mathematical skills that you should be able to perform. These include:

» adding, subtracting, multiplying and dividing
» understanding map scale and the use of scale line and representative fractions
» understanding the terms 'mean', 'mode', 'median' and 'range'
» using averages, decimals, fractions, percentages and ratios
» using standard index notation (standard form), including both positive and negative indices
» understanding significant figures and using them appropriately
» recognising positive and negative relationships shown by scatter graphs
» drawing and interpreting graphs from given data
» selecting suitable scales and axes for graphs
» using a ruler and protractor.

Mathematical skills

Map scale

If the map scale is 1:50,000, this means that every 1 cm on the map represents 50,000 cm on the ground (in reality). The representative fraction is expressed as 1/50,000 – no units are given, as it could be cm, m, km, etc.

Mean, mode, median and range

The **mean** is a type of average. It is found by totalling (Σ) the values for all observations (Σ x), and then dividing by the total number of observations (n):

$$\frac{\Sigma x}{n}$$

For example, the mean number of services in eight villages was found to be:

$$\frac{(5 + 4 + 7 + 8 + 1 + 1 + 2 + 4)}{8} = \frac{32}{8} = 4$$

Activities

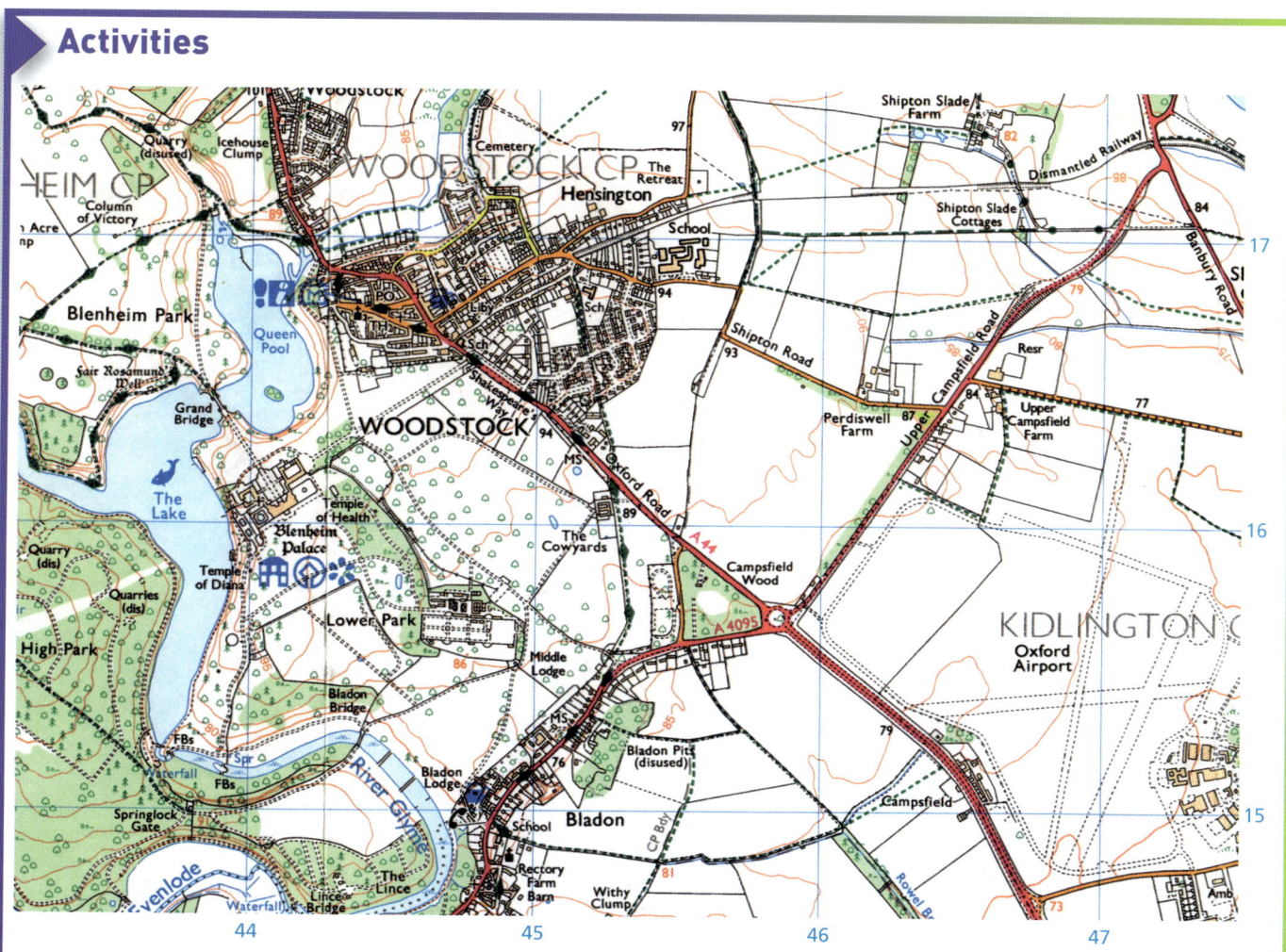

▲ **Figure 11.29** A 1:25,000 map of Woodstock, UK © Crown copyright and database rights 2025 Ordnance Survey (100036470). For Educational Use only

Study Figure 11.29.
1 State the scale of the map.
2 Measure, in centimetres, the distance from the roundabout by Campsfield Wood (458157) to the junction of Upper Campsfield Road and Banbury Road (473173).
3 State the distance from the roundabout by Campsfield Wood to the junction of Upper Campsfield Road and Banbury Road:
   a in metres
   b in kilometres.
4 Estimate the length of the River Glyme from the footbridge at 437152 to Lince Bridge (443146).
5 State the distance of the longest runway at Oxford Airport.

GEOGRAPHICAL SKILLS

There are other types of average. The **mode** refers to the group or value that occurs most often. A pattern with one peak is **unimodal**. In the example above, there are two modes: 1 and 4. Such a pattern, with two peaks (or two modes), is called **bimodal**. A pattern with three or more peaks is called multimodal.

The **median** is the middle value when all the data are placed in either ascending or descending order. In this case, we have:

8, 7, 5, 4, 4, 2, 1, 1

When there are two middle values (as in this example or whenever there is an even number of values) we take the average of the two values, which here is very easy as both middle values are 4, hence the median is 4.

The **range** is the difference between the highest and lowest values. In this example it is:

8 − 1 = 7

Decimals, fractions, percentages and ratios

A **decimal** is a number between two whole numbers. For example, the world's population in 2024 is around 8.2 billion, which means that it is more than 8 billion but less than 9 billion.

A **fraction** is part of a whole – for example, 1/3 (one-third) of Borneo's rainforest has been deforested.

A **percentage** is a number or ratio expressed as a fraction of 100, using the symbol %.

- To convert a fraction to a percentage, first convert the fraction to a decimal and then multiply by 100.
- To convert a fraction to a decimal, divide the number above the line (the numerator) by the number below the line (the denominator). You can use a calculator for this.

Thus, in Borneo 1/3, or 0.33 × 100 = 33%, of the rainforest has been deforested.

A **ratio** is a method of comparing relative sizes or proportions. The deforested area in Borneo compared with the forested area is 1:2 (one-third has been deforested and there are two-thirds left).

> **Activity**
>
> The world's population is approximately 8.2 billion, and China's population is approximately 1.4 billion. Express China's population as a percentage of the world's population.

Standard notation

Standard notation is the number that we would normally write, for example 567. The expanded standard index notation shows that 567 is 5.67×10^2.

A **positive index** is a power value that is positive, for example $2^2 = 2 \times 2 = 4$, or $3^3 = 3 \times 3 \times 3 = 27$.

Negative indices are powers that have a minus sign, for example $2^{-3} = 1/2^3 = 1/8$.

Positive and negative relationships shown by scatter graphs

Relationships between values can be investigated using scatter graphs. Figure 11.30 shows a positive relationship (a), a negative relationship (b), a curved relationship (c), and a positive relationship when all values are considered but no relationship when a subset is used (d).

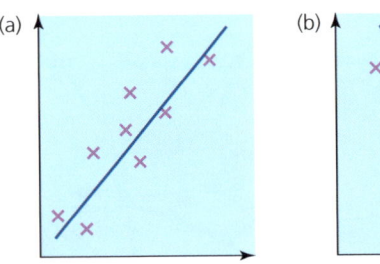
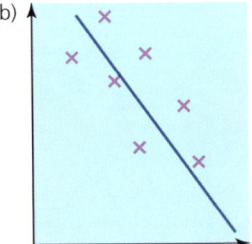
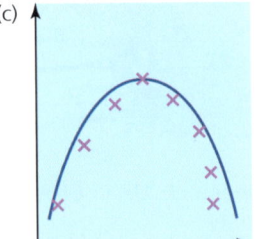
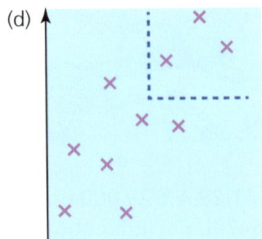

▲ **Figure 11.30** Relationships on a scatter graph

GIS and image skills

Visual images

Geographers use a wide range of visual images to record geographical features as part of fieldwork and research, and to interpret and analyse geographical issues. Photographs are an example of primary data we often collect ourselves, whereas aerial photographs and satellite images are usually secondary data. Figure 11.31 and Table 11.6 compare different types of image.

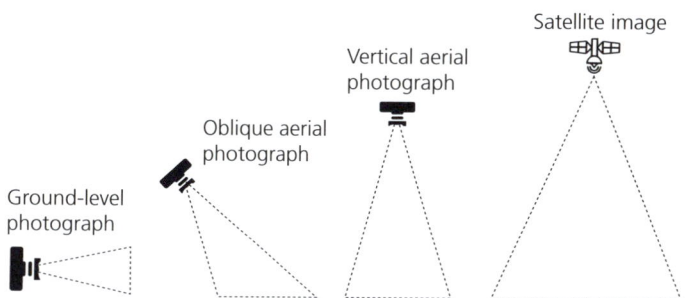

▲ Figure 11.31 Photographs and satellite images

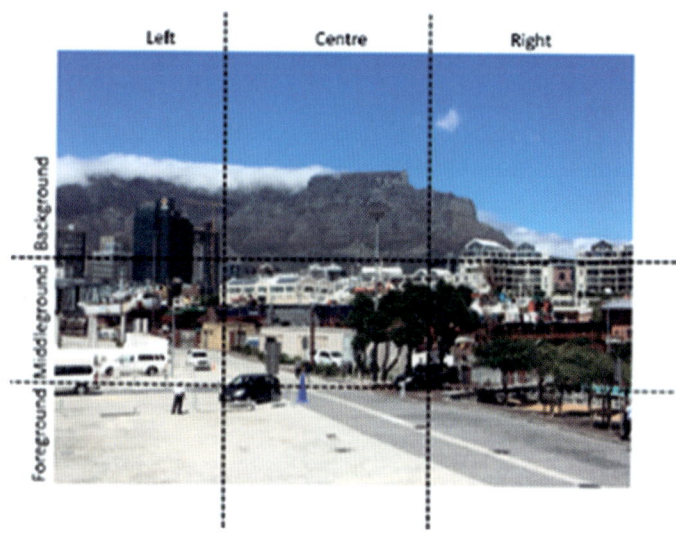

▲ Figure 11.32 An image analysis grid (Cape Town, South Africa)

▼ Table 11.6 Types of photographs

Ground level photographs	Oblique aerial photographs	Vertical aerial photographs	Satellite images
Used to record small-scale features; cheap and easy using digital cameras and smartphones	Used to record large-scale landscapes, using an aircraft or drone	They produce a visual image similar to a map; can show change over time, e.g. deforestation patterns	Taken from space, using visible light or infrared: tracking weather systems or monitoring planet health

Images can be complex, which makes deconstructing and interpreting them challenging. A useful method is to visualise a grid over the image to make analysis easier (Figure 11.32). Specific parts of the image can be referred to and interpreted using the grid.

GIS

GIS stands for Geographic Information System. GIS can be thought of as a very powerful digital map that can be changed and manipulated. This contrasts with a paper map that is static, and outdated from the time it is printed. GIS computer databases contain:

» multiple map layers such as base maps, satellite imagery, vertical aerial photographs and digitised paper maps

» geolocated data such as photographs, data points and measurements.

Data collected during fieldwork can be added to a GIS database (Figure 11.33). GIS databases contain tools that can measure distance, area, altitude and many other features. This makes GIS a tool for analysis. Layers of information can be added and removed from a GIS to analyse patterns, trends and relationships. Both Google Maps and Google Earth are examples of GIS.

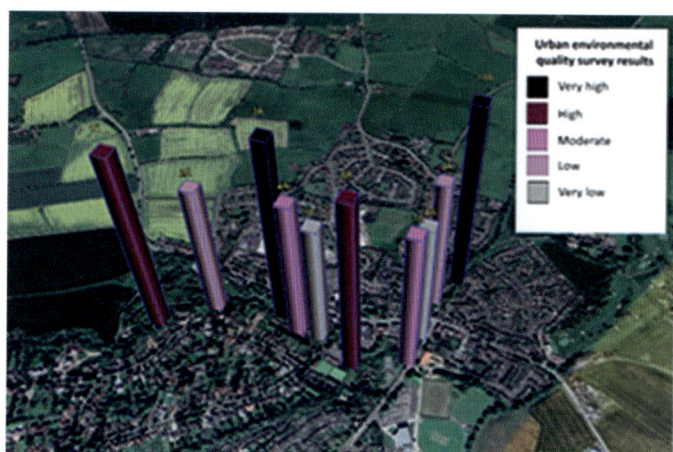

▲ Figure 11.33 Fieldwork data results added to a GIS as coloured bars, showing the pattern of urban environmental quality in Ponteland, Northumberland

GIS has benefits compared to a paper map because it can be constantly updated and added to. It can be used to measure trends in deforestation, ocean warming and sea ice loss due to climate change.

GEOGRAPHICAL SKILLS

GIS is much lower cost than it was 20 years ago, and can even be used on smartphones. Limitations include the need to have some knowledge of ICT to input and analyse data, plus the scale and complexity of some GIS databases.

Cartoons

A cartoon is a diagram that may exaggerate a feature to draw attention to a particular point. What does Figure 11.34 suggest about the increasing similarity of urban areas?

▲ Figure 11.34 An example of a cartoon

Analysing text and sources

Care must be taken to think critically about sources of information used in Geography, be that text, images or data. An example is shown in Table 11.7. This might seem to be an accurate data source but it is not:

1 The data is from 2012, so is very out of date.
2 The source of the data is not shown.
3 Only nine countries have been used, whereas Africa has 54 countries.
4 It is not clear why these nine countries have been chosen.

If all 54 countries are used, average income in 2012 was not $4278 but $2657. This is because the data chosen is biased towards wealthier countries in North Africa. There were 22 African countries with per person incomes under $1000 in 2012, but only one is included.

▼ Table 11.7 Average per capita income in Africa

Income per person, 2012	US$
Angola	5083
Algeria	6096
Egypt	3059
Gabon	9348
Kenya	1289
Malawi	563
Morocco	3164
South Africa	8173
Zambia	1729
Mean income per person	**4278**

Fieldwork skills

Fieldwork is a very important part of Geography because it involves collecting primary data, and using secondary data sources, to investigate a geographical issue or problem. Fieldwork will help you succeed with written assessments and coursework.

Fieldwork enquiry skills, called the route to geographical enquiry, involve a structured, seven stage process to allow you to successfully complete a fieldwork investigation:

1 Identification of issue, question or problem – the hypothesis
2 Objectives of the study are defined
3 Collection of data
4 Selection and collation of data
5 Presentation and recording of the results
6 Analysis and interpretation
7 Making effective conclusions, evaluation and suggestions for further work.

Identification of issue, question or problem – the hypothesis

This section is the introduction to your investigation. You need to identify a topic for investigation through observation, discussion, reading or previous study.

Geographical investigations begin by stating one or more hypothesis used to test the issue, question or problem. Hypotheses are the ideas you intend to test. Before you can set out your hypotheses with confidence you need to ensure that you have a good understanding of the topic (for example, sand dunes) under consideration. Studying the geographical background should ensure that you have clear knowledge and understanding of the theories or models that are used to try to explain your enquiry. You will refer back to these theories and models in your conclusion.

Examples of hypotheses:

- Pedestrian density is highest at the centre of the CBD, and declines with increasing distance from the centre.
- The sphere of influence of settlements increases with settlement size.
- The pH of sand dunes decreases with distance inland.
- Population density is higher in inner urban areas than in the suburbs.
- Sediment size in a river decreases downstream.

For each hypothesis you investigate, you should describe what you expect to find and explain why. Within this section of the investigation you should also justify the geographical location of your inquiry. It should clearly be a good location to address the issue, question or problem you intend to investigate. Make sure that you include the area's site as well as its regional situation. Include clearly labelled location maps. Give each map or diagram used in this section a figure number. For example, a map showing the location of your study area within its wider region might be labelled 'Figure 1: The location of Studland in Dorset'. Follow this procedure for all illustrations used throughout your investigation. Also make sure that you refer to each 'Figure' in your text.

Objectives of the study are defined

Now move on to define the objectives of the study in specific terms, refining the content of the previous section. You will also make decisions concerning:

- what data are relevant to the study
- how the data can be collected.

It is useful in this section to briefly state the sequence of investigation you are going to follow. This should ensure that you are clear about the remaining stages of the investigation and that you tackle the route to geographical enquiry in a logical manner.

> **Activities**
> 1 Draw a diagram to illustrate the route to geographical enquiry.
> 2 What is a hypothesis? Give two examples.
> 3 Why is it important to study the geographical background of your coursework topic?
> 4 How would you go about justifying the geographical location of your enquiry?

Risk assessment

Before you go outside and do fieldwork you must complete a risk assessment. The purpose of this is to identify risks to you and anyone else who is with you and identify ways in which the risks can be reduced. This must happen before you venture outside. Some risks are generic, such as crossing roads and being prepared for bad weather. Others are specific to an environment, such as tide times at the coast or collecting data in a river. Table 11.8 is an example risk assessment.

▼ Table 11.8 An example risk assessment (coastal fieldwork)

Risk	Cause	Consequences	Control measures	Responsibility
Slips, trips and falls	Slippery rocks and paths, uneven surfaces, algae, seaweed	Minor injuries	Suitable footwear	Group leader (planning)
		Major injuries	Avoid dangerous routes /locations	Whole group
Road accidents	Busy roads, dangerous roads and junctions	Minor injuries	Use designated crossings and safe parking	Whole group
		Major injuries		
		Fatalities		
Adverse weather	Cold conditions and rain, strong winds	Exposure	Take bad-weather clothing	Whole group
		Hypothermia		

GEOGRAPHICAL SKILLS

▼ Table 11.8 (Continued)

Tides	Rising tides, getting wet and /or cut-off	Minor injuries	Check tide times	Group leader (planning)
		Major injuries	Stick to timetable	
		Fatalities	Carry mobile phone	
Rockfalls	Cliff collapse / landslides	Minor injuries	Avoid all cliffs and steep slopes	Group leader (planning)
		Major injuries		
		Fatalities		

Collection, selection and collation of data

Data can be collected on a group or individual basis, which may include:

- fieldwork to collect primary data, such as taking measurements and undertaking questionnaires
- gathering data from secondary sources, such as from census information and published maps, newspapers and the internet.

Make sure that you clearly explain the difference between primary data and secondary data. Use as many different techniques as possible to gather information, for example interviews, observations, surveys, questionnaires, maps and looking at figures. Describe and justify each method. Describe the use of primary fieldwork methods and in particular the method and/or equipment used to collect each type of information. Equally, describe and explain the use of secondary sources, for example parish records.

Explain clearly how you decided to use your figures, maps, answers to questions, etc. Some reasoning is necessary here – that is, justify why you used that method or source. For example, explain in detail how you questioned people, collected census figures or obtained maps. Write this up almost like the method for a scientific experiment. You can use a planning sheet here, stating when you collected data, where from, at what time, places you visited, observations you made, interviews you conducted. If you are using a questionnaire then you must justify the questions that you use, for example explain why you have recorded the age and gender of respondents in a shopping survey. Use a range of methods.

To collect data in a sound and logical way so that valid conclusions can be drawn, you should be aware of the characteristics and importance of:

- sampling
- pilot surveys
- questionnaires and interviews
- methods of observing, counting and measuring
- health and safety, and other restrictions.

Sampling

The reasons for sampling

For many geographical investigations it is impossible to obtain 'complete' information. This is usually because it would just take too long in terms of both time and cost. For example, if you wanted to study the shopping habits of all 1000 households in a suburban area by using a doorstep questionnaire, it would be a huge task to visit every household.

However, it is valid to take a 'sample' or proportion of this total 'population' of 1000 households, providing you follow certain rules. The idea is that you are selecting a group that will be representative of the total population.

You might decide to take a 5 per cent or 10 per cent sample, which would involve talking to 50 or 100 of the 1000 households in the area. But how do you decide which 50 or 100 households to sample? There are three recognised methods of sampling that are considered scientifically valid. All three methods avoid bias, which would make results unreliable.

Sampling types

Before selecting the sampling method you need to consider how you are going to take a sample at each location. There are three ways to do this:

- Point sampling – making an observation or measurement at an exact location, such as an individual house or at a precise six-figure grid reference.

Fieldwork skills

- Line sampling – taking measurements along a carefully chosen line or lines, such as a transect across a sand dune ecosystem.
- Quadrat (or area) sampling – quadrats are mainly used for surveying vegetation and beach deposits. A quadrat is a gridded frame.

All three sampling types are shown in Figure 11.35. Here all of the sampling types are illustrated using the systematic method of sampling. When you have read the next section you might think how these diagrams would look using random and stratified sampling.

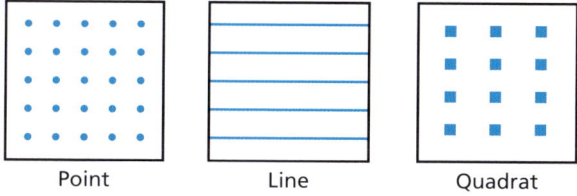

▲ Figure 11.35 Point, line and quadrat sampling

Sampling methods

- **Random sampling** – This method involves selecting sample points by using random numbers (Figures 11.36 and 11.37). Tables of random numbers can be used or the numbers can be generated by most calculators. The use of random numbers guarantees that there is no human bias in the selection process.

61	89	04	24	98	65	96	96
33	79	53	35	51	56	11	78
96	84	68	33	84	15	08	10
28	34	05	81	54	02	60	18
19	35	37	56	39	97	66	15
37	21	22	09	18	99	33	03
46	77	77	83	19	39	43	48
12	44	97	58	79	57	42	30
08	91	47	87	38	21	74	24
98	17	54	62	62	21	06	90
73	53	29	99	11	76	30	00
35	28	06	62	12	99	48	48
50	34	68	74	61	42	19	63
95	49	75	96	49	81	93	10
22	30	86	92	56	79	71	50
68	83	63	59	30	55	37	20
69	67	64	05	14	37	16	36
04	43	66	24	01	62	72	98
03	40	89	99	66	22	11	32
95	44	09	92	08	41	49	27

▲ Figure 11.36 Section of a table of random numbers

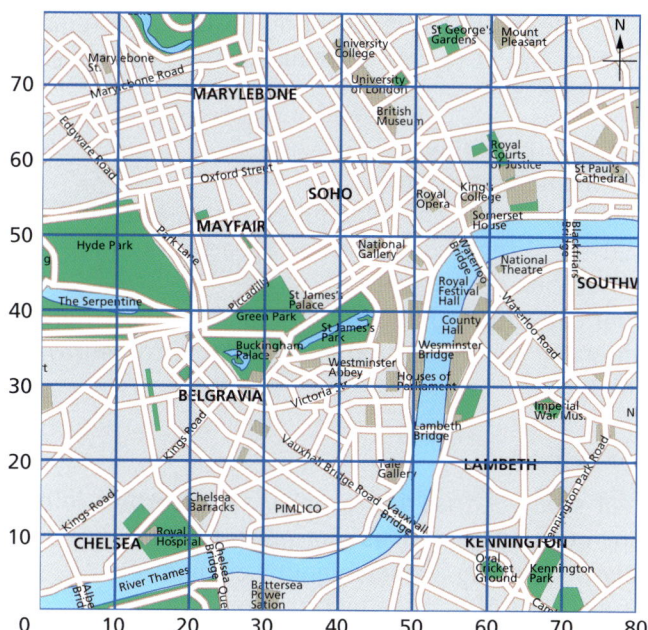

▲ Figure 11.37 How a grid and map can be used for random sampling

- **Systematic sampling** – With this method the sample is taken in a regular way. It might, for example, involve every tenth house or person. When using an Ordnance Survey map it might mean analysing grid squares at regular intervals.
- **Stratified sampling** – Here the area under study divides into different natural areas. For example, rock type A may make up 60 per cent of an area and rock type B the remaining 40 per cent. If you were taking soil samples for each type, you should ensure that 60 per cent of the samples were taken on rock type A and 40 per cent on rock type B (Figure 11.38).

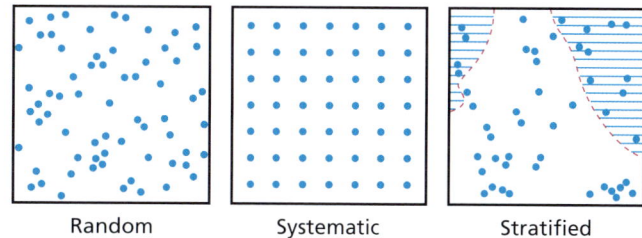

▲ Figure 11.38 Random, systematic and stratified sampling

Deciding on the size of the sample

The larger the sample the more likely you are to obtain a true reflection of the total population. However, this could happen with a very small sample by chance. But, equally, a small sample could give a very misleading picture of the total population.

GEOGRAPHICAL SKILLS

A good rule to follow at Cambridge IGCSE with regard to sample size is to take as many samples as possible with regard to:

» the time available
» the available resources
» the number of samples required for a particular statistical technique, such as Spearman's rank correlation coefficient
» your capacity to handle the data collected (there are many computer programs available to help with this).

Pilot surveys

A pilot study or trial run can play an important role in any geographical investigation. A pilot study involves spending a small amount of time testing your methods of data collection. For example:

» If you are using equipment, does all the equipment work and can everyone in the group use it correctly?
» If your data collection involves a questionnaire, can the people responding understand all the questions clearly?
» If a method of sampling is used, does everyone know how to select the sample points accurately?

A small-scale pilot study allows you to make vital adjustments to your investigation before you begin the main survey. This can save a great deal of time in the long run.

Environmental quality surveys

Environmental quality surveys are used to make judgements about urban, rural or natural environments. They break down an environment into its component parts and then each component is given a score. Often a bipolar scoring system is used, which has both positive and negative numbers. Table 11.9 is an example that could be used in an urban area. It scores six different component parts of the urban environment. This can be repeated in multiple locations. The scores can be averaged, and then presented in graph form or mapped using a GIS or isoline mapping.

Questionnaires

Questionnaire surveys involve both setting questions and obtaining answers. The questions are pre-planned and set out on a specially prepared form. This method of data collection is used to obtain opinions, ideas and information from people in general or from different groups of people. The questionnaire survey is probably the most widely used method to obtain primary data in human geography. In the wider world, questionnaires are used for a variety of purposes, including market research by manufacturing and retail companies and to test public opinion prior to political elections.

Activities

1. Why is sampling so important in geographical investigation?
2. Work in groups to provide outlines of different geographical investigations that would involve:
   a random sampling
   b systematic sampling
   c stratified sampling.
3. Why might it be beneficial to conduct a pilot study prior to beginning a geographical investigation?

▼ Table 11.9 An example of an urban bipolar environmental quality survey

Ward name: Date/time:	Location 1:	Location 2:	Location 3:	Scoring system
Building design, maintenance and condition	+2	-1		
Street furniture design, maintenance and condition	+2	0		+2 Very good
Pavements design, maintenance and safety	+1	-2		+1 Good
				0 Neutral
Road congestion and safety	0	-2		-2 Poor
Litter and graffiti presence / absence	+2	0		-1 Very poor
Vegetation quantity and variety	+1	-2		
Average score:	+1.3	-1.17		

One of the most important decisions you are going to have to make is how many questionnaires you are going to complete. The general rules to follow here are similar to those for sampling, set out in the previous section. Remember, if you have too few questionnaire results, you will not be able to draw valid conclusions. For most types of study, 25 questionnaires is probably the minimum you would need to draw reasonable conclusions. On the other hand, it is unlikely you would have time for more than 100 unless you were collecting data as part of a group.

A good questionnaire (Figure 11.39, page 316):

» has a limited number of questions that take no more than a few minutes to answer
» is clearly set out so that the questioner can move quickly from one question to the next – people do not like to be kept waiting; the careful use of tick boxes can help this objective
» is carefully worded so that the respondents are clear about the meaning of each question
» follows a logical sequence so that respondents can see 'where the questionnaire is going' – if a questionnaire is too complicated and long-winded people may decide to stop halfway through
» avoids questions that are too personal
» begins with the quickest questions to answer and leaves the longer/more difficult questions to the end
» reminds the questioner to thank respondents for their cooperation.

The disadvantages of questionnaires are as follows:

» Many people will not want to cooperate for a variety of reasons. Some people will simply be too busy while others may be uneasy about talking to strangers. Some people may be concerned about the possibility of identity theft.
» Research has indicated that people do not always provide accurate answers in surveys. Some people are tempted to give the answer that they think the questioner wants to hear or the answer they think shows them in the best light.

As with other forms of data collection, it is advisable to carry out a brief pilot survey first. It could be that words or questions you find easy to understand cause problems for some people. Amending the questionnaire in the light of the pilot survey before you begin the survey proper will make things go much more smoothly.

Delivering the questionnaire

There are really three options here:

» Approach people in the street or in another public environment.
» Knock on people's doors.
» Post questionnaires to people. With this approach you could either collect the questionnaire later or enclose a stamped addressed envelope. This last method is costly and experience shows that response rates are rarely above 30 per cent. Another disadvantage is that you will be unable to ask for clarification if some responses are unclear.

If you are conducting a survey of shopping habits you may want to find out if there are significant differences between males and females or between different age groups. In this case you would use a stratified sample divided by gender and the percentage of population in each age group.

The time of day may also be important. In the example given above, very few people in some age groups may be around at a certain time of day. For example, most teenagers will be in school or college at mid-morning on a weekday.

Interviews

Interviews are more detailed interactions than questionnaires. They generally involve talking to a relatively small number of people. A study of an industrial estate might involve interviews if you were trying to find out why companies chose to locate on the estate. An interview is much more of a discussion than a questionnaire, although you should still have a pre-planned question sheet. It can be a good idea to record interviews but you should ask the interviewee's permission first.

Health and safety and other restrictions

It may be sensible to work in pairs when conducting questionnaires as some people can act in an unfriendly manner when approached in the street. Working in pairs can also speed the process up considerably, with one person asking the questions and the other noting the answers. Also, be aware that shopping malls, individual shops and other private premises may not allow you to conduct questionnaires without seeking permission beforehand.

GEOGRAPHICAL SKILLS

A good questionnaire
Introduction: 'Excuse me, I am doing a school geography project. Could I ask you one or two quick questions about where you go shopping?'

1 How often do you come shopping in this town centre?
   More than once a week ☐
   Weekly ☐   Occasionally ☐
2 How do you travel here?
   Walk ☐ Car ☐ Bus ☐ Train/Tube ☐
   Other _____
3 Roughly where do you live? _____
4 Why do you come here rather than any other shopping centre?
   Near to home ☐   Near to work ☐
   More choice ☐   Pleasant environment ☐
   Other _____
5 What sort of things do you normally buy here?
   Groceries ☐   Clothes/shoes ☐
   Everything ☐
   Other _____
6 Do you shop anywhere else, and if so where?
   _____
7 Why do you go shopping there?
   _____
8 What do you buy there?
   _____
9 Sex: M ☐ F ☐   Age (estimate) under 20 ☐
                  20–30 ☐ 30–60 ☐ Over 60 ☐
'Thank you very much for you help'

A bad questionnaire
Introduction: 'Excuse me, but I wonder if I could ask you some questions?'

1 Where do you live?
2 How do you get here?
3 Do you come shopping here often?
4 Why do you come here?
5 Do you buy high- or low-order goods here?
6 Is this a good shopping centre and if so, why?
7 Where else do you go shopping?
8 Do you shop there because it is cheaper or nearer to your home?
9 How old are you? _____
'Right, that's it then.'

▲ **Figure 11.39** Two questionnaires, one good and one bad

> **Activities**
> 1 Design a questionnaire that might be used as part of an investigation into tourism in a small resort.
> 2 Briefly outline a geographical investigation in your local area that could involve the use of interviews.

Observations, counts and measurements

Field sketches

Personal observations or perceptions may form an important element of a coursework investigation. A field sketch is a hand-drawn summary of an environment you are looking at. In both urban and rural environments field sketching is a very useful way of recording the most important aspects of a landscape and noting the relationships between elements of such landscapes. The action of stopping for a period of time to sketch the landscape in front of you will often reveal details that may not have been immediately apparent.

Figure 11.40 is an example of a good field sketch. This sketch highlights the important geographical features of the landscape. Key features should be clearly labelled but make sure that your sketch is not too cluttered. This will detract from the really important details. A good field sketch is a higher-level technique.

Fieldwork skills

Annotated photographs

Annotated photographs should be seen as complementing field sketches rather than just being an alternative to them. Like field sketches, good, fully annotated photographs are regarded as a higher-level skill. Always record the precise location and conditions of the photographs you take. This should include a grid reference, the direction the photograph was taken in, weather conditions and time of day. Such information will make annotation quicker and easier in the long run.

An annotated photograph shows your key perceptions about a location you have visited on fieldwork. A series of such photographs might show:

» how the type and quality of housing varies in an inner city or suburban area

» how a river and its valley changes from source to mouth.

Annotations should be in the form of short, sharp sentences (Figure 11.41, page 318). Moderate abbreviation is fine providing the meaning of the comment remains clear. Some annotations will be just descriptive, but where the opportunity arises some explanation should also be included. Annotation can be most effective when the photograph is placed on the page in landscape format, which will allow more space for annotations on all four sides. As with field sketches, a series of annotated photographs could form a very effective part of your analysis. You should look to correlate annotated photographs with the tables and graphs showing your data analysis. Photographs are also useful to show how you carried out surveys and field measurements.

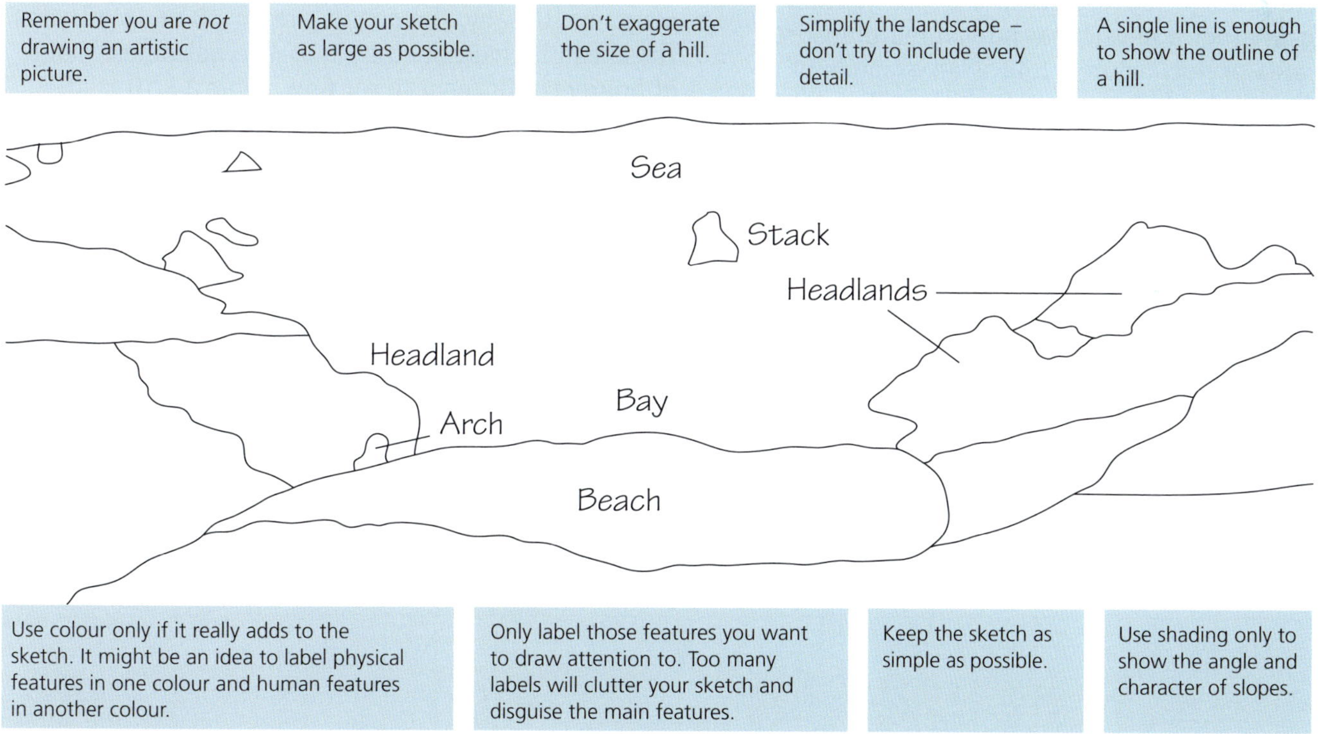

| Remember you are *not* drawing an artistic picture. | Make your sketch as large as possible. | Don't exaggerate the size of a hill. | Simplify the landscape – don't try to include every detail. | A single line is enough to show the outline of a hill. |

[Field sketch showing coastal features labelled: Sea, Stack, Headlands, Headland, Arch, Bay, Beach]

| Use colour only if it really adds to the sketch. It might be an idea to label physical features in one colour and human features in another colour. | Only label those features you want to draw attention to. Too many labels will clutter your sketch and disguise the main features. | Keep the sketch as simple as possible. | Use shading only to show the angle and character of slopes. |

▲ **Figure 11.40** An example of a field sketch

Activities

1. Draw a field sketch of an urban or rural environment within easy reach of your school.
2. Suggest why this location has geographical interest.
3. Annotate a photograph of a location of interest you have visited.

GEOGRAPHICAL SKILLS

▲ Figure 11.41 A river has overflowed its banks: example of an annotated photograph

Recording tables

The most straightforward method of observation is noting whether a physical or human feature exists in an area or not. Figure 11.42 is an example of a recording table showing park facilities. The objective here is to compare the facilities in four parks before attempting to explain the differences between them. Recording is done by placing a tick in the appropriate square. Notice that there is a final column to accommodate any unexpected findings.

Pedestrian and traffic counts

It can be useful to quantify the amount of people and/or traffic in an area. This can be an indicator of how popular an area is, perhaps because it has been regenerated. It can also indicate problems such as congestion and even air pollution. When counting traffic, safety is a key consideration. Traffic and pedestrian counts can be carried out in multiple locations and the results used to construct flow diagrams. Usually, the counts are for 10 or 15 minutes. These can be multiplied by 5 or 4 to get an estimate of hourly traffic/pedestrian flow. Table 11.10 is an example of a traffic count tally recording sheet.

▼ Table 11.10 An example of a traffic count tally

Ward name:	Location /street name:	
Date /time:	Traffic moving right	Traffic moving left
Cars		
Vans		
Buses		
Lorries and other large vehicles		
Motorbikes		
Bicycles		

Park	Size (ha)	Woodland	Children's playground	Sports pitches	Tennis/basket-ball courts	Picnic site	Restaurant/café	Boating/fishing lake	Ornamental gardens	Pavilion/bandstand	Toilets	Car park	Info centre/gift shop	Other
High Lodge Forest Park	120	✓	✓			✓	✓				✓	✓	✓	Maze, jungle gym
Ditchingham Estate Park	25	✓					✓				✓			
Long Stratton Park	4	✓	✓	✓	✓					✓	✓	✓		Skate ramp
Castle Mall Gardens	2					✓	✓		✓					Viewpoint

▲ Figure 11.42 An example of a recording table showing park facilities

Land use surveys

If your fieldwork investigation has an urban focus, particularly the CBD of a town or city, a land use survey is a useful way to measure how economic activity changes along a transect line (or lines) radiating from the CBD. Land use mapping is simple but time consuming. You will need to choose whether to only record ground-floor land use, or first-floor too (which is more time consuming). You need:

1. A detailed base map showing individual buildings at a scale of 1:1250 or less (such as OpenStreetMap).
2. A transect line running from the centre of the CBD outward.
3. A land use key, such as RICEPOTS (see Figure 11.43) and coloured pencils.

Some investigations focus on delimiting the CBD, in other words finding the edge of the CBD to determine its spatial extent or sphere of influence. This can be done by:

- using a land use survey to identify where residential land use takes over from commercial and retail land use
- using pedestrian and traffic counts to map changes in volume
- measuring building height (number of storeys) to find where land values decrease so most buildings are 1–3 storeys, not higher
- using a noise meter, which measures noise in decibels, to measure traffic and other noise.

> **Activity**
>
> Produce a recording table that could be used as part of a geographical investigation in your local area.

Presentation and recording of results

A wide variety of graphical techniques can be used to present geographical data. The skill is in choosing the best type of graph for the particular data set under consideration. You should also consider the size of any graph or diagram you use. It is important that the labels of axes and all other information can be clearly read.

It is important to integrate all maps, graphs, photographs and diagrams with the text. The most elementary way of doing this is to use a sentence such as 'Figure X is a line graph showing temperature change in my garden'.

RICEPOTS land use	Specific use
■ Residential	F = flat /apartment, T = terraced
■ Industrial	M = manufacturing, C = construction
■ Commercial	F = food, C = clothes, P = personal services
■ Entertainment	P = pub, C = café, R = Restaurant
■ Public buildings	E = education, H = health
■ Open space	P = park, S = sports
■ Transport	B = Bus station, T = Train station
■ Services	B = business, E = estate agent, BK = bank

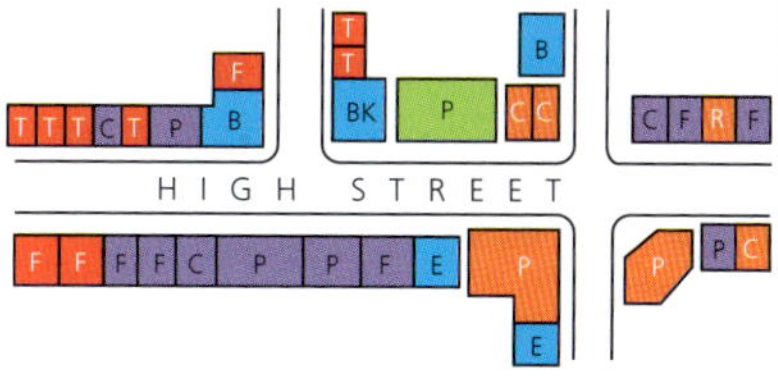

▲ Figure 11.43 A RICEPOTS land use survey key and example

Analysis and interpretation

Here you will analyse and interpret your findings on the issue/question/problem set out in the first section of your investigation. This will be done with reference to relevant geographical concepts.

You need to describe the patterns in data presented in your graphs and tables of results.

- After each graph or technique, describe fully the results or trend or association (using simple descriptive statistics). What do your results tell you? Describe your findings in detail by quoting the evidence from your methods of analysis.
- Do the graphs and other techniques used help to answer the question set?
- Make comments to link the data. For example, show how one diagram, graph or map relates to others.
- Where relevant, consider the values and attitudes of people involved.

Making effective conclusions

Using the evidence from the data you should be able to make judgements on the validity of the original hypothesis or aims of the assignment. Compare the results of the data analysis against standard models and theories.

Link what you have discovered in your enquiry to what you have studied in the syllabus. For example, if you have looked at shopping, have you talked about high-order and low-order shops, shopping hierarchies and other relevant concepts? If your investigation is on leisure, have you linked this to the amount of leisure time people have, spheres of influence of leisure centres, how accessible places are and other key aspects of the topic? These things should be mentioned briefly in the first section of your assignment and need to be discussed now in relation to your findings.

Having described your results, you need to explain and discuss them. Why have you arrived at such results? Do they confirm (accept) or refute (reject) your hypotheses?

Evaluation and suggestions for further work

Your investigation is likely to be slightly less than perfect, and you need to show the examiner that you are aware of this. It is important to show an awareness of the limitations of methods, results and conclusions.

There may be limits to where and when you could carry out a survey, or the number of people you could interview. There may be constraints in terms of expense, in terms of the equipment you are able to use, or how often you can visit the fieldwork location(s). The last item can also be due to a busy school timetable, which can mean your time at a fieldwork location is limited. Investigations in physical geography may be hampered by unexpected conditions. For example, river measurements may be affected by floods or a drought.

You need to make an assessment and state whether or not such limitations have impaired your investigation. If they have, then your results are likely to be compromised, as will any conclusions that you have drawn. For example, a survey of tourists to Oxford on a Tuesday afternoon in April found that most people were attracted to the university buildings and the colleges, and listed the poor weather as their main criticism. Had the survey taken place in another kind of tourist destination, the attractions listed would have been very different. Similarly, had the survey taken place in summer the weather might not have been mentioned as a problem. On the other hand, congestion due to too many tourists might have been mentioned. This example shows that the methods (including the date and time of any survey) produce results that can affect your conclusions and, therefore, your evaluation.

The final part of your evaluation can lead you to suggest future lines of enquiry from the insights you have developed by following the route to geographical enquiry in your investigation.

Coursework assessment

Before you begin your coursework enquiry you should be aware of how your coursework will be assessed. The mark scheme for the assessment of coursework is in the syllabus document. This is in the form of a matrix, which shows what you need to do to meet each of the five assessment criteria.

Fieldwork skills

> **Activities**
> 1. What is the difference between analysis and interpretation?
> 2. Write an analysis and interpretation of the isoline pattern shown in Figure 11.9 (page 297).
> 3. What might you include in the evaluation of your coursework?

Case studies

Analysing sand dunes

▲ **Figure 11.44** Sand dunes at Studland, Dorset, UK

Sand dunes provide an interesting and manageable ecosystem for study at this level. This is because significant changes can be identified over a relatively small area. In a sand dune system, the most recently formed dunes are by the sea. The dunes become older with distance inland. Sand dunes form a series of ridges with intervening 'slacks' between them.

A useful starting point is to survey the morphology (size and shape) of the dunes. Figure 11.45a (page 322) shows how measurements can be taken across a sand dune ecosystem using a tape measure and a clinometer. The transect line should be at right-angles to the coast. The first ranging pole is carefully placed where there is a distinct break in slope from the back of the beach, marking the beginning of the sand dunes. The second ranging pole is placed at the next break of slope. The angle of slope is read from the clinometer. This process is repeated for each break of slope. With about four people working as a team, all the measurements required to draw a cross-section of the dunes (Figure 11.45b) can be taken in a couple of hours. If a larger group of people is available a number of transects could be taken across the sand dunes. Transects could be compared and any differences discussed.

As the dune survey proceeds, other measurements can also be taken. At regular locations across the sand dune system the following can be measured:

- vegetation cover, with the dominant plant species noted
- maximum height of vegetation
- wind speed
- soil moisture content
- soil organic content
- soil pH.

Figure 11.45c provides an example of a recording sheet that could be used for such a survey. While the first few readings might take a little time, once you become familiar with what is required the process should speed up considerably.

All these measurements can be used to test the standard theories about sand dunes presented in textbooks, which can be set out in a series of hypotheses to be tested:

- Vegetation density increases with distance inland.
- The number of species increases with distance inland.
- The height of vegetation increases with distance inland.
- Organic content of soil increases with distance inland.
- Soil pH decreases with distance inland.
- Wind speed decreases with distance inland.

GEOGRAPHICAL SKILLS

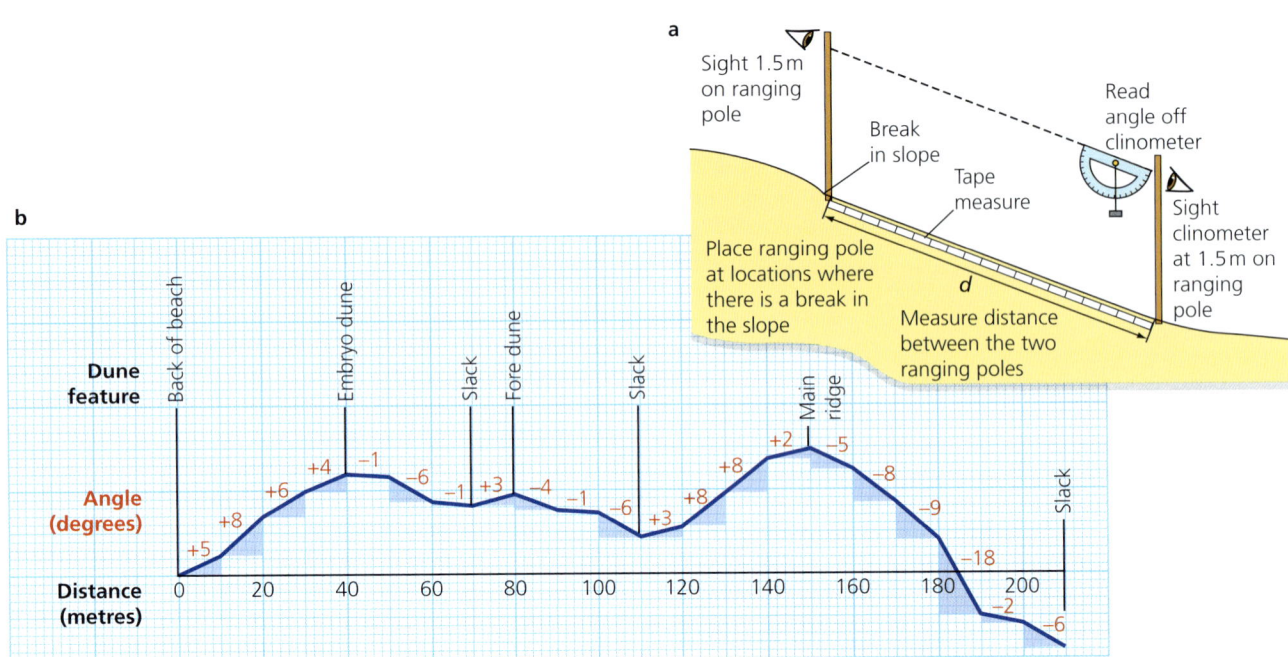

Site	Distance from sea (m)	Angle (°) (uphill)	Vegetation cover (%)	Number of species	Maximum vegetation height (m)	Dominant plant species	Wind speed (m/s)	Soil moisture content (%)	Soil organic content (%)	Soil pH
1	10	+5	50	2	0.51	Marram grass	7.3	8.7	0.2	6.9
2	10	+8	92	3	0.56	Sea couch grass	4.4	25.9	0.4	6.8
3	20	+6	86	3	0.48	Marram grass	3.4	11.2	1.8	6.8
4	30	+4	95	3	0.51	Sea couch grass	3.0	25.0	3.3	6.8
5	40	−1	96	2	0.69	Marram grass	2.4	54.4	3.6	6.8
6	50	−6	94	3	0.71	Marram grass	3.2	9.1	3.7	6.7
7	60	−1	89	5	0.75	Sea couch grass	3.0	24.1	4.6	6.9

▲ **Figure 11.45** Conducting a survey of a sand dune ecosystem

Investigating rivers

Streams and small rivers are a popular focus for geographical investigation because most schools will not be too far from a suitable example. Figure 11.46 shows some of the measurements that can be taken at various locations along the course of a river. For safety reasons it is best to avoid working in streams above the height of your knees.

Fieldwork skills

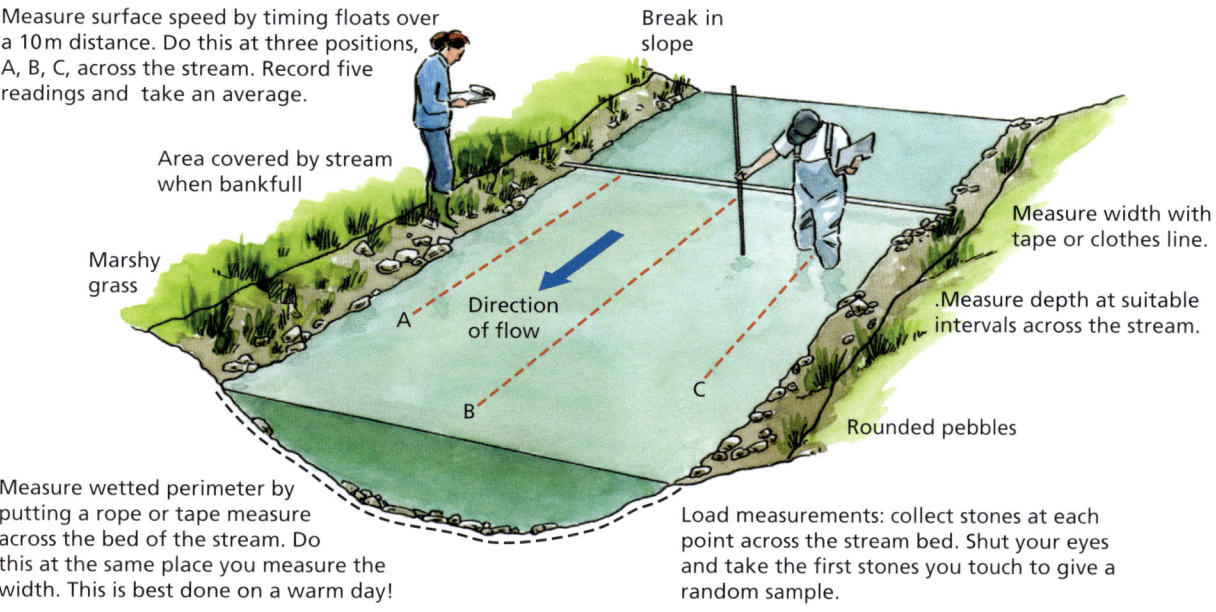

▲ Figure 11.46 Taking river measurements

For most river studies you will want to produce a cross-section of the river channel. The method is as follows:

» Use a tape measure to assess the channel width. This should be done at right-angles to the course of the river. If you want to produce a cross-section of the river when discharge is at its highest, you should look for evidence of the highest point the water reaches on each bank. This will give the bankfull width.

» Channel depth should be measured at regular intervals across the river using a metre stick or ranging pole. Every 20 cm or 30 cm should provide an adequate sampling interval.
» A cross-section can then be drawn using graph paper (Figure 11.47). As with all cross-sections, careful choice of scale is important.
» Use a stopwatch or timer to time how long it takes an orange to travel 10 m in the stream. Repeat at least three times.

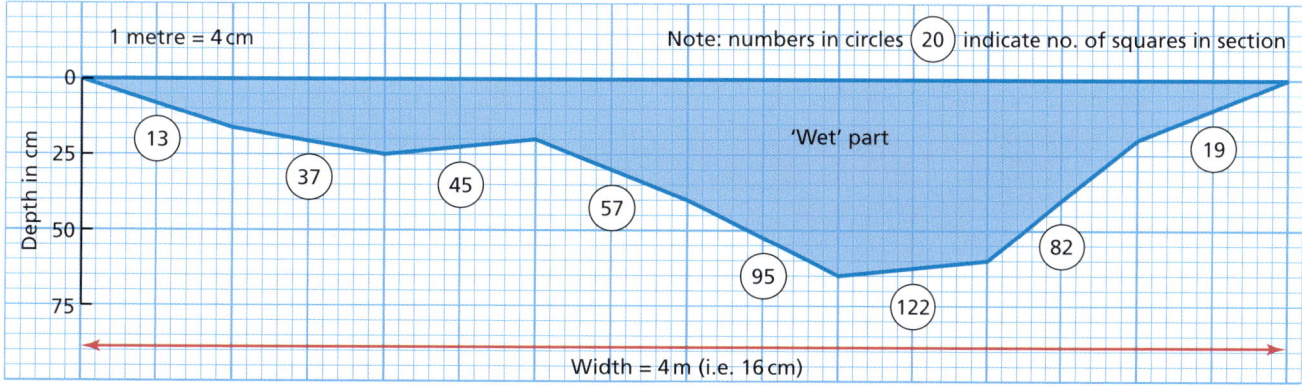

Area = 470 squares. This number has to be converted, according to the scale. The scale 4 cm = 1 m means that each 2-millimetre square on the graph paper represents 0.0025 m² in reality ($1/400$). If the scale had been 1 cm = 1 m then the scaling factor would be $1/100$ = 0.01 m². In this example 470 × 0.0025 = 1.175 m².

▲ Figure 11.47 A river cross section

GEOGRAPHICAL SKILLS

Although various types of float can be used to measure river velocity, it is best to use a flow meter. The impeller (screw device) is pointed upstream at the same points across the river used to calculate the depth intervals. You will be able to see how velocity varies with distance from the banks and how velocity varies with depth. You could also calculate the mean flow rate for this stage of the river.

The discharge of the river can be calculated by multiplying the velocity by the cross-sectional area (Figure 11.47). The gradient of a river can be measured using ranging poles and a clinometer, in the same way that sand dune measurements were carried out in the previous example.

Bedload measurements can also be taken to assess the impact of attrition with increasing distance downstream. Ensure that the samples of bedload are selected randomly by a ranging pole or metre stick at intervals across the river. Collect the stones that are touching the pole or stick. Measure the long axis, shape and radius of curvature of each stone.

As in the sand dune case study, a series of hypotheses based on textbook theory can be set up to be tested.

Use Power's scale of roundness (Figure 11.48) to see whether pebble shape becomes rounder downstream.

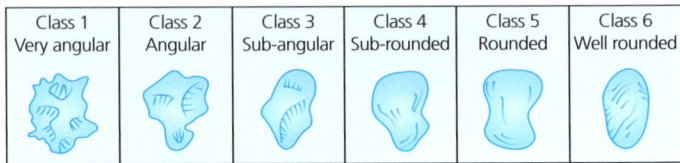

▲ **Figure 11.48** Power's scale of roundness

Command words

The table below includes command words used in the assessment for this syllabus.

Command word	What it means
Assess	Make an informed judgement.
Calculate	Work out from given facts, figures or information.
Compare	Identify/comment on similarities and/or differences.
Define	Give precise meaning.
Describe	State the points of a topic/give characteristics and main features.
Devise	Create a questionnaire to present other information according to specific requirements.
Discuss	Write about issue(s) or topic(s) in depth in a structured way.
Estimate	Use judgement to give a unit value to a distance or area.
Evaluate	Judge or calculate the quality, importance, amount or value of something.
Explain	Set out purposes or reasons/make the relationships between things clear/say why and/or how and support with relevant evidence.
Give	Produce an answer from a given source or recall/memory.
Identify	Name/select/recognise.
Justify	Support a case with evidence/argument.
Locate	Indicate the position of a place, feature or entity from/in a resource.
Plan	Create a method to obtain or present certain information (such as a questionnaire) according to specific requirements.
Plot	Mark point(s) on a graph/diagram/map.
Predict	Suggest what may happen based on available information.
Sketch	Make a simple freehand drawing showing the key features, taking care over proportions.
State	Express in clear terms.
Suggest	Apply knowledge and understanding to situations where there are a range of valid responses in order to make proposals/put forward considerations.

Phrases such as 'How far do you agree...?' and 'To what extent...?' may also be seen in the assessment of this syllabus.

The information in this section is taken from the Cambridge International Education syllabus.

You should always refer to the appropriate syllabus document for the year of examination to confirm the details and for more information. The syllabus document is available on the website: www.cambridgeinternational.org.

Glossary

Abrasion/corrasion The wearing away of the river bed and banks by the river's load hitting them repeatedly.

Accessibility The ease with which a place can be reached. An area with high accessibility will generally have a well-developed transport network and be centrally located.

Active volcano A volcano that is currently erupting or has recently erupted and is likely to erupt again.

Adaptation (to climate change) Methods of learning to live with climate change, e.g. sea walls and malaria tablets.

Adult literacy rate The percentage of an adult population with basic reading and writing skills.

Aeroponics A system where plants are never placed into water, with crops growing suspended in the air. Nutrients are supplied from a mist that is sprayed on to their roots. The entire process is enclosed, so the nutrient mist is able to remain around the plants for longer.

Agglomeration economy The advantages that firms gain when they are located near other similar and related firms, such as cost savings from shared infrastructure.

Agricultural productivity The ratio of agricultural inputs to outputs.

Agricultural region An area with relatively uniform physical characteristics (altitude, climate, soil type), where because of these characteristics a particular type of farming dominates, such as wheat, corn or cotton.

Agricultural technology The application of techniques to control the growth and harvesting of animal and vegetable products.

Agriculture The production of crops and livestock, aquaculture and forest products, e.g. rice, fish and cotton, for food and non-food products.

Agro-industrialisation Industrialised farming that is typically large scale and capital intensive.

Air pressure The pressure at any point on the Earth's surface that is due to the weight of the air above it; it decreases as altitude increases. At sea level the average pressure is 1013 millibars (mb). Areas of relatively high pressure are called anticyclones; areas of low pressure are called depressions.

Albedo The reflectivity of a surface – snow and ice are very reflective, whereas dark surfaces, such as soil, are not.

Alluvium Material deposited by rivers in a channel or on a floodplain.

Altitude Measurement of height, usually given in metres above sea level. Temperature declines, on average, 1°C for every 100 m (and therefore rises 1°C with every decrease in altitude of 100 m).

Annotation Descriptive or explanatory labelling of a photo or diagram to show the main characteristics/features.

Anti-natalist policy A policy that is discouraging and unsupportive of increasing fertility.

Aquaponics A system that combines fish farming (aquaculture) with hydroponics. Fish waste is naturally converted into nutrients for plants, while the plants filter the water for the fish. This creates a closed-loop ecosystem.

Aquifer A rock that allows water to move through it, such as a layer of sandstone.

Arable farming The cultivation of crops.

Arch A natural bridge-like feature formed by erosion. Arches are formed from the erosion of a headland where two caves meet and break through the headland.

Archipelagic country A country made up of a group of islands, interconnecting waters, and other natural features that are closely interrelated.

Arid area Usually defined as an area that receives less than 250 mm of rainfall each year.

Ash fall Very small, hard, jagged material of volcanic origin. It is the most frequent and widespread volcanic hazard.

Atlas maps A collection of maps usually showing different countries, continents, and physical and human geography.

Attrition The process by which particles of rock being transported by a river are rounded and become smaller in size by being struck against one another. Particles near the shoreline become smaller and more rounded due to more frequent attrition.

Backwash The movement of water back down the beach due to the effect of gravity.

Bar A depositional feature – a long ridge of sand or pebbles running parallel to a coastline that is submerged at high tide. Some bars develop as offshore bars when waves disturb sediments on the sea bed and form them into a submarine ridge or bar, while others form from the development of a spit across the whole of a small bay.

Base map A very simple map, often just an outline, which can then be filled in to show some characteristics of the place in question.

Battery energy storage system (BESS) A battery system that enables excess energy from renewables to be stored and then released when the power is needed most.

Bay A wide, open, curving indentation of the sea.

Beach A feature of coastal deposition, consisting of pebbles on exposed coasts or sand on sheltered coasts. It is usually defined by the high and low water marks.

Bedload Material rolled along the river bed.

Big data Extremely large data sets analysed computationally to reveal patterns, trends and associations.

Bimodal A distribution of a population that has two modes (most frequent values).

Biodiversity Biological diversity is a measure of the variety of the Earth's plant and animal species, of genetic differences within species, and of the ecosystems that support those species.

Biofuel Any fuel produced from organic matter, either directly from plants or indirectly from industrial, commercial, domestic or agricultural wastes. For example fossil fuel substitute can be made from a range of crops, including oilseeds, wheat and sugar.

Glossary

Birth rate The number of live births per thousand population in a year.

Body wave A seismic wave that travels through the Earth's interior.

Bradshaw model A geographical model that suggests how a river's characteristics change from the source to the mouth of the river.

Braiding/braided channel The sub-division of a river into several channels, caused by deposition of sediment as small islands in the main channel.

Break-of-bulk point A location, such as a sea port, where freight has to be transferred from one mode of transport to another.

Business/retail park Similar to an industrial estate, but dominated by tertiary activities.

Capacity (rivers) The total amount of sediment a river can carry.

Capital intensive A high level of investment in plant and machinery compared to labour.

Capitalism The social and economic system that relies on the market mechanism to distribute the factors of production (land, labour, capital) in the most efficient way.

Carrying capacity (of a populated area) The largest population that the resources of a given environment can support.

Carrying capacity (of a tourism destination) The maximum number of people that may visit a tourism destination at the same time without causing destruction of the physical, economic or sociocultural environment and an unacceptable decrease in the quality of visitor satisfaction.

Cartoon A diagram that may exaggerate a feature to draw attention to a particular point.

Catchment management The regulations and policies that govern the oversight of surface and groundwater resources within the area of a river catchment.

Cave Coastal caves are large holes formed where relatively soft rock containing lines of weakness is exposed to severe wave action.

Channel flow The movement of water within the river channel.

Channel network (river system) The pattern of a main river and its tributaries within a drainage basin.

Channel roughness The frictional force of a river bed. A rough bed (with boulders, pebbles, potholes) exerts more friction than a smooth channel.

Child mortality rate The number of deaths of children under five years of age per thousand live births per year.

Cinder cone volcano A steep-sided volcano made of fragments of volcanic materials, e.g. cinders and tephra.

Circular economy An economic system based on the reuse and regeneration of materials and products, especially as a means of continuing production in a sustainable way.

Clean air transition The period of change from heavy reliance on fossil fuels to most of people's energy needs being supplied by renewable sources of energy.

Cliff A rock-face along a coastline where coastal erosion, weathering and mass movements are active and the slope rises steeply (over 45°) and for some distance. The nature of the cliff depends on the nature of the rocks, their hardness and their jointing pattern.

Climate The combination of weather conditions at a particular place over a period of time – usually a minimum of 30 years. Climate thus includes the averages, extremes and frequencies of all meteorological elements, such as temperature, atmospheric pressure, precipitation, wind, humidity and sunshine.

Climate loss Pre-harvest food losses from: not planting a crop, or at least not the desired crop; changing crops due to weather; extreme weather reducing or wiping out a crop before harvest; increased predation, pests and/or diseases resulting from climate change.

Closed system A system unconnected to other entities. It has no inputs from, or outputs to, elsewhere.

Cloud Water vapour condensed into minute water particles that float in the atmosphere. Clouds are formed by the cooling of air containing water vapour, which generally condenses around tiny dust or ice particles.

Coastal erosion The wearing away of coastlines by moving agents, such as waves and wind, to create coastal features such as cliffs and wave-cut platforms.

Coastal management strategy A measure taken to prevent coastal erosion and/or flooding. To reduce erosion, several different forms of coastal protection are used. These can be divided into hard engineering and soft engineering.

Coastline The area of contact between land and sea.

Colonialism The establishment of colonies in one or a number of territories by people from another territory.

Commercial farming Farming for profit, where food is produced for sale in the market.

Commodity chain The stages involved in manufacturing a finished product (commodity) for sale to consumers. These stages may occur in factories in different countries.

Community energy Energy produced close to the point of consumption.

Community tourism A form of tourism that aims to include and benefit local communities, particularly in developing countries.

Competence (rivers) The maximum size of particle a river can carry.

Composite cone volcano A steep-sided volcano formed of alternating layers of lava and volcanic fragments, e.g. cinders and tephra.

Concordant coastline A relatively straight coastline in which the relief and geology are parallel to the coast.

Condensation The process by which water vapour changes into water droplets.

Confluence The point where a tributary stream/river joins a main river.

Conservation Allowing for developments that do not damage the character of a location.

Glossary

Conservation of living resources The management of the human use of natural resources to provide the maximum benefit to current generations while maintaining capacity to meet the needs of future generations.

Conservative/transform boundary A plate boundary in which plates move past each other in different directions or at different speeds but no material is destroyed or created. They are associated with earthquake activity.

Constructive wave A wave with a long wave length and a low height. Constructive waves help to build up beaches by deposition.

Consumer culture A society in which patterns of consumption are a key basis for personal identity and status differentiation.

Containerisation A system of intermodal freight transport using standard international containers.

Contour ploughing Prevents or diminishes the downslope movement of water and soil by ensuring that ridges and furrows are at right angles to the slope.

Convectional rainfall Rainfall associated with hot climates, resulting from the rising of convection currents of warm air. Air that has been warmed by the extreme heating of the ground surface rises to great heights and is cooled quickly. The water vapour carried by the air condenses and rain falls heavily. Convectional rainfall is usually associated with thunderstorms.

Convergent/collision zone A plate boundary where two plates meet. It may lead to the creation of fold mountains, e.g. the Himalayas.

Convergent/destructive boundary A plate boundary where two plates meet and the denser oceanic plate subducts/plunges underneath the less dense continental plate. They are associated with creating ocean trenches and fold mountains.

Coral Living organisms that may form large reefs. Coral reefs provide a habitat for a wide diversity of living organisms.

Corruption Dishonest or illegal behaviour by people in powerful positions, typically involving bribery.

Counterurbanisation The movement of people away from larger urban areas to smaller urban areas and/or rural settlements.

Country of destination The country where a migration is completed.

Country of origin The country from which a migration begins.

Crater A bowl-shaped depression at the summit of a volcano, formed by a volcanic eruption.

Cultural diffusion The spreading of cultural traits from one place to another.

Cultural identity The definition of a group or individual (by themselves or others) in terms of ethnicity, nationality, language, religion and gender.

Culture The inherited ideas, beliefs, values and knowledge of a society.

Death rate The number of deaths per thousand population in a year.

Debt Money owed by a country to another country, to private creditors (e.g. commercial banks), or to international agencies such as the World Bank or IMF.

Debt relief The cancellation of debts owed by developing nations to developed countries or to institutions such as the World Bank, in order to allow the government to shift funds toward social development.

Debt service ratio The proportion of a country's GDP required to pay off its debts each year.

Decentralisation The outward movement of economic activity from CBDs and inner cities.

Decimal A way of expressing 'inbetween' numbers. Decimal comes from the Latin word *decimus*, meaning *tenth*. Thus, decimals refer to tenths between whole numbers.

Deep-focus earthquake An earthquake that occurs at a depth of over 70 km beneath the Earth's surface.

Deep-sea port A port that has the depth of water, natural or dredged, and the technology to handle very large ocean-going ships, particularly ultra-large container ships.

Deforestation The removal of forest cover or destruction of forest for timber, fuel or charcoal burning, or clearing for agriculture and extractive industries such as mining. It can be complete or partial. It causes fertile soil to be blown away or washed into rivers, leading to soil erosion, drought, flooding and loss of wildlife.

Deindustrialisation The long-term absolute decline of employment in manufacturing.

Delta A flat, low-lying deposit of sediment that is found at a river's mouth.

Demographer A person who studies human populations.

Demographic dividend The growth in an economy resulting from a change in the age structure of a country's population.

Demographic preparedness The idea that there is time to enact policies and encourage behaviours to limit the potential adverse effects of demographic changes.

Demographic transition model A model illustrating the historical shift of birth and death rates from high to low levels in a population.

Dependency ratio The ratio of the number of people aged under 15 years and over 64 years to those in the 15–64 age group.

Dependency theory A theory that blames the relative underdevelopment of the developing world on exploitation by the developed world, first through colonialism and then by neo-colonialism.

Deposition The laying down of material carried by rivers or the sea because of a reduction of velocity or discharge (both causing a loss of energy), often caused by increased friction with vegetation or coarse particles.

Desert A dry area with limited vegetation. Deserts can be either hot or cold. Characteristics common to all deserts include irregular rainfall of less than 250 mm per year.

Glossary

Desertification The transformation of land that could once support human life into desert.

Desire lines A type of flow line that shows direction of travel.

Destination footprint The environmental impact caused by an individual tourist on holiday in a particular destination.

Destructive wave A wave with a high height and a short wave length, which helps erode beach materials and cliffs.

Detritivores Organisms that feed on dead organic matter (detritus).

Development The use of resources to improve the quality of life in a country.

Development gap The difference in income and quality of life in general between the highest and lowest-income countries in the world.

Diaspora The dispersion of a people, originally belonging to one country, around the world.

Diet The kinds of foods that a person, animal or community habitually eats.

Digital divide The gulf between those who have ready access to computers and the internet and those who do not.

Direct channel precipitation Rainfall and snow falling directly into a river or stream.

Discharge The amount of water passing a specific point at a given time (the volume times the velocity). It is measured in cubic metres per second.

Discordant coastline A coastline in which the relief and geology are at right angles to the coastline.

Disruptive technology An innovation that significantly alters the way that consumers, industries and businesses operate.

Distribution maps Maps that show the distribution of a geographic element in a given area. They can range from small scale, e.g. the distribution of types of shops along a street, to very large scale, e.g. the distribution of global biomes/ecosystems.

Divergent/constructive boundary A plate boundary where new material is created and two plates move away from each other, e.g. the North American and European Plates moving away from the Mid-Atlantic Ridge.

Do minimum A form of managing natural processes that involves very limited human intervention but may aim at protecting human lives.

Do nothing A form of managing natural processes that involves taking no action but allowing natural processes to operate without any human intervention.

Dormant volcano A volcano that has not erupted for a long time, but could erupt in the future.

Doughnut graphs Graphs that show the relationship of parts to the whole features, e.g. different types of energy contributing to total energy. The numbers can be relative (per cent) or absolute and are added to the graph. Doughnut graphs may have different rings to show different factors or show changes over time, e.g. employment structure.

Drainage basin The area of land drained by a river system.

Drainage density The total length of all the streams and rivers in a drainage basin divided by the total area of the drainage basin.

Drought An extended period of dry weather leading to conditions of extreme dryness. Absolute drought is a period of at least 15 consecutive days with less than 0.2 mm of rainfall. Partial drought is a period of at least 29 consecutive days during which the average daily rainfall does not exceed 0.2 mm.

Economic core region(s) The most highly developed region(s) in a country or in the world.

Economic growth An increase in real GDP (%) after inflation is taken into account.

Economic leakage The part of the money a tourist pays for a foreign holiday that does not benefit the destination country because it goes elsewhere.

Economic periphery Parts of a country (or the world) outside of the economic core region(s).

Economic sustainability Conducting economic activities in a way that preserves and promotes long-term economic wellbeing.

Economy of scale The reduction in unit cost as the scale of an operation increases.

Ecosystem An integrated unit consisting of a community of living organisms (animals and plants) and the physical environment (air, soil, water and climate) that they inhabit. Individual organisms interact with each other and with their habitat.

Ecotourism A specialised form of tourism where people experience relatively untouched natural environments, and ensure that their presence does no further damage to these environments. Ecotourism is at the leading edge of sustainable tourism.

Elderly dependency ratio The ratio of the number of people aged 65 and over to those in the 15–64 age group.

Emigration rate The number of emigrants per thousand population leaving a country of origin in a year.

Energy conservation Reducing the consumption of energy by using it more efficiently.

Energy gap The difference between a country's demand for energy and its ability to satisfy this demand from its own resources.

Energy intensity The amount of energy used to produce a given level of output.

Energy ladder The transition from fuelwood and animal dung to 'higher-level' sources of energy such as electricity.

Energy mix The relative contribution of different energy sources to a country's energy consumption.

Energy pathway A supply route between energy producers and consumers, which may be through pipelines, shipping routes or electricity cables.

Energy policy The actions taken by a government to affect the demand for energy as well as the supply of it.

Energy poverty A lack of access to modern energy services, particularly electricity.

Glossary

Energy security The uninterrupted availability of energy sources at an affordable price.

Enhanced greenhouse effect (global warming) The additional heat that is trapped as a result of the increased levels of greenhouse gases in the atmosphere. This is largely as a result of human activities and is causing the world climate to become hotter and more extreme.

Environmental impact assessment A study carried out before a development project is undertaken to assess the possible damage to the environment.

Environmental impact survey A document required by law detailing all the impacts on the environment of a construction project above a certain size.

Environmental Performance Index (EPI) An indicator that uses 58 performance indicators across 11 categories to score and rank countries around the globe on climate change performance, environmental health and ecosystem vitality.

Environmental sustainability The preservation and protection of the natural environment over time. It involves appropriate practices and policies to meet present-day needs without compromising the availability of resources in the future.

Epicentre The point on the Earth's surface immediately above an earthquake's focus.

Erosion The wearing away of the Earth's surface by a moving agent, such as a river, glacier or the sea. In a river, there are several processes of erosion, including hydraulic action, abrasion, attrition and solution. In coastal areas, hydraulic action is the most potent form of erosion.

Eutrophication The depletion of oxygen from water in streams and rivers due to increased levels of nitrates (from fertilisers) or phosphates (from detergents).

Evaporation The process in which a liquid turns to a vapour.

Evapotranspiration A combination of the processes of evaporation and transpiration. Vegetation takes in moisture through its root system. It loses some of this into the air by transpiration. Surface water is also lost by evaporation.

Extensive farming Where a relatively small amount of agricultural produce is obtained per hectare of land, so such farms tend to cover large areas of land. Inputs per unit of land are low.

Extinct volcano A volcano that erupted in the past but will not erupt in the future.

Fairtrade A system designed to help small-scale producers in developing countries achieve sustainable and fair trade relationships.

Fertility rate The number of live births per 1000 women aged 15–44 years in a given year.

Fetch The distance of open water over which wind can blow to create waves. The greater the fetch, the more potential power waves have when they hit the coast.

Flash flood A large but temporary increase in channel discharge.

Flood A discharge great enough to cause a body of water to overflow its channel and submerge (flood) the surrounding area.

Flood abatement Reducing the amount of runoff in a drainage basin through actions such as preservation of natural water stores.

Flood diversion The practice of allowing certain areas, such as wetlands and floodplains, to be flooded to a greater extent.

Flood hydrographs Line graphs that show how the amount of flow in a river (discharge) varies over time following a storm.

Flood impact assessment A process that identifies the impact of a proposed development on the flood risk to the area.

Flood peak The highest discharge of a river flood.

Floodplain The area of periodic flooding along the course of a river valley.

Focus The exact point within or on the Earth's surface where an earthquake occurs.

Food bank A place where stocks of food, typically basic provisions and non-perishable items, are supplied free of charge to people in need.

Food miles The distance food travels from the farm where it is produced to the plate of the final consumer.

Food security When all people at all times have access to sufficient, safe, nutritious food to maintain a healthy and active life.

Food waste Perfectly edible foodstuffs that are thrown away at the retail and consumer level.

Forced migration When people are made to move against their will due to human or environmental factors.

Foreign direct investment (FDI) Investment in physical capital in other countries by transnational corporations.

Fossil fuel A non-renewable resource, such as coal, oil or natural gas, formed from the remains of prehistoric carbon compound- or hydrocarbon-containing materials.

Fraction A fraction is a part of a whole, e.g $\frac{1}{3}$ of Borneo's rainforest has been deforested.

Four-figure square references A four-figured number that shows the location of a large feature. The first two numbers refer to the eastings and the second two numbers refer to the northings

Freeport A special economic zone in which imported goods can be held or processed free of customs duties before re-export.

Friction The resistance encountered when one body moves relative to another body with which it is in contact.

Fuel poverty When a person on a low-income lives in a home that cannot be kept warm at a reasonable cost.

Fuelwood Wood and charcoal used to supply energy.

Gabion A wire basket filled with rocks or stones, used for stabilising slopes and protecting the base of cliffs in areas of coastal erosion.

Gender dividend The increase in economic growth that can result from greater investment in women and girls, including promoting secondary and tertiary education and female workforce participation.

Geopolitics Political relations among nations, particularly relating to disputes about borders, territories and resources.

Geothermal energy The natural heat found in the Earth's crust in the form of steam, hot water and hot rock.

Glacial period The cold phase in an ice age, i.e. when ice sheets and glaciers advance.

Global city A city that is an important nodal point in the global economic system. Global cities are major financial and decision-making centres.

Global civil society Organisations or individuals, independent from the state, whose aim is to transform policies through communal efforts at a national or global scale.

Global value chain (GVC) A phenomenon whereby production is subdivided into activities carried out in different countries.

Globalisation The increasing interdependence of the world economically, culturally and politically.

GNI per capita The total GNI of a country divided by the total population.

Gorge A narrow, steep-sided valley that may or may not have a river at the bottom.

Green belt An area of largely undeveloped land around a city or town upon which there are wide-ranging planning restrictions on building development.

Green Revolution The introduction of high-yielding seeds and modern agricultural techniques in developing countries.

Greenhouse effect Some atmospheric gases, such as water vapour, carbon dioxide, methane and nitrous oxide, are greenhouse gases. This means that they allow short-wave radiation to pass through the atmosphere and reach the Earth's surface, but they absorb some of the outgoing long-wave radiation thereby heating the Earth's atmosphere. This is a natural process.

Groundwater Water stored underground in a permeable rock, e.g. chalk or sandstone.

Groundwater flow The flow of water through permeable rock.

Growth collapse A significant decline in economic activity brought about by factors such as a global economic shock or by the impact of a major natural hazard.

Groyne A wooden or concrete barrier built at right angles to a beach in order to block the movement of material along the beach by longshore drift.

Guest worker A foreigner who is permitted to work in a country on a temporary basis, for example a farm labourer.

Hard engineering Any coastal or river protection scheme that involves altering the natural environment with concrete, stone, steel, metal, etc. – for example the use of sea walls, gabions, groynes and revetments. Artificial structures are built in order to protect the natural environment from erosion.

Headland A point of land projecting into the sea; also known as a cape or a promontory.

Holocene The most recent geological epoch, beginning 11,700 years ago.

Host culture The culture of a country where a person is currently living or has lived.

Human Development Index The composite UN measure of the disparities between countries, using indicators of health, education and income.

Humidity The quantity of water vapour in a given volume of air (absolute humidity), or the ratio of the amount of water vapour in the atmosphere to the maximum amount the air can hold (relative humidity). At dew point the relative humidity is 100 per cent and the air is said to be saturated. Condensation (the conversion of vapour to liquid) may then occur.

Hydraulic action The erosive force exerted by water alone, such as the sheer force of river water removing loose material from the bed and banks of the river. It is particularly effective on jointed rocks, especially during storm conditions.

Hydraulic radius A measure of the efficiency of a stream's shape – that is, the cross-sectional area divided by the wetted perimeter. The higher the ratio, the more efficient the stream and the smaller the frictional loss. The ideal form is semi-circular.

Hydrological cycle The movement of water between air, land and sea.

Hydroponics A system where plants may be suspended in nutrient water all the time or fed by an intermittent flow of water.

Immigration rate The number of immigrants per thousand population entering a receiving country in a year.

Impermeable Rocks that do not allow water to pass through.

Import dependency The proportion of energy consumption supplied by imports from other countries.

Industrial estate An area zoned and planned for the purpose of industrial development. Manufacturing industry has a significant presence.

Industrial Revolution The transformation in the late eighteenth and nineteenth centuries of first the UK and then other European countries and the USA from agricultural into industrial nations.

Infant mortality rate The number of deaths of children under one year of age per 1000 live births per year.

Infiltration The initial movement of water from the surface into the upper level of the soil.

Input An element that is required for a process to take place.

Institutional quality The effectiveness, transparency and strength of the institutions within a country, such as the police and the courts.

Intensive farming Agriculture characterised by high inputs per unit of land to achieve high yields per hectare.

Interception The precipitation that is collected and stored by vegetation.

Glossary

Interglacial period The warm phase during an ice age, i.e. when ice sheets and glaciers retreat.

Interlocking spurs Projecting areas of high ground in the upper course of a river that interlink.

Intermediate technology Aid supplied by a donor country whereby the level of technology and the skills required to service it are properly suited to the conditions in the receiving country.

Intermodal transport hub A facility able to move containers between different modes of transport, with storage space for containers waiting for the next leg of the journey.

Internally displaced person A person who is forced to flee their home due to human or environmental factors, but who remains in the same country.

International aid The giving of resources (money, food, goods, technology, etc.) by one country or organisation to another, lower-income country. The objective is to improve the economy and quality of life in that country.

International tourist arrivals The number of international tourists travelling worldwide or to particular destination countries, usually measured over the course of a year.

International tourist receipts Defined by the WTO as expenditure by international inbound visitors, including their payments to national carriers for international transport.

International trade The exchange of goods and services between countries.

Internet penetration rate The percentage of the total population that use the internet.

Intra-firm trade The international trade of intermediate goods and services within the same firm (company).

Irrigation Supplying dry land with water by systems of ditches or by more advanced means.

Jet stream A narrow band of strong winds that flow from west to east in the Earth's atmosphere.

Kyoto Protocol (1997) An international agreement that aimed to reduce carbon dioxide emissions and the presence of atmospheric greenhouse gases.

Labour intensive A production process where labour costs are high compared to expenditure on capital (factory, machinery, etc.).

Labour migration A migration from one country to another with the primary purpose of seeking employment.

Ladder of agricultural technology Distinct levels of investment in and use of agricultural technology, resulting in an increase in agricultural productivity.

Ladder of development The idea that countries sit on different rungs of a ladder, each associated with a different set of economic activities. As countries develop, they become more capable of producing higher-level goods and services, and move up the ladder.

Lagoon A coastal body of shallow saltwater, usually with limited access to the sea. The term is normally used to describe a shallow sea area cut off by a coral reef or a bar.

Lahar A mudflow consisting of volcanic ash, rain and/or snow.

Landlocked country A country that does not have direct access either to an ocean, or to a sea that is not landlocked.

Land tenure The ways in which land is or can be owned. In the past inheritance laws have had a huge impact on the average size of farms.

Land-use zoning The segregation of land use into different areas for each type of use.

Lateral erosion Sideways erosion of a river bank that results in its valley getting wider.

Latitude Lines parallel to the equator that indicate how far north or south of the equator a place is, e.g. Cairo is located at approximately 30°N.

Lava flow A slow-moving flow of molten lava, typically moving at 1–10 km/hour and with temperatures of over 700°C.

Least low-income countries The very poorest countries.

Levée A raised banks found along the side of a river channel.

Life expectancy at birth The average number of years a newborn infant can expect to live under current mortality levels.

Line of best fit A line that summaries the main features of a scatter graph. It does not have to pass through the origin or all points, but should show the main relationship shown by the graph.

Load The particles of sediment and dissolved matter carried along by a river.

Long profile A longitudinal section of the course of a river drawn along the river from source to mouth.

Longitude Lines running from the north pole to the south pole that indicate how far east or west of the Greenwich Meridian (London) a place is, e.g. Cairo is located at approximately 31°E.

Longshore drift The movement of material along a beach by wave action. When a wave breaks obliquely (at an angle to the beach), pebbles are carried up the beach in the direction of the wave (swash). The wave returns to the sea (backwash) at right angles to the beach (direction of steepest slope), carrying material with it.

Magma Molten or semi-molten material found beneath the Earth's surface.

Magma chamber An underground reservoir of molten magma.

Malnutrition Insufficiency in one or more of the nutritional elements necessary for health and wellbeing.

Managed retreat When the coastline is allowed to retreat (erode) in certain areas where the population density or the value of land is low, so that nature takes its course.

Mangrove A salt-tolerant forest of trees and shrubs that grows in the tidal estuaries and coastal zones of tropical areas.

Marginalisation The process of being pushed to the edge of economic activity.

Marine protected area (MPA) An area of coastline or ocean that is given some form of protection or conservation, e.g. activities such as fishing, mining or drilling may be banned or restricted.

Mass migration The migration of a large group of people from one geographical area to another.

Median The middle value when all the data are placed in ascending or descending order.

Mean A type of average; it is the sum of all the values divided by the number of values.

Meander neck A thin piece of land between two successive outer bank meanders, likely to be eroded in the future to form an oxbow lake.

Mechanisation The replacement of human or animal forms of power by a mechanical system such as a tractor in agriculture.

Mercalli Scale A scale that measures the effects of an earthquake based on its intensity of shaking and the observed damage.

Microcredit Tiny loans and financial services to help people, mostly women, start businesses and escape poverty.

Microgeneration A generator that produces electricity with an output of less than 50 KW.

Migrant culture The attitudes and values of a particular society to the process of migration.

Migration The movement of people across a specified boundary, national or international, to establish a new permanent place of residence.

Mitigation (of climate change) Measures used to prevent or limit global climate change, e.g. burning less fossil fuel and/or planting more trees.

Mixed farming Cultivating crops and keeping livestock together on a farm.

Mode The most common value in a population.

Moment Magnitude Scale A scale that measures the amount of energy released by an earthquake.

Monsoon The season of heavy rain during the summer in hot Asian countries.

Mouth The point at which a river flows into a much larger body of water – an ocean, sea or lake.

Multiplier effect The idea that an initial amount of spending or investment causes money to circulate in an economy, bringing a series of economic benefits over time.

Natural hazard A natural event that puts people, property and livelihoods at risk.

Negative indices Powers that have a minus sign, e.g. $2^3 = \left(\frac{1}{2}\right)^{-3} = \frac{1}{8}$

Negative relationship A negative relationship occurs when one value decreases while the other increases.

Neo-colonialism The dominance of lower-income countries by wealthy ones, not by direct political control (as in colonialism), but by economic power and cultural influence.

Net migration The difference between the numbers of immigrants and emigrants in a year.

New Green Revolution An approach to increasing agricultural production that draws upon the best of the agricultural technologies currently available. It emphasises improved farm management in order to minimise environmental damage from external inputs and benefits farmers and marginal areas bypassed by the original Green Revolution.

New international division of labour (NIDL) Divides production into different skills and tasks that are spread across countries rather than taking place within a single country.

Newly industrialised country A country that since the 1960s has experienced very rapid growth in manufacturing industry.

Non-renewable energy The fossil fuels (coal, oil, natural gas) and nuclear fuel. These resources are finite so that as they are used up the supply that remains is reduced. Eventually, these non-renewable resources could become completely exhausted.

Nutrient cycling The re-use of nutrients over time. They are recycled between the atmosphere, plants, animals and the soil.

Open economy A country with few investment and trade barriers, which encourages business with other countries.

Organic farming Agriculture that does not use manufactured chemicals such as chemical fertilisers, pesticides, insecticides and herbicides.

Output A finished product that is sold to customers or consumed, for example what a farm produces, such as milk, eggs, meat and crops.

Overgrazing The grazing of natural pastures at stocking intensities above the livestock carrying capacity.

Overland flow Water flowing over the surface under the influence of gravity. It occurs when the soil is saturated.

Oxbow lake A curved lake found on the floodplain of a river. Oxbows are caused by the loops of meanders being cut off at times of flood and the river subsequently adopting a shorter course.

Package tour The most popular form of foreign holiday, where travel, accommodation and meals may all be included in the price and booked in advance.

Paris Agreement A legally binding international treaty on climate change. It was adopted by 196 Parties in 2015 and came into force in 2016. Its aim is to keep the increase in the global average temperature to well below 2°C above pre-industrial levels.

Pastoral farming The rearing of livestock, such as dairy cattle, sheep and pigs.

Per capita calorie supply The main international indicator used to show variations in global access to food supply. It has limitations – a wholesome diet requires more than just energy and supply figures do not include consumption-level waste (food that is wasted at household, restaurant and retail levels). However, it is the best indicator available. It is measured in kilocalories per person per day.

Percentage A number or ratio expressed as a fraction of 100 (%).

Percolation The downward, vertical movement of water within soil or rock.

Glossary

Permeable Rocks that allow water to pass through, either due to porosity (large volume of pore spaces) or due to being pervious (having joints and cracks/fissures).

Plantation A large farm or estate where one crop is produced commercially, such as palm oil in Malaysia or tea in Sri Lanka. Plantations are usually owned by large companies, often multinational corporations. Many plantations were established in countries under colonial rule, using slave labour.

Pleistocene The geological era lasting from about 2.6 million years ago to around 11,700 years ago, characterised by multiple glacial (cold) periods and interglacial (warm) periods.

Pollution Contamination of the environment.

Population momentum When a country's fertility rate declines to or below the replacement level, but the population continues to grow because of its age structure.

Population projection A prediction of the future population, usually based on current and past trends.

Population pyramid A bar chart, arranged vertically, that shows the distribution of a population by age and gender.

Population structure The composition of a population, the most important elements of which are age and sex (gender).

Positive index A power value that is positive, e.g. $2^3 = 2 \times 2 \times 2 = 8$.

Positive relationship A positive relationship exists between two values when both increase/decrease together.

Post-industrial society A country where the tertiary sector dominates employment and where considerable deindustrialisation has occurred.

Pothole A circular depression in a river bed formed by the abrasive action of rock particles caught up in vortices in turbulent river flow.

Precipitation Water that falls to the Earth from the atmosphere. It is part of the hydrological cycle. Forms of precipitation include rain, snow, sleet, hail, dew and frost.

Preservation Maintaining a location exactly as it is and not allowing development.

Primary product dependent (commodity dependent) When a country relies on one or a small number of primary products for most of its export earnings.

Primary sector The economic sector that exploits raw materials. Farming, fishing, forestry and mining make up most jobs in this sector.

Primary wave The fastest type of seismic wave, which causes the ground to shake.

Process An operation that take places on a farm, such as ploughing and harvesting, or the industrial activities that take place in a factory to make a finished product.

Product chain The full sequence of activities needed to turn raw materials into a finished product.

Productive capability The productive resources, entrepreneurial capabilities, and production links that together determine a country's capability to produce goods and services, and enable it to grow and develop.

Pro-natalist policy A policy that is encouraging and supportive of increasing the fertility rate.

Pro-poor tourism Tourism that results in increased net benefits for lower-income people.

Proportional symbols Symbols such as squares, circles and cubes that are drawn relative to the features being looked at, e.g. proportional circles to show the size of world megacities.

Protectionism The practice of shielding a country's domestic industries from foreign competition by taxing imports.

Protozoans A group of single-celled organisms, either free-living or parasitic, that feed on organic matter or organic debris.

Pull factor A positive factor at the destination.

Purchasing power parity (PPP) Takes into account variations in the cost of living between countries.

Push factor A negative factor at the place of origin.

Pyroclastic flow A superheated cloud of volcanic ash, cinder, tephra and other fragments. It can reach temperatures of over 700°C and is capable of travelling at speeds of over 500 km/hour.

Quaternary sector The economic sector that uses high technology to provide information and expertise to all sectors of an economy.

Rainfall A form of precipitation in which drops of water fall to the Earth's surface from clouds. The drops are formed by the accumulation of fine droplets that condense from water vapour in the air. The condensation is usually brought about by the rising and subsequent cooling of air.

Range The difference between the highest value and the lowest value.

Range How far people will travel to obtain a good or service, e.g. a short distance for a newspaper (a low-order good), or a large distance for a car/computer (a high-order good).

Rapid Where very thin alternating bands of hard and soft rock cross the course of a river, creating an uneven river bed and a zone of turbulent water.

Rate of natural change The difference between the birth rate and death rate. If it is positive, it is termed natural increase. If it is negative, it is known as natural decrease.

Ratio A method of comparing relative size or proportions, e.g. the area deforested in Borneo compared to the forested areas is 1:2 (i.e. one-third has been deforested and two-thirds remain).

Refugee A person who is forced to flee their home due to human or environmental factors and who crosses an international border into another country.

Remittance Money sent back by a migrant to their family in their home community.

Renewable energy Sources of energy such as solar and wind power that are not depleted as they are used.

Replacement level fertility A fertility rate of 2.1 children per woman is required to maintain a population at its current level. Below this level, a population will eventually decline.

Glossary

Repowering Replacing first-generation wind turbines with modern multi-megawatt turbines, which give a much better performance.

Reurbanisation The increase in population in the inner city of an urban area following a period of urban decline.

Richter Scale A scale to measure the magnitude of an earthquake. The scale is logarithmic, so an earthquake of 5.0 magnitude is ten times greater than one of 4.0 magnitude.

River cliff A steep slope that may be found on the outer bend of a meander.

River terrace A relatively flat river deposit marking the position of a former floodplain.

Root zone The area of soil and oxygen surrounding the roots of a plant.

Route maps Maps that show the main transport elements in an area, e.g. roads, rail lines, or maps that show the fastest/shortest route between two places.

Saffir-Simpson Scale A scale to measure the intensity, windspeed and likely damage caused by tropical cyclones.

Saltation When larger particles of a river's load (sand, gravel, very small stones) are transported in a series of 'hops' or bounces along the river bed. It is the means by which bedload (material that is too heavy to be carried in suspension) is transported downstream.

Sand dune A mound or ridge of wind-drifted sand common on coasts and in deserts. In coastal areas, sand is trapped by vegetation, notably sea couch grass and marram grass, to form stable dunes.

Sanitary landfill Municipal waste facilities designed to control leachate and methane, thus minimising the risk of land pollution from solid waste disposal. Sanitary landfill sites are prepared with impermeable bottom liners to collect leachate and prevent contamination of groundwater.

Saturated When the soil is so full of water that it cannot absorb any more.

Scales Scales refer to the relationship between the distance on a map and the corresponding real-life distance on the ground, e.g. on a 1:25,000 map, every 4 cm on the map refers to 1 km on the ground, and on a 1:50,000 map every 2 cm on the map refers to 1 km on the ground.

Sea floor spreading The gradual increase in the size of the oceans as new material is added to the oceans at constructive plate boundaries/mid-ocean ridges.

Sea level An average level of the sea, between the high water mark and low water mark.

Seasonality Fluctuation in demand in the tourism industry over the course of a year, due mainly to seasonal variations in the weather.

Secondary cone A smaller cone found on the side of a volcano's main cone.

Secondary sector The economic sector that manufactures primary materials into finished products.

Secondary wave A secondary or shear wave is a seismic wave that travels through solids, with an undulating/wavelike movement (up and down).

Seismic wave A wave-like motion through the Earth caused by an earthquake, volcanic eruption or landslide.

Sequester To capture and remove CO_2 from the Earth's atmosphere and to store it permanently.

Shallow-focus earthquake An earthquake that occurs up to a depth of 70 km.

Shelterbelt A barrier of trees and shrubs that protects against the wind and reduces erosion.

Shield volcano A low-angle volcano formed of very hot, runny, basaltic lava.

Shifting cultivation A farming system in which farmers move on from one place to another when the land becomes exhausted. The most common form is slash-and-burn agriculture, where land is cleared by burning so that crops can be grown. After a few years the soil fertility is reduced and the land is abandoned. A new area is cleared while the old land recovers its fertility.

Six-figure references A set of six numbers that identifies the exact location of a place on a map. The first three numbers refer to the eastings and the second three numbers refer to the northings.

Slip-off slope The gentle slope found on the inside bank of a river meander.

Slow travel Visitors seeking a deeper connection to the culture, history and nature of the place they are visiting. They usually stay for longer at a destination while avoiding crowds of tourists and appreciating lesser-visited attractions.

Smart meter An internet-capable device that records the detailed use of household energy.

Smart power grid An electricity network that uses digital and other advanced technologies to monitor and manage the transport of electricity from all generating sources to meet the varying electricity demands of end users.

Social business A form of business that seeks to profit from investments that generate social improvements and serve a broader human development purpose.

Social norm The general attitude of a population to an important issue, such as family size, contraception, religion, politics, etc.

Social sustainability The wellbeing of people and the communities they live in.

Soft engineering Any form of coastal or river protection that involves the use of natural means, e.g. sand dunes, saltmarshes, tree planting and/or beach replenishment.

Soil The outermost layer of the Earth's solid surface, consisting of weathered rock, air, water and decaying organic matter overlying the bedrock. Soil comprises minerals, organic matter (humus) derived from decomposed plants and organisms, living organisms, air and water.

Soil erosion The wearing away and redistribution of the Earth's soil. It is caused by the action of water, wind and ice, and also by unsustainable methods of agriculture.

Glossary

Solution (or corrosion) The process by which the minerals in a rock, notably calcium ions, are dissolved in acid water. Solution is one of the processes of erosion.

Source The origin or starting point of a river.

Sphere of influence maps Maps that show the influence of a particular function, i.e. how far people will travel to a place/service such as a settlement or a shop. Some spheres of influence are very small, e.g. a small hamlet, whereas others may be international, e.g. the sphere of influence of a world city.

Spit A ridge of sand or shingle connected to the land at one end and the open sea at the other end. It is formed by the interruption of longshore drift due to wave refraction, river currents, secondary winds and/or changes in the shape of the coastline.

Stack An isolated, upstanding pillar of rock that has become separated from a headland by coastal erosion. It is usually formed by the collapse of an arch.

Standard notation The number that we would normally write, e.g. 281.

Stemflow Water that trickles down plant stems and tree trunks.

Steric effect The expansion of sea water as its temperature increases.

Store (of water) A body of water that receives, holds and releases volumes of water. On land, these include rivers, lakes, reservoirs and aquifers.

Strategic petroleum reserve A large reserve of oil held by a country to tide it over for a few months or so if normal oil supplies are disrupted.

Stratovolcano A cone-shaped volcano formed of viscous, sticky lava.

Structural transformation The transition of an economy from low productivity and labour-intensive activities to higher productivity and skill-intensive activities.

Stump An eroded stack that is exposed only at low tide.

Subduction zone An area where a dense oceanic plate plunges underneath a less-dense continental plate.

Subsistence agriculture The most labour-intensive and generally the smallest scale of agriculture, where the produce is consumed entirely or mainly by the family who work the land or tend the livestock.

Subsistence farming The most basic form of agriculture, in which the produce is consumed entirely or mainly by the family who work the land or tend the livestock. If a small surplus is produced, it may be sold or traded.

Suburbanisation The growth of residential areas at the edges of urban areas; the movement of people into the suburbs (outer parts of the city).

Suburb A residential area found towards the edge of a city and located relatively far from the central areas.

Supervolcano A volcano with a Volcanic Explosivity Index of 8.0, i.e. an eruption of over 1000 km³.

Surface runoff The unconfined flow of water over the ground surface. It occurs when excess precipitation can no longer infiltrate into the soil.

Surface wave A seismic wave at the Earth's surface. Love waves cause the ground to move sideways, and Rayleigh waves cause the ground to move up and down.

Suspension When the smallest and lightest particles of a river's load (silt and clay) are carried in suspension by the moving water.

Sustainable development Development that meets the needs of the present without compromising the ability of future generations to meet their own needs.

Sustainable Development Goals (SDGs) The UN's 17 Sustainable Development Goals with their 169 targets were adopted by UN Member States in 2015. The overall objectives were to end poverty, protect the planet, and ensure all people enjoy peace and prosperity.

Sustainable drainage system (SuDS) A natural drainage method for the drainage of surface water to nearby watercourses in urban areas in particular.

Sustainable tourism Tourism organised in such a way that its level can be sustained in the future without creating irreparable environmental, social and economic damage to the receiving area.

Swash The movement of material up the beach in the direction of the prevailing wind.

System A situation in which there are recognisable inputs, processes and outputs, such as raw materials, labour, energy and capital. For example, a farm requires a range of inputs, such as labour and energy, before anything else can happen.

Technology transfer The transfer of new technology, especially from developed to developing countries, in an attempt to boost their economies.

Tephra All volcanic fragments ejected into the air during an eruption.

Terracing Created when a steep slope is converted into a series of flat steps with raised outer edges (bunds). The monsoon regions of Southeast Asia exhibit widespread terracing.

Tertiary sector The economic sector that provides services to people and to businesses.

Thermal limit for crop growth The temperature (low or high) beyond which a plant is unable to grow to a healthy state.

Threshold characteristic The number of people needed to support a service, e.g. a small number for a newsagent, a large number for a car salesroom.

Throughflow The flow of water through the soil under gravity.

Tied aid Foreign aid that must be spent in the country providing the aid (the donor country).

Tiger economy An economy that undergoes rapid economic growth, usually accompanied by an increase in the standard of living.

Glossary

Tipping point A critical threshold when even a small change can have dramatic effects and cause a disproportionately large response.

Tombolo A bar or ridge that links an island to the mainland.

Total fertility rate The average number of children born to a woman in a country during her lifetime.

Tourism Travel away from the home environment for leisure, recreation and holidays, or to visit friends and relatives, or for business and professional reasons.

Tourist-generating country A country from which many people take holidays abroad.

Traction The sliding or rolling of sediment along the river bed or sea floor.

Trade bloc A group of countries that share trade agreements between each other.

Transfer (of water) The movement of water between stores in the hydrological cycle.

Transnational corporation (TNC) A firm that owns or controls productive operations in more than one country through foreign direct investment.

Transpiration The loss of moisture from vegetation into the atmosphere.

Transport hub A location where routes converge, allowing for the efficient movement of goods and people. The most important transport hubs are often intermodal.

Transportation The movement of a river's load by the processes of traction, saltation, suspension and solution.

Travel motivator The reasons that people travel. The three major categories are leisure, business and visiting friends and relatives.

Triangular graph An equilateral triangle with three axes. Points can be plotted by reading the correct value off each of the axes and placing a point on the triangular grid.

Tributary A river or stream flowing into a larger river or lake.

Tropical cyclone (hurricane, typhoon) A region of very low atmospheric pressure in tropical regions. Hurricanes originate in latitudes between 5° and 20° north or south of the equator, when the surface temperature of the ocean is above 27°C. A central calm area, called the eye, is surrounded by inwardly spiralling winds (anticlockwise in the Northern Hemisphere) of up to 320 km/hr.

Tropical rainforest Dense forest usually found on or near the equator, where the climate is hot and wet. The vegetation in tropical rainforests typically includes a canopy formed by high branches of tall trees that provides shade for lower layers, an intermediate layer of shorter trees, tree roots and lianas, and a ground cover of mosses and ferns.

Tsunami A large and unusual wave in the water generated by a submarine earthquake, volcanic eruption or landslide.

Ultra Low Emission Zone (ULEZ) A scheme in London that has significantly reduced the number of older, high-polluting vehicles entering the city's urban area.

Ultra-large container ship A vessel with a capacity of at least 14,000 TEUs (twenty-foot equivalent units).

Undercut Erosion taking place at the base of a river bank or waterfall and leading to the formation of a notch and an overhang.

Unicorn A recently established business (start-up) valued at US$1 billion or over.

Unimodal A distribution of a population where there is one clear most common size.

United Nations Commission for Trade and Development (UNCTAD) The UN's commission that deals with trade, investment and development issues, assisting countries to integrate into the world economy on an equitable basis. It is based in Geneva, Switzerland.

United Nations Development Programme (UNDP) The UN's development agency working to eradicate poverty. It plays a crucial role in helping countries achieve the Sustainable Development Goals.

United Nations Framework Convention on Climate Change (UNFCCC) The UN process for reaching an agreement to limit dangerous climate change. Its main objective is the stabilisation of greenhouse emissions to prevent dangerous human-caused climate change.

Urban growth The absolute growth of urban areas in population size and/or the physical growth of urban areas.

Urbanisation An increase in the proportion of people living in urban areas.

US Geological Survey A United States scientific agency that studies the landscape of the USA, its natural resources, and the natural hazards that threaten it.

Velocity The speed of a river's water, measured in metres per second.

Venn diagrams Diagrams that show the relation (or overlap) between different sets of data, e.g. reasons for rate of development in LICs.

Vent A fracture on the summit of a volcano through which lava can be ejected.

Vertical erosion Downward erosion occurring in the upper course of the river, where the river cuts down into its bed, deepening the valley.

Volcanic block A fragment of rock measuring over 64 mm in diameter that is erupted in a solid condition.

Volcanic Explosivity Index A measure of the amount of volcanic material ejected in an eruption, the height the material is thrown into the atmosphere, and how long the eruption lasts. The scale is logarithmic, so a VEI 6.0 is ten times greater than a VEI 5.0.

Volcanic gas A gas given off by an active volcano, including water vapour, carbon dioxide, sulphur dioxide, hydrogen sulphide, nitrogen, methane and carbon monoxide.

Volume The amount of water in the river.

Voluntary migration When individuals have a free choice about whether to migrate or not.

V-shaped valley Rivers in upland areas contain large boulders that can erode the bed rapidly when the river is in flood. This results in the river cutting downward into its bed by vertical erosion to form steep V-shaped valleys.

Glossary

Water quality The suitability of water to sustain various uses or processes. Any use will have certain requirements for the physical, chemical or biological characteristics of the water.

Water table The upper level of groundwater saturation in permeable rocks.

Water treatment Improving water quality to make it acceptable for a particular end use, such as human consumption or irrigation.

Waterfall A steep fall of river water where its course is suddenly and significantly interrupted.

Watershed A ridge of high land that forms the boundary between two drainage basins.

Wave The circular or elliptical movement of water near the surface of the sea.

Wave refraction The way in which a wave changes shape and loses speed as it comes into contact with the sea bed. If refraction is complete, waves break parallel to the coastline. If refraction is not complete, longshore drift occurs.

Wave-cut platform A gently sloping rock surface found at the base of a coastal cliff. It is covered by water at high tide but is exposed at low tide. It is formed by the erosion (by waves) of a former cliff face.

Weight gain Where the finished product is heavier than the weight of the raw materials required to manufacture it.

Weight loss Where the finished product is lighter than the weight of the raw materials required to manufacture it.

Wellfield The land above and surrounding wells drilled into an aquifer.

Wetted perimeter The total length of the cross-section at the interface between a channel bed and the stream water that occupies it.

Wind roses Diagrams that show the direction and wind flow for a certain time period. More complex ones can also show variations in wind over the same time.

Youth dependency ratio The ratio of the number of people aged 0–14 years to those in the 15–64 age group.

Acknowledgements

Maps included by kind permission of Ordnance Survey (OS) on **p.291** (r) and **p.307**. Ordnance Survey (OS) is the national mapping agency for Great Britain, and a world-leading geospatial data and technology organisation. As a reliable partner to government, business and citizens across Britain and the world, OS helps its customers in virtually all sectors to improve quality of life.

p.2 Table 1.1 *Geography for CSEC*, 2nd edition, Nelson Thornes, 2016; **p.4** Figure 1.5 G. Nagle, *AS & A2 Geography for Edexcel B* (Oxford University Press, 2003), copyright © Garrett Nagle 2003; **p.15** Figure 1.29 G. Nagle, *ORG GCSE Geography* (Through Diagrams) (Oxford University Press, 1998), copyright © Garrett Nagle 1998; **p.17** Figure 1.33 IGN map Provence-Alpes- Cote d'Azure 2013, © IGN France 2012; **p.21** Table 1.3 Data from *State of Global Water Resources 2022 Report*, World Meteorological Organization, pp.26–28; **p.23** Table 1.4 Data from UK Environment Agency (Defra), 2007, *Improving the Flood Performance of New Buildings*; **p.26** Figure 1.42 www.mississippiriverdelta.org/5-reasonswhy-2019s-mississippi-river-flood-is-the-mostunprecedented-of-our-time; **p.27** Figure 1.43 H.L. Schramm and B.S. Ickes, 'The Mississippi River: A Place for Fish' p.6, www.researchgate.net/publication/267897610; **p.32** Figure 1.47 Baker, S. et al. 1996, *Pathways in Senior Geography*, Nelson, Figure 5.3, p.88; **p.33** Figure 1.48 www.researchgate.net/figure/Water-flow-from-Blue-Nile-and-Nile-River-before-and-after-the-GERDTesfa-2013_fig7_336591659; **p.37** Figure 2.5 G. Nagle, *AS & A2 Geography for Edexcel B* (Oxford University Press, 2003) copyright © Garrett Nagle 2003; **p.40** Figure 2.14 (t) Republic of South Africa/Department of Land Affairs/Surveys and Mapping; **p.41** Figure 2.15 G. Nagle, *ORG GCSE Geography* (Through Diagrams) (Oxford University Press, 1998), copyright © Garrett Nagle 1998; **p.59** Figure 2.35 adapted from S. Warn and C. Roberts, *Coral Reefs: Ecosystem in Crisis?* (Field Studies Council, 2001); **p.63** Figure 2.40 https://old.mpatlas.org/campaign/sulu-sulawesi-seascape/; **p.64** Figure 2.41 Nagle, G. and Spencer, K., 1997, *Geographical Enquiries: Skills and Techniques for Geography*, Stanley Thornes, Figure 6.39, p.100; **p.65** Figure 2.42 Nagle, G. and Spencer, K., 1997, *Geographical Enquiries: Skills and Techniques for Geography*, Stanley Thornes, Figure 6.42, p.102; Figure 2.44 Guinness, P. and Nagle, G., *Geography for Pearson Edexcel International GCSE*, Figure 2.25, p.58; **p.67** Figure 3.1 Dodds, K., 2012, *The Antarctic: A Very Short Introduction*, Oxford, Figure 1, p.3; **p.68** Figure 3.2 Sugden, D., 1982, *Arctic and Antarctica: A modern geographic synthesis*, Blackwell, Figure 3.5, p.50; Table 3.1 Sugden, D., 1982, *Arctic and Antarctica: A modern geographic synthesis*, Blackwell, Table 3.1, p.60; **p.69** Table 3.2 www.coolantarctica.com/Antarctica%20fact%20file/antarctica%20environment/vostok_south_pole_mcmurdo.php; **p.75** Figure 3.9 www.bas.ac.uk/wp-content/ uploads/2018/09/NTS-Rothera-Wharf-Reconstruction.pdf; **p.77** Figure 3.12 (part) G. Nagle, *ORG GCSE Geography* (Through Diagrams) (Oxford University Press, 1998), copyright © Garrett Nagle 1998; **p.82** Figure 3.20 https://news.mongabay.com/2023/11/deforestation-in-the-brazilian-amazon-falls-22-in-2023/; Figure 3.21 https://news.mongabay.com/2023/11/deforestation-in-the-brazilian-amazon-falls-22-in-2023/; **p.85** Figure 3.23 Sugden, D., 1982, *Arctic and Antarctica*, Blackwell, p.147, Figure 6.14; **p.95** Figure 4.11 (part) GeoFactsheets, 121, www.curriculum-press.co.uk; **p.108** Figure 4.26 Armstrong, D. et al., *Geology*, 2008, Heinemann, p.32, Figure 1; **p.110** Figure 5.1 www.climate.gov/news-features/understanding-climate/climate-change-global-temperature; **p.111** Figure 5.2 www.bas.ac.uk/data/our-data/publication/ice-cores-and-climate-change/; Figure 5.3 www.carbonbrief.org/guest-post-piecing-together-arctic-sea-ice-history-1850/; **p.116** Figure 5.10 www.ipcc.ch/site/assets/uploads/2018/02/WG1AR5_Chapter12_FINAL.pdf; **p.120** Figure 5.13 www.researchgate.net/figure/Flood-map-of-Bangladesh-Source-WARPO_fig4_235707918; **pp.124–5** Table 6.1 UN *World Population Prospects*, 2024; **p.126** Table 6.2 UN *World Population Prospects*, 2022; Table 6.3 2023 World Population Data Sheet, Population Reference Bureau; **p.127** Table 6.4 2023 World Population Data Sheet, Population Reference Bureau; Table 6.5 2023 World Population Data Sheet, Population Reference Bureau; **p.130** Figure 6.9 (Earth Policy Institute, 2001); **p.134** Figure 6.15 M. Harcourt and S. Warren, *Tomorrow's Geography* (Hodder Murray, 2012); **p.135** Figure 6.16 CIA World Factbook; **p.136** Table 6.7 2022 World Population Data Sheet, Population Reference Bureau; **p.138** Table 6.8 2023 World Population Data Sheet, Population Reference Bureau; **p.139** Figure 6.18 ECB Occasional Paper Series No 296, June 2022; **p.141** Table 6.9 World Bank and Statista; **p.143** Figure 6.22 http://ourworldindata.org/migration; Table 6.10 World Migration Report, 2022; **p.144** Table 6.11 Data from World Development Report, 2023; **p.148** Figure 6.26 World Bank; **p.153** Table 6.14 www.migrationpolicy.org/article/mexicanimmigrants-united-states; **p.155** Figure 6.32 Davis, A. and Nagle, G., 2024, *Environmental Science and Societies, Third Edition*, Pearson; Figure 6.33 Davis, A. and Nagle, G., 2024, *Environmental Science and Societies, Third Edition*, Pearson; **p.162** Figure 7.8 CC-BYSA-3.0, http://en.wikipedia.org/wiki/File:2009Blikkiesdorp.JPG; **p.164** Figure 7.12 GeoFactsheets, 121, www.curriculum-press.co.uk; **p.180** Table 8.2 Selected data from the 2023 World Population Data Sheet; Per capita kilocalorie data from UN FAO; **p.181** Table 8.3 Selected data from the 2023 World Population Data Sheet; **p.183** Figure 8.6 http://hdr.undp.org/en/statistics/hdi/; Figure 8.7 Wikipedia; **p.184** Figure 8.8 P. Guinness, *Geography for the IB Diploma: Patterns and Change* (Cambridge University Press, 2010), © Cambridge University Press 2010; Figure 8.9 https://datatopics.worldbank.org/worlddevelopment-indicators/the-world-by-income-and-region.html; **p.185** Figure 8.10 https://datatopics.worldbank.org/world-development-indicators/stories/the-classification-of-countries-by-income.html; Table 8.6 World Bank; **p.188** Figure 8.13 P. Guinness, *Geography for the IB Diploma: Patterns and Change* (Cambridge University Press, 2010), © Cambridge University Press 2010; **p.191** Figure 8.15 P. Guinness and B. Walpole, 2015, *Environmental Systems and Societies for the IB Diploma, Second edition*, CUP, p.33, Figure 1.18; **p.200** Figure 8.25 Park, J.D., 'Assessing the role of foreign aid donors and recipients', www.researchoutreach.org; **pp.202–3** Table 8.10 2023 Population Data Sheet, Population Reference Bureau; **p.203** Figure 8.29 www.ft.com; **p.204** Figure 8.30 www.ft.com; **p.212** Table 9.1 World Bank; Table 9.2 Selected data from the 2023 World Population Data Sheet, Population Reference Bureau; Figure 9.11 World Bank data for 2022; **p.218** Figure 9.19 www.statista.com/statistics/268750/global-gross-domestic-product-gdp/; **p.223** Table 9.4 Data from www.lojistico.com/post/where-are-the-largest-logistics-hubs-around-the-world; **p.224** Figure 9.25 P. Guinness, *Geography for the IB Diploma: Global Interactions* (Cambridge University Press, 2011), © Cambridge University Press 2011; **p.228** Table 9.6 P. Guinness, *Geography for the IB Diploma: Global Interactions* (Cambridge University Press, 2011), © Cambridge University Press 2011; **p.229** Figure 9.29 P. Guinness, 'Mongolia: a smaller developing country', GeoFile No. 826, Sept 2022, OUP; **p.232** Figure 9.34 www.statista.com; **p.233** Table 9.7 www.statista.com; Table 9.8 www.statista.com; **p.236** Figure 9.38 (Global Insight: Tourism Satellite Accounting); **p.243** Figure 9.46 www.climatestotravel.com/climate/jamaica; Figure 9.47 Jane Dove et.al., *OCR AS Geography* (Heinemann Educational, 2008); **p.248** Figure 10.5 D. Waugh, *Geography: An Integrated Approach*, 1st Edition (Nelson Thornes, 1990); **p.251** Figure 10.9 www.ourworldindata.org/food-supply; **p.252** Table 10.1 www.ourworldindata.com; **p.254** Figure 10.13 P. Guinness and B. Walpole, *Environmental Systems and Societies for the IB Diploma* (Cambridge University Press, 2012), © Cambridge University Press 2012; **p.256** Figure 10.15 The State of Agricultural Commodity Markets 2022: Part 1 – Global and regional trade networks, Figure 1.2, page 4/12; **p.257** Figure 10.16 *Geography for the IB Diploma: Global Interactions*, CUP, Figure 1, page 156; **p.259** Table 10.2 'The Water Crisis: A Matter of Life and Death', Understanding Global Issues; **pp.260–1** Table 10.3 P. Guinness and B. Walpole, *Environmental Systems and Societies for the IB Diploma* (Cambridge University Press, 2012), © Cambridge University Press 2012; **p.263** Figure 10.20 2023 Global Hunger Index; **p.268** Figure 10.25 https://234intel.com/economics/navigating-through-the-shadows-the-food-crisis-in-nigeria/; **p.273** Figure 10.28 Energy Institute Statistical Review of World Energy 2023; **p.274** Figure 10.29 Energy Institute Statistical Review of World Energy

Acknowledgements

2023; Table 10.6 Data from the 2024 Statistical Review of World Energy; **p.275** Figure 10.30 2023 Statistical Review of World Energy; Table 10.7 Data from the 2024 Statistical Review of World Energy; Table 10.8 Data from the 2024 Statistical Review of World Energy; **p.282** Table 10.10 Data from the 2024 Statistical Review of World Energy; **p.288** Figure 10.39 © BP Statistical Review of World Energy via Our World in Data/ Creative Commons Attribution 4.0 International (CC BY 4.0); **p.290** Davis, A and Nagle, G., 2024 *Environmental science and societies*, Third Edition, Pearson, p. 674 Figure 7.26; **p.294** Figure 11.5 (Directorate of Overseas Surveys/Department for International Development), © Crown copyright; **p.296** Figure 11.8 © Crown Copyright; **p.298** Figure 11.12 Garrett Nagle; **p.301** Figure 11.16 Korea *Statistical Yearbook 2000*; **p.302** Figure 11.18 B. Lenon and P. Cleves, *Fieldwork Techniques and Projects in Geography* (Collins Educational, 2001); **p.303** Figure 11.22 B. Lenon and P. Cleves, *Fieldwork Techniques and Projects in Geography* (Collins Educational); **p.316** Figure 11.39 B. Lenon and P. Cleves, *Fieldwork Techniques and Projects in Geography* (Collins Educational, 2001).

Photo credits

Paul Guinness: p.3 *both*, p.8, p.9, p.10 *both*, p.12 *both*, p.13 *both*, p.14, p.15, p.16 *both*, p.18 *both*, p.23, p.28, p.30, p.43 *br*, p.65 *r*, p.123 (and p.149), p.124, p.129, p.130, p.132 *both*, p.134, p.141, p.142, p.145, p.148, p.149 *both*, p.153, p.195, p.196 *both*, p.198, p.201, p.202, p.177 (and p.183), p.178, p.181, p.186 *both*, p.207 (and p.210), p.208 *t*, p.209 *tl*, p.211 *both*, p.213, p.214, p.215 *both*, p.226, p.229, p.230, p.231, p.233, p.237, p.238 *both*, p.239, p.240, p.242 *both*, p.246 (and p.270), p.247 *both*, p.248 *l*, p.249 *l*, p.254, p.259, p.261, p.265 *both*, p.266, p.270 *both*, p.279, p.280, p.281, p.282, p.283, p.284 *both*, p.309 *both*, p.321

Garrett Nagle: p.1 (and p.6), p.22 *both*, p.34 (and p.40), p.35 *both*, p.36, p.39, p.40 *bl* and *tr*, p.41 *l*, p.42, p.44 *tl*, *cl* and *bl*, p.46, p.47, p.49 *both*, p.50 *all except third from top*, p.51, p.55 *br* and *bl*, p.59 *both*, p.61 *both*, p.65 *l*, p.66 (and p.77), p.78 *both*, p.80, p.82, p.86 (and p.90), p.90 *all*, p.92, p.93, p.96 *l*, p.97, p.99 *tl*, p.105, p.107, p.108 *both*, p.114 *both*, p.158, p.160, p.166, p.168 *both*, p.170, p.171, p.295, p.298, p.299

Other photos reproduced by permission of: **p.31** © Songquan Deng/ Alamy Stock Photo; **p.40** *br* © Bennymarty/stock.adobe.com; **p.41** *r* © Mrks_v/stock.adobe.com; **p.43** *tl* and *bl* © Used with permission from Sean Molloy, **p.44** *cr* © Mudassar/stock.adobe.com; **p.50** *third from top* © Mark Sunderland Photography/Alamy Stock Photo; **p.52** © Used with permission from Robert A Rohde, Berkeley Earth.; **p.53** © Used with permission of NOAA Atlantic Oceanographic and Meteorological Laboratory; **p.55** *t* © Danvis Collection/Alamy Stock Photo; **p.70** © RLS Photo – stock.adobe.com; **p.74** Trent Schindler, NASA Goddard Space Flight Center Scientific Visualization Studio; **p.75** © Frans lemmens/Alamy Stock Photo; **p.84** © GFC Collection/ Alamy Stock Photo; **p.96**, *r* © Almannavarnir/Iceland Civil Protection/ Alamy Stock Photo; **p.99** *br* © Anna/stock.adobe.com; **p.106** *t* Used with permission from Esri. Copyright © 2024 Esri. All rights reserved, *br* © Odelyn Joseph/Associated Press/Alamy Stock Photo; **p.109** (and **p.112**) © Ian Dagnall Computing/Alamy Stock Photo; **p.114** *t* © Dimitrios/stock.adobe.com; **p.119** © Siddharth/stock.adobe.com; **p.126** © robertharding/Alamy Stock Photo; **p.138** © Antony SOUTER/Alamy Stock Photo; **p.151** "World Bank. 2023. World Development Report 2023: Migrants, Refugees, and Societies. © Washington, DC : World Bank. http://hdl.handle.net/10986/39696 License: CC BY 3.0 IGO."; **p.156** © mauritius images GmbH / Alamy Stock Photo; **p.161** © Steve Vidler/Alamy Stock Photo; **p.162** *t* © Used with the permission of Dr. Laura Nkula-Wenz; *b* Africa Media Online/Alamy Stock Photo; **p.163** © Airmaria/stock.adobe.com; **p.165** Piper33/stock.adobe.com; **p.167** © Mark & Audrey Gibson/Stock Connection Blue/Alamy Stock Photo; **p.173** © Imaginechina-Tuchong/Imaginechina Limited/Alamy Stock Photo; **p.176** *t* Observer Research Foundation; *b* Garrett Nagle; **p.182** © Multiple sources compiled by World Bank (2024) – processed by Our World in Data. "Medical doctors per 1,000 people" [dataset]. Data compiled from multiple sources by World Bank, "World Development Indicators" [original data]. Retrieved October 16, 2024 from https:// ourworldindata.org/grapher/physicians-per-1000-people; **p.192** United Nations Sustainable Development Goals web site: https:// www.un.org/sustainabledevelopment/"The content of this publication has not been approved by the United Nations and does not reflect the views of the United Nations or its officials or Member States".; **p.194** Ellen Macarthur Foundation (circular economy system diagram, 2019); **p.206** From The least developed countries in the post-COVID world: Learning from 50 years of experience, by Rolf Traeger. et al.. © 2021 United Nations. Reprinted with the permission of the United Nations.; **p.208** *b* © Romaset/stock.adobe.com; **p.209** *bl* © Kevpix/Alamy Stock Photo, *r* © Vadim/stock.adobe.com; **p.244** © Konstantin Kulikov/ stock.adobe.com; **p.248** *r* © Robert Harding Productions/Alamy Stock Photo; **p.249** *r* © Mustbeyou/stock.adobe.com; **p.264** © US Air Force Photo / Alamy Stock Photo; **p.291** *r* © Crown copyright and database rights 2024 Ordnance Survey (100036470). For Educational Use only.; **p.307** © Crown copyright and database rights 2024 Ordnance Survey (100036470). For Educational Use only.; **p.310** Louis Hellman via RIBA; **p.318** © Jez/stock.adobe.com.

Index

A
abrasion 10, 35
Accumulated Cyclone Energy (ACE) 54
Achill Island 43
active volcanoes 95
adaptation 118–19
adult literacy rate 180
advection 7
age, population structures 134, 136, 137–9
agglomeration 216–17
agriculture
   and climate change 114
   and coastal areas 48
   subsistence 210
   sustainable 194
   and urbanisation 166
   see also farming
agro-industrialisation 253
aid 198–200, 263–4
air freight 223
albedo 68
alluvium 11
Amazon River 2–3
Angel Falls 14
annotation 292–3, 297, 317
Antarctic Circle 67
Antarctic Treaty 1959 74
Antarctica 7, 67
   abiotic factors 70–1
   biotic factors 70–1
   climate 67–8
   climate change 72, 73
   ecosystem 67, 69–75
   fishing 72–3
   ice sheets 67, 72, 73
   management strategies 74–5
   pollution 73
   threats to 72–3
   tourism 73
aquaculture 48
aquaponics 249
aquifers 8–9
arable farming 247, 253–4
area, estimating 293
artificial intelligence (AI) 225
ash fallout 96
asylum seekers 142, 152
atlas maps 291–2
atoll reefs 59
attrition 10, 36
Australia, offshoring asylum seekers 152
averages 307–8

B
backwash 36
Bangladesh, impacts of climate change 120–1
bar graphs 300–1
barrier reefs 59
base maps 292–3
battery energy storage systems (BESS) 286
bays 39
beaches 41–3, 45
   see also coastal areas
bedload 11, 36
bilharzia 29
biodiversity 77
biofuels industry 80
biomass 283–4
birth rate 127, 128–33
Bradshaw model 5
braided channels 16
Brazil, deforestation 81–3
break-of-bulk points 214
Butler Model 235–6
butterflies 79

C
calcium carbonate 36
calorie intake 181
carbon dioxide 111, 114
Caribbean Sea 53
cartographic skills 291–8
cartoons 310
cattle ranching 80
caves 39–40
chlorofluorocarbons 114
cholera 29
choropleth maps 297
cinder cone volcanoes 95
circular economy 193–4
Clark-Fisher economic sector model 210
clean air transition 271
cliff drainage 50
cliff regrading 51
cliffs
   coastal areas 39, 41
   rivers 14
climate
   Antarctica 67–8
   impact on rivers 11
   tropical rainforests 76–7
climate change
   adaptation and mitigation 118–19
   in Antarctica 72–3
   evidence for 110–12
   and flooding 20–1
   global warming 72, 110, 115
   greenhouse effect 113–14, 115–16
   human influences 113–14
   impacts of 115–17
   management strategies and responses 118–21
   natural causes 112–13
   orbital changes 112–13
   tipping point 115
   and volcanic activity 113
climate loss 261
coal 214, 274, 279
coastal areas
   bars and spits 43–4
   beaches 41–3, 45
   concordant/discordant coastlines 46
   erosion 35–8, 49, 57
   fieldwork 63–4
   hazards 48–9, 54–5
   landforms 35, 39
   living near 47–9
   management 49–51
   pollution 49
   sand dunes 45
   stacks and stumps 39–40
   waves 35, 37–40
colonialism 188
commercial farming 247–8
commodity chains 221–2
communications systems 224
condensation 8
cone volcanoes 95
conservation 74
construction industry 195–6
constructive waves 37
consumer culture 224
convection currents 88–9
convectional rains 76–7
coral reefs 58–60, 62–3
Coral Triangle 62–3
corrasion 10, 35
corrosion 13, 36
cotton crops 259
counterurbanisation 158
crater 96
cultural diffusion 223–4
cultural identity 151
culture
   and development 190
   and globalisation 223–4

Index

D

Danube River 3
Danum Valley Conservation Area 84
data collection 312
death rate 127, 128, 132–3
debt relief 200
decentralisation 167
decimals 308
deforestation 20, 81–2
deindustrialisation 211
deltas 3, 16–17
demographic change 125
demographic dividend 137
dependency theory 188–9
deposition 6, 11, 15, 16, 36
desertification 266–7
deserts 9
destructive waves 37
development 178
   categories of 179
   and culture 190
   dependency theory 188–9
   economic indicators 179–82
   gap 187–8
   high-income countries (HICs) 184–5, 187
   Human Development Index (HDI) 182–3
   ladder of 184
   low-income countries (LICs) 184–6
   middle-income countries MICs) 184–5, 186–7
   sustainable 191–6
   and technology 190
   trade and investment 189–90
   uneven 184–90, 196–205
diaspora 149
digital divide 190
digital twins 225
dispersion graphs 303
disruptive technology 225
distribution maps 298
doctors, per thousand people 181–2
dormant volcanoes 95
doughnut graphs 305
drainage basins 2–3, 7

E

Earth
   crust 87
   inner core 87
   layers and characteristics 87
   lithosphere 87
   mantle 87
   outer core 87
   plate boundaries 89–94
   tectonic plates 88
earthquakes 91–4, 98–106
economic development 179–82
economic migration 142, 144, 146
economic sectors 208–10, 212
economic sustainability 194
Ecosystem Approach to Fisheries Management (EAFM) 63
ecotourism 241
education 165, 180
electric vehicles 195
emigration rate 127
employment 163
employment structures 210
energy
   conservation 286
   consumption 272–5
   gap 276
   mix 271, 274
   non-renewable 270–1, 279–81
   poverty 271
   production 272–5, 277–8, 279
   renewable 270–1, 274, 281–5
   security 276
   storage 286
   supply 276–7, 285–6
   surplus and deficit 276–7
enhanced greenhouse effect 114, 115–16
environment
   and tourism 239–41
   and urbanisation 170
environmental impact surveys 74
Environmental Performance Index (EPI) 192–3
environmental sustainability 193
equatorial areas 76–7
erosion
   coastal 35–8
   rivers 5, 10–11, 14–15
European Union (EU) 222
eutrophication 29
evaporation 7–8
evapotranspiration 9
extinct volcanoes 96

F

Fairtrade 198
farming
   aquaponics 249
   arable 247, 253–4
   commercial 247–8
   extensive 249
   hydroponics 249
   intensive 249
   mixed 247
   no-till 266
   organic 249
   pastoral 247, 253–4
   subsistence 247–8
   sustainable 264–5
   systems 250
   see also agriculture
fault lines 13

Favela Barrio Project 164–5
fertility 128–31
field sketches 316–17
fieldwork
   coastal areas 63–4
   skills 310–20
fisheries 48
fishing 62, 72–3
flash floods 20, 57
flood hydrographs 305
flooding 115, 121
   causes of 19–20, 26
   and erosion 11
   flood peak 82
   impacts of 18–19, 20–1, 26, 57–8
   lake formation 15
   Mississippi River 25–7
   precipitation 26–7
   prediction and prevention 22–4
   risk management 21
floodplains 15–16, 20
flow-line diagrams 302–3
food aid 263–4
food production 116–17, 252–5
food security 261–4, 268–9
food supply
   challenges 258–64, 268–9
   global variations 251–3
   globalisation of 255–7
   strategies 254–5
food waste 252
Foote, Eunice Newton 111
fractions 308
Frank, A.G. 188–9
freeports 216
friction 9–10, 14
fringing reefs 58
fuel poverty 271
fuelwood 270–1, 278

G

gabions 50
Gangotri Glacier 3
gender dividend 141
gentrification 168
Geographic Information System (GIS) 309–10
geothermal electricity 284
glacial period 113
glaciation 13
glaciers 3
global cities 218
global civic society 197
Global Hunger Index (2023) 262–3
global tourism 232
global value chains (GVCs) 226
global warming 72, 110, 115
globalisation 218–31, 255–7
gorges 13
gradient 294–5

Index

graphical skills 299–305
graphs 299–305, 308
gravitational sliding 88
Green Revolution 254–5
greenbelts 167, 213
greenhouse effect 113–14, 115–16
greenhouse gases 113–14, 117
Greenland ice sheet 7
gross domestic product (GDP) 179, 182, 232
gross national income (GNI) 179–80, 184–5, 212
gross national product (GNP) 179
groundwater 7–9
groundwater flow 3
Group of 20 (G20) 222
groynes 36, 50

H

hard engineering 22, 49
hazard-resistant design 23
headlands 38, 39
heatwaves 115–16
Heavily Indebted Poor Countries (HIPC) Initiative 200
High Force waterfall 5–6
high-income countries (HICs)
   development 184–5, 187
   employment structures 211–12
histograms 301–2
Holocene 112
housing 162–3, 168, 172
Human Development Index (HDI) 182–3
humidity 76
Hurricane Andrew (1992) 58
Hurricane Gilbert (1988) 55
Hurricane Luis (1995) 55
Hurricane Severity Index 54
hurricanes *see* tropical cyclones
hydraulic action 10, 13, 35
hydraulic radius 4
hydroelectric power 82, 274, 281
hydrogen fuel cells 285
hydrological cycle 7–8
hydroponics 249

I

ice ages 112–13
ice core analysis 110–11
ice sheets 111
   Antarctica 7–8, 67, 72, 73
   Greenland 7
immigration rate 127
impermeable rock 9
Indonesia 201–5
Industrial Revolution 210, 213, 214
industry
   and coastal areas 48
   factors affecting 212–17

managing growth 169
   sustainable 196
inequality 162
infant mortality rate 180–1
infiltration 9
interception 9
interglacial period 113
interlocking spurs 12
internal displacement 142
international aid 198–200
International Federation of Robotics (IFR) 225
International Organization for Migration (IOM) 150
internet penetration rate 190
irrigation 259
isoline diagrams 297

J

Jamaica, tourism 243–5

K

kite graphs 302
krill 70
Kyoto Protocol (1997) 118

L

labour 215
labour migration 152, 221
ladder of development 184
lahars 93, 96
lakes 15
land tenure 260
landfills 196
landforms
   coastal 35, 39
   lowland 14–17
   upland 12–14
lava flows 96
leisure 169
levées 22
life expectancy 180
limestone 10
line graphs 299
literacy 180
longshore drift 36–7, 44
low-income countries (LICs)
   development 184–6
   employment structures 210, 212
   food production 260–1
   population structures 126, 136
   water quality 28
low-pressure systems 52–3

M

magma 88
magma chamber 96
malaria 29, 62
malnutrition 262

mangroves 58, 60–2
maps
   atlas maps 291–2
   base maps 292–3
   choropleth maps 297
   direction 294
   distribution maps 298
   gradient 294–5
   isoline diagrams 297
   references 293–4
   route maps 298
   scales 291, 293, 307
   sketch maps 297
   sphere of influence maps 298
   symbols 295
marine ecosystems 70–1
marine protected area (MPA) 62
marram grass 45–6
mathematical skills 306–8
mean 307–8
meander neck 15
median 308
Mercalli Scale 100
methane 111, 114
Mexico, migration 152–4
microcredit 201
microplastics 28
middle-income countries MICs)
   development 184–5, 186–7
   employment structures 210–11, 212
   population structures 126, 136
migration 142–54, 161
   barriers to 144–5
   categories of 142–3
   causes of 144
   impacts of 147–50
   managing 150–2
   Mexico 152–4
   push and pull factors 145–6
Milankovitch, Milutin 113
Milankovitch cycle theory 113
mineral exploitation 72
mineral extraction 195
mining 72, 82, 195
Mississippi River 2–3, 22, 25–7
mitigation 118–19
mixed farming 247
mobile homes 58
mode 308
Moment Magnitude Scale 101
Mongolia, transnational corporations (TNCs) 228–31
Mount Etna 107–8
multiculturalism 161
multiplier effects 189

N

natural gas 274, 280
neo-colonialism 188
net migration 127

Index

new international division of labour (NIDL) 220
New York 161, 167
Niagara Falls 13
Nigeria, food insecurity 268–9
Nike 221–2
Nile River 2
non-renewable energy 270–1, 279–81
North American Free Trade Agreement (NAFTA) 222
no-till farming 266
nuclear energy 274, 280–1
nutrient cycling 82

O

obesity 252
oil 274, 279–80
one-child policy 130–1
open economies 187
organic farming 249
overgrazing 22
overland flow 3
oxbow lakes 15

P

Pacific Ring of Fire 91
package tours 233–4
Palisadoes 37, 44
Paris Agreement (2015) 118
pastoral farming 247, 253–4
Patagonian toothfish 70–1
penguins 70–2
per capita calorie supply 251–2
percentages 308
percolation 9
permeable rock 9
phytoplankton 70–1
pictograms 299
pie charts 300–1
pilot surveys 314
pitcher plants 78
plantations 80
Pleistocene 112
pollution
    Antarctica 73
    coastal areas 49
    rivers 28–32
    vehicles 195
population pyramids 134–5
populations
    decline 127
    demographic change 125
    demographic transition model 131–3
    dependency ratio 136
    diaspora 149
    distribution 125–6
    growth 124–9
    net migration 127
    structures 134–41

    see also migration
post-industrial societies 211
potholes, river bed 12–13
poverty 188, 271
precipitation 7–9
    Antarctica 68
    and climate change 117
    see also rainfall
primary sector 208, 212
product chain 209
protectionism 189
pyroclastic flows 93, 97

Q

quality of life 178, 216
quarrying 195
Quaternary Period 112–13
quaternary sector 209
questionnaires 314–16

R

radial graphs 303–4
rainfall 7, 19–20, 52, 76–7, 81–2
    see also precipitation
rainforests *see* tropical rainforests
range 308
rapids 5, 13, 14
ratios 308
raw materials 213
ray diagrams 303
recording tables 318
reforestation 22
refugees 142–4
renewable energy 270–1, 274, 281–5
revetments 50
Rhône River 17
Richter Scale 99
risk assessment 311–12
river cliff 14
river terrace 16
rivers
    Bradshaw model 5
    braiding 16
    channel roughness and shape 4–5
    characteristics 2–6
    deltas 3, 16–17
    deposition 6, 11, 15, 16
    discharge 4–5, 10
    drainage basins 2–3, 7
    erosion 5, 10–11
    investigating 322–3
    landforms 12–17
    living near 18–21
    meanders 14
    mouth 3
    pollution 28–32
    processes 9–11
    source 3
    transportation 11
    velocity 4–5, 10

robotics 225
Rocinha 165
rock
    impermeable 9
    permeable 9
    river erosion 5, 10–11
rock armour 50
route maps 298
rural-to-urban fringe 165–7

S

Saffir-Simpson Scale 53–4
saltation 11
sampling 312–14
sand dunes 45, 321–2
sandbanks 17
sanitary landfills 196
scales 291, 293, 307
scatter graphs 304, 308
sea breeze 37–8, 44
sea couch grasses 45–6
sea floor spreading 93
sea ice 111–12
    see also ice sheets
sea level change 35, 57–8, 115–16
sea temperatures 53
sea walls 49
seals 70–1
secondary sector 208, 212
sediment 16–17, 36, 38, 43–4
seismic waves 94
seismograph 99
settlements, unplanned 164
sewage 28
sex, and population structures 134, 137
shale oil 280
Shanghai, urbanisation 170–4
shelterbelts 265
shield volcanoes 95
shipping 256
Sitges 235
sketch maps 297
sloths 78
smart meters 285–6
Smoke Control Areas (SCAs) 195
social businesses 201
social investment 189–90
social norms 132
social sustainability 193
soft engineering 22, 51
soil
    erosion 82, 264–5
    salt-affected 266
    saturated 9
    tropical rainforests 78, 80–1
solar power 282–3
solution (erosion) 10
Southern Ocean 70
sphere of influence maps 298
spit 43

springs 9
standard notation 308
stemflow 9
steric effect 112
storm surges 57
stormwater 30
stratovolcanoes 95
subsidence 58
subsistence agriculture 210, 247–8
suburbanisation 158
sunspots 113
supervolcano 101
surface runoff 9
suspended load 36
sustainable development 191–6, 286–7
Sustainable Development Goals 178, 191–2
sustainable drainage systems (SuDS) 24
Suzhou Creek 31–2
Sweden, energy management 287–9

T

tarrifs 197, 255–6
technology
  and development 190
  disruptive 225
  and globalisation 224–5
  and industry 217
tectonic activity
  earthquakes 91–4, 98–106
  hazards of 98–9
  impacts of 98
  measuring 99–101
  plate boundaries 89–94
  predicting 103
  responses to 102
  volcanoes 91–7, 99–103, 107–8, 113
Tees River 5–6
temperature, Antarctica 68, 73
tephra 96
tertiary sector 208–9, 212
Thomas Cook 233–4
throughflow 3
thunderstorms 52
tidal power 283
tied aid 199
tomobolos 44
tourism
  Antarctica 73
  Butler Model 235–6
  coastal areas 47–8
  economic impacts 236–7
  ecotourism 241

environmental impacts 239–41
  global 232–45
  Jamaica 243–5
  package tours 233–4
  protected areas 241–2
  seasonality 237
  social and cultural impacts 237–9
  sustainable 240–1
  travel motivators 234
traction 11
trade 189, 197–8, 221, 236, 255–6
trade blocs 221–2
trade winds 37
Trans-Amazonian Highway 80
transnational corporations (TNCs) 225–31
transpiration 9
transport hubs 222–3
transportation 11, 36, 47–8, 80, 161, 163–4, 166, 168–9, 172, 216, 256
triangular graphs 304–5
tributaries 2–3
tropical cyclones 37, 44, 48, 51–7, 96, 120
  management strategies 55–6, 58
  Saffir-Simpson Scale 53–4
tropical rainforests
  climate 76–7
  ecosystem 77
  flora and fauna 78–9
  food chain 79
  global distribution 76
  management strategies 83–4
  threats to 80
tsunamis 48, 99
Typhoon Haiyan (2012) 56–7

U

Ukraine, migration 144
Ultra Low Emission Zone (ULEZ) 195
United Nations Conference on Trade and Development (UNCTAD) 197
United Nations Development Programme (UNDP) 178
United Nations Framework Convention of Climate Change (UNFCCC) 118
upland landforms 12
urban fringe 165–7
urban growth 157
urban sprawl 165–7
urbanisation 20, 47–8, 157–74, 253
  challenges 162–5

environmental issues 170
  managing growth 168–74
  opportunities 161
  rural-to-urban fringe 165–7
  social issues 169–70
  sustainable 195

V

Venn diagrams 305
visual images 309–10
volcanic block 97
Volcanic Explosivity Index 101
volcanic gasses 97
volcanoes 91–7, 99–103, 107–8, 113
Volga River 3
V-shaped valleys 12

W

waste management 164, 173, 196
waste water treatment 28–30
water
  cycle 7–8
  fresh 7–8
  quality 28–9
  stores 7
  stormwater 30
  table 9
  transfers 8
waterfalls 5, 13–14
watershed 2
wave power 283
waves 35, 37–40
weather, impacts of climate change 115
weight loss (manufacturing) 213
weight-gain industries 216
wellbeing 178
wellfields 21
wetted perimeter 4
whales 70–1
wind power 282
wind speeds 52–3, 68
World Trade Organization (WTO) 197–8, 221

Y

Yangtze River 2
yellow fever 29

Z

zooplankton 70–1